Lecture Notes in Artificial Intelligence 8482

Subseries of Lecture Notes in Computer Science

T0212737

Moonis Ali   Jeng-Shyang Pan
Shyi-Ming Chen   Mong-Fong Horng (Eds.)

Modern Advances in Applied Intelligence

27th International Conference
on Industrial Engineering and Other Applications
of Applied Intelligent Systems, IEA/AIE 2014
Kaohsiung, Taiwan, June 3-6, 2014
Proceedings, Part II

 Springer

Volume Editors

Moonis Ali
Texas State University, San Marcos, TX, USA
E-mail: ma04@txstate.edu

Jeng-Shyang Pan
National Kaohsiung University of Applied Sciences, Taiwan
E-mail: jspan@cc.kuas.edu.tw

Shyi-Ming Chen
National Taiwan University of Science and Technology, Taipei, Taiwan
E-mail: smchen@mail.ntust.edu.tw

Mong-Fong Horng
National Kaohsiung University of Applied Sciences, Taiwan
E-mail: mfhorng@ieee.org

ISSN 0302-9743   e-ISSN 1611-3349
ISBN 978-3-319-07466-5   e-ISBN 978-3-319-07467-2
DOI 10.1007/978-3-319-07467-2
Springer Cham Heidelberg New York Dordrecht London

Library of Congress Control Number: 2014939236

LNCS Sublibrary: SL 7 – Artificial Intelligence

Typesetting: Camera-ready by author, data conversion by Scientific Publishing Services, Chennai, India

Printed on acid-free paper

Springer is part of Springer Science+Business Media (www.springer.com)

Preface

On behalf of the Organizing Committee of the 27th International Conference on Industrial, Engineering and Other Applications of Applied Intelligent Systems (IEA-AIE 2014), it is our pleasure to present the proceedings of the conference. IEA-AIE 2014 was hosted by the National Kaohsiung University of Applied Sciences, Taiwan, Chaoyang University of Technology, Taiwan, and Harbin Institute of Technology University, China. IEA/AIE 2014 continued the tradition of emphasizing applications of applied intelligent systems to solve real-life problems in all areas including engineering, science, industry, automation and robotics, business and finance, medicine and biomedicine, bioinformatics, cyberspace, and human–machine interactions. IEA/AIE 2014 included oral presentations, invited speakers, special sessions, and a computer vision competition.

We receive 294 submissions from 27 countries in Africa, Asia, Australia, Europe, North America, and South America. Due to the page limit, only 106 papers were accepted and are published in the proceedings. Each paper was reviewed by at least two reviewers. The reviewers are mainly researchers from universities and industry. We congratulate and thank all the authors of the accepted papers. In addition to the paper presentations, four plenary talks were kindly delivered by Prof. John Roddick from Flinders University, Australia, Prof. Ce Zhu, University of Electronic Science and Technology of China, Prof. Daniel S. Yeung, South China University of Technology, China, and President Han-Chien Chao, Ilan University, Taiwan. We hope that these high-quality and valuable keynotes are beneficial for your future research.

Finally, we sincerely appreciate the contribution of the authors, reviewers, and keynote speakers. We would also like to express our sincere appreciation to the Program Committee members and local committee members for their valuable effort that contributed to the conference's success.

June 2014

Ali Moonis
Jeng-Shyang Pan
Shyi-Ming Chen
Mong-Fong Horng
Rung-Ching Chen
Ngoc Thanh Nguyen
Ali Selamat

Organization

Honorary Chairs

Cheng-Hong Yang   National Kaohsiung University of Applied
Sciences, Taiwan

Jen-Chin Chung   Chaoyang University of Technology, Taiwan

General Chairs

Ali Moonis   Texas State University-San Marcos, USA

Jeng-Shyang Pan   National Kaohsiung University of Applied
Sciences, Taiwan

Shyi-Ming Chen   National Taiwan University of Science and
Technology, Taiwan

Program Committee Chairs

Mong-Fong Horng   National Kaohsiung University of Applied
Sciences, Taiwan

Rung-Ching Chen   Chaoyang University of Technology, Taiwan

Ngoc Thanh Nguyen   Wroclaw University of Technology, Poland

Ali Selamat   Universiti Teknologi Malaysia

Advisory Committee Board

Ajith Abraham   Technical University of Ostrava,
Czech Republic

Bin-Yi Liao   National Kaohsiung University of Applied
Sciences, Taiwan

Yau-Hwang Kuo   National Cheng kung University, Taiwan

Kuan-Ming Hung   National Kaohsiung University of Applied
Sciences, Taiwan

Publication Chairs

Chin-Shiuh Shieh   National Kaohsiung University of Applied
Sciences

Muhammad Khurram Khan   King Saud University, Saudi Arabia

Invited Session Chairs

Jeng-Shyang Pan	National Kaohsiung University of Applied Sciences, Taiwan

Publicity Chairs

Tien-Tsorng Shih	National Kaohsiung University of Applied Sciences, Taiwan

Keynote Speeches

John Roddick	Dean of the School of Computer Science, Engineering and Mathematics, Flinders University, Australia
Ce Zhu	Professor, University of Electronic Science and Technology of China, China
Daniel S. Yeung	Chair Professor, South China University of Technology, China
Han-Chieh Chao	President, National Han University, Taiwan
Tzung-Pei Hong	Department of Computer Science and Information Engineering, National University of Kaohsiung, Taiwan

Invited Sessions Organizers

Shyi-Ming Chen	National Taiwan University of Science and Technology, Taiwan
Yuh-Ming Cheng	Shu-Te University, Taiwan
Tsong-Yi Chen	National Kaohsiung University of Applied Sciences, Taiwan
Lijun Yan	Harbin Institute of Technology Shenzhen Graduate School, China
Linlin Tang	Shenzhen Graduate School of Harbin Institute of Technology, China
Shu-Chuan Chu	Flinders University, Australia

International Program Committee

Aart Van Halteren	Vrije University Amsterdam, The Netherlands
Ah-Hwee Tan	Nanyang Technological University, Singapore
Alexander Felfernig	Technische Universität Graz, Austria
Alfons Salden	Almende BV, The Netherlands
Al-Mubaid Hisham	University of Houston Clear Lake, USA
Altincay Hakan	Eastern Mediterranean University, Turkey

Amruth N. Kumar — School of Theoretical and Applied Science, USA
Andrea Orlandini — Institute of Cognitive Sciences and Technologies, National Research Council, Italy
Andres Bustillo — University of Burgos, Spain
Antonio Peregrin — Universidad de Huelva, Spain
Anwar Althari — Universiti Putra Malaysia, Malaysia
Ariel Monteserin — Universidad Nacional del Centro de la Provincia de Buenos Aires, Argentina
Arzucan Özgür — University of Michigan, USA
Ashok Goel — Georgia Institute of Technology, USA
Aydogan Reyhan — Delft University of Technology, The Netherlands
Azizi Ab Aziz — Universiti Utara, Malaysia
Bae Youngchul — Chonnam National University, Korea
Barber Suzanne — University of Michigan, USA
Benferhat Salem — Université d'Artois, France
Bentahar Jamal — Concordia University, Canada
Bipin Indurkhya — International Institute of Information Technology, India
Björn Gambäck — Norwegian University of Science and Technology, Norway
Bora İsmail Kumova — Izmir Institute of Technology, Turkey
César García-Osorio — Universidad de Burgos, Spain
Catherine Havasi — Massachusetts Institute of Technology, USA
Chang-Hwan Lee — Dongguk University, South Korea
Chao-Lieh Chen — National Kaohsiung First University of Science and Technology, Taiwan
Chen Ling-Jyh — Academia Sinica, Taiwan
Chen Ping — University of Houston, USA
Cheng-Seen Ho — National Taiwan University of Science and Technology, Taiwan
Chien-Chung Wang — National Defense University, Taiwan
Chih Cheng Hung — Southern Polytechnic State University, USA
Colin Fyfe — Universidad de Burgos, Spain
Correa da Silva, F.S. — University of São Paulo, Brazil
Dan E. Tamir — Texas State University, USA
Daniel Yeung — South China University of Technology, China
Dariusz Krol — Wrocław University of Technology, Poland
Darryl Charles — University of Ulster, UK
Dianhui Wang — La Trobe University, Australia
Dilip Kumar Pratihar — Indian Institute of Technology
Dirk Heylen — University of Twente, The Netherlands
Djamel F. H. Sadok — Universidade Federal de Pernambuco, Brazil
Don-Lin Yang — Feng Chia University, Taiwan
Dounias Georgios — University of the Aegean, Greece

Duco Nunes Ferro   Almende BV, The Netherlands
Enrique Herrera Viedma   Universidad de Granada, Spain
Esra Erdem   Sabanci University, Turkey
Evert Haasdijk   Universiteit van Amsterdam, The Netherlands
Fariba Sadri   Imperial College London, UK
Fernando Gomide   Universidade Estadual de Campinas, Brazil
Fevzi Belli   University of Paderborn, Georgia
Floriana Esposito   Università' di Bari, Italy
Fran Campa   University of the Basque Country, Spain
Francesco Marcelloni   University of Pisa, Italy
Francisco Fernández
  de Vega   University of Extremadura, Spain
Francisco Herrera   Universidad de Granada, Spain
François Jacquenet   UFR des Sciences et Techniques, France
Frank Klawonn   University of Applied Sciences
                              Braunschweig/Wolfenbüttel, Germany
Gérard Dreyfus   Laboratoire d'Électronique École Supérieure
                              de Physique et de Chimie Industrielles,
                              France
Greg Lee   National Taiwan Normal University, Taiwan
Gregorio Ismael
  Sainz Palmero   University of Valladolid, Spain
Hamido Fujita   Iwate Prefectural University, Japan
Hans Guesgen   University of Auckland, New Zealand
Hasan Selim   Dokuz Eylül Üniversitesi, Turkey
He Jiang   Dalian University of Technology, China
Henri Prade   Université Paul Sabatier, France
Hiroshi G. Okuno   Kyoto University, Japan
Honggang Wang   University of Massachusetts Dartmouth, USA
Huey-Ming Lee   Chinese Culture University, Taiwan
Humberto Bustince   Public University of Navarra, Spain
Hyuk Cho   Sam Houston State University, USA
Jean-Charles Lamirel   Lorrain de Recherche en Informatique et
                              ses Applications, France
Jeng-Shyang Pan   National Kaohsiung University of
                              Applied Sciences, Taiwan
Jing Peng   Montclair State University, USA
João Paulo Carvalho   INESC-ID, Instituto Superior Técnico,
                              Technical University of Lisbon, Portugal
João Sousa   Technical University of Lisbon, Portugal
Joaquim Melendez Frigola   Universitat de Girona, Spain
John M. Dolan   Carnegie Mellon University, USA
José Valente de Oliveira   University of the Algarve, Portugal

Jose-Maria Peña	DATSI, Universidad Politecnica de Madrid, Spain
Jun Hakura	Iwate Prefectural University, Japan
Jyh-Horng Chou	National Kaohsiung First University of Science and Technology, Taiwan
Kaoru Hirota	The University of Tokyo, Japan
Kazuhiko Suzuki	Okayama University, Japan
Khosrow Kaikhah	Texas State University, USA
Kishan G. Mehrotra	Syracuse University, USA
Kuan-Rong Lee	Kun Shan University, Taiwan
Kurosh Madani	University of Paris, France
Lars Braubach	University of Hamburg, Germany
Leszek Borzemski	Wrocław University of Technology, Poland
Maciej Grzenda	Warsaw University of Technology, Poland
Manton M. Matthews	University of South Carolina, USA
Manuel Lozano	University of Granada, Spain
Marcilio Carlos Pereira De Souto	Federal University of Rio Grande do Norte, Brazil
Marco Valtorta	University of South Carolina, USA
Mark Last	Ben-Gurion University of the Negev, Israel
Mark Neerincx	Delft University of Technology, Netherlands
Mark Sh. Levin	Institute for Information Transmission Problems Russian Academy of Sciences, Russia
Martijn Warnier	Delft University of Technology, The Netherlands
Michele Folgheraiter	Nazarbayev University, Georgia
Miquel Sànchez i Marrè	Technical University of Catalonia, Spain
Murat Sensoy	University of Aberdeen, UK
Nashat Mansour	Lebanese American University, Lebanon
Natalie van der Wal	Vrije Universiteit Amsterdam, The Netherlands
Ngoc Thanh Nguyen	Wroclaw University of Technology, Poland
Nicolás García-Pedrajas	University of Córdoba, Spain
Niek Wijngaards	Delft Cooperation on Intelligent Systems, The Netherlands
Oscar Cordón	Granada University, Spain
Paolo Rosso	Universidad Politecnica de Valencia, Spain
Patrick Brézillon	Laboratoire d'Informatique de Paris 6, France
Patrick Chan	South China University of Technology, China
Paul Chung	Loughborough University, UK
Prabhat Mahanti	University of New Brunswick, Canada
Qingzhong Liu	New Mexico Institute of Mining and Technology, USA

Richard Dapoigny   Université de Savoie, France
Robbert-Jan Beun   University Utrecht, The Netherlands
Rocío Romero Zaliz   Universidad de Granada, Spain
Safeeullah Soomro   Sindh Madresatul Islam University, Pakistan
Sander van Splunter   Delft University of Technology,
                              The Netherlands
Shaheen Fatima   Loughborough University, UK
Shogo Okada   Tokyo Institute of Technology, Japan
Shyi-Ming Chen   National Taiwan University of Science and
                              Technology, Taiwan
Simone Marinai   University of Florence, Italy
Snejana Yordanova   Technical University of Sofia, Bulgaria
Srini Ramaswamy   University of Arkansas at Little Rock, USA
Sung-Bae Cho   Yonsei University, Korea
Takatoshi Ito   Nagoya Institute of Technology, Japan
Tetsuo Kinoshita   Tohoku University, Japan
Tibor Bosse   Vrije Universiteit Amsterdam, The Netherlands
Tim Hendtlass   Swinburne University of Technology, Australia
Tim Verwaart   Wageningen University and Research Centre,
                              The Netherlands
Valery Tereshko   University of the West of Scotland,
                              UK
Vincent Julian   Universidad Politecnica de Valencia, Spain
Vincent Shin-Mu Tseng   National Cheng Kung University, Taiwan
Vincenzo Loia   University of Salerno, Italy
Walter D. Potter   The University of Georgia
Wei-Tsung Su   Aletheia University, Taiwan
Wen-Juan Hou   National Taiwan Normal University, Taiwan
Xue Wang   University of Leeds, UK
Yo-Ping Huang   Asia University, Taiwan
Yu-Bin Yang   Nanjing University, China
Zhijun Yin   University of Illinois at Urbana-Champaign,
                              USA
Zsolt János Viharos   Research Laboratory on Engineering &
                              Management Intelligence, Hungary

Reviewers

Aart Van Halteren   Al-Mubaid Hisham   Antonio Peregrin
Adriane B. de S.   Altincay Hakan   Anwar Althari
  Serapião   Amruth Kumar   Ariel Monteserin
Ah-Hwee Tan   Andrea Orlandini   Arzucan Özgür
Alexander Felfernig   Andres Bustillo   Ashok Goel
Alfons Salden   Andrzej Skowron   Aydogan Reyhan

Azizi Ab Aziz
Bae Youngchul
Barber Suzanne
Belli Fevzi
Benferhat Salem
Bentahar Jamal
Bipin Indurkhya
Björn Gambäck
Bora İsmail Kumova
César García-Osorio
Catherine Havasi
Chang-Hwan Lee
Chao-Lieh Chen
Chen Ling-Jyh
Chen Ping
Cheng-Seen Ho
Cheng-Yi Wang
Chien-Chung Wang
Chih Cheng Hung
Ching-Te Wang
Chun-Ming Tsai
Chunsheng Yang
Ciro Castiello
Colin Fyfe
Correa da Silva, F.S.
Dan E. Tamir
Daniel Yeung
Dariusz Krol
Darryl Charles
Dianhui Wang
Dilip Kumar Pratihar
Dirk Heylen
Djamel F.H. Sadok
Don-Lin Yang
Dounias Georgios
Duco Nunes Ferro
Enrique Herrera Viedma
Esra Erdem
Evert Haasdijk
Fariba Sadri
Fernando Gomide
Fevzi Belli
Flavia Soares Correa da
    Silva
Floriana Esposito

Fran Campa
Francesco Marcelloni
Francisco Fernández de
    Vega
Francisco Herrera
François Jacquenet
Frank Klawonn
Gérard Dreyfus
Greg Lee
Gregorio Ismael
    Sainz Palmero
Hamido Fujita
Hannaneh Najd Ataei
Hans Guesgen
Hasan Selim
He Jiang
Henri Prade
Herrera Viedma Enrique
Hiroshi G. Okuno
Honggang Wang
Huey-Ming Lee
Hui-Yu Huang
Humberto Bustince
Hyuk Cho
Jaziar Radianti
Jean-Charles Lamirel
Jeng-Shyang Pan
Jesús Maudes
Jing Peng
João Paulo Carvalho
João Sousa
Joaquim Melendez
    Frigola
John M. Dolan
José Valente de Oliveira
Jose-Maria Peña
Julian Vicent
Jun Hakura
Jyh-Horng Chou
Kaoru Hirota
Kazuhiko Suzuki
Khosrow Kaikhah
Kishan G. Mehrotra
Kuan-Rong Lee
Kurash Madani

Lars Braubach
Lei Jiao
Leszek Borzemski
Linlin Tang
Li-Wei Lee
Lin-Yu Tseng
Maciej Grzenda
Manabu Gouko
Manton M. Matthews
Manuel Lozano
Marcilio Carlos Pereira
    De Souto
Marco Valtorta
Mark Last
Mark Neerincx
Mark Sh. Levin
Martijn Warnier
Michele Folgheraiter
Miquel Sànchez i Marrè
Mong-Fong Horng
Murat Sensoy
Nashat Mansour
Natalie van der Wal
Ngoc Thanh Nguyen
Nicolás García-Pedrajas
Niek Wijngaards
Oscar Cordón
Paolo Rosso
Patrick Brezillon
Patrick Chan
Paul Chung
Pinar Karagoz
Prabhat Mahanti
Qian Hu
Qingzhong Liu
Reyhan Aydogan
Richard Dapoigny
Robbert-Jan Beun
Rocio Romero Zaliz
Rosso Paolo
Safeeullah Soomro
Sander van Splunter
Shaheen Fatima
Shogo Okada
Shou-Hsiung Cheng

Table of Contents – Part II

Samrt Data-Mining in Industrial Applications

Innovations in Intelligent Systems and Applications

Media Processing

Media Processing

Smart Living

Smart Living

Information Retrieval

Intelligence Systems for E-Commerce amd Logistics

Table of Contents – Part I

Data Mining and QA Technology

Optimization in Industrial Applications

Pattern Recognition and Machine Learning

Innovations in Intelligent Systems and Applications

Machine Learning

Intelligent Industrial Applications and Control Systems

Multi-objective Optimzation

A Framework to Evolutionary Path Planning for Autonomous Underwater Glider

Shih Chien-Chou[1], Yang Yih[1], Mong-Fong Horng[2], Pan Tien-Szu[2], and Jeng-Shyang Pan[3]

[1] Taiwan Ocean Research Institute, NARL, N0. 219, Sec. 1, Dongfang Rd., Qieding Dist., Kaohsiung City, Taiwan (R.O.C)
{raymondshih,yy}@narlabs.org.tw
[2] National Kaohsiung University of Applied Sciences, N0. 415, Chienkung Rd., Sanmin Dist., Kaohsiung City, Taiwan (R.O.C)
{mfhorng,tpan}@cc.kuas.edu.tw
[3] Harbin Institute of Technology Shenzhen Graduate School, HIT Campus of ShenZhen University Town, XiLi, ShenZhen, China
jengshyangpan@gmail.com

Abstract. In recent decade years, AUG has been attached importance to oceanographic sampling tool. AUG is a buoyancy driven vehicle with low energy consumption, and capable of long-term and large-scale oceanographic sampling. However, ocean environment is characterized by variable and severe current fields, which jeopardizes AUG cruise. Therefore, an efficient path planning is a key point that can assist AUG to arrive at each waypoint and reduces the energy consumption to prolong AUG sampling time. To improve AUG cruise efficiency, a path planning framework with evolutionary computation is proposed to map out an optimal cruising path and increases AUG mission reachability in this work.

Keywords: We would like to encourage you to list your keywords in this section.

1   Introduction

Autonomous Underwater Glider (AUG) was painted by Henry Stommel. Initially, AUG was acquired to establish the World Ocean Observing System (WOOS), a facility capable of monitoring the global ocean, using a fleet of small neutrally-buoyant floats called Slocum that draw their power from the temperature stratification of the ocean [1]. In 2001, a variety of Autonomous Underwater Vehicles (AUV) were developed and supported by a communication and power infrastructure, comprise the basic observation system, applied in Autonomous Ocean-Sampling Network (AOSN) [2]. Among these AUVs, a slow moving, buoyancy driven with low power consumption, called AUG, was employed for long-term and large-scale ocean sampling processes. Three types of AUG were employed in AOSN mission: Seaglider, Spray and Slocum, respectively. Seaglider is designed to glide from the ocean surface to a programmed depth and back while measuring temperature, salinity, depth-averaged

A. Moonis et al.   (Eds.): IEA/AIE 2014, Part II, LNAI 8482, pp. 1–11, 2014.

current, and other quantities along a saw-tooth trajectory through the water [3]. Spray is an autonomous profiling float that uses a buoyancy engine to cycle vertically and wings to glide horizontally while moving up and down [4]. Slocum is operated 40,000 kilometers range which harvests its propulsive energy from the heat flow between the vehicle engine and the thermal gradient of the temperate and tropical ocean [5]. These three types of AUG are small and reusable, designed for the missions of several thousand kilometers and many months.

AUG is designed for deep water where the vehicle can traverse large areas with minimal use of energy and are specifically designed for the needs of the Blue Water scientist, which require greater control over the vehicle [6]. However, ocean environment is variant and unpredictable. The cruise of AUG is highly influenced by ocean current so that AUG is difficult to procure the mission. Therefore, AUG is equipped satellite transceiver such as Iridium terminal so that it communicates with onshore or offshore station to receive mission commands or transmit its position. AUG is also equipped Global Positioning System (GPS) receiver to fix its location. However, radio or Electromagnetic (EM) communication between AUG and satellite can only be established when AUG surfacing because propagation of EM wave is severe attenuation in water. Therefore, AUG can't rely on EM communication when AUG is diving into water.

To overcome this problem, some algorithms are applied in AUG path planning. In traditional, AUG path planning is based on dead-reckoning. Dead-reckoning is a process of calculating one's current position by using a previously determined position and advancing that position based upon known or estimated speeds over elapsed time and course [7]. In recent decade years, simulated evolution is quickly becoming the method of choice for complex problem solving, especially when more traditional methods cannot be efficiently applied or produce unsatisfactory solutions [8]. Evolutionary computation (EC) is the field of research devoted to the study of problem solving via simulated evolution [9].

In this paper, a framework to evolutionary path planning for AUG is proposed. Before discussion of framework, characteristic of AUG is reviewed in Section 2. Session 3 introduces Evolutionary computation. The framework is based on object-oriented approach and consisted of four importance objects: ocean current model, TSP algorithm, AUG and waypoint, is represented in Section 4. Section 5 demonstrates the simulation, and finally, conclusion and feature work are discussed in Section 6.

2   Autonomous Underwater Glider

AUG is autonomous vehicle that profile vertically by controlling buoyancy and move horizontally on wings. AUG propels themselves by changing buoyancy and using wings to produce forward motion. Buoyancy is changed by varying the vehicle volume to create a buoyancy force. Wing lift balances the across-track buoyant force while the forward buoyant force balances drag. Table 1 summarizes the specifications of three AUGs [10]. Without active propeller, the disadvantage of AUG is difficult to overcome ocean current when ocean current exceeds half of AUG velocity [11].

In general, a path planning is proposed according to mission of AUG sampling before AUG operation. AUG operation may include release, positioning, diving, surfacing and retrieve procedures [12], as shown in Fig. 1. Release is to push AUG into sea surface from onshore or offshore station, and then AUG starts to obtain GPS signal from GPS satellite. Once AUG fixes its position and transmits its position backup to station via satellite communication such as Iridium, then AUG is ready to perform ocean sampling mission and slips into water. During ocean sampling period, AUG adjusts the buoyancy engine to dive or climb so that trajectory of AUG is like a saw-toothed pattern. Normally, one dive-climb AUG will float on sea surface and correct its position and transmit its position back to station again. Sometimes AUG's path needs to be changed, AUG will receive a new command from station via satellite communication. After AUG arrives all of waypoints or ocean sampling mission is accomplished, AUG floats on sea surface and waits for retrieving.

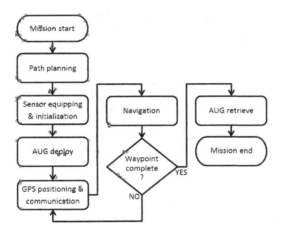

Fig. 1. AUG operation

AUG reserves a space to place scientific devices such as oceanic sensors and communication modules. Usually, communication modules are equipped in AUG can be classified two types: EM and acoustic waves. Normally, EM and acoustic waves are used in AUG surfacing and diving, respectively. Satellite communications such as GPS receiver and Iridium transceiver use EM wave to establish communication links while underwater communications such as sonar and transponder use acoustic wave to establish communication links.

EM wave cannot be adapted in underwater environment is because of the velocity of EM waves in water is more than 4 orders faster than acoustic waves so the channel latency is greatly reduced. The primary limitation of EM wave propagation in water is the high attenuation due to the conductivity of water [12]. Fig. 2 illustrates that EM wave transports in fresh water has highly propagation loss.

Table 1. Specifications of Seaglider, Spray and Slocum

Seaglider	
Hull	Length 180 cm, diameter 30 cm, mass 52 kg, payload 4 kg
Lift surfaces	Wing span 100 cm, vertical stabilizer span 40 cm
Batteries	81 D Lithium cells in 2 packs, energy 10 MJ, mass 9.4 kg
Communication	Iridium, 180 byte/s net, GPS navigation
Operating	Max P 1000 dbar, max U 45 cm/s, control on depth+position+attitude+vertical W
Endurance	U = 27 cm/s, 16° glide, buoyancy 130 gm, range 4600 km, duration 200 days
Spray	
Hull	Length 200 cm, diameter 20 cm, mass 51 kg, payload 3.5 kg
Lift surfaces	Wing span 120 cm, vertical stabilizer length 49 cm
Batteries	52 DD Lithium CSC cells in 3 packs, energy 13 MJ, mass 12 kg
Communication	Iridium, 180 byte/s net, GPS navigation
Operating	Max P 1500 dbar, max U 45 cm/s, control on depth+altitude+attitude+vertical W
Endurance	U = 27 cm/s 18° glide, buoyancy 125 gm, range 7000 km, duration 330 days
Slocum	
Hull	Length 150 cm, diameter 21 cm, mass 52 kg, payload 5 kg
Lift surfaces	Wing span 120 cm, stabilizer length 15 cm
Batteries	250 Alkaline C cells, energy 8 MJ, mass 18 kg
Communication	Freewave LAN or Iridium, GPS navigation
Operating	Max P 200 dbar, max U 40 cm/s, control on depth+altitude+attitude+vertical W
Endurance	U = 35 cm/s, 25° glide, buoyance 230 gm, range 500 km, duration 20 days

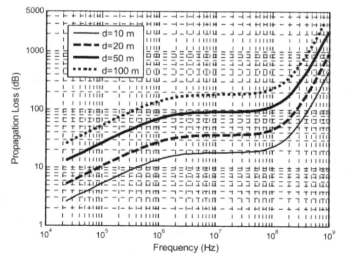

Fig. 2. Propagation loss at deep fresh water [12]

As mentioned above shows that two main weaknesses are limited AUG cruise: motion control and positioning. Therefore, an efficient path planning applied in AUG cruise can assist AUG to improve mission reachability and accomplish the ocean sampling goal.

3   Evolutionary Computation

Over the past 30 years the field of EC has itself been evolving. Originally, the first generation of EC consisted of three evolution-based paradigms: evolution strategies, evolutionary programming and genetic algorithms (GAs). Each of these evolutionary techniques was developed for solving distinct problems [9]. The evolutionary computation is a subfield of computational intelligence which is used to solve NP-hard combinatorial optimization problem. Architecture of computational intelligence is shown in Fig. 2. Any problem belongs to the NP-class can be formulated as Travelling Salesman Problem (TSP) which is to visit each city exactly once and find the shortest tour or path. The well-known algorithms of evolutionary computation are such as Genetic Algorithm (GA), Ant Colony Optimization (ACO), Particle Swarm Optimization (PS) and so on. These algorithms possess the meta-heuristic or stochastic character to find a global optimal solution.

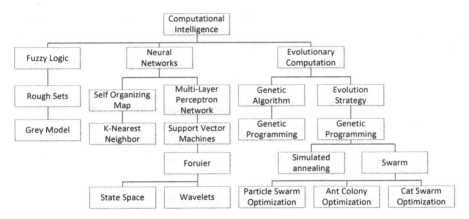

Fig. 3. Architecture of computational intelligence

Algorithm belongs to EC is characterized by a meta-heuristic or stochastic optimization character with a global optimization method. EC adapts iterative process to evolve the offspring in populations which are randomly generated using specified method, and compares each solution to achieve the desired end. Programming for EC is called Evolutionary Programming (EP). EP is introduced by Lawrence J. Fogel, who is known as the father of EP and a pioneer in EP. The first EP is realized by John Henry Holland while his method is called Genetic Algorithm [13]. EC is wildly applied in a lot of area and some applications are listed in Table 2.

Table 2. Applications of EC

EC	Application	Article
GA, PSO	Electromagnetic	[15, 28]
GA, ACO, CSO	Wireless Sensor Networks	[16, 22, 29]
GA, PSO	Image Process	[17, 18]
GA	Service Scheduling	[19]
GA, ACO	Power System	[20, 26]
GA	IC Design	[21]
ACO	Optical Networks	[23]
ACO	Routing Protocol	[24, 25]
PSO	Path Planning	[27]

4   Framework

In this paper, the framework is called Autonomous Underwater Glider Path Planning (AUGPP). Four main objects are constructed the framework: AUG, ocean current model, waypoint and TSP algorithm, as shown in Fig. 4. Waypoint object consists of coordinate (latitude, longitude and depth) and tour. Tour is a sequence of randomized waypoints, also called a path where AUG travels on. In GA, a tour is a population and several tours construct a solution space. After iterative evolution, the optimal solution may be found, namely, an optimal tour is suitable for a specified problem or path planning. Ocean current model object mimics an oceanic environment for AUG cruising with fixed or variant current fields. AUG object models a physical AUG parameters such as motion or attitude (heading, pitch and roll), and position. Some modules of scientific device can be implemented here. TSP algorithm object is a container which composes of EC algorithm. EC algorithm is applied in AUGPP to find an optimal solution, namely, an optimal path. An optimal path can be a shortest distance path or a lowest cost path. Cost can be formulated a function to represent time or power consumption, or travelling distance. Cost is a variable if ocean current field is not stationary.

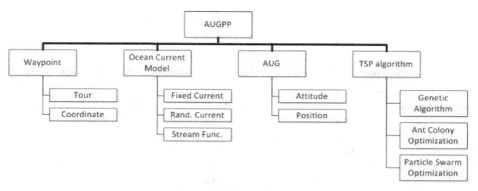

Fig. 4. AUGPP architecture

A framework to evolutionary path planning for AUG is depicted in Fig. 5. Navigation path and navigation control indicate two stages: offline path planning and online path control. In navigation path stage, a path is offline planned when an ocean sampling mission is initiated. Path planning is highly depending on mission due to different mission has a different path planning. In navigation control stage, AUG needs to control its motion by specified algorithm. Dead-reckoning algorithm is a basic path planning method that is applied in Seaglider [3], Spray [4] and Slocum [5] gliders. Parameters such as AUG parameters, attitude and altimeter of AUG, and current fields are input parameters in both stages. Waypoint is only needed in navigation path stage. When evolutionary path planning is finished and an optimal path is found from waypoints, an optimal path is became an ocean sampling path for a specified mission. AUG will load this optimal path into internal memory and follow it to complete the mission. Kinematic and energy models is implemented in AUG to cyclic control its motion in real-time. Except for EC algorithm, other method such as graph theory can also be implemented in AUGPP. In graph theory, waypoints can be indicated a node and converted to an adjacent matrix or reachable matrix.

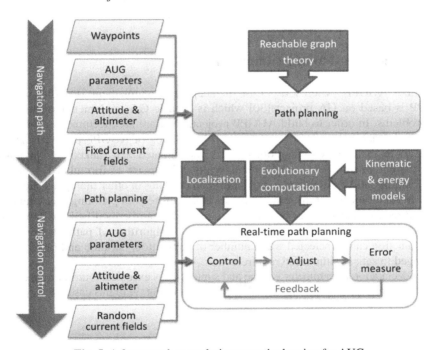

Fig. 5. A framework to evolutionary path planning for AUG

To model an ocean environment, an Extensible Ocean Model (EOM) [12], as shown in Fig. 6, can be implemented in AUGPP to simplified ocean current model. EOM is discretized in space over $n_x \times n_y \times n_z$ regular grids along the three Cartesian directions. Ocean space of OEM is scalable according to waypoints. Once ocean space is created, ocean current fields such as fixed current field, random current field and stream function current field can be applied in OEM.

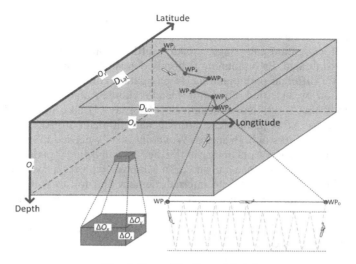

Fig. 6. Extensible Ocean Model

5   Demonstration of Framework

To realize AUGPP, Microsoft VC++ is adapted to implement AUGPP project. AUGPP is based on GA project [30] which is a basic GA implementation to solve TSP problems. In order to build AUGPP project, software architecture of GA project has been restructured to add more features. Main dialog of AUGPP project is shown in Fig. 7. In main dialog, four objects are implemented and ranked by waypoint, ocean current model, AUG and TSP algorithm. Waypoint is formulated as a text file (including latitude, longitude and depth) and drawing in other dialog, as shown in Fig. 8. In each object, parameters can be declared before path planning. For example, current direction and velocity of ocean current model can be modified and attributes of AUG can be adjusted. In TSP algorithm object, algorithm of path planning and fitness function can be selected while number of population, iteration and run can also be changed to satisfy simulation requirement.

After declaration of parameter is done and simulation is executed, tour information and simulation result are shown in main dialog, as shown in Fig. 9. In this case, 48 waypoints

Waypoints	Current Model	Autonomous Under Glider	Algorithm
Open WP	Fixed	Number of glider 1	Path planning Offline
	Constrained Random		Algorithm GA
0 -119.793323 22.119149 -1579.00000	Streamfunction	Nominal velocity (m/s) 0.350000	Fitness function Distance
1 -119.567207 21.935395 -2419.00000			
2 -119.868687 21.844302 -2270.00000	Direction East-Northward	Max. pitch angle 25.000000	Populations 50
3 -119.845847 21.589262 -3102.00000			
4 -119.551827 21.677659 -2852.00000	U (m/s) 0.200000 UU Range 0.100000	Control cycle (s) 60.000000	Iterations 500	Run
5 -119.265520 21.998427 -1997.00000			
6 -119.477042 21.338638 -3181.00000	V (m/s) 0.100000 VV Range 0.100000	Surfacing Time(s) 650	Run Number:
7 -120.069950 21.251239 -3015.00000			1 Reset counter Reset List2
8 -119.236749 21.600301 -2743.00000	Z (m/s) 0.000000 ZZ Range 0.100000		

Fig. 7. Main dialog

are planned in this mission. Current direction and velocity are defined as fixed, and GA is applied in TSP algorithm. Simulation result shows a sequence of optimal path followed by distance between waypoints, and displays total distance of optimal path.

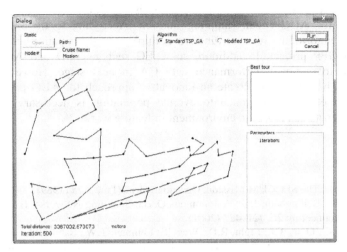

Fig. 8. Waypoint drawing

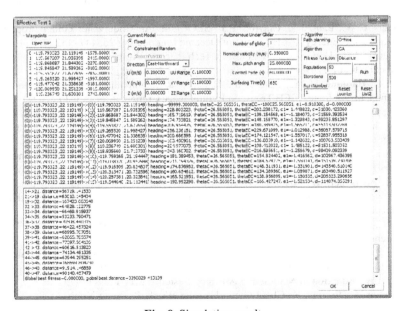

Fig. 9. Simulation result

6 Conclusion and Feature Work

In this paper, a framework of evolutionary path planning for AUG is proposed. To successfully accomplish an ocean sampling mission, a key point of path planning is necessary to guide AUG on the right way. EC is an efficient technique to find an optimal path for AUG ocean sampling mission. Accroding to the framework of AUGPP, GA algorithm with cost function is implemented for path planning in first stage of AUGPP project. In addition, some EC such as PSO and ACO will be implemented to compare performance with GA in next stage. However, appling present EC is insufficient to create an innovative approach for AUG path planning. Therefore, an effecitive approach to evolute population is necessary to estimate variable path cost such as ocean environment in feature work.

References

1. Stommel, H.: The SLOCUM Mission. Oceanography 2(1), 22–25 (1989)
2. Curtin, T.B., Bellingham, J.G.: Autonomous Ocean-Sampling Networks. IEEE Journal of Oceanic Engineering 26, 421–423 (2001)
3. Eriksen, C.C., Osse, T.J., Light, R.D., Wen, T., Lehman, T.W., Sabin, P.L., Ballard, J.W.: Seaglider: A Long-Range Autonomous Underwater Vehicle for Oceanographic Research. IEEE Journal of Oceanic Engineering 26, 424–436 (2001)
4. Sherman, J., Davis, R., Owens, W.B., Valdes, J.: The Autonomous Underwater Glider "Spray". IEEE Journal of Oceanic Engineering 26, 437–446 (2001)
5. Douglas, C.W., Paul, J.S., Clayton, P.J.: SLOCUM: An Underwater Glider Propelled by Environmental Energy. IEEE Journal of Oceanic Engineering 26, 447–452 (2001)
6. Alexander, V.I.: Underwater Vehicles. In-Tech, Vienna, Austria (2009)
7. Wikipedia, http://en.wikipedia.org/wiki/Dead_reckoning
8. Bäck, T., Hammel, U., Schwefel, H.P.: Evolutionary Computation: Comments on the History and Current State. IEEE Trans. on Evolutionary Computation 1, 3–17 (1997)
9. Gerry, D., Abdollah, H., Edward, T., Darryl, B.: An Introduction to Evolutionary Computation: Intelligent Control Systems Using Soft Computing Methodologies, ch. 17. CRC Press (2001)
10. Rudnick, D.L., Davis, R.E., Eriksen, C.C., Fratantoni, D.M., Perry, M.J.: Underwater Gliders for Ocean Research. Marine Technology Society Journal 38, 48–59 (2004)
11. Garau, B., Bonet, M., Álvarez, A., Ruiz, S., Pascual, A.: Path planning for autonomous underwater vehicles in realistic oceanic current fields: Application to gliders in the Western Mediterranean sea. Journal of Maritime Research 6, 5–21 (2009)
12. Shih, C.-C., Horng, M.-F., Pan, J.-S.: 3-D Adaptive Bearing Sampling for Route Planning in Extensible Ocean Model. In: COUTA 2012, pp. 71–84 (2012)
13. Jiang, S., Georgakopoulos, S.: Electromagnetic Wave Propagation into Fresh Water. Journal of Electromagnetic Analysis and Applications 3, 261–266 (2011)
14. John, H.H.: Adaption in Natural and Artificial Systems. MIT Press (1975)
15. Youngjun, A., Jiseong, P., Cheol-Gyun, L., Jong-Wook, K., Sang-Yong, J.: Novel Memetic Algorithm implemented With GA (Genetic Algorithm) and MADS (Mesh Adaptive Direct Search) for Optimal Design of Electromagnetic System. IEEE Transactions on Magnetics 46, 1982–1985 (2010)

16. Yin, W., Wenbo, L.: Routing protocol based on genetic algorithm for energy harvesting-wireless sensor networks. IET Wireless Sensor Systems 3, 112–118 (2013)
17. GuangJian, T., Yong, X., Zhang, Y., Dagan, F.: Hybrid Genetic and Variational Expectation-Maximization Algorithm for Gaussian-Mixture-Model-Based Brain MR Image Segmentation. IEEE Transactions on Information Technology in Biomedicine 15, 373–380 (2011)
18. Brinkman, W., Thayaparan, T.: Focusing inverse synthetic aperture radar images with higher-order motion error using the adaptive joint-time-frequency algorithm optimised with the genetic algorithm and the particle swarm optimisation algorithm - comparison and results. IET Signal Processing 4, 329–342 (2010)
19. Ip, W.H., Dingwei, W., Cho, V.: Aircraft Ground Service Scheduling Problems and Their Genetic Algorithm With Hybrid Assignment and Sequence Encoding Scheme. IEEE Systems Journal 7, 649–657 (2013)
20. Xiangning, L., Shuohao, K., Zhengtian, L., Hanli, W., Xionghui, H.: A Fault Diagnosis Method of Power Systems Based on Improved Objective Function and Genetic Algorithm-Tabu Search. IEEE Transactions on Power Delivery 25, 1268–1274 (2010)
21. Hsien-Chie, C., Chung, I.C., Wen-Hwa, C.: Thermal Chip Placement in MCMs Using a Novel Hybrid Optimization Algorithm. IEEE Transactions on Components, Packaging and Manufacturing Technology 2, 764–774 (2012)
22. Xuxun, L.: Sensor Deployment of Wireless Sensor Networks Based on Ant Colony Optimization with Three Classes of Ant Transitions. IEEE Communications Letters 16, 1604–1607 (2012)
23. Triay, J., Cervello-Pastor, C.: An ant-based algorithm for distributed routing and wavelength assignment in dynamic optical networks. IEEE Journal on Selected Areas in Communications 28, 542–552 (2010)
24. Karia, D.C., Godbole, V.V.: New approach for routing in mobile ad-hoc networks based on ant colony optimisation with global positioning system. IET Networks 2, 171–180 (2013)
25. Li-Ning, X., Rohlfshagen, P., Ying-Wu, C., Xin, Y.: A Hybrid Ant Colony Optimization Algorithm for the Extended Capacitated Arc Routing Problem. IEEE Transactions on Systems, Man, and Cybernetics, Part B: Cybernetics 41, 1110–1123 (2011)
26. Young-Min, K., Eun-Jung, L., Hong-Shik, P.: Ant Colony Optimization Based Energy Saving Routing for Energy-Efficient Networks. IEEE Communications Letters 15, 779–781 (2011)
27. Yangguang, F., Mingyue, D., Chengping, Z.: Phase Angle-Encoded and Quantum-Behaved Particle Swarm Optimization Applied to Three-Dimensional Route Planning for UAV. IEEE Transactions on Systems, Man and Cybernetics, Part A: Systems and Humans 42, 511–526 (2012)
28. Xiongbiao, L., Mori, K.: Robust Endoscope Motion Estimation Via an Animated Particle Filter for Electromagnetically Navigated Endoscopy. IEEE Transactions on Biomedical Engineering 61, 85–95 (2014)
29. Temel, S., Unaldi, N., Kaynak, O.: On Deployment of Wireless Sensors on 3-D Terrains to Maximize Sensing Coverage by Utilizing Cat Swarm Optimization With Wavelet Transform. IEEE Transactions on Systems, Man, and Cybernetics: Systems 44, 111–120 (2014)
30. Genetic Algorithms Applied to Travelling Salesman Problems in C++, http://www.technical-recipes.com/2012/genetic-algorithms-applied-to-travelling-saleman-problems-in-c/

A Three-Stage Decision Model Integrating FAHP, MDS and Association Rules for Targeting Smartphone Customers

Shiang-Lin Lin[*] and Heng-Li Yang

Department of Management Information Systems
National Chengchi University, Taipei, R.O.C.
{102356505,yanh}@nccu.edu.tw

Abstract. Human living consists of a series of decision making that involves a lot of decision objectives and factors. Decision makers would be concerned with the best selection among different alternatives. In this study, we propose a three-stage decision making model combining with Fuzzy Analytic Hierarchy Process (FAHP), Multidimensional Scaling (MDS) and Association Rules (AR). The proposed model enables decision makers to find out the importance of each decision criterion and select the optimal alternative for solving decision problems. We applied the proposed model to understand the consumer's evaluation factors in their smartphone purchasing. In stage I, the FAHP analysis results indicate the decision criteria as follows. When customers intend to buy a smartphone, the order of importance of their decision criteria is as follows: display size, thickness, and hardware efficiency. Further, in stage II, several interesting AR analysis rules reveal that consumers of high self-glory would consider the hardware efficiency and the pixel of lens of the product when they purchase a smartphone. Finally, through the perceptual map of MDS in stage III, we can classify consumers into four groups, namely, "students", "IT professionals", "house brokers and bankers & insurers", and "teachers and healthcare workers". Customers in the same group have similar evaluation criteria in their smartphone purchasing. The integrated decision model proposed in this study can help the suppliers in making right marketing strategies for different consumer groups, and can also be applied in various fields for decision marking in the future.

Keywords: Multi-Criteria Decision Making (MADM), Association Rules, Fuzzy Analytic Hierarchy Process (FAHP), Multidimensional Scaling (MDS).

1 Introduction

Human living consists of a series of decision making that ranging from trivial things to important business issues. Facing all these decision problems, decision makers need to consider with so many objects and factors. When making selections among multiple alternatives, most people become hesitant and unable to make the correct decision due to insufficient information and unclear outcome. In view of this, the principal concern for the decision makers is how to pick and act on the optimal plan out of the others.

[*] Corresponding Author.

A. Moonis et al. (Eds.): IEA/AIE 2014, Part II, LNAI 8482, pp. 12–21, 2014.

Some decision methods assist decision makers to resolve difficulty of decision making, as well as make the best decision. Keeney and Raiffa proposed Multiple Criteria Decision Making (MCDM) that facilitated the decision maker to consider correctly and make the rational decision when multiple objects need to be evaluated and selected [1]. The primary part of this method is to find out the optimal decision during the decision making process based on multiple criteria. MCDM have been widely applied in many fields, including industrial R&D [2], Environmental Engineering [3] and financial analysis [4].

Among various decision making methods, Analytic Hierarchy Process (AHP) is a MCDM method which is most commonly used [5]. It could structuralize the complicated decision making issues and privide criteria and weight for assessment, then verifies the consistency of each criterion [6], and builds a hierarchical decision-making structure   [7]. Via this structure, decision maker can see the relation between hierarchies and take the priorities of the factors into consideration. However, the traditional AHP is unable to well represent the human thought by fuzziness, result in the experts determine the priority of the factors, whose may cause deviation of the consequence [8], [9]. Therefore, Buckley proposed the Fuzzy Analytic Hierarchy Process (FAHP) integrating the Fuzzy theory with AHP [10]. It primarily improved the shortcomings of fuzziness and subjective opinion, when conducting the AHP questionnaire, which might result in the deviation of the evaluation results.

In addition to the AHP, Association Rule (AR) analysis is the most common method in decision science area, which could find out the hidden correlating rules among a mass of datum. These rules can effectively provide the results of prediction and decision making. AR has been widely applied in the fields of production process [11], marketing solution [12] and supply chain [13] and so on. Furthermore, recent research devoted to represent the alternative solution for decision problem solving, e.g., Multidimensional Scaling (MDS). MDS is a method to simplify the data, which presents the complicated relation between the alternatives in a visualized way. The principal purpose of MDS is to allow decision makers to observe the similarities and differences between the alternatives quickly from the Perceptual Map. MDS has been widely applied in various fields such as psychology or marketing [14].

Overall, FAHP can facilitate the decision maker to find out the optimal solution from evaluative factors. If combining with MDS and AR, it would assist decision makers to make decisions more accurately, and to understand the relation among all alternatives more clearly. Based on above reasons, this study combines with these three methods, FAHP, MDS, and AR to propose a three-stage integrated decision model. First, it applies FAHP and MDS to assist decision makers to select appropriate plans for decision making issues. Moreover, the proposed model would apply AR to reveal the criteria with correlation, to provide foundation for decision makers when carrying out the alternative.

2   Literature Review

2.1   Fuzzy Analytical Hierarchy Process

Analytic Hierarchy Process (AHP) was proposed by Saaty in 1977, and has been widely applied in ranking, evaluation and prediction. The application of AHP assumes the hypothesis that a system could be divided into multiple classes and components; and forms the hierarchical structure among all components with independence [15].

In 1985, Buckley proposed the Fuzzy Analytic Hierarchy Process (FAHP) method that integrated the Fuzzy theory with AHP. FAHP can improve the disadvantage of fuzziness and subjective opinions brought by the component pairwise comparison when conducting the AHP expert questionnaire survey, Compared with the traditional AHP, FAHP requires more complicated calculation steps, which are detailed as below:

- Build the hierarchical structure of evaluation. Confirm the decision making issues, select dimensions, criteria and alternatives for the target decision making.
- Conduct component pairwise comprison: After the hierarchical structure of evaluation is built, we adopt the 9-level pairwise comparison scale proposed by Saaty between the components as shown in Table 1, to conduct pairwise comparison for each component. Through the fuzzy semantic membership functions, we convert the nine levels into triangular fuzzy numbers [16].

Table 1. Pairwise comparison scale and triangular fuzzy numbers conversion

Scale	Definition	$(\widetilde{M}_{ij}) = (L_{ij}, M_{ij}, R_{ij})$
1	Equally important	$\tilde{1}=(1, 1, 3)$
3	Weakly important	$\tilde{3}=(1, 3, 5)$
5	Essentially important	$\tilde{5}=(3, 5, 7)$
7	Very Strongly important	$\tilde{7}=(5, 7, 9)$
9	Absolutely important	$\tilde{9}=(7, 9, 9)$

- Build pairwise comparison matrix: On the upper triangular part of the pairwise comparison matrix A, shown in equation (1), we place the evaluation value of the comparison result for a group of components made up of $A_1, A_2, ..., A_n$ obtained by an expert, where a_{ij} represents the relative priority of component i to component j.

$$A = \begin{bmatrix} 1 & a_{12} & .. & .. & a_{1j} \\ 1/a_{12} & 1 & . & . & a_{2j} \\ . & . & 1 & . & . \\ . & . & . & 1 & . \\ 1/a_{1j} & 1/a_{2j} & . & . & 1 \end{bmatrix} \quad (1)$$

- Build the fuzzy positive reciprocal matrix: Through the above matrix A, we convert the matix value of the compenents into the triangular fuzzy numbers $\widetilde{M}_{ij}$ by using Table1. And we build the fuzzy positive reciprocal matrix M as shown in equation (2), where , $\widetilde{M}_{ij}$ is the relative fuzzy number of factor i to factor j, and $\widetilde{M}_{ji}=1/\widetilde{M}_{ij}$:

$$M = \left[\widetilde{M}_{ij}\right] = \begin{bmatrix} 1 & \widetilde{M}_{12} = (L_{12}, M_{12}, R_{12}) & \widetilde{M}_{13} = (L_{13}, M_{13}, R_{13}) & . & . & \widetilde{M}_{1j} = (L_{1j}, M_{1j}, R_{1j}) \\ \widetilde{M}_{21} = 1/\widetilde{M}_{12} & 1 & \widetilde{M}_{23} = (L_{23}, M_{23}, R_{23}) & . & . & \widetilde{M}_{2j} = (L_{2j}, M_{2j}, R_{2j}) \\ \widetilde{M}_{31} = 1/\widetilde{M}_{13} & \widetilde{M}_{32} = 1/\widetilde{M}_{23} & 1 & . & . & \widetilde{M}_{3j} = (L_{3j}, M_{3j}, R_{3j}) \\ . & . & . & 1 & . & . \\ . & . & . & . & 1 & \widetilde{M}_{ij} = (L_{ij}, M_{ij}, R_{ij}) \\ \widetilde{M}_{j1} = 1/\widetilde{M}_{1j} & \widetilde{M}_{j2} = 1/\widetilde{M}_{2j} & \widetilde{M}_{j3} = 1/\widetilde{M}_{3j} & . & \widetilde{M}_{ji} = 1/\widetilde{M}_{ij} & 1 \end{bmatrix} \quad (2)$$

- Calculate the max of eigenvalue (λ_{max}) and: through the eigenvalue solution method in the value analysis, it further finds out the eigenvector and the maximum eigenvalue λ_{max}, and calculates the weights of the factors in each hierarchy.
- Consistency verification: It may be very difficult for decision makers to achieve consistency when conducting the pairwise comparison. Therefore, it needs to verify the consistency to confirm the reasonable degree of the decision makers'

judgment during the evaluation process. Consistency index (C.I.) and consistency ratio (C.R.) can be used to verify the consistency of the weight.

o Consistency Index (C.I.): Saaty suggests $C.I.\leqq0.1$ is more ideal. It is calculated by $(\lambda_{max}-n)/(n-1)$, where n is the number of the factors.

o Consistency Ratio (C.R.): It is more difficult to verify the consistency when research problem becomes complicated. For this issue, Saaty proposes Random Index (R.I.), to adjust the changes of differnt C.I. values in different orders, namely, Consistency Ratio (C.R.), it is calculated by $C.I. / R.I.$. Saaty also suggests that the $C.R.\leqq0.1$ is better.

• Calculate fuzzy weights: Calculate the triangular fuzzy weights (W_i) based on the overall triangular fuzzy numbers (L_i, M_i, R_i) and the sum up of triangular fuzzy numbers (L_i'', M_i'', R_i''). The calculation is in equation (3):

$$W_i = (WL_i, WM_i, WR_i) = (L_i'/{L_i''}, M_i'/{M_i''}, R_i'/{R_i''}) \qquad (3)$$

• Defuzzification and Normalization: First, we de-fuzzy the triangular fuzzy weights obtained in the previous step, then convert them into a real number DW_i, and set the sum value DW_i of all components as 1. It then needs to conduct normalization of each component, as shown in equation (4):

$$DW_i = \frac{[(WR_i-WL_i)+(WM_i-WL_i)]}{3}+WL_i \qquad (4)$$

• Priority of evaluation factors: Through above steps, it could obtain the weight NW_i of dimension i under the goal, and the weight NW_{ij} of criterion j under dimension i. Further, the absolute weight NW_k of criterion k is figured out by $NW_i * NW_{ij}$. Through these weights, each criteria could be used to further rank the priorities of importance.

2.2   Multidimensional Scaling

Torgerson [17] proposed the Multidimensional Scaling (MDS) in 1952. MDS provides a visualized presentation through conducting pairwise comparison of the complex relation between n different components, and presents a perceptual map in low-dimension [18]. For obtaining this map, we must compute the Euclidean distance between each two components), further, the Euclidean distance matrix of the components will be generated via these values. The Euclidean distance function is shown as follows:

$$D_{ij} = \sqrt{(x_{i1}-x_{j1})^2 + (x_{i2}-x_{j2})^2 + \cdots + (x_{ik}-x_{jk})^2} \qquad (5)$$

where X_i denotes the percived value of component a, and X_j denotes the coordinate of alternative b.

2.3   Association Rules Analysis

Association Rules (AR) analysis is a data mining method proposed by Agrawal et al. in 1990s [19]. It is defined set $I = \{i_1, i_2, i_3 \cdots i_m\}$ as a set of a group of specific items, while business database D is the set of all transactional events T, where each data T is the subset of any item in I, namely, $T \subseteq I$.

AR is defined as $X \rightarrow Y$, where $X \subseteq I$, $Y \subseteq I$ and $X \cap Y = \emptyset$. Each criterion needs two parameters of "support" *Supp* and "confidence" *Conf* to judge whether this criterion has existence meaning. Assume the possibility of T in D containing X and Y is s%, then it presented $Supp(X \rightarrow Y) = Supp(X \cup Y)$; moreover, assume the possibility of T in D containing Y when it contains X is c%, then it presented $Conf(X \rightarrow Y) = Supp(X \cup Y)/Supp(X)$.

For a valid criterion, the parameters of Support ($Supp(X \rightarrow Y)$) and Confidence ($Conf(X \rightarrow Y)$) must be bigger or equal the minimum support (minsupp) and minimum confidence (minconf) which set by the user. Only the criterion meets the condition, will it be meaningful and representative.

3 The Proposed Model with Case Study

This study proposes a 3-stage integrated decision model as shown in Fig. 1 that consists of three stages.

- **Stage I.** Decision problem definition: Decision makers need to choose the goal of decision problem, further, select the decision dimensions, criteria and alternatives. Then, in order to build the hierarchical structure of evaluation, those will be taken as the foundation to design an expert questoinnaire.
- **Stage II.** Adopting FAHP, decision maker can acquire the relative fuzzy-weights for each decision component. Furthermore, via multiplying weights in each layer, the important priority of each criterion and alternative could be obtained. On other hand, applying AR, the proposed model could find out the correlation among the decision criteria and some interesting AR-rules can be revealed.
- **Stage III.** Conduct the integration analysis of relative fuzzy-weights via FAHP by using MDS. Through calculating the Euclidean distance matrix of alternatives, the proposed model can bring out the perceptual map for all alternatives allocation. According to this map, decision makers can get the similarity and difference among the alternatives of solutions.

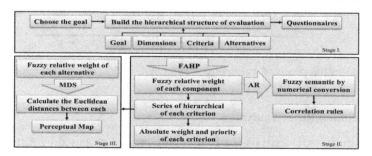

Fig. 1. Integrated decision model

4 Apply the Proposed Model to Understand the Evaluation Factors of Consumer's Smartphone Purchasing

In this study, we applied the proposed integrated decision model to understand the key factors of smartphone purchasing for different consumer groups. The hierarchical structure of evaluation consists of multiple decision dimensions, decision criteria and consumer groups for each layer.

4.1 Fuzzy Analytical Hierarchy Process

To build the hierarchical structure of evaluation as the decision making target confirmed, a decision maker needs to compare the evaluation components and the groups of consumers.

Build the Hierarchical Structure of Evaluation
Based on the selected components, the proposed model builds the hierarchical structure of evaluation, in which the levels from left to right are goal, dimensions (D), and criteria (C), as shown in Fig. 2. Six groups of consumers (G) are listed in the right-hand of Fig. 2. We referred to the structure of evaluation for designing the AHP expert-questionnaire; and invited the vendors of smartphone as subjects for the expert interview. Each subject was required to compare the components pairwise and also asked to compare six groups of consumers pair by pair. Finally, we received 40 expert questionnaires, and calculated the relative weight of the components in each questionnaire.

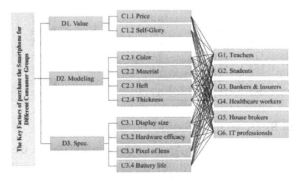

Fig. 2. Hierarchical structure of evaluation

The Relative Fuzzy-Weight
After all the questionnaires have retrieved, the pairwise comparison matrix for each question could be obtained. Afterward, it needs to be furhter convert into fuzzy pairwise comparison matrix, which can facilitate the calculation of the relative fuzzy-weight. Through converting the matrix, the triangular fuzzy number requires the calculation of its geometric mean first. Subsequently, via defuzzification and normalization, we could get the relative fuzzy-weight between all components, as shown in Table 2.

Table 2. Relative fuzzy-weight of all components

Components	Triangular fuzzy numbers	Defuzzification	Relative fuzzy-weight
D1	**(0.043, 0.072, 0.145)**	**0.086**	**0.066**
C1.1	(0.097, 0.125, 0.174)	0.132	0.128
C1.2	(0.649, 0.875, 1.168)	0.897	0.872
D2	**(0.144, 0.279, 0.682)**	**0.368**	**0.303**
C2.1	(0.106, 0.263, 0.696)	0.355	0.281
C2.2	(0.050, 0.118, 0.325)	0.164	0.130
C2.3	(0.029, 0.055, 0.145)	0.076	0.060
C2.4	(0.237, 0.564, 1.205)	0.669	0.529
D3	**(0.291, 0.649, 1.268)**	**0.736**	**0.632**
C3.1	(0.260, 0.576, 1.114)	0.650	0.536
C3.2	(0.107, 0.253, 0.643)	0.334	0.276
C3.3	(0.026, 0.044, 0.094)	0.055	0.045
C3.4	(0.066, 0.128, 0.327)	0.174	0.143

Priority of Evaluation Criteria

As getting the relative fuzzy-weight of all components, we further calculate the absolute fuzzy-weight in order to obtain the priority of criteria, as shown in Table 3.

Table 3. Absolute fuzzy-weight and priority of criteria

Criteria	Tiangular fuzzy numbers	Defuzzification	Absolute fuzzy-weight	Priority
C3.1	(0.004, 0.009, 0.025)	0.621	0.321	1
C2.4	(0.028, 0.063, 0.169)	0.338	0.175	2
C3.2	(0.015, 0.073, 0.474)	0.337	0.174	3
C2.1	(0.007, 0.033, 0.222)	0.188	0.097	4
C3.4	(0.004, 0.015, 0.099)	0.172	0.089	5
C2.2	(0.034, 0.157, 0.822)	0.087	0.045	6
C1.2	(0.076, 0.374, 1.413)	0.086	0.045	7
C3.3	(0.031, 0.164, 0.816)	0.052	0.027	8
C2.3	(0.008, 0.028, 0.119)	0.040	0.020	9
C1.1	(0.019, 0.083, 0.415)	0.013	0.007	10

4.2   Association Rules Analysis

Through FAHP, we further apply the Apriori algorithm to conduct the AR analysis for exploring whether there are meaningful rules among the criteria. In actual application, it needs to convert the continuous fuzzy weight of each criterion into the discrete fuzzy semantic based on the fuzzy semantic by numerical conversion as shown in Table 4 [20].

Table 4. Fuzzy semantic by numerical conversion

Scale	No. of terms used	Value												
		0.077	0.154	0.231	0.308	0.385	0.462	0.539	0.616	0.693	0.770	0.847	0.924	1.000
1	2							●			●			
2	3		Ignored			●	Minor		●	Important		●	Very Important	
3	5		●			●			●			●		●
4	5				●	●			●			●		●
5	6			●		●	●			●		●	●	
6	7			●	●			●		●	●		●	
7	9			●	●		●	●			●	●	●	●
8	11	●		●		●	●	●	●	●	●	●		●	●

Source: [20]

In this study, the importance is divided by five scales (i.e. four ranges) according to the triangular fuzzy number conversion in Table 1. Therefore, when conducting   data conversion, we adopt the scale 2 in Table 4, to also divide the relative fuzzy-weight of the criteria from each questionaire into four fuzzy semantics, namely, Ignored (0-0.307), Minor (0.308-0.538), Important (0.539-0.769) and Very Important (0.770-1). As converting fuzzy semantic, we conduct the AR analysis, and further obtain some meaningful rules among the criteria as shown in Table 5.

Table 5. An example of assoication rules

Left-hand Side	Right-hand Side	Confidence
C1.2 Self-glory (Important)	C3.2 Hardware efficiency (Very Important)	0.97
C1.2 Self-glory (Very Important)	C3.3 Pixel of lens (Very Important)	0.91
C2.4 Thickness (Very Important)	C3.4 Battery life (Important)	0.88

4.3 Multidimensional Scaling Analysis

Through FAHP, the relative fuzzy-weight of all criteria of consumer groups can be achieved, as shown in Table 6.

Table 6. The weight of consumer groups

Criteria	Relative fuzzy-weight					
	G1. Teachers	G2. Students	G3. Bankers & Insurers	G4. Healthcare workers	G5. House brokers	G6. IT professionals
C1.1	0.27	0.43	0.07	0.15	0.05	0.02
C1.2	0.04	0.03	0.19	0.06	0.23	0.45
C2.1	0.03	0.43	0.08	0.03	0.15	0.27
C2.2	0.04	0.08	0.15	0.03	0.27	0.43
C2.3	0.05	0.16	0.39	0.07	0.31	0.03
C2.4	0.05	0.03	0.27	0.06	0.44	0.15
C3.1	0.14	0.03	0.27	0.05	0.43	0.09
C3.2	0.03	0.04	0.25	0.06	0.17	0.45
C3.3	0.19	0.42	0.05	0.23	0.09	0.03
C3.4	0.03	0.08	0.39	0.03	0.3	0.16

According to these weights, we further calculated the Euclidean distances between each two alternatives, and built the Euclidean distance matrix, as shown in Table 7.

Table 7. Euclidean distance matrix between groups of consumers

Alternatives	G1.	G2.	G3.	G4.	G5.	G6.
G1. Teachers	0	-	-	-	-	-
G2. Students	1.66	0	-	-	-	-
G3. Bankers & Insurers	2.21	2.76	0	-	-	-
G4. Healthcare workers	0.51	1.73	2.12	0	-	-
G5. House brokers	2.41	2.93	1.01	2.39	0	-
G6. IT professionals	2.64	2.98	2.19	2.49	2.22	0

By the Euclidean distance matrix, we could generate a perceptual map, as shown in Fig. 3, which indicates four different purchasing groups of consumers.

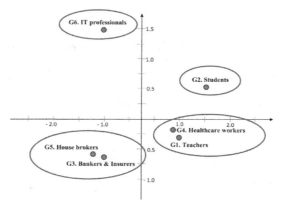

Fig. 3. Perceptual Map of two dimensions space

5   Conclusions

People always face with the decision making problem and alternative selection issues in daily life. This study proposes a three-stage integrated decision model and applied to explore the evaluation factors of smartphone purchasing. According to the result of FAHP, evaluation factor emphasized by the consumer is "display size", "thickness" and "hardware efficiency" when purchasing smartphone. Further, through the AR analysis, we found that consumers of high "self-glory" would put more emphasis on "hardware efficiency" and "pixel of lens". On the other hand, most consumers who put more emphasis on "thickness" would also pay more attention to "battery life". In addition, appling with the result of percuptual map in MDS, we can realize that there are four target consumer groups: "students", "IT professionals", "house brokers and bankers & insurers", "teachers and healthcare workers", while considering the smartphone purchase. Therefore, according to our model, supplier can set the target customers and select the promotion programs. Moreovr, the proposed model can indeed help the supplier make the right marketing strategies for different consumer groups. For example, when suppliers intend to release new products, they could choose the most appropriate consumer groups as the main target for new product promotion.

The integrated decision model proposed in this tudy can be applied to the decision making analysis of various fields in the future, such as SCM, BPM and so on. It could also be integrated with other decision making methods, such as DEMATEL and TOPSIS, to improve the accuracy and reliability of the study results.

References

1. Keeney, R.L., Raiffa, H.: Decision with Multiple Objectives: Preferences and Value Tradeoffs. The Press Syndicate of the University of Cambridge (1976)
2. Oral, M., Kettani, O., Lang, P.: A Methodology for Collective Evaluation and Selection of
   Industrial R & D Projects. Management Science 37(7), 871–885 (1991)

3. Teng, J.Y., Tzeng, G.H.: Multicriteria Evaluation for Strategies of Improving and Controlling Air Quality in the Super City: A Case Study of Taipei City. Journal of Environmental Management 40(2), 213–229 (1994)
4. Hallerbach, W., Spronk, J.: The Relevance of MCDM for Financial Decisions. J. Multi-Crit. Decis. Anal. 11(4), 187–195 (2002)
5. Saaty, T.L.: A Scaling Method for Priorities in Hierarchical Structures. Journal of Mathematical Psychology 15(3), 231–281 (1977)
6. Mahdavi, I., Fazlollahtabar, H., Heidarzade, A., Mahdavi-Amiri, N., Rooshan, Y.I.: A Heuristic Methodology for Multi-Criteria Evaluation of Web-Based E-Learning Systems Based on User Satisfaction. Journal of Applied Sciences 8(24), 4603–4609 (2008)
7. Shee, D.Y., Tzeng, G.H., Tang, T.I.: AHP, Fuzzy Measure and Fuzzy Integral Approaches for the Appraisal of Information Service Providers in Taiwan. Journal of Global IT Management 6(1), 8–30 (2003)
8. Belton, V., Gear, T.: The Legitimacy of Rank Reversal-A Comment. Omega 13(3), 143–144 (1985)
9. Millet, I., Harkear, P.T.: Globally Effective Questioning in the Analytic Hierarchy Process. European Journal of Operational Research 48(1), 88–97 (1990)
10. Buckley, J.J.: Fuzzy Hierarchical Analysis. Fuzzy Sets and Systems 17(3), 233–247 (1985)
11. Kamsu-Foguem, B., Rigal, F., Mauget, F.: Mining Association Rules for the Quality Improvement of the Production Process. Expert Systems with Applications 40(4), 1034–1045 (2013)
12. Lee, D., Park, S.H., Moon, S.: Utility-Based Association Rule Mining: A Marketing Solution for Cross-Selling. Expert Systems with Applications 40(7), 2715–2725 (2013)
13. Fang, J., Jie, Z., Hanlin, G.: Research on the Application of Improved Association Rule Algorithm in Supply Chain Management. Advanced Materials Research 121-122, 309–313 (2010)
14. Huang, J.J., Tzeng, G.H., Ong, C.S.: Multidimensional Data in Multidimensional Scaling Using the Analytic Network Process. Pattern Recognition Letters 26(6), 755–767 (2005)
15. Chiou, H.K., Tzeng, G.H., Cheng, D.C.: Evaluating Sustainable Fishing Development Strategies Using Fuzzy MCDM Approach. Omega 33(3), 223–234 (2005)
16. Saaty, T.L.: How to Make a Decision: The Analytic Hierarchy Process. European Journal of Operational Research 48(1), 9–26 (1990)
17. Torgerson, W.S.: Multidimensional Scaling: I. Theory and Method. Pstchometrika 17(4), 401–419 (1952)
18. Chen, M.F., Tzeng, G.H., Ding, C.G.: Combining Fuzzy AHP with MDS in Identifying the Preference Similarity of Alternatives. Applied Soft Computing 8(1), 110–117 (2008)
19. Agrawal, R., Lmielinski, T., Swami, A.: Mining Association Rules between Sets of Items in Large Databases. Proc. of the ACM SIGMOD Int. Conf. on Management of Data 22(2), 207–216 (1993)
20. Chen, S.J., Hwang, C.L.: Fuzzy Multiple Attribute Decision Making: Methods and Applications. Springer (1992)

Variable Ordering and Constraint Propagation for Constrained CP-Nets

Eisa Alanazi and Malek Mouhoub

Department of Computer Science
University of Regina
Regina, Canada
{alanazie,mouhoubm}@uregina.ca

Abstract. In many real world applications we are often required to manage constraints and preferences in an efficient way. The goal here is to select one or more scenarios that are feasible according to the constraints while maximizing a given utility function. This problem can be modeled as a constrained Conditional Preference Networks (Constrained CP-Nets) where preferences and constraints are represented through CP-Nets and Constraint Satisfaction Problems respectively. This problem has gained a considerable attention recently and has been tackled using backtrack search. However, there has been no study about the effect of variable ordering heuristics and constraint propagation on the performance of the backtrack search solving method. We investigate several constraint propagation strategies over the CP-Net structure while adopting the most constrained heuristic for variables ordering during search. In order to assess the effect of constraint propagation and variable ordering on the time performance of the backtrack search, we conducted an experimental study on several constrained CP-Net instances randomly generated using the RB model. The results of these experiments clearly show a significant improvement when compared to the well known methods for solving constrained CP-Nets.

1 Introduction

Managing both constraints and preferences is often required when tackling a wide variety of real world applications [1,2,3]. For instance, one of the important aspects of successful deployment of autonomous agents is the ability to reason about user preferences. This includes representing and eliciting user preferences and finding the *best* scenario for the user given her preference statements [1]. Moreover, many agents work in a constrained environment where they must take into consideration the feasibility of the chosen scenario [4]. For example, consider a PC configuration application, the user has some preferences over different attributes (i.e. screen size, brand and memory . . . etc) while the constraints could be the manufacturer compatibility constraints among the attributes. Moreover, the user, whose the agent is acting on behalf, might have other requirements like budget limit. Therefore, the agent must look for a scenario (solution or

A. Moonis et al. (Eds.): IEA/AIE 2014, Part II, LNAI 8482, pp. 22–31, 2014.
© Springer International Publishing Switzerland 2014

outcome) that satisfies the set of constraints while maximizing its utility. This problem can be viewed as a constrained CP-Net problem where the goal is to find one or more solutions that are feasible and not dominated by any other feasible solution [5,6]. We refer to such set of solutions as the Pareto optimal set where a feasible solution is Pareto optimal if it is not dominated by any other feasible solution. Finding the set of Pareto solutions for such problems is known to be an NP-hard problem in general [6].

In this paper, we demonstrate that variable ordering heuristics in addition to constraint propagation techniques play an important role for efficiently solving the CP-Net problem when a backtrack search method is used. Note that, while several attempts have been made for solving constrained CP-Nets [5,6,7], variable ordering as well as constraint propagation have been neglected during the solving process. More precisely we extend Search-CP algorithm proposed in [5] with several constraint propagation techniques in addition to a variable ordering heuristics. In order to evaluate the performance of our extended Search-CP with the plain Search-CP we conducted several tests on several constrained CP-Net instances randomly generated based on the RB model. The results demonstrate the superiority of our method over Search-CP, thanks to the constraint propagation and variable ordering.

The rest of the paper is structured as follows. The next section introduces CP-Nets and constraint satisfaction. Then, the related work is presented in section 3. Section 4 is devoted to our proposed search method. The experimental evaluation is then reported in section 5. Finally concluding remarks as well as future directions are listed in section 6.

2   Background

2.1   Conditional Preference Networks (CP-Nets)

A Conditional Preferences network (CP-Net) [8] is a graphical model to represent qualitative preference statements including conditional preferences such as: "*I prefer A to B when X holds*". A CP-Net works by exploiting the notion of preferential independency based on the *ceteris paribus* (with all other things being without change) assumption. Ceteris Paribus (CP) assumption provides a clear way to interpret the user preferences. For instance, I prefer A more than B means I prefer A more than B if there was no change in the main characteristics of the objects. A CP-Net can be represented by a directed graph where nodes represent features (or variables) along with their possible values (variables domains) and arcs represent preference dependencies among features. Each variable X is associated with a ceteris paribus table (denoted as $CPT(X)$) expressing the order ranking over different values of X given the set of parents $Pa(X)$. An outcome for a CP-Net is an assignment for each CP-Net variable from one its domain values. Given a CP-Net, one of the main queries is to find the best outcome given the set of preferences. We say outcome o_i is better than outcome o_j if there is a sequence of worsening flips going from o_i to o_j [8]. A worsening flip is a change in the variable value to a less preferred value according to the variable's CPT. The relation between different outcomes for a CP-Net can be captured through an induced graph constructed as follows. Each node in the graph represents an outcome of the network. An edge going from o_j to o_i exists if there is an improving flip according to the CPT of one of the variables in o_j all else being equal. Consider a simple CP-Net and its induced graph shown in Figure 1 which is similar to the example in [8]. The CP-Net has three variables A, B and C where A and B unconditionally prefer a and b to $\bar{a}$ and $\bar{b}$ respectively. However, the preference values over C depend on A and B values. For instance when $A = a$ and $B = \bar{b}$, the preference order for C is $\bar{c} \succ c$. The induced graph represents all the information we need to answer regarding the different dominance relations between outcomes. An outcome o_i dominates another outcome o_j if there is a path from o_j to o_i in the induced graph otherwise they are incomparable (denoted as $o_i \bowtie o_j$). Since the induced graph is acyclic, there is only one optimal outcome which resides at the bottom of the graph. For example, the optimal outcome for CP-Net in Figure 1 is abc.

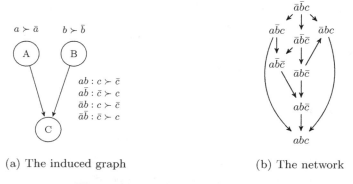

(a) The induced graph   (b) The network

Fig. 1. A CPNet and its induced graph

2.2   Constraint Satisfaction Problem (CSP)

A Constraint Satisfaction Problem (CSP) [9] consists of a set of variables each defined on a set of possible values (variable domain) and a set of relations restricting the values that each variable can take. A solution to a CSP is a complete assignment of values to variables such that all the constraints are satisfied.

A CSP is known to be an NP-Hard problem. In order to overcome this difficulty in practice, several constraint propagation techniques have been proposed [9]. The goal of these techniques is to reduce the size of the search space before and during the search for the solution to the CSP. One of the well-known constraint propagation techniques is called Arc Consistency (AC) [10]. The aim of AC is to enforce a 2 consistency over the constraint problem. More precisely, the 2 consistency consists in making sure that for each pair of variables (X, Y) sharing a constraint, every value x_i from X's domain has a corresponding value in Y's domain such that the constraint between X and Y is satisfied, otherwise a is eliminated. Forward Checking (FC) [9] maintains AC during the search space as whenever a new assignment x_i is made to variable X, we enforce AC between X and any non assigned variable connected with it through a given constraint. This results in removing inconsistent values from X's connected variables domains.

3   Constrained CP-Nets

Several approaches have been proposed for solving constrained CP-Nets [5,6,7]. The work in [7] seeks a generalised framework to solve the problem based on how different parameters are defined. The method in [6] transforms (a possibly cyclic) CP-Net into a CSP. Then, the solutions for the CSP are the Pareto solutions of the corresponding constrained CP-Net problem. In [5], an anytime algorithm (called Search-CP) to handle the problem has been proposed. Search-CP iteratively calls a subnetwork of the original CP-Net and strengthen the set of constraints to the current instantiated variables. However, Search-CP adopts solely the CP-Net ordering and therefore not considering any other variable ordering heuristics. In addition Search-CP did not explicitly consider any propagation technique.

The work in [11] shows some pruning conditions while adopting a relaxed notion of the dominance task. The paper shows through experimental tests that such conditions may result in saving the computational time of the problem. In [6] a new method has been proposed where the constrained (possibly cyclic) CP-Net is transformed into a CSP problem. Then, the solutions of the CSP are the optimals of the corresponding constrained CP-Net problem. The paper also proposes an approximation for the set of pareto optimals by assuming a chain of feasible outcomes. In [12] an approximation method has been proposed to approximate the CP-net into a soft constraint problem which can then be solved by different soft constraint techniques [13]. Finally, in [14] a solving approach, using Arc Consistency (AC) in the preprocessing phase, has been applied to solve the constrained CP-Net. The result is a new CP-Net (called Arc Consistent CP-Net (ACCPnet)) where some variables domain values (and thus preference statements) been removed from the constrained CP-Net problem.

4   The Proposed Solving Approach

The hardness of the constrained CP-Net problem comes from the fact that the most preferred outcome may not be feasible with respect to the constraints. One solution to overcome this difficulty in practice is to enforce constraint propagation techniques that will detect sooner any possible inconsistency. This will prevent the backtrack search algorithm to go over reconsider some decisions if such inconsistencies have not been detected earlier. The propagation techniques we are considering are: Arc Consistency (AC) before search and Forward Checking (FC) or Maintaining Arc Consistency (MAC) during search [9]. Note that since FC and MAC do not eliminate feasible solutions from the search space, so do our solving method. Also, our method preserves the anytime property following the CP-Net semantics.

Our ordering heuristic is based on the Most Constrained Heuristic (MCH) variable ordering [9] and works as follows. Given an acyclic CP-Net $\mathcal{N}$ and its related CSP representing the constrained variables, the typical order is a topological sorting over $\mathcal{N}$ from top to bottom. An important dependency condition here is that any variable v must be ordered before any of its descendants. In the context of constrained CP-Net, different orders might result in different performance of the chosen algorithm. Therefore, we adopt the most constrained variable heuristic to sort the set of variables in $\mathcal{N}$. More precisely, we first order the variables according to the related CSP structure. Afterwards, we iterate over this order and for every variable, we position its parents before it. We stop when every variable meets the dependency condition. The resulted order respects the CP-Net structure while taking into consideration the most constrained variable heuristic.

4.1   Algorithm Details

As we mentioned earlier, our algorithm is is an extension of Search-CP [5] with constraint propagation and variable ordering heuristics. This allows the algorithm to quickly detect inconsistencies and return a Pareto if exists in a very efficient running time.

We maintain a *fringe* containing the set of nodes to be expanded. The fringe acts as a stack to expand the nodes. Whenever a new node is expanded, we assign its current variable A to its most preferred value according to $CPT(A)$ given its parent values. We order the variables based on the most constrained heuristic. The ACCPnet procedure [14] is applied to the problem before the search starts. This results in reducing the number of domain values and thus the size of the search space before the backtrack search method. Algorithm 1 shows the procedure to find one Pareto optimal. Upon the assignment of the current variable, $X = x$ (line 19), we apply Forward Checking (FC) propagation technique as described earlier. The result is a new CSP where some domain values of the non assigned variables have been removed. The search stops either when the fringe is empty (in which case an inconsistency is detected) or when the algorithm finds a solution (line 5).

Notice the power of CP-Net semantics. For finding one Pareto optimal, we do not need to perform dominance checking over outcomes. If for any variable X in the search we assign x such that x is the most preferred consistent value of X, then the first complete solution s is guaranteed to be Pareto. That is, there is no other solution $\hat{s}$ dominating s. However, when looking for more than one solution, dominance checking is required [5].

Algorithm 1. Algorithm to find one pareto optimal

---

input : N : CP-Net
         con : CSP
output:
         S : A Pareto outcome

1 $\pi \leftarrow$ MCH(N,con)   // variables ordering heuristic
2 fringe $\leftarrow$ {root}
3 S $\leftarrow \emptyset$
4 $flag \leftarrow true$
5 **while** (fringe $\neq \emptyset \wedge flag$) **do**
6 | $n \leftarrow$ *pop first item in* fringe
7 | **if** isConsistent(n, con) **then**
8 | | **if** isComplete(n) **then**
9 | | | $S = n$
10 | | | $flag \leftarrow false$
11 | | **end**
12 | | **else**
13 | | | let X $\leftarrow$ nextVariable(π)
14 | | | let $\succ$ be the order given $Pa($X$)$
15 | | | let $x \leftarrow$ nextValue(X,$\succ$)
16 | | | **if** $x \neq \emptyset$ **then**
17 | | | | $con \leftarrow$ FC(X $= x, con$)
18 | | | | $n \leftarrow n\cup$ X $=x$
19 | | | | add n to fringe
20 | | | **end**
21 | | **end**
22 | **end**
23 **end**
24 Return S

---

5   Experimentation

The goal of the experiments conducted in this section is to evaluate the time performance of our proposed approach and therefore assess the effect of constraint propagation and variable ordering heuristics on the backtrack search. We compare our extended Search-CP to the plain Search-CP algorithm proposed in [5]. The reason for choosing Search-CP is because, unlike other methods, it is a decoupled approach and does not require transforming the CP-Net into another structure.

The experiments were conducted on a MAC OS X version 10.8.3 with 4GB RAM and 2.4 GHz Intel Core i5. Since there is no known library for constrained CP-Net instances, we randomly generate these instances based on the RB model as follows. Note that the choice for the RB model is motivated by the fact that it has exact phase transition and is capable of generating very hard instances that are close to the phase transition. We generate random acyclic instances of

CP-Nets. Each instance has 50 variables where each variable may take between 0 and 2 parents. The constraint part has 25%, 50% or 75% of these variables ($n = 13, 25 and 25$ respectively) and has been generated by the RB Model [15] with $r = 0.6$ and $\alpha = 0.5$ where r and α $(0 < \alpha < 1)$ are two positive constants used by the model RB. Given these parameters settings and according to the RB model, the domain size of the variables and the number of constraints are respectively equal to $d = n^\alpha$ and $nc = rn \ln n$. The phase transition will then be: $pt = 1 - e^{-\alpha/r} = 0.5$

We vary the tightness p value between 0.05 and 0.4 with 0.05 increment in each iteration. The constraint tightness is defined here as the number of non eligible tuples over the cartesian product of the two domains of the variables sharing the constraint. In every iteration, we fix p and generate random constrained CP-Net instances. Finally, we take the average running time (over 20 run) needed to find one Pareto solution. We choose a random number of CP-Net variables n' where $3 \leq n' \leq 20$. Then we create a random acyclic CP-Net containing n' variables with equal domain size. Any variable can take up to 2 variables as immediate ancestors or parents.

The charts in Figure 2 show the average running time in seconds required to find a single Pareto solution when the number of CSP variables is respectively equal to 13, 25 and 38 (which respectively corresponds to 25%, 50% and 75% of the CP-Net variables). The tightness p varies from 0.05 to 0.4. The following 5 methods are used for these comparative tests.

Plain Search-CP . Algorithm proposed in [5]
FC . Search-CP with FC during the search phase.
FC + MCH . FC using the most constrained variable ordering heuristics.
AC + FC + MCH . FC + MCH using AC in a preprocessing phase to reduce
    the size of the variables domains before search.
AC + MAC + MCH . Previous method when using MAC instead of FC.

As we can notice from the top chart in figure 2, the time spent by AC before and during the search through FC or MAC does not really pay off as the plain Search-CP provides the best results. Indeed, for these under constrained problems (with only 25% of the CP-Net variables being constrained) constraint propagation does not do much.

The situation is however different when the percentage of CSP variables is larger. The middle chart in figure 2 shows that the running time for Search-CP without any propagation and ordering heuristic (the Plain Search-CP) is substantially large compared to the other methods. We can also clearly see that applying FC during search improves considerably the time performance of the search especially if the variable ordering heuristics is used. Finally, the best result is obtained when adding a preprocessing phase before the search where AC is applied to reduce the size of the search space. We notice however that MAC is slightly better than FC in most of the cases. Finally, the superiority of MAC (with AC in the preprocessing phase) is noticeable for a larger number of CSP variables as we can easily see from the bottom chart of figure 2 especially for instances close to the phase transition.

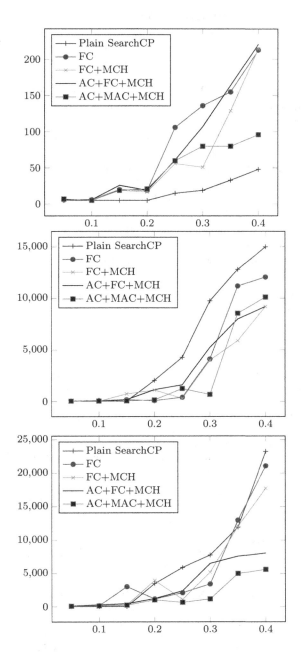

Fig. 2. Average running time for finding an optimal with 13 CSP variables (top chart), 25 CSP variables (middle chart) and 38 variables (bottom chart)

6   Conclusion and Future Work

In this work, we showed that the Search-CP algorithm achieves significant improvements in terms of process time when constraint propagation and variable ordering heuristics are added. Indeed, the experimental results conducted on randomly generated constrained CP-Nets clearly show the superiority of extending Search-CP with constraint propagation before and during search in addition to the most constrained variable heuristics.

The proposed extension of the Search-CP has been integrated into an online shopping system based on user's preference ellicitation [16]. In the near future we plan to study closely those constrained CP-Net instances that are near the phase transition. We will as well explore other variable ordering heuristics based on learning and metaheuristics [17,18]. Another important line of work is finding the $k-$Pareto set of optimal solutions (when looking for more than one solution) as well as the theoretical justification for choosing the number k. Finally, we plan to extend this work to the case of temporal constraints using the temporal constraint solving techniques we have proposed in the past [19,20,21].

Acknowledgements. Eisa Alanazi's research is supported by the Ministry of Higher Education, Saudi Arabia.

References

1. Jiang, W., Sadaoui, S.: Evaluating and ranking semantic offers according to users'interests. Journal of Electronic Commerce Research 13(1) (2012)
2. Mouhoub, M., Sukpan, A.: Managing temporal constraints with preferences. Spatial Cognition & Computation 8(1-2), 131–149 (2008)
3. Mouhoub, M., Sukpan, A.: Managing dynamic csps with preferences. Applied Intelligence 37(3), 446–462 (2012)
4. Chen, B., Sadaoui, S.: A generic formal framework for constructing agent interaction protocols. International Journal of Software Engineering and Knowledge Engineering 15(01), 61–85 (2005)
5. Boutilier, C., Brafman, R.I., Hoos, H.H., Poole, D.: Preference-based constrained optimization with cp-nets. Computational Intelligence 20, 137–157 (2001)
6. Prestwich, S.D., Rossi, F., Venable, K.B., Walsh, T.: Constraint-based preferential optimization. In: AAAI, pp. 461–466 (2005)
7. Boerkoel, Jr., J.C., Durfee, E.H., Purrington, K.: Generalized solution techniques for preference-based constrained optimization with cp-nets. In: AAMAS, pp. 291–298 (2010)
8. Boutilier, C., Brafman, R.I., Domshlak, C., Hoos, H.H., Poole, D.: Cp-nets: A tool for representing and reasoning with conditional ceteris paribus preference statements. J. Artif. Intell. Res. (JAIR) 21, 135–191 (2004)
9. Dechter, R.: Constraint processing. Elsevier Morgan Kaufmann (2003)
10. Mackworth, A.K.: Consistency in networks of relations. Artificial Intelligence 8(1), 99–118 (1977)
11. Wilson, N., Trabelsi, W.: Pruning rules for constrained optimisation for conditional preferences. In: Lee, J. (ed.) CP 2011. LNCS, vol. 6876, pp. 804–818. Springer, Heidelberg (2011)

12. Domshlak, C., Rossi, F., Venable, K.B., Walsh, T.: Reasoning about soft constraints and conditional preferences: complexity results and approximation techniques. CoRR abs/0905.3766 (2009)
13. Bistarelli, S., Montanari, U., Rossi, F.: Semiring-based constraint satisfaction and optimization. Journal of the ACM 44(2), 201–236 (1997)
14. Alanazi, E., Mouhoub, M.: Arc consistency for cp-nets under constraints. In: FLAIRS Conference (2012)
15. Xu, K., Li, W.: Exact phase transitions in random constraint satisfaction problems. Journal of Artificial Intelligence Research 12, 93–103 (2000)
16. Mohammed, B., Mouhoub, M., Alanazi, E., Sadaoui, S.: Data mining techniques and preference learning in recommender systems. Computer & Information Science 6(5) (2013)
17. Mouhoub, M., Jashmi, B.J.: Heuristic techniques for variable and value ordering in csps. In: [22], pp. 457–464
18. Abbasian, R., Mouhoub, M.: An efficient hierarchical parallel genetic algorithm for graph coloring problem. In: [22], pp. 521–528
19. Mouhoub, M.: Reasoning about Numeric and Symbolic Time Information. In: The Twelfth IEEE International Conference on Tools with Artificial Intelligence (ICTAI 2000), pp. 164–172. IEEE Computer Society, Vancouver (2000)
20. Mouhoub, M.: Analysis of Approximation Algorithms for Maximal Temporal Constraint Satisfaction Problems. In: The 2001 International Conference on Artificial Intelligence (IC-AI 2001), Las Vegas, pp. 165–171 (2001)
21. Mouhoub, M.: Systematic versus non systematic techniques for solving temporal constraints in a dynamic environment. AI Communications 17(4), 201–211 (2004)
22. Krasnogor, N., Lanzi, P.L. (eds.): Proceedings of 13th Annual Genetic and Evolutionary Computation Conference, GECCO 2011, Dublin, Ireland, July 12-16. ACM (2011)

A Three Stages to Implement Barriers in Bayesian-Based Bow Tie Diagram

Ahmed Badreddine, Mohamed Aymen Ben HajKacem, and Nahla Ben Amor

LARODEC, Université de Tunis, Institut Supérieur de Gestion de Tunis,
41 Avenue de la liberté, cité Bouchoucha, 2000 Le Bardo,Tunisia
{badreddine.ahmed,aymen_1717}@hotmail.com, nahla.benamor@gmx.fr

Abstract. The preventive and protective barriers in Bow tie diagrams are often defined by experts that ignore the real aspect of the system. Thus implementing barriers in Bow tie diagrams in automatic way remains a real challenge. This paper proposes a new approach to implement preventive and protective barriers. This approach is based on the multi-objective influence diagrams which are a graphical model to solve decision problems with multiple objectives.

Keywords: Bow tie diagram, Preventive barriers, Protective barriers, Multi-objective influence diagrams.

1 Introduction

The identification of possible accidents in an industrial process is a key point of risk analysis. Thus, several tools have been proposed such as, *Bow tie diagrams* [3], *fault and event trees* [7] and *barriers block diagrams* [14]. However, the Bow tie diagram is the most appropriate to represent the whole scenario of a given risk [2]. In fact, this tool represents the causes and the consequences of the given risk, also called *the top event* TE, via two parts namely *fault tree* and *event tree*. In addition, this technique proposes a set of *preventive* and *protective* barriers to reduce respectively the frequency and the severity of TE.

Generally, the implementation of these barriers is based on the experts knowledge. In this context, we can mention in particular Delvosalle et al. who have proposed to implement preventive and protective barriers by examining the structure of Bow tie by using survey [4]. However, defining preventive and protective barriers using experts knowledge is not an easy task since it depends on four criteria namely, *effectiveness*, *reliability*, *availability* and *cost* [3].

In the literature, few researches have been proposed to implement barriers by using the Bayesian networks [1,9,10]. For instance, Léger et al. have proposed to quantify the barriers and evaluate their impact on the global system performance [10]. Recently, Baderddine et al. have proposed a Bayesian approach to implement preventive and protective barriers. This approach is based on two phases. Firstly, they studied the impact of events on the TE to detect the most critical branch. Secondly, they applied the Analytical Hierarchical Process (AHP) [8] in order to deal with the different criteria to select the appropriate preventive

A. Moonis et al. (Eds.): IEA/AIE 2014, Part II, LNAI 8482, pp. 32–41, 2014.

and protective barriers [1]. Indeed, the AHP method [11] is based on pair-wise comparison between different barriers according to each criteria to select the appropriate preventive and protective barriers. However, this comparison is based on the experts knowledge without considering the system behavior.

To overcome this weakness, we propose in this paper a new approach to implement the appropriate preventive and protective barriers. This approach is based on the multi-objective influence diagrams [5] which are a graphical model to solve decision problems with multiple objectives. This paper is organized as follows: Section 2 describes the basics of Bow tie diagrams. Section 3 gives a brief recall on the Bayesian approach to construct Bow tie diagram. Section 4 presents the basics of multi-objective influence diagram. The new approach to implement preventive and protective barriers in Bow tie diagrams is discussed in section 5 followed by the conclusion in section 6.

2 Basics of Bow Tie Diagrams

In the early nineties, Bow tie diagrams were developed by the SHELL company to represent the whole scenario of an accident for risk analysis. For each identified risk R also named *the top event* TE, the Bow tie diagram represents the whole scenario of TE via two parts. The first part corresponds to *fault tree*(FT) to represent all possible causes of the TE. These causes can be classified into two types namely: *the initiator events IE* which define the principal causes of TE, and *the undesired and critical events IndE and CE* which are the causes of IE. The second part corresponds to *event tree*(ET) to represent all possible consequences of TE. These consequences can be classified into three types namely: *second events* (SE) which define the primary consequences of TE, *dangerous effects*(DE) which are the dangerous consequences of SE and *major events*(ME) of each DE.

Bow tie diagrams allow us to implement *preventive* and *protective* barriers to reduce respectively the frequency and the severity of TE. In fact, this implementation is constrained by several criteria namely, *effectiveness*, *reliability*, *availability* and *cost* [3]. To select the appropriate preventive and protective barriers, Baderddine et al. [1] have proposed a Bayesian approach which will be detailed in what follows.

3 Brief Recall on the Bayesian Approach to Construct Bow Tie Diagrams

The approach proposed by Baderddine et al. is based on two algorithms. First, they proposed an algorithm based on Bayesian approach to generate automatically the graphical component of Bow tie diagram (i.e fault and event tree which are considered as Bayesian networks) and the numerical component (i.e a set of conditional probability table and a set of severity degree) from a given training set (i.e a set of observation). Second, on the basics of these two components, they proposed a second algorithm to implement preventive and protective barriers. This implementation is based on two phases. Firstly, they studied the impact of events on TE by using the propagation algorithm in Bayesian networks [11] to detect the most critical branch. Secondly, they applied the AHP method [8] to deal with the different criteria to select the appropriate preventive and protective barriers [1]. In fact, the AHP method [11] is based on pair-wise comparison between different barriers according to each criteria. However, this comparison is based on the experts knowledge without considering the numerical component of Bow tie diagram which can lead to a dissociation between the construction Bow tie diagram algorithm and barriers implementation algorithm.

In order to improve the Bayesian approach [1], we propose in this paper a new approach to implement preventive and protective barriers by using the multi-objective influence diagram [5]. Before detailing the proposed approach, we will give a brief recall on this model.

4   Basics of Multi-objective Influence Diagrams

Multi-objective influence diagrams [5], denoted by MID, are an extension of classical influence diagrams [12] to solve the decision problem with multiple objectives. Formally, MID can be defined as an assembly of graphical and numerical components.

The graphical component is devoted to consider the multi-objective influence diagram as a directed acyclic graph (DAG), denoted by $G = (N,A)$, while N represents the nodes and A represents the arcs between them. The set of nodes N is composed of three subsets C, D and V defined as follows: *Chance nodes*: $C_i \in C$, represents a set of random uncertain variables of decision problem. *Decision nodes*: $D_i \in D$, represents a set of decision that must be taken by the decision maker. *Value node*: $V_i \in V$, denoted a *multi objective value node* which contains the objectives of the problem [5]. The set of arcs A is composed of two types according to their target: *Conditional arcs (into chance and value nodes)* and *Informational arcs (into decision nodes)*. The graphical component of MID is required to satisfy some property to be regular namely: there is no cycles, the value node has no children and there is a directed path containing all decision nodes, known as the *no-forgetting* property [5].

The numerical component is devoted to quantify the different arcs. Each conditional arc which has a chance node C_i as target, is quantified by conditional probability distribution of C_i in the context of its parents $Pa(C_i)$ [12]. Each conditional arc which has a value node V as target, is quantified by utility function of V in the context of its parents [5].

Once the graphical and the numerical components of MID are defined, an evaluation algorithm was proposed by Micheal et al., to generate the optimal strategy (i.e decisions) satisfying all the objectives [5].

5   A New Approach to Implement Preventive and Protective Barriers in Bow Tie Diagram

To implement preventive and protective barriers in Bow tie diagram, we propose in this paper a new approach which is divided into three phases (see Figure 1): a building phase where we propose a mapping algorithm from the graphical component of Bow tie diagram [1] and the possible barriers into a qualitative multi-objective influence diagram MID, a quantitative phase where we propose to quantify this qualitative MID by using the numerical component of Bow tie diagram [1] and an evaluation phase where we propose to apply the evaluation algorithm of MID [5] to generate the optimal strategy of preventive and protective barriers implementation. In what follows, we will detail these three phases.

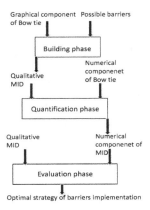

Fig. 1. A new approach to implement preventive and protective barriers in Bow tie diagram

5.1   Building Phase

- Let BT be the graphical component of Bow tie diagram [1], which is composed of:$\{IE, IndE, CE, SE, DE, ME\}$ such that IE (resp. IndE, CE, SE, DE) is the set of initiator (resp. critical, undesired, second, dangerous, major) events. All these events can take two states (i.e present=(True) or absent=(False)).
- Let Bar be the set of the possible preventive and protective barriers which can be implemented to reduce the frequency and the severity of TE. These barriers are represented as couple (Bar_i, E_i), where Bar_i is the barrier relative to the event E_i. All possible barriers can take two states (i.e implemented=(True) or non implemented=(False)).
- Let $\{C_1 \ldots C_4\}$ be the criteria of the barriers selection which are respectively *effectiveness, reliability, availability* and *cost*.
- Let C and D be respectively the set of chance and decision nodes of MID.

- Let V be the value node of MID.
- Let A be the set of arcs of MID.
- Let $Arc(X_i, Y_i)$ is a function which returns all arcs between X_i and Y_i events in BT.
- Let ord be the order between the decision nodes.

The building phase is outlined in Algorithm 1.

Algorithm 1.1: Building phase

Data: BT; Bar; $\{C_1 \dots C_4\}$; ord

Result: Qualitative MID

begin

    $MID.V \leftarrow \emptyset \;\; MID.C \leftarrow \emptyset \;\; MID.D \leftarrow \emptyset \;\; MID.A \leftarrow \emptyset$

    % *Collect the criteria in the same value node V*

    $MID.V \leftarrow C_1 \cup C_2 \cup C_3 \cup C_4$

    % *Add events and their arcs*

    $MID.C \leftarrow MID.C \cup BT.IE \cup BT.Ind \cup BT.CE \cup BT.TE \cup BT.SE \cup BT.DE \cup BT.ME$

    $\forall IE_i \in BT.IE, \forall Ind_i \in BT.Ind, \forall CE_i \in BT.CE$

    $MID.A \leftarrow MID.A \cup Arc(IE_i, CE_i) \cup Arc(IE_i, Ind_i) \cup Arc(IE_i, TE)$

    $\forall SE_i \in BT.SE, \forall DE_i \in BT.DE, \forall ME_i \in BT.ME$

    $MID.A \leftarrow MID.A \cup Arc(TE, SE_i) \cup Arc(SE_i, DE_i) \cup Arc(DE_i, ME_i)$

    % *Add barriers*

    $\forall Bar_i \in Bar$

    $MID.D \leftarrow Bar_i$

    $MID.A \leftarrow MID.A \cup (Bar_i \rightarrow E_i)$

    $MID.A \leftarrow MID.A \cup (Bar_i \rightarrow MID.V)$

    $l = |MID.D|$

    for $t \leftarrow 1 \dots (l-1)$ **do**

        for $t \leftarrow (t+1) \dots (l)$ **do**

            $MID.A \leftarrow MID.A \cup (MID.D_{ord} \rightarrow MID.D_{ord+1})$

end

As outlined in Algorithm 1, first our idea is to gather all the barriers selection criteria in the same value node. Second, we propose to add the events of Bow tie diagram and their arcs to the MID. To this end, we suggest to represent these events as chance nodes. Then, we propose to add all the possible preventive and protective barriers as decision nodes and to connect each barrier with its relative event. Next, we suggest connecting all the possible barriers with the value node. Finally, we propose to connect all the barriers together by respecting an order between them, to satisfy the *no-forgetting* property of MID. It is important to note that this algorithm generates a regular MID.

Example 1. *Let us consider the graphical component of Bow tie diagram relative to a major fire and explosion on tanker truck carrying hydrocarbon TE which is constructed by the Bayesian algorithm [1]. The events relative to fault* *tree are: (hydrocarbon gas leak HGL, source of ignition SI, tank value failure TVF and exhaust failure EF) and the events relative to event tree are: (pool fire PF, thermal effects THE, toxic effects TOE and production process in stop PPS, Damage to the other trucks DT, Toxic damage to persons TODP and Late delivery LD). The graphical component is illustrated in Fig 2.(a). Then, we consider two preventive barriers which are: Fire simulation FS to minimize EF event and Education and training ETT to reduce HGL event. After that, we assume two protective barriers which are: Blast protection window film BP to reduce PF event and Setting up equipment SUEH to limit THE event.*

First, the building algorithm groups all criteria in the same value node V_C. Then, it maps the graphical component of the Bow tie diagram into MID by adding the events (i.e TVF, EF, HGL, SI, PF, THE, TOE, PPS, TDP, DT, LD and TODP) as chance nodes and connecting them (i.e (TVF to SI), (EF to HGL), (HGL to TE), (SI to TE), (PF to THE), (PF to TOE), (PF to PPS), (THE to DT), (PPS to LD) and (TOE to TODP)). After that, this algorithm adds the barriers (i.e FS, ETT, BP and SUEH) as decision nodes, then it connects these barriers with their relative events (i.e (FS to EF), (ETT to HGL), (BP to PF) and (SUEH to THE)). Next, it connects all these barriers with the value node (i.e (FS to V_C), (ETT to V_C), (BP to V_C) and (SUEH to V_C)). Finally, it connects all the barriers together to respect the order between them (i.e (FS to ETT), (ETT to BP) and (BP to SUEH)). The qualitative MID is depicted in Fig 2.(b).

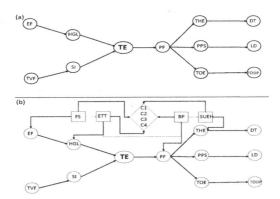

Fig. 2. (a): The graphical component of Bow tie diagram. (b): The qualitative MID.

5.2   Quantification Phase

This phase consists on generating the numerical component of qualitative MID. First, we quantify each event (i.e chance node) in the context of its parents, by using the conditional probability table CPT and the severity degree S of Bow tie diagram. In fact, this quantification varies according to situations that depend on the way the events and the barriers are connected in the MID (see Figure 3).

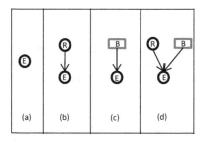

Fig. 3. Connection between events and barriers

•**Situation (a)**: We assign the conditional probability table relative to the cause event E (i.e $P(E)$) by transferring these values from the CPT of Bow tie. We can mention that this situation concerns only the causes events.

•**Situation (b)**: We assign the conditional probability table relative to the event E in the context of the event R (i.e $P(E|R)$) which differs from the causes and consequences events. Regarding the causes events, this quantification can be performed by using the CPT of Bow tie diagram. For the consequences events, this quantification can be carried out by using the severity degree S of the Bow tie diagram. So that, we compute the $P(E = T|R = T)$ and $P(E = T|R = F)$ by applying the Pearl's propagation algorithm in Bayesian networks [11].

•**Situation (c)**: We assign the conditional probability table relative to the cause event E in the context of the preventive barrier B (i.e $P(E|B)$). The quantification of $P(E|B = T)$ can be performed on three steps. First, we propose a set of intervention denoted by $I_B=\{I_{B1},\ldots,I_{Bn}\}$, relative to the implementation of the barrier B. Second, we propose to study the impact of each intervention denoted by $I_{Bi}=\{I_{Bi1},\ldots,I_{Bin}\}$ on the top event TE, in order to detect the most efficient intervention that reduces the occurrence of TE. Thus, we compute $P(TE = T|I_{Bi1} = e_1,\ldots,I_{Bin} = e_n)$, where ej is the probability relative to the event I_{Bij} (i.e $I_{Bij}=1$ (resp. $I_{Bij}=0$) means that $I_{Bij}=$T (resp. $I_{Bij}=$F) otherwise, if $e_j \neq 0$ or $e_j \neq 1$ means that $P(I_{Bij})=e_j$). Finally, we compute the impact of the efficient intervention on the event E.

The quantification of $P(E|B = F)$ is equivalent to quantifying $P(E)$ which is performed in similar way to Situation (a). We note that this situation concerns only the causes events and all the computational are performed by applying the Pearl's propagation algorithm [11].

•**Situation (d)**: We assign the conditional probability table relative to the consequence event E in the context of the protective barrier B and the event R (i.e $P(E|B,R)$). The quantification of $P(E|B = T,R)$ is made on three steps. First, we propose a set of intervention concerning the implementation of the protective barrier B. Then, we suggest studying the impact of each intervention on the major events ME to detect the most efficient intervention that reduces the occurrence of ME. Finally, we compute the impact of the efficient intervention I_{Bi}^* on the event E i.e we are interested in the value $P(E|I_{Bi}^*)$.

The quantification of $P(E\,|B = F, R)$ is equivalent to quantify $P(E\,|R)$ which can be carried out in similar manner than Situation (b).

Example 2. *Once the qualitative MID is constructed by the building algorithm (see Figure 2.(b)), we can quantify it by using the CPT and the severity degree S of Bow tie diagram. On the basis of these information, we assign the conditional probability table relative to each event of MID. We can take the case of the CPT relative to event EF in the context of barrier FS (i.e $P(EF\,|FS)$). The quantification of $P(EF\,|FS = T)$ is first made by proposing two scenarios of intervention regarding the implementation of barrier FS which are:*

- *$I_{FS1}=\{I_{FS11} : source\,of\,ignition, I_{FS12} : tank\,value\,failure\}, \{e_1=0.8, e_2 = 0.6\}$, and*
- *$I_{FS2}=\{I_{FS21} : source\,of\,ignition, I_{FS22} : tank\,value\,failure\}, \{e_1=0.6, e_2 = 0.2\}$.*

Then the propagation process leads to:

- *$P(TE{=}T\,—I_{FS11}{=}0.8, I_{FS12}{=}0.6){=}0.18$, and*
- *$P(TE{=}T\,—I_{FS21}{=}0.6, I_{FS22}{=}0.2){=}0.29$.*

On the basics of these values, we can deduce that I_{FS1} is more efficient, then we compute its impact on EF event. The propagation process leads to:

- *$P(EF{=}T\,—I_{FS11}{=}0.8, I_{FS12}{=}0.6){=}0.55$.*

The quantification of $P(EF\,|FS = F)$ is performed by using the CPT of Bow tie diagram. The conditional probability table of EF is listed in Table 1. Due to the lack of space in this paper we cannot show the numerical component of Bow tie diagram and MID.

Table 1. Conditional probability table of EF event

FS	EF	$P(EF\,|FS)$
T	T	0.55
T	F	0.45
F	T	0.38
F	F	0.62

Then, we assign the utility function of the value node in the context of its parents (i.e preventive and protective barriers). To this end, we propose to define for each possible barriers two indicators which are described as follows:

- $MTBF$: is defined as the mean operating time between failures of barrier [6].
- $MTTR$: is defined as the mean time to repair when the barrier has failed [6].

The value node that contains a vector of four objectives namely, *effectiveness, reliability, availability* and *cost*, can be quantified as follows:

- The effectiveness relative to strategy of preventive and protective barriers implementation is quantified by using the MTBF and MTTR indicators of these barriers. We note that the maximization of effectiveness means by increasing MIBF and reducing MTTR.

- The reliability relative to strategy of preventive and protective barriers implementation is assigned by using the MTBF indicator of these barriers.
- The availability relative to strategy of preventive and protective barriers implementation is quantified by using the MTBF and MTTR indicators of these barriers.
- The cost relative to strategy of preventive and protective barriers implementation is based on the cost of their maintenance. Hence the maintenance can be estimated by using the MTTR indicator of these barriers.

Example 3. *Let us consider the MTBF and MTTR relative to possible preventive and protective barriers which are listed in Table 2. On the basics of these information, we assign the utility function of the value node V_C.*

Table 2. MTBF and MTTR of preventive and protective barriers

Barriers	MTBF	MTTR
FS	0.85	0.19
ETT	0.67	0.57
BP	0.22	0.69
SUEH	0.73	0.54

5.3 Evaluation Phase

Once the graphical and the numerical components of MID are obtained, we can apply the evaluation algorithm which is proposed by Micheal et al., to generate the optimal strategy of preventive and protective barriers implementations satisfying all criteria [5].

Example 4. *Once the qualitative MID is defined which is depicted in Fig 2.(b), and its numerical component is generated, we can apply the evaluation algorithm [5] to generate the optimal strategy of preventive and protective barriers implementations. The optimal strategy of our example is:*

- *Strategy 1=$\{FS = T, ETT = T, BP = F, SUEH = T\}$.*
- *Strategy 2=$\{FS = T, ETT = F, BP = F, SUEH = T\}$.*

6   Conclusion

This paper proposes a new approach to implement preventive and protective barriers in Bow tie diagram. This approach is based on the multi-objective influence diagram [5]. First, we have proposed a mapping algorithm from the graphical component of the Bow tie diagram and possible barriers into qualitative MID. Second, we have suggested quantifying this MID by using the numerical component of Bow tie diagram. Finally, we have applied the evaluation algorithm

[5] to generate the optimal strategy of preventive and protective barriers implementation. As future work, we propose to use the possibility theory [13] instead of the probability theory in generating the numerical component of Bow tie diagram. Indeed, the probability theory is only appropriate when all numerical data are available. In fact, the possibility theory offers an appropriate framework to handle uncertainty arises in data.

References

1. Badreddine, A., Amor, N.B.: A Bayesian approach to construct bow tie diagrams for risk evaluation. Process Safety and Environmental Protection 91(3), 159–171 (2013)
2. Cockshott, J.E.: Probability bow-ties: A transparent risk management tool. Process Safety and Environmental Protection 83(4), 307–316 (2005)
3. Couronneau, J.C., Tripathi, A.: Implementation of the new approach of risk analysis in france. In: 41st International Petroleum Conference, Bratislava (2003)
4. Delvosalle, C., Fievez, C., Pipart, A., Debray, B.: ARAMIS project: A comprehensive methodology for the identification of reference accident scenarios in process industries. Journal of Hazardous Materials 130(3), 200–219 (2006)
5. Diehl, M., Haimes, Y.Y.: Influence diagrams with multiple objectives and tradeoff analysis. IEEE Transactions on Systems, Man and Cybernetics, Part A: Systems and Humans 34(3), 293–304 (2004)
6. Availability, Reliability, SIL Whats the difference?, http://www.mtl-inst.com
7. Kumamoto, H., Henley, E.J.: Automated fault tree synthesis by disturbance analysis. Industrial Engineering Chemistry Fundamentals 25(2), 233–239 (1986)
8. Saaty, T.L.: Axiomatic foundation of the analytic hierarchy process. Management science 32(7), 841–855 (1986)
9. Khakzad, N., Khan, F., Amyotte, P.: Dynamic safety analysis of process systems by mapping bow-tie into Bayesian network. Process Safety and Environmental Protection 91(1), 46–53 (2013)
10. Léger, A., Duval, C., Weber, P., Levrat, E., Farret, R.: Bayesian network modelling the risk analysis of complex socio technical systems. In: Workshop on Advanced Control and Diagnosis, ACD 2006 (2006)
11. Pearl, J.: Fusion, propagation, and structuring in belief networks. Artificial intelligence 29(3), 241–288 (1986)
12. Shachter, R.D.: Evaluating influence diagrams. Operations Research 34(6), 871–882 (1986)
13. Zadeh, L.A.: Fuzzy sets as a basis for a theory of possibility. Fuzzy Sets and Systems 100, 9–34 (1999)
14. Sklet, S.: Comparison of some selected methods for accident investigation. Journal of Hazardous Materials 111(1), 29–37 (2004)

Multi-criteria Utility Mining Using Minimum Constraints

Guo-Cheng Lan[1], Tzung-Pei Hong[2,3], and Yu-Te Chao[2]

[1] Department of Computer Science and Information Engineering,
National Cheng-Kung University, Tainan, Taiwan
[2] Department of Computer Science and Information Engineering,
National University of Kaohsiung, Kaohsiung, Taiwan
[3] Department of Computer Science and Engineering,
National Sun Yat-Sen University, Kaohsiung, Taiwan
rrfoheiay@gmail.com, tphong@nuk.edu.tw, ny152_david@hotmail.com

Abstract. Different from existing utility mining techniques with a single minimum utility criterion, this work presents a new utility-based framework that allows users to specify different minimum utility thresholds to items according to the characteristics or importance of items. In particular, the viewpoint of minimum constraints in traditional minimum utility mining is extended to decide the proper criterion for an itemset in mining when its items have different criteria. In addition, an effective mining approach is proposed to cope with the problem of multi-criteria utility mining. The experimental results show the effectiveness of the proposed viewpoint and approach under different parameter settings.

Keywords: Data mining, utility mining, minimum constraint, multiple thresholds.

1 Introduction

In the existing studies related to utility mining [4][6][8], all items in the utility-based framework are treated uniformly since a single minimum utility threshold is used as the utility requirement for all items in a database. However, a single minimum utility is not easily used to reflect the natures of the items, such as the significances of items. For example, in real world, since the profit of the item "LCD TV" is obviously higher than that of "Milk", only a utility requirement is not easily used to reflect the importance of the two items. As this example notes, developing a utility-based framework with multiple minimum utilities is then a critical issue. In addition, since the existing utility mining approaches cannot directly be applied to handle this problem, designing a proper mining method for avoiding any information losing case in the problem is also another critical issue.

Based on the above reasons, this work presents a new research issue named multi-criteria utility mining, which allows users to define different minimum utilities for all items in a database. In particular, the viewpoint of minimum constraints is adopted in

A. Moonis et al. (Eds.): IEA/AIE 2014, Part II, LNAI 8482, pp. 42–47, 2014.

the work to find interesting patterns information. Based on the existing upper-bound model, the minimum-utility sorting strategy is designed to keep the downward-closure property in mining under the minimum constraints. Finally, the experimental results show the proposed approach has good performance in execution efficiency when compared with the state-of-the-art mining approach, *TP* [6]. To our best knowledge, this is the first work on mining high utility itemsets with the consideration of the multi-criteria in the field of utility data mining.

The remaining parts of this paper are organized as follows. The related works are reviewed in Section 2. The problem to be solved and its definitions are described in Section 3. The proposed approach is stated in Section 4. The experimental results are showed in Section 5. Finally, the conclusions are stated in Section 6.

2   Review of Related Works

Data mining techniques are used to extract useful information from various types of data. Among the published techniques, association-rule mining is one of important issues in the field of data mining due to the consideration of the co-occurrence relationship of items in transactions [1][2]. However, association rule mining [1][2] only uses a single minimum support threshold to determine whether or not an item is frequent in a database. However, in practical application, items may have different criteria to assess their importance [7]. That is, different items should be different support requirements. To address this problem, Liu *et al.* presented a new issue, namely association-rule mining with multiple minimum supports [5], which agreed the users to assign different minimum requirements for items by the significance of the items, such as profit or cost. To achieve this goal, Liu *et al.* designed a minimum constraint to determine the minimum support of an itemset [5]. The minimum constraint was that the minimum value of the minimum supports of all items in an itemset was regarded as the minimum support of that itemset. Different from Liu *et al.*'s study [5], Wang *et al.* [7] then presented a bin-oriented, non-uniform support constraint, which allowed the minimum support value of an itemset to be any function of the minimum support values of items contained in the itemset.

In real applications, however, a transaction usually involves quantities and profits of items other than the item information. Yao *et al.* thus proposed a utility function [8] which considered not only the quantities of the items but also their individual profits in transactions, to find high utility itemsets from a transaction database. According to Yao *et al.*'s definitions [8], local transaction utility (quantity) and external utility (profit) are used to measure the utility of an item. By using a transaction dataset and a utility table together, the discovered itemset is able to better match a user's expectations than if found by considering only the transaction dataset itself. To effectively reduce the search space in mining, Liu et al. then proposed a two-phase approach (abbreviated as *TP*) to efficiently handle the problem of utility mining [6]. In particular, an upper-bound model was developed to keep the downward-closure property in mining [6]. Afterward, most of existing approaches were based on the framework of the *TP* algorithm to copy with various applications with the viewpoint of utility mining, such as on-shelf utility mining [4], and so on.

As association-rule mining [1][2], however, one of main limitations for utility mining is that all the items are treated uniformly [4][6][8]. Accordingly, designing a utility-based framework with multiple minimum utilities is a critical issue.

3   Problem Statement and Definitions

In this section, to illustrate the multi-criteria utility mining problem to be solved clearly, a simple example is given. Assume there are ten transactions in a quantitative transaction database (D), as shown in Table 1, and each of which consists of three features, transaction identification (TID) and items with sold quantities. There are six items in the transactions, respectively denoted from A to F. The value attached to each item in the corresponding slot is the sold quantity in a transaction. The individual profit values of the six items are 3, 10, 1, 6, 5 and 2, respectively, and their minimum utility criteria are 0.20, 0.40, 0.25, 0.15, 0.20 and 0.15.

Table 1. The ten transactions in this example

TID	*A*	*B*	*C*	*D*	*E*	*F*
Trans₁	1	0	2	1	1	1
Trans₂	0	1	25	0	0	0
Trans₃	0	0	0	0	2	1
Trans₄	0	1	12	0	0	0
Trans₅	2	0	8	0	2	0
Trans₆	0	0	4	1	0	1
Trans₇	0	0	2	1	0	0
Trans₈	3	2	0	0	2	3
Trans₉	2	0	0	1	0	0
Trans₁₀	0	0	4	0	2	0

In this study, an itemset X is a subset of the items I, $X \subseteq I$. If $|X| = r$, the set X is called an r-itemset. $I = \{i_1, i_2, ..., i_n\}$ is a set of items may appear in the transaction. In addition, a transaction $(Trans)$ consists of a set items purchased with their quantities. A database D is then composed of a set of transactions. That is, $D = \{Trans_1, Trans_2, ..., Trans_y, ..., Trans_z\}$, where $Trans_y$ is the y-th transaction in D.

Based on Yao *et al.*'s utility function, the utility u_{yi} of an item i in $Trans_y$ is the external utility s_i multiplied by the quantity q_{zj} of i in $Trans_y$, and the utility u_{yX} of an itemset X in $Trans_y$ is the summation of the utilities of all items in X in $Trans_y$. Furthermore, the actual utility au_X of an itemset X in a quantitative transaction database D is the summation of the utilities of X in the transactions including X of D. For example, the utility of $\{AE\}$ in the first transaction can be calculated as $3*1 + 5*1$, which is 8, and then the actual utility $au_{\{AE\}}$ of $\{AE\}$ in Table 1 can be calculated as $8 + 16 + 19$, which is 43.

For an itemset, its actual utility ratio aur_X is the summation of the utilities of X in the transactions including X of QDB over the summation of the transaction utilities of all transactions Q. For example, in Table 1, the summation of the transaction utilities of all transactions can be calculated as $18 + 35 + 12 + 22 + 24 + 12 + 8 + 45 + 12 +$ 14, which is 202. Since the actual utility $au_{\{AE\}}$ of $\{AE\}$ in Table 3.1 is 43, the actual utility ratio $au_{\{AE\}}$ of $\{AE\}$ in Table 3.1 can be calculated as 43/202, which is 0.2128. Finally, let λ_i be the predefined individual minimum utility threshold of an item i. Note that here a minimum constraint is used to select the minimum value of minimum utilities of all items in X as the minimum utility threshold λ_X of X. Hence, an itemset X is called a high utility itemset (abbreviated as HU) if $au_X \geqq \lambda_X$. For example, in Table 3.1, the actual utility ratio of the itemset $\{AE\}$ is a high utility itemset under the minimum constraint due to its actual utility ratio (= 0.2128).

However, to keep the downward-closure property in this problem the existing transaction-utility upper-bound model (abbreviated as $TUUB$) model [6] is introduced to solve this problem. The main concept is that the transaction utility of a transaction is used as the upper-bound of any subsets in that transaction, and the transaction-utility upper-bound $tuubr_X$ of an itemset X in D is the summation of the transaction utilities of the transactions including X in D over the summation of transaction utilities of all transactions in D. If $tuubr_X \geqq \lambda_X$, the itemset X is called a high transaction-utility upper-bound itemset (abbreviated as $HTUUB$). For example, in Table 1, since the itemset $\{AE\}$ appears in the three transactions, $Trans_1$, $Trans_5$ and $Trans_8$, and their transaction utilities are 18, 24 and 45, and the total utility of D is 202, the transaction-utility upper-bound ratio of $\{AE\}$ can be calculated as 87/202, which is about 0.4307. Accordingly, $\{AE\}$ is a $HTUUB$ due to its $tuubr$ (= 0.4307) and the minimum utility threshold (= 0.2).

4   The Proposed Mining Algorithm

The execution process of the TPM_{min} is then stated as follows.

INPUT: A set of items, each with a profit value and a minimum utility threshold, a transaction database D, in which each transaction includes a subset of items with quantities.
OUTPUT: A final set of high utility itemsets (HUs) satisfying their minimum utilities.

Phase 1: Finding High Transaction-Utility Upper-Bound Itemsets
STEP 1: Sort the items in transactions in ascending order of their minimum utility values.
STEP 2: For each transaction $Trans_y$ in D, do the following substeps.
   (a)   Find the utility u_{yz} of each item i_z in $Trans_y$.
   (b)   Find the transaction utility tu_{yz} of $Trans_y$.
STEP 3: Find the total transaction utility of transaction utilities of all transactions in D.
STEP 4: For each item i in D, calculate the transaction-utility upper-bound ratio $tuubr_i$ of item i.
STEP 5: Find the smallest value of the minimum utility constraints of all items in D, and denote it as $\lambda_{min,1}$.

STEP 6: For each item i in D, if the transaction-utility upper-bound ratio $tuubr_i$ of i is larger than or equal to the corresponding minimum utility threshold $\lambda_{min,1}$, put it in the set of high transaction-utility upper-bound 1-itemsets, $HTUUB_1$.

STEP 7: Set $r = 1$, where r represents the number of items in the current set of candidate utility r-itemsets (C_r) to be processed.

STEP 8: Generate from the set $HTUUB_r$ the candidate set C_{r+1}, in which all the r-sub-itemsets of each candidate must be contained in the set of $HTUUB_r$.

STEP 9: For each candidate $(r+1)$-itemset X in the set C_{r+1}, find the transaction-weighted utility ratio $twur_X$ of X in QDB.

STEP 10: Find the first one of itemsets sorted in the set C_{r+1} in ascending order of the minimum utility values of the itemsets, and find the smallest value of the minimum utility constraints of the items in the first itemset, and denote it as $\lambda_{min,r+1}$.

STEP 11: For each candidate utility $(r+1)$-itemset X in set C_{r+1}, check whether the transaction-weighted utility ratio $twur_X$ of X is larger than or equal to the threshold $\lambda_{min,r+1}$ under the minimum constraint. If it is, put it in set $HTUUB_{r+1}$.

STEP 12: If $HTUUB_{r+1}$ is null, do STEP 13; otherwise, set $r = r + 1$ and repeat STEPs 8 to 12.

Phase 2: Finding High Utility Itemsets (HUs) Satisfying their Minimum Utilities

STEP 13: Scan D once to the actual utility ratio aur_X of X in all $HTUUB$ sets.

STEP 14: For each itemset X in all $HTUUB$ sets, do the following substeps.

   (a) Find the minimum value λ_i among all items in X as the minimum utility constraint λ_X of X under the minimum constraint concept.

   (b) Check whether the actual utility ratio aur_X of X is larger than or equal to the minimum utility threshold λ_X. If it is, put it in set HU_{r+1}.

STEP 15: Output the final set of high utility itemsets satisfying their own criteria, HUs.

5   Experimental Evaluation

A series of experiments were conducted to compare the performance of the proposed two-phase multi-criteria approach using minimum constraints (abbreviated as TPM_{min}) and traditional two-phase utility mining approach in Liu *et al.*'s study [5] in terms of the execution efficiency. The experiments were implemented in J2SDK 1.7.0 and executed on a PC with 3.3 GHz CPU and 2GB memory.

In the experiments, the public *IBM* data generator [3] was used to produce the experimental data. Figure 1 showed the performance of the proposed TP_{min} and the traditional *TP* approach under various λ_{min}. Here the symbol λ_{min} represented the minimum value of minimum utility thresholds of all items in databases, and λ_{min} was regarded as the minimum utility threshold in traditional utility mining. As shown in the figure, the proposed TPM_{min} approach did not need to generate a huge number of high utility itemsets than the traditional *TP* approach using only the single minimum utility threshold. Hence, the proposed utility-based framework using minimum constraints might be a proper framework in both effectiveness of minimum constraint viewpoint and efficiency when items had different minimum utilities.

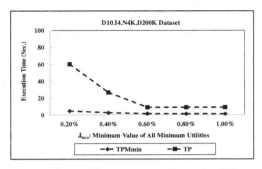

Fig. 1. Efficiency comparison of the two approaches under different thresholds λ_{min}

6   Conclusion

This work has proposed a new research issue, namely multi-criteria utility mining with the minimum constraints, to define the minimum utilities of itemsets in a database when items have different criteria. In particular, this work also presents an effective strategy, minimum-utility sorting, to keep the characteristic of downward-closure property in mining, such that a two-phase mining (TPM_{min}) approach can effectively avoid any information lose case in mining. The experimental results show that proposed TPM_{min} has good performance on execution efficiency.

References

1. Agrawal, R., Imielinksi, T., Swami, A.: Mining Association Rules between Sets of Items in Large Database. In: ACM SIGMOD International Conference on Management of Data, pp. 207–216 (1993)
2. Agrawal, R., Srikant, R.: Fast Algorithm for Mining Association Rules. In: International Conference on Very Large Data Bases, pp. 487–499 (1994)
3. IBM Quest Data Mining Project, Quest synthetic data generation code, http://www.almaden.ibm.com/cs/quest/syndata.html
4. Lan, G.C., Hong, T.P., Tseng, V.S.: Discovery of High Utility Itemsets from On-Shelf Time Periods of Products. Expert Systems with Applications 38(5), 5851–5857 (2011)
5. Liu, B., Hsu, W., Ma, Y.: Mining Association Rules with Multiple Minimum Supports. In: International Conference on Knowledge Discovery and Data Mining, pp. 337–341 (1999)
6. Liu, Y., Liao, W.K., Choudhary, A.: A Fast High Utility Itemsets Mining Algorithm. In: International Workshop on Utility-Based Data Mining, pp. 90–99 (2005)
7. Wang, K., He, Y., Han, J.: Mining Frequent Itemsets Using Support Constraints. In: The 26th International Conference on Very Large Data Bases, pp. 43–52 (2000)
8. Yao, H., Hamilton, H.J., Butz, C.J.: A Foundational Approach to Mining Itemset Utilities from Databases. In: The 4th SIAM International Conference on Data Mining, pp. 482–486 (2004)

A Bayesian Learning Automata-Based Distributed Channel Selection Scheme for Cognitive Radio Networks

Lei Jiao[1], Xuan Zhang[1], Ole-Christoffer Granmo[1], and B. John Oommen[1,2]

[1] Dept. of ICT, University of Agder, Grimstad, Norway
[2] School of Computer Science, Carleton University, Ottawa, Canada*
{lei.jiao,xuan.zhang,ole.granmo}@uia.no, oommen@scs.carleton.ca

Abstract. We consider a scenario where multiple Secondary Users (SUs) operate within a Cognitive Radio Network (CRN) which involves a set of channels, where each channel is associated with a Primary User (PU). We investigate two channel access strategies for SU transmissions. In the first strategy, the SUs will send a packet directly without operating Carrier Sensing Medium Access/Collision Avoidance (CSMA/CA) whenever a PU is absent in the selected channel. In the second strategy, the SUs implement CSMA/CA to further reduce the probability of collisions among co-channel SUs. For each strategy, the channel selection problem is formulated and demonstrated to be a so-called "Potential" game, and a Bayesian Learning Automata (BLA) has been incorporated into each SU so to play the game in such a manner that the SU can adapt itself to the environment. The performance of the BLA in this application is evaluated through rigorous simulations. These simulation results illustrate the convergence of the SUs to the global optimum in the first strategy, and to a Nash Equilibrium (NE) point in the second.

Keywords: Cognitive radio, Channel access, Multiple users, Potential game, BLA.

1 Introduction

In Cognitive Radio Networks (CRNs) [1], Secondary Users (SUs) can access the spectrum allocated to Primary Users (PUs) opportunistically whenever the respective channel is not occupied by the PU. Due to the randomness of PU's behavior, different channels may have various probabilities of being idle as far as the SUs are concerned. In order to utilize the channels efficiently, the SUs are supposed to learn the properties of the channels and to thereafter "intelligently" select a channel that possesses a higher idle probability. To achieve this, Learning Automata (LA) [2,3,8] have been applied to SUs for channel selection in CRNs [4–7]. The advantage of using LA for channel selection is that the SUs can then learn from the environment and adjust themselves accordingly. Besides, the process of learning and making decisions happens simultaneously without the system requiring any prior knowledge of the environment.

According to the number of SU communication pairs in the CRNs, the existing work in channel selection of CRNs using LA can be classified into two categories. In the

---

* *Chancellor's Professor*; *Fellow: IEEE* and *Fellow: IAPR*. The Author also holds an *Adjunct Professorship* with the Dept. of ICT, University of Agder, Norway.

A. Moonis et al. (Eds.): IEA/AIE 2014, Part II, LNAI 8482, pp. 48–57, 2014.

first category, one assumes that only a single communication pair exists in a multi-channel network [4–6]. In such a scenario, an SU communication pair equipped with LA searches for the best channel for transmission, and will converge to the best channel through learning. In [4], the authors utilized the Discretized Generalized Pursuit Algorithm (DGPA) in which the scheme's optimal learning speed (achieved by selecting the optimal learning parameter) had to be pre-determined in order to achieve the best trade-off between the learning speed and the accuracy. However, due to the time-variant characteristics of PU traffic in different channels, the idle probabilities of the various channels could vary with time. Consequently, it is not an easy task for the user to find the optimal learning parameter *a priori*, or for the scheme to adapt the learning parameter so as to follow a dynamic environment. The authors of [5] discussed the issue of determining the circumstances under which a new round of learning had to be triggered, and the learning-parameter-based DGPA was adopted to model the SU pairs again.

To avoid the limitations of the DGPA in such applications, our recent work [6] suggested the incorporation of the Bayesian Learning Automata (BLA) [8] into channel selection, where the channel's switching functionality was enabled. The advantage of the BLA over the DGPA and other learning-parameter-based LA is that one requires no learning parameters to be pre-defined so as to achieve a reasonable (if not ideal) trade-off between the learning speed and the associated accuracy. Besides, the switching functionality allows the SU to move to another channel if the current channel is occupied by a PU, further facilitating the transmission task of the SU.

In the second category, multiple SU communication pairs are considered, as in [7]. To achieve a successful packet delivery among different SU communication pairs, SUs do not only need to avoid collisions with PUs, but also have to avoid colliding with *other* SUs. This scenario is much more complicated than the above-mentioned single SU pair case, and has been analyzed in depth in [7]. In [7], the channel selection problem was formulated as a game, and thus more mathematical insights were provided from the aspects of both the game itself and its potential solutions. To allow the SU communication pairs to converge in a distributed manner, a Linear Reward-Inaction (L_{R-I}) LA was utilized to play the game. As the L_{R-I} has the same drawback as the DGPA in that it requires a pre-defined learning parameter, and as the efficiency of the BLA in game playing was earlier demonstrated in solving the Goore game [9], we were motivated to incorporate the BLA to the multi-SU scenario in CRNs, with the ultimate hope that the system's overall performance could be further improved by its inclusion.

In this work, we consider two channel access strategies, i.e., channel access with and without Carrier Sensing Medium Access/Collision Avoidance (CSMA/CA). Based on the strategies and system models, we formulate and analyze their corresponding game models. Thereafter, the BLA learning scheme has been demonstrated as a novel solution to this problem, and its efficiency has been validated through simulations.

The rest of this article is organized as follows. In Section 2, we describe and analyze the system model and the channel selection problems. In Section 3, we proceed to present the BLA-based distributed channel access scheme. Section 4 provides extensive simulation results that demonstrate the advantage of the BLA in channel selection. We conclude the paper in Section 5.

2   System Model and Problem Formulation

2.1   System Model and Assumptions

Two types of radios, PUs and SUs, operate in a spectrum band consisting of N channels allocated to PUs. PUs access the spectrum in a time-slotted fashion and the behavior of the PUs is independent from one channel to another. The SUs are synchronized with the PUs and the supported data rate for the SUs are the same in all the channels. There are M SU communication pairs in the network, and each of them needs to select, out of N channels, a channel for the purpose of communication. Without loss of generality, unless otherwise stated, we utilize the term "SUs" to refer to these SU pairs.

To model the behavior of the PU in the ith channel, $i \in \{1, \ldots, N\}$, we adopt a two-state Markov chain model, as shown in Fig. 1. State 0 stands for the channel being idle. State 1 represents the case when the channel is being occupied by a PU. d_i and b_i are the transition probabilities between these two states in channel i. Thus one can verify that the steady state probability of the channel being idle is given by $p_i = \frac{b_i}{b_i + d_i}$.

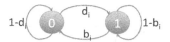

Fig. 1. The On/Off channel model for the PU's behavior

To avoid collision with PUs, at the beginning of each time slot, there is a quiet period for the SUs to sense the channel. If the channel is found unoccupied by the PU, the SUs can access the channel following a strategy prior to the packet transmission. There are two strategies considered in this work, depending on whether the operation of CSMA/CA is carried out or not. If the strategy being utilized does not support CSMA/CA, the SUs will transmit packets directly. However, if the strategy permits the CSMA/CA operation, the SUs compete for the accesses to the channels in the following manner. In each round of contention, an SU picks an integer randomly and individually from a contention window that has a fixed size, and then initiates counting down. During the counting-down period, the SUs always listen to the channel and check if there is an SU transmission in the channel. An SU will transmit a packet when the integer counts down to zero, and at the same time, no other SU has initiated transmitting in this channel within the current round of competition. Obviously, the SU that selects the smallest integer among all the integers selected by the co-channel SUs will reach zero first and consequently win the contention. If two or more contending SUs select the same integer that happens to be the minimum one, we say that a collision has occurred. The SUs who lose in the contention or who collide with one another, will give up their rights to transmit during this round.

The SU packet size is adjusted to be *one* packet transmission per time slot for the given data rate. We assume that channel sensing is ideal, and that due to the advanced coding scheme, interruption because of channel fading for SUs will not occur at the given rate. We also assume that there is a background protocol supporting channel access, and the detailed signaling process is outside the scope of this work. We also assume that SUs always have packets ready for transmission.

In what follows, we formulate the problems for channel access with or without CSMA/CA, as games, and in particular, as "Potential" games.

2.2   Channel Access without CSMA/CA

Consider the strategy where CSMA/CA is not utilized by the SUs after detecting an idle channel. In this case, collision among the SUs will happen if two or more SUs appear on the same channel after spectrum sensing. To avoid such a collision, we require that at most one SU can exist in each channel. If $M > N$, i.e., if there are more SUs than channels, collision among the SUs cannot be avoided because at least one channel will have more than one SU. For this reason, the first strategy is applicable only when $M \leq N$. The case when $M > N$ will be studied in the next subsection where CSMA/CA is enabled.

We formulate the channel selection problem without CSMA/CA as a game denoted by $\mathscr{G} = [\mathscr{M}, \{A_m\}_{m \in \mathscr{M}}, \{u_m\}_{m \in \mathscr{M}}]$, where $\mathscr{M} = \{1, \ldots, M\}$ is the set of SUs, $A_m = \{1, \ldots, N\}$ is the set of possible action/channel selections of the SU m, and u_m is the utility function of the SU m. The utility function of the SU m is defined as

$$u_m(a_m, \boldsymbol{a}_{-m}) = p_{a_m} f(h(a_m)), \tag{1}$$

where $a_m \in A_m$ is the action/channel selected by the SU m, and $\boldsymbol{a}_{-m} \in A_1 \times A_2 \times \ldots \times A_{m-1} \times A_{m+1} \ldots \times A_M$ represents the channels selected by all the other SUs, and where the symbol $\times$ represents the Cartesian product. The function $f(k) = 1$ if $k = 1$ and 0 otherwise. We denote $h(k)$ as the number of SUs that have selected channel k. $f(h(a_m))$ represents the event that a successful transmission has occurred, and this happens if and only if exactly a single SU exists in channel a_m.

We define the potential function for $\mathscr{G}$ as $\delta(a_m, \boldsymbol{a}_{-m}) = \sum_{i=1}^{N} \sum_{j=0}^{h(i)} p_i f(j)$. By definition, $\mathscr{G}$ is an Exact Potential game if we illustrate that when an arbitrary SU m changes from channel a_m to channel $\tilde{a}_m$ while other SUs keep the channel selections unchanged, the change in the potential equals to the change in the utility of the SU m.

Theorem 1. *The game $\mathscr{G}$ is an Exact Potential game that has at least one pure strategy NE point.*

The proof of Theorem 1 is a consequence of the definition of a Potential game, and the details of the proof are not included here in the interest of brevity.

We now define the overall system capacity $C(a_1, a_2, \ldots, a_M)$ as the sum of the utility functions of all the SUs, i.e., $C(a_1, a_2, \ldots, a_M) = \sum_{k=1}^{M} p_{a_k} f(h(a_k))$. Based on the implications of Proposition 1 in [6], it is easy to conclude that the optimal solution for the overall system capacity, $(a_1^*, a_2^*, \ldots, a_M^*) = \arg\max C(a_1, a_2, \ldots, a_M)$, is to select the channels with $p_i, i \in \{1, \ldots, N\}$ from high to low, and each channel being associated with only a single SU. Furthermore, it is not difficult to show that $(a_1^*, a_2^*, \ldots, a_M^*)$ is actually an NE point of $\mathscr{G}$, as a unilateral change of any SU will result in either a collision, i.e., if it tunes onto a channel where another SU resides, or an usage of a channel with a lower idle probability, i.e., if it tunes onto a channel where there is no SU. Thus, the global optimal solution is, indeed, an NE point of $\mathscr{G}$.

2.3   Channel Access with CSMA/CA

Similar to the analysis above, when CSMA/CA is enabled, the utility function of SU m can be defined as

$$u_m(a_m, \boldsymbol{a}_{-m}) = p_{a_m} f'(h(a_m), c), \tag{2}$$

with a_m, $\boldsymbol{a}_{-m}$, and $h(k)$ being the same as defined in (1). The function $f'(h(a_m), c)$ represents the probability of a successful transmission of SU m given $h(a_m)$ co-channel SUs that have a contention window of size $c > 0$. In more details, $f'(h(a_m), c)$ is expressed by

$$f'(h(a_m), c) = \begin{cases} 0, & \text{for } h(a_m) = 0, \\ 1, & \text{for } h(a_m) = 1, \\ \frac{\sum_{i=1}^{c-1}(c-i)^{(h(a_m)-1)}}{c^{h(a_m)}}, & \text{for } 1 < h(a_m) \leq M. \end{cases} \tag{3}$$

Although it is not shown explicitly here, the game can be demonstrated to be an Exact Potential game when CSMA/CA is utilized, implying that it has at least one pure strategy NE point.

It is worth mentioning that the channel selection which results in a global optimal system capacity is not necessarily an NE point when CSMA/CA is utilized. For example, consider a scenario that has two SUs communicating in a network in which there are two channels with idle probabilities $p_1 = 0.2$ and $p_2 = 0.8$, respectively. The optimal system capacity is $p_1 + p_2 = 1$ and this is achieved when one SU utilizes channel 1 and the other utilizes channel 2. However, the selection is not an NE when CSMA/CA is enabled because the one that selects channel 1 can move to channel 2 to achieve a higher capacity for itself. In this case, the NE point is the selection that requires both of the SUs to adopt channel 2.

3   BLA-Based Distributed Channel Access Scheme

Based on the analyses and the comprehension of the games as presented in the previous section, we propose a BLA-based *distributed* scheme for the SU channel access. Since at least one pure strategy NE point exists in a Potential game, we expect that after learning, each SU can converge to an NE point. The proposed scheme requires neither pre-information about the system, nor information being exchanged between the SUs. The scheme is described, in detail, below.

For each SU packet transmission, in the quiet period at the beginning of each slot, the SU selects a channel and senses its status. If the channel is occupied by its PU, it receives a penalty. Otherwise, the SU will proceed with one of the following two distinct steps depending on whether CSMA/CA is supported or not.

1. When CSMA/CA is not supported, the SU will start transmitting after the quiet period on this idle channel. If the transmitting SU is the only SU that selected this channel, the transmission succeeds and a reward is received. Otherwise, the transmission fails due to collision and each of the colliding SUs received a penalty.

2. When CSMA/CA is supported, if there is only a single SU that wins the contention, the transmission succeeds and the SU receives a reward. For an SU who experiences a collision or who loses in the channel contention, the transmission fails and it received a penalty.

Because of the randomness of the activities of the PUs and the unpredictable behaviors of the other SUs, it is challenging to determine the best channel for any specific SU. However, by equipping each of the SUs with a BLA described as follows, the problem can be solved both elegantly and efficiently.

Algorithm: BLA − Channel Access

Initialization:

- $\alpha_i := \beta_i := 1$, where $i = 1, 2, ..., N$, and N is the number of channels. These hyperparameters, α_i and β_i, count the respective rewards and penalties that each channel has received.
- $\gamma := 0$. γ counts the number of time slots.

Loop:

1. $\gamma := \gamma + 1$. Draw a value x_i randomly from the following Beta distributions:

$$f(x_i; \alpha_i, \beta_i) = \frac{\int_0^{x_i} v^{(\alpha_i-1)}(1-v)^{(\beta_i-1)} dv}{\int_0^1 u^{(\alpha_i-1)}(1-u)^{(\beta_i-1)} du}.$$

Then list x_i, $i \in [1, N]$ in their descending orders.
2. Select the channel associated with the greatest element in the list of x_i. Suppose the selected channel is channel j.
3. Sense the channel status and try to transmit in channel j.
   If transmission succeeds, it receive a reward; Update $(\alpha_j := \alpha_j + 1)$.
   If transmission fails, a penalty occurs; Update $(\beta_j := \beta_j + 1)$.

End Algorithm: BLA − Channel Access

4 Simulation Results

In this section, we present the results we have obtained by simulating the above solution. In the first subsection, we mainly focus on the convergence of the BLA in the Potential games. In the second subsection, we compare the convergence speed of the BLA and the L_{R-I} scheme. In the third subsection, we study the performance of the BLA and the L_{R-I} in terms of the normalized system capacities in different system configurations, where the normalized capacity is the result of dividing the total number of packets that are transmitted for all SUs in the system by the total number of simulated time slots.

Extensive simulations were conducted based on various configurations, a subset of which are reported here. Table 1 shows three groups (Conf. 1-3) of transition probabilities, each of which has nine channels, i.e., from channel 1 (c1) to channel 9 (c9). We will use below the configuration number and the channel number to indicate a specific simulation setting. For example, we use Conf. 1 (c1-c3) to represent that the simulation environment consists of Channel 1, Channel 2 and Channel 3 from configuration one.

Table 1. The transition probabilities used in different channels and configurations

Index	Configurations	c1	c2	c3	c4	c5	c6	c7	c8	c9
d_i	Conf.1	.9	.8	.7	.6	.5	.4	.3	.2	.1
b_i		.1	.2	.3	.4	.5	.6	.7	.8	.9
d_i	Conf.2	.4	.4	.4	.4	.5	.5	.5	.5	.5
b_i		.6	.6	.6	.6	.5	.5	.5	.5	.5
d_i	Conf.3	.2	.1	.8	.2	.2	.5	.1	.3	.1
b_i		.8	.9	.2	.3	.6	.5	.4	.9	.3

4.1   Convergence of the BLA in the Potential Games

We tested the BLA-based algorithm in various environment settings, in each of which we conducted an ensemble of 100 independent experiments, and where each of the experiments had 80,000 iterations/time slots. The results show that the BLA is able to converge to an NE point with a high probability. As more results will be presented in Section 4.3, to avoid repetition, we highlight the cases in Conf. 1 (c1, c2 and c9) to demonstrate that the BLA-equipped SUs converge to an NE point of the game.

Consider Conf. 1 (c1, c2 and c9) with 3 SUs. When CSMA/CA are utilized, the NE point of the game is that all SUs converge to Channel 9, while the global optimal is obtained when each of the SUs converge to a different channel. The values of the theoretical system capacity at the NE point and the global optimal point are 0.9 and 1.2 respectively. The theoretical capacity value is calculated by summing up the static idle probabilities of the channels at a specific point, which can be considered as the upper bound of the capacity at this point in reality.

For an SU, if its probability of selecting one channel surpasses 95%, we say that the SU has converged to this specific channel. If all the SUs converge, we consider that the game has converged. Further, if the game converges to an NE point, we say that the game has converged correctly.

In all the 100 experiments conducted, the game converged correctly. Also, all the SUs converged to Channel 9, which is an NE point, before 80,000 iterations. The normalized capacity achieved was 0.8203, which is fairly close to the theoretical value of 0.9. The gap between the normalized and the theoretical capacities is mainly due to the deviation from the NE point during the learning process before all the SUs converged.

4.2   Convergence Speed of the BLA and the L_{R-I}

In this subsection, we compare the convergence speed of the BLA and the L_{R-I} in playing the games. In the configurations where we have tested the BLA, we did the same simulations for the L_{R-I}, i.e., with the same number of experiments and the same number of iterations. As mentioned earlier, for the L_{R-I}, one has to define a learning parameter, λ, before the algorithm can be used. If we denote the game convergence accuracy probability as ξ (i.e., a game is required to converge correctly with probability ξ), we will see from the following results that different values of λ yield different values of ξ when the L_{R-I} is used to play the games.

From the simulation results, we observe that the BLA converged faster than the L_{R-I} given the same or even higher accuracy probabilities. For example, in Conf. 1 (c1-c4) with two SUs and where CSMA/CA were not permitted, the L_{R-I} achieved a convergence accuracy of 99% with the learning parameter, $\lambda = 0.047$. The average number of steps it took for the game to converge was $5,956.7$. In the same configuration, the convergence accuracy of the BLA was 100% and the average number of steps for convergence was $4,646.6$. Similarly, in Conf. 3 (c1-c4) with 2 SUs and where CSMA/CA was not permitted, the BLA converged with an accuracy of $\xi = 100\%$ within, on average, $1,327.7$ steps, while the L_{R-I} had a convergence accuracy of 96%, for which it needed an average of $4,899.3$ steps with the learning parameter $\lambda = 0.032$.

It is easy to see that the learning parameter for the L_{R-I} differs for distinct configurations and for various requirements for the convergence accuracy. Moreover, the convergence accuracy, in one sense, conflicts with the scheme's convergence speed. A larger learning parameter accelerates the convergence but may decrease the accuracy, while an unnecessarily small learning parameter gains more accuracy but at the same time, yields to a loss in the rate of convergence. In CRNs, it is quite challenging to find a reasonable or universal learning parameter that compromises well between the accuracy and the speed for all different scenarios. We conclude, therefore, that the L_{R-I} is not totally suitable for CRNs. On the contrary, the Bayesian nature of the BLA provides a much better tradeoff between the learning speed and its accuracy, and more importantly, these tradeoff are achieved automatically, i.e., without any parameter tuning.

It is also worth mentioning that the L_{R-I} can always achieve a higher convergence accuracy by reducing its learning parameter, λ, at the expense of a slower speed of convergence. For instance, in Conf. 3 (c1-c4) with 2 SUs and where CSMA/CA was not permitted, by tuning λ to be 0.0055, the L_{R-I} was able to achieve a convergence accuracy of 99%, where the average number of steps for convergence was 28,954 steps. In other words, if an extremely high accuracy of convergence is required, the L_{R-I} can always achieve this goal, because there is a provision for the learning parameter to be tuned to be smaller while it is still positive[1]. However, on the other hand, the cost can be expensive too, as a smaller learning parameter results in a slower convergence speed.

The above-mentioned arbitrarily high convergence accuracy can be achieved by the L_{R-I}. This cannot be expected from the BLA as the latter has no tunable learning parameter. Fortunately, though, the advantage of the BLA is not degraded, as it can, for almost all the scenarios, yield a competitive learning accuracy and speed.

4.3 Comparison of the Capacities of the BLA and the L_{R-I}

In this section, we compare the performance of the BLA and the L_{R-I} by examining their normalized capacities. We organize the numerical results in two parts according to whether CSMA/CA is utilized or not, as shown in Table 2 and Table 3 respectively. All the results are the averaged values obtained from 100 independent experiments each of which had 80,000 iterations. The learning rate for the L_{R-I} was fixed as 0.01.

In Tables 2 and 3, GO and NE stand for global optimal and Nash equilibrium, respectively. The values in the rows for GO/NE represent the theoretical capacities at those points. Note that when CSMA/CA is not enabled, the GO point is an NE point; but when it is enabled, the GO point does not necessarily have to be an NE point.

As can be seen from Table 2, the capacity achieved by the L_{R-I} is generally lower than that obtained by the BLA, showing that the BLA approach is generally superior

---

[1] One can refer to [7] for a proof of this statement.

to the L_{R-I}. Besides, the capacity of the BLA is quite close to the theoretical upper bound, which means that in the scenarios where CSMA/CA is disabled, the BLA is quite efficient in terms of the transmission of the packets.

Table 2. The capacity of the BLA with different number of SUs in 9-channel configurations, where CSMA/CA is not permitted

Conf.	Alg.	$M=2$	$M=4$	$M=6$	$M=8$
Conf.1	BLA	1.6982	2.9952	3.8891	4.3743
	L_{R-I}	1.6844	2.9565	3.8203	4.2260
(c1-c9)	NE/GO	1.7	3	3.9	4.4
Conf.2	BLA	1.1957	2.3828	3.3859	4.3835
	L_{R-I}	1.1943	2.3207	3.3276	4.2817
(c1-c9)	NE/GO	1.2	2.4	3.4	4.4
Conf.3	BLA	1.6976	3.2441	4.7383	5.8346
	L_{R-I}	1.6865	3.2067	4.6405	5.7234
(c1-c9)	NE/GO	1.7	3.25	4.75	5.85

Table 3. The capacity of the BLA with different number of SUs in 9-channel configurations where CSMA/CA is permitted

Conf.	Alg.	$M=4$	$M=8$	$M=12$	$M=16$
Conf.1	BLA	2.9926	3.7788	3.9582	3.9146
	L_{R-I}	2.9592	3.6754	3.8980	3.8602
(c1-c9)	NE/GO	3/3	3.9/4.4	4.2/4.5	4.2/4.5
Conf.2	BLA	2.3872	4.3725	4.7637	4.6432
	L_{R-I}	2.3534	4.3153	4.7107	4.6058
(c1-c9)	NE/GO	2.4/2.4	4.4/4.4	4.9/4.9	4.9/4.9
Conf.3	BLA	3.2438	5.8188	5.6333	5.4860
	L_{R-I}	3.2165	5.5121	5.5868	5.4582
(c1-c9)	NE/GO	3.25/3.25	5.85/5.85	5.85/6.05	5.85/6.05

Table 3 illustrates the simulation results when CSMA/CA is enabled. Again, the BLA outperforms the L_{R-I} as the normalized capacity achieved by the former is generally higher than that of the L_{R-I}. Note that in this table, the theoretical capacity at the GO point does not necessarily equal to that at an NE point, and all the normalized capacities achieved by both algorithms tend to approach the NE capacity instead of the GO capacity. This is because both the BLA and the L_{R-I} tend to converge to an NE point, even if there exists a GO point that may yield a superior capacity.

One can also see from Table 3 that in the cases where there are more SUs, the achieved capacity is relatively further away from the NE theoretical capacity. The reasons for this are twofold. Firstly, a larger number of SUs implies a more complicated environment requiring more steps for the SUs to converge. Secondly, CSMA/CA cannot be considered as ideal when the number of SUs increases. Indeed, understandably, more collisions could occur if there are more SUs contending for transmission.

We finally investigate the issue of fairness in CRNs. In cases in which CSMA/CA is not enabled, the SUs tend to stay in the NE point after the game has converged. In other words, the SU which has converged to a channel with a higher static idle probability, will always have a better chance for communication, resulting in an unfairness among the SUs. This can be resolved by re-initiating the learning process after a specific time interval. In this way, SUs can take turns to use the different channels and the fairness can be achieved statistically. But when CSMA/CA is enabled, the fairness among SUs is improved because co-channel SUs will share channel access opportunities.

5   Conclusion

This paper studies the channel selection problem in CRNs when multiple SUs exist. The problem includes two channel access strategies, i.e., when CSMA/CA is enabled, and when it is not. Each of the strategies has been formulated as a Potential game, and a BLA-based approach is presented to play these games. Simulation results show the advantages of the BLA scheme from four aspects. Firstly, the BLA is able to converge to the NE point of the game with a high accuracy. Secondly, the cost paid by the BLA before converging to the NE point is less than what the L_{R-I} algorithm demands in most cases. Thirdly, with regard to the learning parameter based LA, e.g., the L_{R-I}, the BLA does not have to tune any parameter to achieve a good tradeoff between its learning speed and accuracy. Finally, the BLA has been shown to be able to achieve a normalized system capacity close to the theoretical capacity value at an NE point.

References

1. Liang, Y.C., Chen, K.C., Li, G.Y., Mahönen, P.: Cognitive radio networking and communications: An overview. IEEE Trans. Veh. Technol. 60(7), 3386–3407 (2011)
2. Narendra, K.S., Thathachar, M.A.L.: Learning automata: An introduction. Prentice-Hall (1989)
3. Lakshmivarahan, S.: Learning algorithms theory and applications. Springer, New York (1981)
4. Song, Y., Fang, Y., Zhang, Y.: Stochastic channel selection in cognitive radio networks. In: IEEE Global Telecommunications Conference, Washington DC, USA, pp. 4878–4882 (November 2007)
5. Tuan, T.A., Tong, L.C., Premkumar, A.B.: An adaptive learning automata algorithm for channel selection in cognitive radio network. In: IEEE International Conference on Communications and Mobile Computing, Shenzhen, China (April 2010)
6. Zhang, X., Jiao, L., Granmo, O.C., Oommen, B.J.: Channel selection in cognitive radio networks: A switchable Bayesian learning automata approach. In: IEEE International Symposium on Personal, Indoor and Mobile Radio Communications, London, UK (September 2013)
7. Xu, Y., Wang, J., Wu, Q., Anpalagan, A., Yao, Y.-D.: Opportunistic spectrum access in unknown dynamic environment: A game-theoretic stochastic learning solution. IEEE Trans. on Wirel. Commun. 11(4), 1380–1391 (2012)
8. Granmo, O.C.: Solving two-armed Bernoulli bandit problems using a Bayesian learning automaton. International Journal of Intelligent Computing and Cybernetics 3(2), 207–234 (2010)
9. Granmo, O.C., Glimsdal, S.: Accelerated Bayesian learning for decentralized two-armed bandit based decision making with applications to the Goore game. Applied Intelligence 38(4), 479–488 (2013)

Towards the Design of an Advanced Knowledge-Based Portal for Enterprises: The KBMS 2.0 Project

Silvia Calegari, Matteo Dominoni, and Emanuele Panzeri

Department of Informatics, Systems and Communication (DISCo),
University of Milano-Bicocca,
v.le Sarca 336/14, 20126 Milano, Italy
{calegari,dominoni,panzeri}@disco.unimib.it

Abstract. This paper presents the KBMS 2.0 Project, a system that is aimed at managing and searching information effectively. The idea is to leverage the information stored in any enterprise to support users during their interactions with the defined portal. The advanced portal has been developed to help users in finding information related to their information needs with the use of a friendly user interface. This task has been done by considering two key aspects: (1) context, and (2) personalization, respectively. Contextual information allows to filter out knowledge not related to users, whereas personalization aspects have been used to support users during their searches by considering user preferences and user activities.

1   Introduction

The huge amount of information on enterprise environments is continuously overwhelming. It is more and more difficult for people to discover information that satisfies their information needs. In fact, for users the search for relevant information is a time consuming activity, as they spend a lot of time on relevant search results. Let us consider the scenario when a user interacts with a search engine where many pages of results are obtained after the evaluation of the query. In this typical example, the relevant result can be on the eighth page of results, with the consequence that the user might never even view it. Indeed, a deep analysis has shown that a user examines up to the fourth page of results [6]. The task is to help people to identify the relevant search results faster and easier. To face this issue, several aspects have to be considered to design an advanced knowledge-based portal for enterprises, such as the need to guarantee a certified quality of documents for both the retrieval and management of information.

The KBMS 2.0 Project[1] is aimed at providing an advanced knowledge-based portal to manage knowledge innovatively, with the objective to support people in all the phases of their searches and also in the emergence of new knowledge

---

[1] KBMS 2.0: Knowledge-Base Management System 2.0 is a research project that has involved the University of Milano-Bicocca, Italy, along with Enel SpA energy company.

A. Moonis et al. (Eds.): IEA/AIE 2014, Part II, LNAI 8482, pp. 58–67, 2014.

in the system. Our objective is to support users during their searches with the aim of retrieving the relevant search results according to the user's interests and preferences. The portal we developed is mainly based on two key aspects: (1) contextualization, and (2) personalization.

The former considers the *concept of context* that has received a great deal of attention in several research areas of Computer Science, such as information seeking and retrieval, artificial intelligence, ubiquitous computing, etc. [5,3]. Several meanings can be associated to the significance of context, and there is not a unique definition that covers all the ways the term is used. In [2] the context has been defined as any type of knowledge characterizing the situation of each entity where an entity can be a person, a place, or an object involved in the interactions between a user and an application, including the user and applications themselves. The latter refers to applications where the *personalization* aspect is based both on modeling the user context from the knowledge on users that represents user preferences and user activities, and on the definition of processes that exploits the knowledge representing the user in order to tailor the search outcome to the user's needs [4,7,8,9].

The KBMS 2.0 Project properly uses the advantage of integrating context and personalization aspects with the objective to define an effective portal to improve the search and management of information with the knowledge of the user. The combined use of these features is very innovative as they can focus the knowledge according to the user's information needs. The context allows to filter the knowledge characterizing the content of the portal, whereas the personalization allows to give to the user only the knowledge that can satisfy his/her interests. In addition, these aspects are supported with the definition of a friendly user interface that helps the user to interact in an easier way with the advanced portal.

The advanced knowledge-base portal developed in the KBMS 2.0 Project is targeted to all the multiple roles that people have in a company in order to have different access to the information. In fact, the individuals who work in a company are logically organized in a hierarchical structure, and for each level in the organization a specific role is assigned with the consequence that not all the knowledge has to be accessible or viewed in the same way for each role. The developed portal considers the roles that users can have in the Enel SpA energy company during the definition and the searches of the users. But in Enel SpA the hierarchical structure is more complex than other enterprises as two hierarchical structures are defined-one for each internal logical division, namely *market*, i.e. *Free-Market* and *Protected Categories Market*. Each market implicitly defines our context information. In fact, every person belongs to a specific market, and when he/she interacts with the advanced portal only specific information is visualized based on the market information. This means that the market acts as a filter on the knowledge stored in the system: if a user belongs to a specific market, he/she will have access only to some parts of the portal, and he/she will interact only with a subset of the information, namely the one related to his/her market. For each market, personalized aspects are considered to support the user in finding appropriate information. In detail, in the KBMS 2.0 Project the personalization aspects have been defined for managing information in several modules: search engine, grid of navigation, urgent news, newsletters, and links to external applications.

This paper describes the combined use of context and personalization issues in the two most important modules that have been developed in the KBMS 2.0 Project that are the search engine and the grid of navigation, respectively.

The paper is organized as follows. Section 2 shows the architecture of the KBMS 2.0 Project, Section 3 describes the search engine and the grid of navigation modules, and Section 4 gives an overview of some evaluation steps performed up to now according to a user-center driven monitoring. Finally, in Section 5 conclusions and future trends of research are given.

2   The KBMS 2.0 Architecture

The KBMS 2.0 architecture is presented in Figure 1. The KBMS 2.0 is based on the open source edition of the Liferay portal[2]. Due to the specific requests of the Enel SpA stakeholders and some limitations of the open source edition of Liferay, several new modules have been defined for developing the KBMS 2.0 Project. The new modules defined are: *Workflow, Search Engine, Grid of Navigation, Newsletter, Spot-News,* and *MyLinks*. From this set of modules only the *Search Engine* has been extended by integrating some functionalities provided by Lucene collocated into the native bundle of Liferay with new features related to: architectural aspects, such as the integration of Lucene with Apache Solr, and usability aspects, such as the addition of new social interactions, re-ranking of results etc. (see Section 3.2). In detail:

Presentation Layer.

- **Workflow:** it guarantees a certification of the documents that have to be inserted in the official repository of the system. The quality of the documents is given by the definition of a new workflow that allows to examine each document from three levels of certification where, for each level, several expert users analyze each document.
- **Newsletter:** it sends the recent published news via email to users according to some processes of contextualization and personalization.
- **Spot-News:** urgent news is visualized in the area-spot in the main page of the system. The visualization of urgent news is performed thanks to some processes of contextualization and personalization.
- **MyLinks:** a user has access to external applications by clicking-on specific links. There is a personalized area called *MyLinks*, where a user can indicate his/her preferred external links.
- **Search Engine:** an advanced search engine has been defined based on the following three features: contextualization, personalization and social activities (for more details see Section 3).

---

[2] Liferay website: `www.liferay.com`

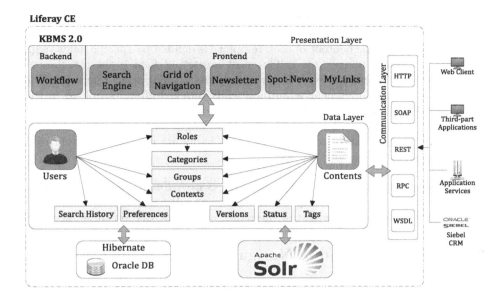

Fig. 1. The KBMS 2.0 architecture

- **Grid of Navigation:** it allows users to access documents that have been categorized according to some specific categories. Aspects of contextualization and personalization have been defined (for more details see Section 3).

Data Layer. Aspects of contextualization and of personalization have an impact on both users and contents. The context is given by analysing the information of the logical division in the enterprise. In the case of the Enel SpA energy company the context is given by the market, i.e. *Free-Market* and *Protected Categories Market*. The personalization aspects are related to: *Search History, Preferences, Roles, Categories,* and *Groups* (see Section 3). The user's roles are grouped by the *Group* field to classify roles under specific categories. Moreover, to each content three other fields have been defined, i.e. the field *Versions* that stores all the versions of a content, the field Status that indicates if a content is published, pending or is a draft, and the field *Tags* that allows to assign labels that indicate the topic of the content.

Communication Layer. The KBMS 2.0 Project can be accessible by several external devices/applications such as *Web Clients, Application Services, CRM (Customer Relationships Management),* etc. For each device/application there is a specific protocol of communication as shown in Figure 1. The KBMS 2.0 also provides a web service of communication between the *Search Engine* module and the CRM in order to provide documents (or their portions) regarding some critical issues for the Enel SpA operators when people call.

Contents. In the KBMS 2.0 Project the contents are grouped in 10 templates that are: script, faq, offers, glossary, models, banks, billing calendar, conventions, news and general. The content of each template is defined by specific fields (e.g., authors, title, date, images, etc.). The most important field is the attachment that allows to associate one or more documents to a specific template. The information of each document is automatically indexed after its upload in the system. It is possible to insert documents in several formats (such as .pdf, .docx, .doc, .pptx, .ppt, .xslx, .xls), and in several languages (such as English, Italian, Spanish, etc.). For sake of readability, in this paper we refer to a content as a generic document.

Search Engine and Grid of Navigation. *Search Engine* and *Grid of Navigation* modules are the most important components developed in the KBMS 2.0 Project. For this reason, the rest of the paper will be focused on these modules.

3   KBMS 2.0: "Grid of Navigation" and "Search Engine"

This section presents the two modules defined to help users during their searches for useful information. The two modules are: (1) grid of navigation, and (2) search engine. They consider aspects of context and personalization. The context is used to identify the user's market in order to filter the knowledge related to it; instead, the personalization aspects are used to highlight user's interests and preferences for make their searches faster.

3.1   Grid of Navigation

The *Grid of Navigation* is the first object that appears to users after the login in the KBMS 2.0 system. Its aim is to allow easy navigation of documents stored in the enterprise repository, by using a taxonomy of categories as support. In fact, when a user selects a category, all the documents categorized with it will be visualized. This way, a user has a general overview of all the documents that are related to the topic of the selected category. Figure 2 shows the interface developed for the *Grid of Navigation module*: the left-hand side is used to visualize the hierarchical structure of categories, whereas the right-hand side is the part of the interface devoted to visualizing all the information of documents that corresponds to each selected category. There are two hierarchical structures of categories, one for each market, in order to contextualize/filter the knowledge that a user can access as previously described. For each market, we have:

1. **Taxonomy of Categories:** A category in the taxonomy identifies a topic of interest. To each category two operations are performed: (1) a category is used to categorize each document stored in the enterprise repository, and (2) a category is associated with the roles defined in the enterprise that are the same used for both users and documents (see Section 2). This means that for each user the following aspects of personalization are enabled: (1) only the documents related to the categories in the hierarchy are visualized

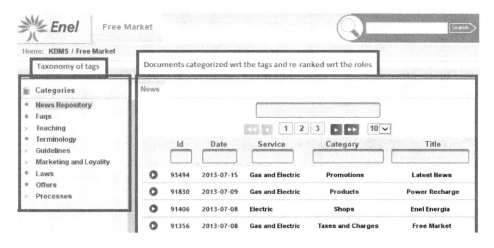

Fig. 2. An example of use of the *Grid of Navigation* module

to the user, and (2) only the categories related to the document roles are visualized in the taxonomy.

2. **Visualization of Documents:** All the information related to the documents will be visualized. As explained in Section 2, there are several templates that identify all the types of documents. Figure 2 shows the example where a user has selected the category *News Repository*, and only the template *News* is active for this category. The user interface has been developed to visualize all the other templates as tabs beside the tab *News.* The interface is a strong point of the system we developed. In fact, in only one page a user can access the information by navigating the *Taxonomy of categories* and by selecting the categories related to the document templates. When a tab is selected, only the documents related to the tab will be visualized.

For this part the following aspect of personalization is enabled: all the documents are re-ranked according to the user's role, with the objective of showing in the first positions only the documents that can be of interest to the user. A user has a trace of which documents have been re-ranked, as they are in bold with respect to the other ones.

3.2 Search Engine

The *Search Engine* module allows to users to perform searches in the documents repository, in order to satisfy the information needs faster and easier. A user has to identify in a few seconds/minutes the documents that can be useful for him/her. This feature has to be available in all the search engines used for any company; but for Enel SpA this activity assumes a key aspect in the system. Indeed, the search engine can be used by technical users for people's practical problems by phone solving. As happens for the *Grid of Navigation*, all the

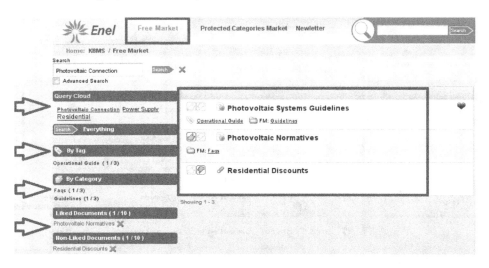

Fig. 3. An example of use of the *Search Engine* module

documents are contextualized/filtered according to the selected market. Figure 3 depicts an example where the selected tab is *Free Market*, so that a user can access only the documents contextualized for the market. The main features of the developed *Search Engine* are:

1. **Query Cloud:** it is used to visualize the keywords searched most by a user. Each keyword in this area has a different size: if a keyword is searched more times with respect to the others, then it will have a greater size. The Query Cloud section is an aspect of personalization of the search engine, as it stores the user's keywords history with the objective of tracking the user's interests, corresponding to what the user searches frequently. In addition, all the keywords can be used to support the user during the formulation of new queries: each keyword selected will appear in the textual area devoted to the formulation of the queries.
2. **Tag:** it is used to identify the topics of each document. This section shows all the tags related to the documents ranked after the evaluation of the query. A user can select one of these tags in order to filter the documents from the list of results.
3. **Category:** it is used to identify the categories of each document. This section shows all the categories assigned to the documents ranked after the evaluation of the query. A user can select one of these categories in order to filter the documents from the list of results. The categories are the ones corresponding to the *Taxonomy of categories* described in Section 3.1.
4. **Liked/Non-Liked Documents:** it corresponds to the social aspect included in the system. A user can attribute a social preference on each documents belonging to the list of results. Each document can have two social values: '*like*' thumb up, and '*non-like*' thumb down. As shown in Figure 3

when a user clicks on the *'like'* value the thumb up assumes a green color. Instead, when a user clicks on the *'non-like'* value the thumb down assumes a red color. If a social value is activated, the corresponding document will be also visualized in one of the two ad-hoc sections called *Preferred Documents* and *Non-Preferred Documents.* The social preferences are logically linked to the queries written by the user. In fact, if the same query is performed, then the past social values assigned to the documents will be visualized. The system has memory on which documents can be of interest (or non-interest) for the user, in order to make his/her searches faster, by focusing the attention on new documents inserted in the repository on the same query.

5. **Re-ranking:** the documents obtained after the evaluation of the query are re-ranked by considering the role information. The documents having the same roles of the user are positioned at the first positions. These documents are highlighted to the user thanks to the adoption of the *heart* icon situated on the right of each ranked document (see Figure 3).

4   Evaluation Issue

The KBMS 2.0 Project adopts a user-centered approach; this means to tailor services that better fit the user's preferences. In our approach a preliminary feedback has been performed to test the released KBMS 2.0 version by interviewing 80 users having a different role in the hierarchical organization in Enel SpA. The current version of the system has received positive judgement from users. They have been very collaborative by proposing ideas for extending the KBMS 2.0, either with new modules or new functionalities. Our goal is to obtain analytic evaluations in order to test if the services provided can really satisfy the user's information needs. The evaluation process should make explicit use of the information provided by the user preferences as an additional constraint in order to obtain relevant information and then to estimate its relevance. Two main techniques can be adopted in order to capture the user's preferences: explicit and implicit[1,9]. With the explicit approach the user must explicitly specify his/her preferences to the system. This can be achieved by filling in questionnaires and/or by providing short textual descriptions in order to specify the topics of interests. With the implicit approach, the user's preferences are automatically gathered by monitoring the user's actions. To this regard, the defined techniques range from click-through data analysis to query and log analysis, etc.

Our approach consists in using the implicit approach by analysing the log files. A log file is created every day and the actions performed by users on the system are monitored in this file. The choice of the information to be stored in the log file is not an easy step as it plays a key role in order to have data for making good evaluations. In fact, the information stored in the log file will be used to make evaluations for comparing our approach with other presented in the literature. At the moment, we have made an in depth study in order to insert specific information related to the aspects of personalization in the *Search Engine* and the *Grid of Navigation* modules. In detail, we have monitored the following information that will be used to generate implicit knowledge among users and documents:

Search Engine.
`general information:=` `<time,IP>` $\mapsto$ these parameters are inserted for all the information monitored. `time` is the time on when the action has been performed by a user, and `IP` is the IP address.
`query:=` `<market, keywords, groupsId, rolesId, userId>` $\mapsto$ `market` is the market indicating the logical division of the enterprise, `keywords` is the set of keywords written by a user during his/her query, `groupsId` is the set of user's groups, `roleId` is the set of user's roles, and `userId` is the code that identifies a user who has performed the query.
`category:=` `<assetCategory>` $\mapsto$ `assetCategory` is the category selected by a user after a first search to filter the ranked list of results.
`tags:=` `<assetTags>` $\mapsto$ `assetTags` is the tag selected by a user after a first search to filter the ranked list of results.
`liked/non-liked ranked documents:=` `<type of template, documentId, keyword, queryHash, market, userId, score, old score>` $\mapsto$ `type of template` indicates the typology/name of the document, `documentId` is the Id of the document selected in the list of results, `keyword` is the set of keywords written by a user during his/her query, `queryHash` is the hash code that identifies unequivocally a query, `market` is the market indicating the logical division of the enterprise, `userId` is the code that identifies a user who has performed the query, `score` is the current social status of the document, i.e. if like or non-like, and `old score` is the previous social status of the document, i.e. if like or non-like.
`liked/non-liked click-action on documents:=` `<url document>` $\mapsto$ `url document` is the url of the document, and it allows to open the page related to this document.
`query cloud:=` `<keyword>` $\mapsto$ `keyword` is the keyword stored in the Query Cloud area that has been selected by a user to formulate his/her new query.
`result:=` `<total results, scoreDocumentId, details scoreDocumentId>` $\mapsto$ `total results` indicates the number of results obtained after the evaluation of a query, `scoreDocumentId` is the score calculated for each ranked document, and `details scoreDocumentId` is a list of detailed information (e.g. the well known *tf* or *idf*) related on how the final score of a document has been obtained.

Grid of Navigation.
`general information:=` `<time,IP>` $\mapsto$ these parameters are inserted for all the information monitored. `time` is the time on when the action has been performed by a user, and `IP` is the IP address.
`grid:=` `<type of template, url document>` $\mapsto$ `type of template` indicates the typology/name of the document, and `url document` is the url of the document, and it allows to open the page related to this document.

5   Conclusions

In this paper, we have presented the KBMS 2.0 Project. This advanced portal is able to satisfy the user's information needs by providing a certificate quality of information and by supporting a user during his/her searches, with the goal of retrieving relevant documents faster. This goal has been reached by jointly considering aspects of context and personalization. Encouraging preliminary evaluations have been obtained to test the quality of KBMS 2.0. In the future, we plan to perform analytic evaluations by gathering the information contained in the defined log files, in order to generate new implicit knowledge among users and documents and to establish a fully collaborative system to support, for example, the learning of new users. This way, we will compare our approach with other presented in the literature.

Acknowledgements. This paper was written within the Enel SpA Project, and we want to thank all the people who allowed us to make an in depth study for the development of the KBMS 2.0 system.

References

1. Claypool, M., Brown, D., Le, P., Waseda, M.: Inferring user interest. IEEE Internet Computing 5, 32–39 (2001)
2. Dey, A.K.: Understanding and using context. Pers Ubiquitous Comput. 5(1), 4–7 (2001)
3. Dourish, P.: What we talk about when we talk about context. Personal and Ubiquitous Computing 8 (2004)
4. Harman, D.: Overview of the fourth text retrieval conference (trec-4). In: TREC (1995)
5. Ingwersen, P., Järvelin, K.: Information retrieval in context: Irix. SIGIR Forum 39, 31–39 (2005)
6. Jansen, B.J., Spink, A.: An analysis of documents viewing patterns of web search engine users. In: Scime, A. (ed.) Web mining: Applications and techniques, pp. 339–354. IGI Publishing, Hershey (2004)
7. Ma, Z., Pant, G., Sheng, O.R.L.: Interest-based personalized search. ACM Trans. Inf. Syst. 25(1) (2007)
8. Micarelli, A., Gasparetti, F., Sciarrone, F., Gauch, S.: Personalized search on the world wide web. In: The Adaptive Web: Methods and Strategies of Web Personalization, ch. 6, pp. 195–230 (2007)
9. Teevan, J., Dumais, S.T., Horvitz, E.: Personalizing search via automated analysis of interests and activities. In: Proceedings of the 28th Annual International ACM SIGIR Conference on Research and Development in Information Retrieval, SIGIR 2005, pp. 449–456. ACM, New York (2005)

A Topic Model for Traffic Speed Data Analysis

Tomonari Masada[1] and Atsuhiro Takasu[2]

[1] Nagasaki University, 1-14 Bunkyo-machi, Nagasaki 8528521, Japan
masada@nagasaki-u.ac.jp
[2] National Institute of Informatics, 2-1-2 Hitotsubashi, Chiyoda-ku,
Tokyo 1018430, Japan
takasu@nii.ac.jp

Abstract. We propose a probabilistic model for traffic speed data. Our model inherits two key features from latent Dirichlet allocation (LDA). Firstly, unlike e.g. stock market data, lack of data is often perceived for traffic speed data due to unexpected failure of sensors or networks. Therefore, we regard speed data not as a time series, but as an unordered multiset in the same way as LDA regards documents not as a sequence, but as a bag of words. This also enables us to analyze co-occurrence patterns of speed data regardless of their positions along the time axis. Secondly, we regard a daily set of speed data gathered from the same sensor as a document and model it not with a single distribution, but with a mixture of distributions as in LDA. While each such distribution is called topic in LDA, we call it patch to remove text-mining connotation and name our model *Patchy*. This approach enables us to model speed co-occurrence patterns effectively. However, speed data are non-negative real. Therefore, we use Gamma distributions in place of multinomial distributions. Due to these two features, Patchy can reveal *context dependency* of traffic speed data. For example, a 60 mph observed on Sunday can be assigned to a patch different from that to which a 60 mph on Wednesday is assigned. We evaluate this context dependency through a binary classification task, where test data are classified as either weekday data or not. We use real traffic speed data provided by New York City and compare Patchy with the baseline method, where a simpler data model is applied.

1 Introduction

Riding on the trend of Big Data, sensor data analysis increases its importance. Especially, traffic speed is one among the most important targets, because the necessity of analyzing this type of data prevails in most urban cities. It is commonplace to see traffic data as a time series and to apply techniques proposed for such type of data [5]. However, we need to notice that sensors and networks are vulnerable to damages from various natural causes and thus often stop sending data unlike the case of e.g. stock market data. Therefore, it is a possible option to regard traffic speed data not as a time series, but as an unordered multiset. This is reminiscent of the bag-of-words philosophy in text mining. We adopt this

A. Moonis et al. (Eds.): IEA/AIE 2014, Part II, LNAI 8482, pp. 68–77, 2014.

philosophy and regard a set of traffic speed data in a time window of a certain width as a bag of traffic speed data. In this paper, we set the window width to one day, showing a clear periodicity, and make windows non-overlapped, though other settings can be chosen depending on applications. Further, this approach also enables us to focus on co-occurrence patterns of traffic speed data regardless of their positions in each window. Many text mining methods also analyze co-occurrence patterns of words regardless of their positions in each document.

Our proposal has another important feature aside from the bag-of-words philosophy. We envision our method to be used for the off-line analysis addressing growing traffic congestion e.g. in need of redesigning traffic signal timing. Therefore, our method can be regarded as a traffic event analysis, not as a traffic prediction by using the dichotomy proposed in [7]. In particular, we aim to mine co-occurrence patterns of traffic speed data and thus propose a Bayesian probabilistic model inheriting *two* features from latent Dirichlet allocation (LDA) [2], which can be used for revealing latent co-occurrence patterns of words. The one feature is the bag-of-words philosophy, as is discussed above. The other is the admixture property. That is, LDA models each document with multiple word multinomial distributions, each of which is called *topic*. This is in contrast with naive Bayes that models each document only with a single multinomial corresponding to the class of the document. Following LDA, we model each set of traffic speed data with multiple Gamma distributions, which are used in place of multinomials, because speed data are non-negative real. To remove text-mining connotation, we call topic in our model *patch* and name our model *Patchy*. We use the patch assignments obtained by Patchy for evaluating similarities among different speed data sets in the same way as we use the topic assignments by LDA for evaluating similarities among documents [3].

The rest of the paper is organized as follows. Section 2 gives related work. Section 3 describes our model and its variational Bayesian inference. Section 4 presents the experimental results. Section 5 concludes the paper.

2   Related Work

While LDA is proposed for word frequency data of documents, we can modify it and propose a new topic model for other types of data. A typical example is Latent Process Decomposition (LPD) [8], which is proposed for analyzing microarray data represented as a real matrix with rows and columns corresponding to samples and genes, respectively. In LPD, samples, corresponding to documents in LDA, are modeled with a multinomial distribution over topics as in LDA. However, microarray data consists of real numbers. Therefore, LPD generates observed data as follows: Draw a topic from the per-document multinomial distribution over topics; and draw a real number from the univariate Gaussian distribution corresponding to the drawn topic for each gene.

Microarray data are similar to traffic speed data in that they consist of continuous values. However, traffic speed data are non-negative. Microarray data can be negative depending on preprocessings performed over raw expression level data. On the other hand, we can access raw data for traffic speed, and the data are non-negative. Therefore, with Gamma distribution, which can effectively model non-negative real data, we propose a new probabilistic model.

Gamma distribution has another advantage. It behaves very similar to Gaussian distribution when the shape parameter is large, which is the case for our experiment. For example, the density in Fig. 1 looks almost identical to that of Gaussian. Therefore, while respecting the non-negativity of speed data, we can also perform an analysis similar to that achieved with Gaussian distribution.

3 Patchy

In this paper, we regard a set of traffic speed data over one day (from 0 a.m. to 12 p.m.) gathered from the same sensor as a document and call such set of speed data *roll*. Patchy reconstructs every roll as a mixture of patches. Let D and K be the numbers of rolls and that of patches, respectively. Let n_d denote the number of traffic speed data in the dth roll. Patchy generates rolls as follows.

1. For each patch $k = 1, \dots, K$, draw a rate parameter β_k of a Gamma distribution Gamma(α_k, β_k) from the gamma prior distribution Gamma(a, b).
2. For each roll $d = 1, \dots, D$,
   (a) Draw a K-dimensional vector $\boldsymbol{m}_d$ from a K-dimensional Gaussian distribution $\mathcal{N}(\boldsymbol{\mu}, \boldsymbol{\Sigma})$ and let $\theta_{dk} \equiv \frac{\exp(m_{dk})}{\sum_j \exp(m_{dj})}$.
   (b) For each $i = 1, \dots, n_d$, draw a latent patch z_{di} from a multinomial distribution Discrete$(\boldsymbol{\theta}_d)$, and draw the ith observed traffic speed data x_{di} from the Gamma distribution Gamma$(\alpha_{z_{di}}, \beta_{z_{di}})$.

The parameter $\boldsymbol{\theta}_d$ of per-roll multinomial distribution over patches is generated not from Dirichlet distribution, but from logistic normal distribution. This is because we would like to make Patchy easily extensible for using metadata, e.g. sensor locations. For example, when two sensors are similar in their locations, we can make the corresponding distributions over patches similar by using the technique proposed in [4]. Further, in this paper, we assume that $\boldsymbol{\Sigma}$ is diagonal in favor of simplicity and denote its kth diagonal entry as σ_k^2. When we allow $\boldsymbol{\Sigma}$ to be non-diagonal, our inference needs to be modified based on [1].

The full joint distribution of Patchy is given as follows:

$$p(\boldsymbol{x}, \boldsymbol{z}, \boldsymbol{\beta}, \boldsymbol{m} | \boldsymbol{\mu}, \boldsymbol{\Sigma}, \boldsymbol{\alpha}, a, b) = p(\boldsymbol{\beta}|a, b) \cdot p(\boldsymbol{m}|\boldsymbol{\mu}, \boldsymbol{\Sigma}) \cdot p(\boldsymbol{z}|\boldsymbol{m})p(\boldsymbol{x}|\boldsymbol{\alpha}, \boldsymbol{\beta}, \boldsymbol{z})$$

$$= \prod_k \frac{b^a \beta_k^{a-1} e^{-b\beta_k}}{\Gamma(a)} \cdot \prod_d \prod_k \frac{1}{\sqrt{2\pi\sigma_k^2}} \exp\left\{ -\frac{(m_{dk} - \mu_k)^2}{2\sigma_k^2} \right\}$$

$$\cdot \prod_d \prod_i \prod_k \left\{ \frac{\exp(m_{dk})}{\sum_j \exp(m_{dj})} \cdot \frac{\beta_k^{\alpha_k} x_{di}^{\alpha_k-1} e^{-\beta_k x_{di}}}{\Gamma(\alpha_k)} \right\}^{\delta(z_{di}=k)}, \tag{1}$$

where $\delta(\cdot)$ is equal to 1 if the condition in the parentheses holds and is equal to 0 otherwise. Note that $\boldsymbol{\mu}$, $\boldsymbol{\Sigma}$, $\boldsymbol{\alpha}$, a, b, and K are free parameters.

After introducing a factorized variational posterior distribution $q(z)q(\beta)q(m)$, a lower bound of the log evidence can be obtained as follows:

$$\ln p(x|\mu, \Sigma, \alpha, a, b) \geq \int \sum_z q(z)q(m) \ln p(z|m)dm$$

$$+ \int q(m) \ln p(m|\mu, \Sigma)dm + \int \sum_z q(z)q(\beta) \ln p(x|\alpha, \beta, z)d\beta$$

$$+ \int q(\beta) \ln p(\beta|a, b)d\beta + H[q(z)q(\beta)q(m)] , \qquad (2)$$

where $H[\cdot]$ denotes the entropy. For variational posterior distributions, we assume the followings: $q(z)$ is factorized as $\prod_{d,i,k} q(z_{di}|\gamma_{di}) = \prod_{d,i,k} \gamma_{dik}^{\delta(z_{di}=k)}$, where each $q(z_{di}|\gamma_{di})$ is a multinomial; $q(\beta)$ is factorized as $\prod_k q(\beta_k|c_k, d_k)$, where each $q(\beta_k|c_k, d_k)$ is a Gamma distribution; and $q(m)$ is factorized as $\prod_{d,k} q(m_{dk}|r_{dk}, s_{dk})$, where each $q(m_{dk}|r_{dk}, s_{dk})$ is a univariate Gaussian.

We explain the terms of the right hand side of Eq. (2) in turn. The first term can be rewritten as $\sum_{d,i,k} \gamma_{dik}r_{dk} - \sum_{d,i,k} \gamma_{dik} \int q(m_d|r_d, s_d) \ln \sum_j \exp(m_{dj})dm_d$. By using the trick proposed in [1], we can obtain a tractable lower bound. That is, since $\ln x \leq \frac{x}{\nu} - 1 + \ln \nu$ for any $\nu > 0$, we obtain the following inequality:

$$\int \sum_z q(z)q(m) \ln p(z|m)dm \geq \sum_{d,i,k} \gamma_{dik}\left\{r_{dk} - \ln \nu_d + 1 - \frac{1}{\nu_d} \sum_j \exp\left(r_{dj} + \frac{s_{dj}^2}{2}\right)\right\} .$$
$$(3)$$

The other four terms of the right hand side of Eq. (2) are rewritten as follows:

$$\int q(m) \ln p(m|\mu, \Sigma)dm = -D \sum_k \ln \sqrt{2\pi\sigma_k^2} - \frac{1}{2} \sum_{d,k} \frac{s_{dk}^2}{\sigma_k^2} - \sum_{d,k} \frac{(r_{dk} - \mu_k)^2}{2\sigma_k^2} ,$$
$$(4)$$

$$\int \sum_z q(z)q(\beta) \ln p(x|\alpha, \beta, z)d\beta$$

$$= \sum_{d,i,k} \gamma_{dik}\left[-\ln \Gamma(\alpha_k) + (\alpha_k - 1)\ln x_{di} + \alpha_k\left\{\Psi(c_k) - \ln d_k\right\} - x_{di}\frac{c_k}{d_k}\right] , \quad (5)$$

$$\int q(\beta) \ln p(\beta|a, b)d\beta$$

$$= -K \ln \Gamma(a) + Ka \ln b + (a - 1) \sum_k \{\Psi(c_k) - \ln d_k\} - b \sum_k \frac{c_k}{d_k} , \qquad (6)$$

$$H[q(z)q(\beta)q(m)]$$

$$= -\sum_{d,i,k} \gamma_{dik} \ln \gamma_{dik} + \sum_{d,k} \ln s_{dk} + \sum_k \{\ln \Gamma(c_k) - (c_k - 1)\Psi(c_k) - \ln d_k + c_k\} ,$$
$$(7)$$

where we omit constant terms.

Table 1. 142 sensors used in our experiment

2, 3, 4, 17, 20, 27, 28, 31, 32, 35, 40, 42, 44, 47, 48, 55, 58, 60, 65, 69, 70, 79, 80, 81, 82, 83, 87, 92, 93, 94, 95, 96, 98, 99, 102, 131, 135, 162, 163, 164, 168, 173, 179, 180, 185, 186, 188, 190, 191, 192, 205, 243, 249, 251, 252, 253, 254, 256, 257, 258, 259, 260, 262, 263, 264, 267, 268, 270, 272, 278, 279, 281, 285, 287, 288, 289, 291, 292, 294, 295, 296, 297, 298, 299, 300, 302, 303, 304, 305, 308, 314, 315, 317, 318, 319, 320, 322, 323, 325, 326, 327, 329, 330, 331, 332, 333, 334, 335, 336, 337, 338, 343, 344, 345, 346, 356, 358, 359, 362, 363, 364, 365, 372, 373, 374, 375, 376, 377, 378, 379, 380, 381, 382, 384, 387, 388, 389, 411, 412, 413, 414, 415

Let $\mathcal{L}$ denote the lower bound of the log evidence resulting from the above discussion. By taking a derivative of $\mathcal{L}$ with respect to each parameter, we can obtain the corresponding update rule. ν_d and γ_{dik} can be updated as follows:

$$\nu_d = \sum_k \exp(r_{dk} + s_{dk}^2/2) \ , \ \gamma_{dik} \propto \exp(r_{dk}) \cdot \exp\{\alpha_k \Psi(c_k)\} \cdot \frac{x_{di}^{\alpha_k-1} e^{-c_k x_{di}/d_k}}{\Gamma(\alpha_k) d_k^{\alpha_k}} \ .$$

$$(8)$$

While we have no closed form update rule for r_{dk}, the derivative is written as:

$$\frac{\partial \mathcal{L}}{\partial r_{dk}} = n_{dk} - \frac{n_d}{\nu_d} \exp\left(r_{dk} + \frac{s_{dk}^2}{2}\right) - \frac{r_{dk} - \mu_k}{\sigma_k^2} \ . \tag{9}$$

Therefore, we can use binary search on the interval $(0, \mu_k + n_{dk}/\sigma_k^2)$ for solving $\partial \mathcal{L}/\partial r_{dk} = 0$. For s_{dk}, we have the following derivative:

$$\frac{\partial \mathcal{L}}{\partial s_{dk}} = -\frac{n_d s_{dk}}{\nu_d} \exp(r_{dk} + s_{dk}^2/2) - \frac{s_{dk}}{\sigma_k^2} + \frac{1}{s_{dk}} \ . \tag{10}$$

We can also use binary search on the interval $(0, \sigma_k)$ for solving $\partial \mathcal{L}/\partial s_{dk} = 0$.

We estimate the free parameters by maximizing $\mathcal{L}$. With respect to α_k, we solve $\partial \mathcal{L}/\partial \alpha_k = 0$ and obtain the following update: $\alpha_k = \Psi^{-1}\left(\frac{\sum_{d,i} \gamma_{dik} \ln x_{di}}{N_k} + \Psi(c_k) - \ln d_k\right)$, where $N_k \equiv \sum_{d,i} \gamma_{dik}$. c_k and d_k can be updated as: $c_k = a + N_k \alpha_k$, $d_k = b + \sum_{d,i} \gamma_{dik} x_{di}$. Further, a and b can be updated as: $a = \Psi^{-1}\left(\ln b + \frac{\sum_k \{\Psi(c_k) - \ln d_k\}}{K}\right)$, $b = \frac{Ka}{\sum_k c_k/d_k}$. Finally, μ_k and σ_k can be updated as: $\mu_k = \frac{\sum_d r_{dk}}{D}$, $\sigma_k^2 = \frac{\sum_d s_{dk}^2 + \sum_d (r_{dk} - \mu_k)^2}{D}$. With respect to the free parameter K, i.e., the number of patches, we use several different values in the experiment.

4 Experiment

4.1 Data set

We downloaded real traffic speed data provided by New York City[1]. The data file, formatted as a tab-separated text file, contains a set of traffic speed data

---

[1] https://data.cityofnewyork.us/Transportation/
Real-Time-Traffic-Speed-Data/xsat-x5sa

in mph originated from sensors scattered all around the city and is updated frequently. We fetched the data file from the Web site every minute. Based on 43,569 text files we fetched, a training set and a test set were composed.

The training set is a set of traffic speed data gathered from May 27 to June 16 in 2013 and is used for the variational inference described in Section 3. Every daily set of traffic speed data gathered from the same sensor is called *roll*. Rolls correspond to documents in LDA. The training set contains 21 rolls for each sensor, because there are 21 days from May 27 to June 16. The test set is a set of speed data collected from July 23 to August 5 in 2013 and contains 14 rolls for each sensor, because there are 14 days from July 23 to August 5. Since the data from some sensors made us suspect malfunction, we only used the rolls coming from the 142 sensors in Table 1. Consequently, we have $21 \times 142 = 2,982$ training rolls and $14 \times 142 = 1,988$ test rolls.

4.2   Context Dependency

We performed an experiment for evaluating the *context-dependency* of patch assignments achieved by Patchy. In case of text data, the same word may mean different things depending on its context. LDA can detect such differences by assigning the same word to different topics. In a similar manner, Patchy can assign the same traffic speed in mph to different patches depending on its context. Below, we give patch assignment examples to make our motivation clear.

Fig. 2 presents the examples, where we set the number K of patches to 60. The top, middle, and bottom panels of Fig. 2 visualize the 21 training rolls gathered from the sensors whose IDs are 289, 315, and 359, respectively. The horizontal axis presents the dates from May 27 to June 16, and the vertical axis presents traffic speed values. Markers of different colors and shapes correspond to different patches. For each traffic speed observation x_{di}, the patch giving the largest probability, i.e., arg $\max_k \gamma_{dik}$, is depicted by the corresponding marker.

The top panel of Fig. 2 shows that the observations of around 60 mph on June 2, 8, 9, 15, and 16, which are all weekends, are assigned to the patch depicted as dark purple circle. Further, the observations of mph values in the same range are assigned to the same patch also on May 27, a holiday in the United States. However, on weekdays, the observations of mph values in the same range are assigned to the patch depicted as light yellow down-pointing triangle. Fig. 1 presents the density functions of the two corresponding Gamma distributions. Based on Fig. 1, we can discuss that the observations of around 60 mph are assigned to the patch depicted as dark purple circle when the observations show less variance in one day. Therefore, it can be said that Patchy assigns the same mph value to different patches depending on daily co-occurrence patterns. The middle panel of Fig. 2 also shows that the observations of around 60 mph are assigned to the patch depicted as dark purple circle on May 27, June 2, and June 16, which are all weekends. This patch is the same with that depicted as dark purple circle in the top panel. However, the observations of mph values in the same range are assigned to other patches on weekdays. A similar discussion can be repeated for this panel.

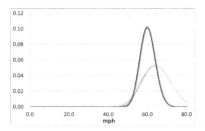

Fig. 1. Two Gamma distributions corresponding to the patches that appear in the top panel of Fig. 2. Each density function is depicted with the same color as in Fig. 2.

The bottom panel of Fig. 2 gives an example of different nature. The observations of around 42 mph are assigned to the patch depicted as light purple square on June 2, 8, 9, and 15, which are all weekends. However, the observations of mph values in the same range are assigned to a wide variety of patches on weekdays, depending on the co-occurrence patterns.

In sum, it is suggested that Patchy can discriminate weekdays from weekends, because Patchy reveals that there is a remarkable difference between the daily co-occurrence patterns of traffic speed data on weekdays and those on weekends. Therefore, we evaluate Patchy through a binary classification task where test rolls are classified as either weekday data or not. From now on, we call the rolls gathered on weekdays as weekday rolls and the others as weekend rolls.

4.3   Predictive Probability

We compare Patchy with the baseline method that uses a simpler probabilistic model. This baseline fits each training roll with a single Gamma distribution, whose parameters are estimated by maximum likelihood [6]. Consequently, we have 21 Gamma distributions for each sensor, because 21 training rolls are obtained from each sensor. Let $\text{Gamma}(\bar{\alpha}_d, \bar{\beta}_d)$ denote the Gamma distribution fitting the dth training roll. For a test roll $\boldsymbol{x}_0 = \{x_{01}, \dots, x_{0n_0}\}$, we calculate its probability with respect to the dth training roll as follows:

$$P_{\text{baseline}}(\boldsymbol{x}_0|\boldsymbol{x}_d) = \prod_{i=1}^{n_0} \frac{\bar{\beta}_d^{\bar{\alpha}_d}}{\Gamma(\bar{\alpha}_d)} x_{0i}^{\bar{\alpha}_d-1} e^{-\bar{\beta}_d x_{0i}} \ . \tag{11}$$

The training roll giving the largest probability, i.e., $\arg\max_{\boldsymbol{x}_d} P_{\text{baseline}}(\boldsymbol{x}_0|\boldsymbol{x}_d)$, is selected. If the selected roll is a weekday roll, we have an answer that the test roll $\boldsymbol{x}_0$ is also a weekday roll. Otherwise, we have an answer that $\boldsymbol{x}_0$ is a weekend roll. May 27 is regarded as a weekend, because it is a holiday in the United States. Since there are 142 sensors as Table 1 shows, we have 142 answers, each corresponding to different sensors. Therefore, the baseline method conducts a voting. When more than half of the 142 sensors give an answer that the test roll is a weekday roll, we classify the test roll as a weekday roll.

Patchy gives an answer to our classification problem in the following manner. When we are given a test roll $\boldsymbol{x}_0 = \{x_{01}, \ldots, x_{0n_0}\}$, we calculate its probability with respect to the dth training roll as below:

$$P_{\text{Patchy}}(\boldsymbol{x}_0|\boldsymbol{x}_d) = \prod_i \sum_k \theta_{dk} \frac{\beta_k^{\alpha_k}}{\Gamma(\alpha_k)} x_{0i}^{\alpha_k-1} e^{-\beta_k x_{0i}} \ . \tag{12}$$

When we calculate the probability in Eq. (12), we do not integrate out the parameter β_k, but simply replace β_k with an estimated mean c_k/d_k in favor of computational efficiency. This is justified by the fact that both c_k and d_k are estimated as very large ($> 10^3$) in most cases and thus give a peaky Gamma posterior. Based on Eq. (12), Patchy classifies the test roll as either weekday or not depending on whether the training roll giving the largest probability, i.e., $\arg \max_{\boldsymbol{x}_d} P_{\text{Patchy}}(\boldsymbol{x}_0|\boldsymbol{x}_d)$, is a weekday roll or not. Also in this case, we have 142 answers, each corresponding to different sensors. Therefore, we conduct a voting.

4.4   Comparison Experiment

The test set is composed from the traffic speed data gathered from July 23 to August 5. Therefore, we have 14 days to be classified. For example, July 25 should be classified as a weekday, because it is Thursday. The dates are widely different from those of the training set, because the similarity intrinsic to the data on contiguous dates may make the classification task unnecessarily easier. We compare Patchy and the baseline by the number of days correctly classified.

Before comparing our method with the baseline, we show the result achieved by the nearest neighbor (NN) to grasp the hardness of our classification. We calculate the distance between a test roll $\boldsymbol{x}_0$ and the dth training roll $\boldsymbol{x}_d$ as $\sum_{i=1}^{n_0}(x_{0i} - x_{di})^2/n_0$, where the difference $x_{0i} - x_{di}$ is calculated for the two observations, i.e., x_{0i} and x_{di}, having the same timestamp i. We then classify the test roll as either weekday or not based on whether the training roll giving the smallest distance is a weekday roll or not. Also in this case, we perform a voting with the answers given by the 142 sensors. We obtained the following result: NN classified 13 among the 14 days correctly and only misclassified July 27 as a weekday. However, NN uses the timestamps of traffic speed data, because the formula $\sum_i(x_{0i} - x_{di})^2/n_0$ for distance calculation is the sum of squares of the difference between the two observations having the same timestamp i. This means that NN accesses a richer amount of information than Patchy and the baseline. Therefore, NN is inefficient in memory usage when compared with Patchy and the baseline, both of which are based on the bag-of-words philosophy.

From now on, we discuss the comparison between Patchy and the baseline. Firstly, we provide the result of the baseline method. The baseline could classify 12 days correctly and misclassified July 26, Friday and August 5, Monday as weekends. Secondly, we provide the result of Patchy. K was varied from 30 to 100 with a step size of 10. Further, we ran the variational inference 20 times after initializing γ_{dik}s randomly, for each of the eight settings of K. Consequently, we obtained 20 different inference instances for each setting of K. Table 2 summarizes classification results. For example, when $K = 100$, one, six, 13, and zero

Table 2. Classification results achieved by Patchy

	# days correctly classified			
K	11	12	13	14
30	0/20	10/20	10/20	0/20
40	0/20	9/20	11/20	0/20
50	0/20	6/20	14/20	0/20
60	0/20	4/20	16/20	0/20
70	0/20	5/20	15/20	0/20
80	0/20	7/20	13/20	0/20
90	0/20	5/20	15/20	0/20
100	1/20	6/20	13/20	0/20

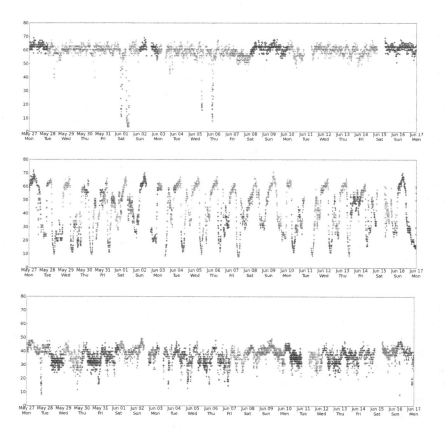

Fig. 2. Examples of patch assignments obtained by Patchy when $K = 60$. The horizontal axis presents the dates, and the vertical axis presents the traffic speed in mph.

inference instances among 20 instances enabled us to correctly classify 11, 12, 13, and 14 days, respectively. For any setting of K, no inference instance could give a perfect classification. The best result was obtained when $K = 60$, where

16 inference instances enabled us to correctly classify 13 days, and only August 5, Monday was misclassified as weekend. It is interesting that this date is different from the date misclassified by NN, i.e., July 27. The other four inference instances correctly classified 12 days, where July 27, Saturday and August 5, Monday were misclassified. Recall that NN correctly classified 13 days. Therefore, it can be concluded as follows: 1) Patchy can give a better accuracy than the baseline; and 2) Patchy can achieve almost the same accuracy as NN even when we adopt the bag-of-words philosophy and further represent data in a more compact form with an LDA-like topic modeling approach.

5   Conclusions

This paper proposed a new topic model called Patchy for realizing a context dependent analysis of traffic speed data. In the evaluation experiment, we used real data set provided by New York City and clarified that Patchy could assign observations of the same traffic speed to different patches depending on daily co-occurrence patterns of traffic speed data. In fact, the context dependency revealed by Patchy led to an effectiveness in our binary classification task.

An important future work is to extend Patchy by considering similarities among sensors. The traffic speed data we used in this paper also contains location information of sensors. Therefore, we can modify topic probabilities e.g. by considering distances among sensors.

References

1. Blei, D.M., Lafferty, J.D.: Correlated topic models. In: NIPS (2005)
2. Blei, D.M., Ng, A.Y., Jordan, M.I.: Latent Dirichlet allocation. JMLR 3, 993–1022 (2003)
3. Drummond, A., Jermaine, C., Vagena, Z.: Topic models for feature selection in document clustering. In: SDM, pp. 521–529 (2013)
4. Hennig, P., Stern, D.H., Herbrich, R., Graepel., T.: Kernel topic models. In: AIS-TATS (2012)
5. Mills, T.C., Markellos, R.N.: The Econometric Modelling of Financial Time Series. Cambridge University Press (2008)
6. Minka, T.P.: Estimating a Gamma distribution (2002), http://research.microsoft.com/en-us/um/people/minka/papers/minka-gamma.pdf
7. Pan, B., Demiryurek, U., Shahabi, C.: Utilizing real-world transportation data for accurate traffic prediction. In: ICDM, pp. 595–604 (2012)
8. Rogers, S., Girolami, M., Campbell, C., Breitling, R.: The latent process decomposition of cDNA microarray data sets. IEEE/ACM Trans. Comput. Biol. Bioinformatics 2(2), 143–156 (2005)

Semi-automatic Knowledge Acquisition through CODA

Manuel Fiorelli, Riccardo Gambella, Maria Teresa Pazienza,
Armando Stellato, and Andrea Turbati

ART Research Group, Dept. of Enterprise Engineering (DII),
University of Rome, Tor Vergata
Via del Politecnico, 1, 00133 Rome, Italy
{fiorelli,pazienza,stellato,turbati}@info.uniroma2.it,
gambella.riccardo@gmail.com

Abstract. In this paper, we illustrate the benefits deriving from the adoption of CODA (Computer-aided Ontology Development Architecture) for the semi-automatic acquisition of knowledge from unstructured information. Based on UIMA for the orchestration of analytics, CODA promotes the reuse of independently developed information extractors, while providing dedicated capabilities for projecting their output as RDF triples conforming to a user provided vocabulary. CODA introduces a clear workflow for the coordination of concurrently working teams through the incremental definition of a limited number of shared interfaces. In the proposed semi-automatic knowledge acquisition process, humans can validate the automatically produced triples, or refine them to increase their relevance to a specific domain model. An experimental user interface tries to raise efficiency and effectiveness of human involvement. For instance, candidate refinements are provided based on metadata about the triples to be refined, and the already assessed knowledge in the target semantic repository.

Keywords: Human-Computer Interaction, Ontology Engineering, Ontology Population, Text Analytics, UIMA.

1 Introduction

Efficient Information Management and Information Gathering are becoming extremely important to derive value from the large amount of available information. While the uptake of Linked Data [1]promoted uniform standards for the publication of information as interlinked datasets, the Web still consists mainly of unstructured content.

Dealing with this heterogeneous content asks for coordinated capabilities of several dedicated tools. In fact, development of knowledge acquisition systems today largely requires non-trivial integration effort and the development of ad hoc solutions for tasks, which could be better defined and channeled into an organic approach.

Platforms such as GATE [2] and UIMA [3] provide standard support for content analytics, while they completely demand to developers tasks concerning data transformation and publication. There have been a few attempts at completing information extraction architectures with facilities for the generation of RDF

A. Moonis et al. (Eds.): IEA/AIE 2014, Part II, LNAI 8482, pp. 78–87, 2014.

(Resource Description Framework) [4] data, e.g. the RDF UIMA CAS Consumer[1] and Apache Stanbol[2]. The former is a UIMA component consuming the analysis metadata to generate RDF statements, while the latter focuses on semantic content management. These projects share the same approach, which consists in a vocabulary-agnostic serialization of metadata inferred by analytics. However, Stanbol provides a component, called Refactor Engine[3], which can refactor the extracted triples, possibly to target a user chosen vocabulary.

In our opinion, these approaches lack an overall perspective on the task, as well as a proposal of a gluing architecture supporting the triplification process as a whole.

In this paper we discuss the benefits of adopting CODA (Computer-aided Ontology Development Architecture), an architecture based on standard technologies (such as UIMA, OSGi[4], and OBR[5]) and data models (e.g. RDF, OWL [5] and SKOS [6]), which defines a comprehensive process for the production of semantic data from unstructured content. This process includes content analysis, generation of RDF out of extracted information, and involvement of humans for validation and refinement of the produced triples.

2   CODA Architecture

CODA (Fig. 1) extends UIMA with specific capabilities for populating RDF datasets and evolving ontology vocabularies with information mined from unstructured content.

Section 2.1 provides an overview of UIMA, while section 2.2 introduces the main components of CODA and details their role in the overall architecture. Finally, section 2.3 describes our approach for a user interface supporting validation and enrichment of the automatically extracted triples.

2.1   UIMA: Unstructured Information Management Architecture

In this section, we introduce some basic concepts about UIMA, in order to show how CODA can work jointly with this architecture.

UIMA is an architecture (OASIS Standard[6] in 2009) and an associated framework[7] supporting development and orchestration of analysis components – called Analysis Engines (AEs) – for the extraction of information from unstructured content of any media type.

UIMA adopts a data-driven approach for component integration, in which Analysis Engines represent and share their results by using a data structure called CAS (*Common Analysis Structure*[7]).

---

[1] http://uima.apache.org/downloads/sandbox/RDF_CC/
RDFCASConsumerUserGuide.html
[2] http://stanbol.apache.org/
[3] https://stanbol.apache.org/docs/trunk/components/
rules/refactor.html
[4] http://www.osgi.org/Specifications/HomePage
[5] http://felix.apache.org/site/
apache-felix-osgi-bundle-repository.html
[6] http://docs.oasis-open.org/uima/v1.0/uima-v1.0.html
[7] http://uima.apache.org

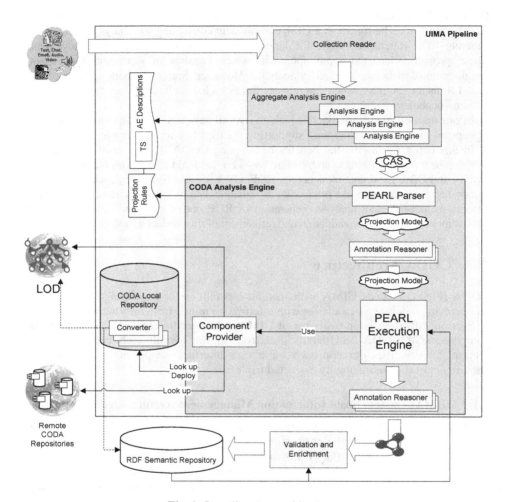

Fig. 1. Overall system architecture

A CAS contains both the information to be analyzed (called Subject of Analysis, or *Sofa*) and the extracted metadata, represented as typed feature structures [8]constrained by a model called *Type System* (TS) in the UIMA terminology. A specific kind of metadata (*annotations*) refers to specific sofa segments, and assigns explicit semantics to them. The CAS also contains an index over the metadata, which guarantees fast access to them, and allows iterating over the annotations, usually according to the order in which they appear in the sofa.

CODA fits the UIMA architecture by providing a concrete Analysis Engine that projects UIMA annotations onto RDF knowledge bases with a user-defined vocabulary.

2.2   CODA Analysis Engine

The *CODA Analysis Engine* represents the real innovative contribution provided by CODA that proves relevant to applications dealing with knowledge acquisition tasks.

While UIMA manages the unstructured content analytics, CODA provides components and their orchestration for projecting the resulting CAS onto an RDF Semantic Repository.

This projection is a complex process including the following activities:

— Selection of relevant information from within the CAS;
— Construction of RDF nodes out of UIMA metadata;
— Instantiation of user defined graph patterns with the prepared RDF nodes.

To support the user in the specification of this process, CODA provides a rule-based pattern matching and transformation language, called PEARL (Proj Ection of Annotations Rule Language) [9].

The *PEARL Execution Engine* orchestrates the triplification process, and executes PEARL rules against the CASes populated by UIMA AEs. These specifications may include annotations, providing various kinds of metadata, which may guide extensions of the execution engine, or, if required, be propagated to the output triples. User-supplied *annotation reasoners* may automatically annotate the PEARL rules, or customize the mechanism for annotating the results.

PEARL syntax covers UIMA metadata selection and RDF graph construction patterns. Conversely, PEARL externalizes the construction of RDF nodes to external components, called *converters*, since this is a varied task, possibly depending on domain/application specific procedures, or on external services(such as OKKAM [10]).

PEARL specifications refer to converters indirectly in the form of URIs that identify the desired behavior (a *contract*) rather than concrete implementations. During the triplification process, the *Component Provider* follows a two-step procedure for resolving these references into suitable converters. By first, the provider lookups a match in well-known repositories (starting from the *CODA Local Repository*). If no candidate is found, the Component Provider activates a procedure to discover additional repositories in the Web. Indeed, complying with the principle of self-descriptiveness [11] and grounding in the Web, contract URIs dereference to RDF descriptions which include a list of authoritative repositories of known implementations.

This architecture enables mostly autonomous configuration of CODA systems, which is valuable when reusing PEARL documents written by third parties, as in the open and distributed scenario described in. In fact, the reuse of PEARL specifications benefits from the lazy resolution of contracts into converters, as this approach allows choosing (among functionally equivalent ones) the implementation best fitting the non-functional requirements of a specific user, including consumption of computational resources, performance with respect to the tasks and even licensing terms.

2.3   CODA Human Interface

Human experts may revise the triples produced by *CODA Analysis Engine*, in order to raise their quality. An experimental user interface, called *CODA-Vis*, hides low-level issues concerning the RDF syntax, and supports more effective interaction modalities with domain experts. CODA-Vis (see later Fig. 3) benefits from the interaction with a Knowledge Management and Acquisition framework for RDF [13] for what concerns the UI and the interaction with the target semantic repository.

The validation aspect is limited to the binary judgment about the acceptability of individual triples. User support consists of additional features:

- Exploration of the source data which originated the suggestion, and of the semantic context of the elements in the triple;
- Alternative ways of representing the acquired data (e.g. list of triples, focus on common subjects, or groups of subject/predicate)

However, users may actively enter into the knowledge acquisition process by enhancing the automatically acquired results, in order to augment their preciseness, while strengthening their binding to the specific domain. For instance, the RDF triple "art:Armando rdf:type ex:Person" is obtained by the projection of a result from a UIMA Named Entity Recognizer into the generic class representing persons. However, a further evolution of the ontology for the Academic domain contains further specializations such as professor, researcher and student. The results would be more relevant, if the individual were assigned to the most specific class. In order to enable this kind of triple modification and improvement, the system has to organize and present the relations between the extracted information and the underlying RDF dataset, thus supporting the user in more effective decisions. In the given scenario, the system might pop up a class tree rooted at the class representing persons, thus providing the user with a focused slice of the underlying ontology, which should bring the user the right specialization.

3   Running Example

In this section, we step through the main phases of the triplification process by using a sample application. In section 3.1, we illustrate the use of PEARL for specifying the extraction and transformation process of information contained in UIMA annotations. For this example, we assume the existence of a UIMA Analysis Engine able to recognize a Person with its working Organization (if present). This means that the resulting CAS may contain several annotations, each one holding the identifiers of the recognized entities (the person and possibly the organization). In section 3.2, we show how CODA-Vis enables humans not just to accept or reject the suggested triples, but also to refine them with the help of a dedicated context sensitive user interface.

```
prefix ontologies: <http://art.uniroma2.it/ontologies#>
annotations = {
@Duplicate
@Target(Subject, Object)
Annotation IsClass.
}
rule it.uniroma2.art.Annotator id:PEARLRule{
nodes = {
person uri personId
organization uri organizationId
}
graph = {
@IsClass(Object).
$person a ontologies:Person .
OPTIONAL { $person ontologies:organization $organization . } .
}
}
```

Fig. 2. A PEARL rule example

3.1   Metadata Matching and Projection

PEARL is a pattern-matching and transformation language for the triplification of UIMA metadata. It allows concise yet powerful specifications of the transformation, while allowing the externalization (see section 2.2) of highly application and/or domain specific procedures.

Since its initial specification [9], PEARL has evolved towards greater expressiveness and support for more complex workflows. In this paper, we will emphasize the features introduced for supporting human validation and enrichment (see the previous section and section 3.2).

Fig. 2 provides an example of a PEARL document consisting of one rule, which drives the projection process in the running example.

A rule definition begins with the keyword *rule*, followed by the matched UIMA type name. The rule applies when an annotation from the index matches the declared UIMA type: the rule execution flows through a series of steps, specified within the rest of the rule definition.

The *nodes* section initializes *placeholders* with RDF nodes, which are constructed out of information that is extracted from the matched annotation using a *feature path* (a sequence of features inside a feature structure).For instance, in the rule in Fig. 2, the value of feature `person Id`is transformed into a URI resource that is assigned to the placeholder `person`. Default conversion heuristics are applied, and are inferred based on the specified node type: in this case, as the target is a URI, the feature value is first normalized to match the restrictions on the URI syntax, and then prefixed with the namespace of the target semantic repository.

The graph section defines a graph pattern[8], which contains mentions (starting with the symbol "$") of placeholders originating from the nodes section. During a rule execution, this graph pattern is instantiated into RDF triples, by substituting the nodes they hold for the placeholders occurring in the pattern. For example, the triple pattern "$person a ontologies:Person" states that CODA will suggest a triple, asserting that the value inside the placeholder person is an instance of the class ontologies:Person. Since the other triple pattern is inside an OPTIONAL graph pattern, it is possible that CODA will not produce any triple, if the annotation triggering the rule has no value for the feature organizationId (so novalue is stored inside the placeholder organization).

PEARL allows providing metadata about the elements of a graph pattern and the rules themselves, by attaching annotations to them. CODA provides a few meta-annotation types, for the further characterization of new annotation types, defined in the *annotations* section. *Target* indicates the elements, which an annotation type applies to: e.g. a triple as a whole, the subject, the predicate or the object. *Duplicate* controls whether annotations of a give type will be propagated to the results that originated from the annotated elements. In fact, with the exception of the aforementioned meta-annotation types, PEARL is almost agnostic with respect to the meaning of the annotations. In fact, annotations within a PEARL specification stand primarily as hints to extensions of the execution engine, which may somehow react to these annotations. Additionally, if an annotation is marked with *Duplicate*, it is subject to a uniform propagation process. Therefore, the output of the PEARL execution engine is a set of triples, possibly annotated with metadata helping their further exploitation.

While remaining consistent with this linear process, it is possible to plug reasoners that apply simple pattern matching techniques, to annotate either a PEARL document, or directly the output triples. These reasoners may handle common and simple cases, while manual annotation remains an option for solving the uncommon and hard ones.

In the example in Fig. 2, we first define the annotation *IsClass*, which is applicable only to the subject and the object of a triple. Then, we annotate the object of the first triple pattern with *IsClass*. As the annotation has been marked as *Duplicate*, CODA will propagate the annotation to the triples instantiated out of the triple pattern.

3.2   Human Validation and Enrichment

The annotated triples produced by the CODA Analysis Engine might not meet the quality requirements for being committed to the target semantic repository as they are. Therefore, we developed an experimental user interface for supporting the involvement of humans in the validation and enrichment of the acquired data. The validation involves a binary judgment, whether to accept or reject the proposed triples, while the enrichment consists in the specialization of the suggested triples to better suit the specific domain model.

Continuing the example before, it could be useful to specialize the class of each recognized person with respect to finer-grain distinctions made by the target vocabulary. To support this use case, the interface provides pluggable enhancers for different enrichment scenarios. In Fig. 3, for instance, a human validator activates the enhancer to assign an instance of the class `Person` resulting from a knowledge acquisition process to a more specific subclass found in the target ontology.

---

[8]http://www.w3.org/TR/rdf-sparql-query/#GraphPattern

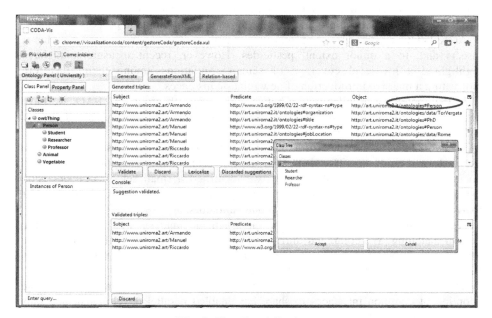

Fig. 3. Class Specialization

To achieve this result the user should select a suggested RDF resource produced by CODA. The enhancer suggests relevant modifications by dynamically constructing a class tree consisting of all the subclasses of the class `Person`. This specific behavior is offered since the selected resource has been annotated with *IsClass*. Further annotation types are associated to other behaviors, e.g. refining a property, refining a SKOS concept, or refining a street address by mashing-up popular online map applications. An extension mechanism in the UI assures that new enhancers can be associated to new annotation types. The selection of the relevant behaviors amounts to suitably annotating the triples, without any constraint on the mechanism to adopt. The primary means for producing these annotations is the duplication of the annotations in the PEARL transformation rules. However, these annotations do not need to be hard-coded in the rules; in fact, dedicated reasoners may automatically annotate the rules, usually by simple pattern matching techniques.

4   A Use Case: AgroIE

AGROVOC[14] [9] is a thesaurus developed and maintained by The Food and Agriculture Organization (FAO)[10] of the United Nations (UN). It contains more than 30000 concepts in up to 22 languages covering topics of interest for the FAO. Recently the vocabulary has been migrated to SKOS [15] and published as Linked

---

[9] http://aims.fao.org/standards/agrovoc/about
[10] http://www.fao.org

Data. In compliance with the fourth Linked Data rule, reliable mappings [16] to other popular datasets have been produced by means of a semi-automatic process.

AGROVOC contains, among others, a number of concepts regarding plants, insects and, to a minor extent, pesticides. However, recent measurements have highlighted a low-level coverage of semantic relations among them, therefore motivating a massive introduction of new relation instances.

In the context of a collaboration between our research group and FAO, we have developed a semi-automatic platform, called *AgroIE,* for the augmentation of AGROVOC, based on the CODA framework.

We worked on web pages from Wikipedia and other specialized sites that tend to use a formal and repetitive language. By experimentation on about a hundred documents from the available sources, we identified a small set of high precision extraction patterns for the desired relations (e.g. an insect is a pest of a plant), in order to avoid an unsustainable load on the human validators.

AgroIE uses a UIMA pipeline for the linguistic analysis and lexico-syntactic pattern matching. We combined third-party analysis engines from DKPro (wrapping Stanford Core NLP) to ad-hoc engines for the recognition of the relevant entities (insects, plants and pesticides) and the relationship among them.

The clear separation between content extraction and triplification issues and the adoption of common interfaces to share has enabled parallel development and thus increased productivity. In our setting, two MsC students have been developing – in the context of their thesis work – the UIMA Information Extraction engine, a PhD with expertise in CODA has written the PEARL document once the UIMA type system had been defined, while a fourth developer familiar with linked data has written the two *converters/identity resolvers.*

5   Conclusions

In the context of a specific application, we confirmed that CODA benefits the development of knowledge acquisition systems in several ways. The use of UIMA for content analysis increases the chances of reusing independently developed solutions, thus possibly raising performance and reducing the development effort. Furthermore, the clear process definition underpinning CODA supports concurrent development on different aspects, while reducing the synchronization needs to the (possibly incremental) definition of shared interfaces. Finally, an experimental user interface completes the envisioned *Computer-aided* Ontology Development Architecture, towards more effective and productive interactions with humans during the validation and enhancement activities.

References

1. Bizer, C., Heath, T., Berners-Lee, T.: Linked Data - The Story So Far. International Journal on Semantic Web and Information Systems, Special Issue on Linked Data (IJSWIS) 5(3), 1–22 (2009)

2. Cunningham, H.: GATE, a General Architecture for Text Engineering. Computers and the Humanities 36(2), 223–254 (2002), doi:10.1023/A:1014348124664

3. Ferrucci, D., Lally, A.: Uima: An architectural approach to unstructured information processing in the corporate research environment. Nat. Lang. Eng. 10(3-4), 327–348 (2004)

4. Manola, F., Miller, E.: RDF Primer. In: World Wide Web Consortium (W3C), http://www.w3.org/TR/rdf-primer/ (accessed February 10, 2004)

5. W3C: OWL 2 Web Ontology Language. In: World Wide Web Consortium (W3C), http://www.w3.org/TR/2009/REC-owl2-overview-20091027/ (accessed October 27, 2009)

6. Isaac, A., Summers, E.: SKOS Simple Knowledge Organization System Primer. In: World Wide Web Consortium (W3C), http://www.w3.org/TR/skos-primer/ (accessed August 18, 2009)

7. Götz, T., Suhre, O.: Design and implementation of the UIMA common analysis system. IBM System Journal 43(3), 476–489 (2004)

8. Carpenter, B.: The Logic of Typed Feature Structures, hardback edn. Cambridge Tracts in Theoretical Computer Science, p. 32. Cambridge University Press (1992)

9. Pazienza, M.T., Stellato, A., Turbati, A.: PEARL: ProjEction of Annotations Rule Language, a Language for Projecting (UIMA) Annotations over RDF Knowledge Bases. In: International Conference on Language Resources and Evaluation (LREC 2012), Instanbul, Turkey (2012)

10. Bouquet, P., Stoermer, H., Bazzanella, B.: An Entity Name System (ENS) for the Semantic Web. In: Bechhofer, S., Hauswirth, M., Hoffmann, J., Koubarakis, M. (eds.) ESWC 2008. LNCS, vol. 5021, pp. 258–272. Springer, Heidelberg (2008)

11. Mendelsohn, N.: The Self-Describing Web. In: World Wide Web Consortium (W3C) Technical Architecture Group (TAG), http://www.w3.org/2001/tag/doc/selfDescribingDocuments.html (accessed February 7, 2009)

12. Diosteanu, A., Turbati, A., Stellato, A.: SODA: A Service Oriented Data Acquisition Framework. In: Pazienza, M.T., Stellato, A. (eds.) Semi-Automatic Ontology Development: Processes and Resources, pp. 48–77. IGI Global (2012)

13. Pazienza, M.T., Scarpato, N., Stellato, A., Turbati, A.: Semantic Turkey: A Browser-Integrated Environment for Knowledge Acquisition and Management. Semantic Web Journal 3(3), 279–292 (2012)

14. Caracciolo, C., Stellato, A., Morshed, A., Johannsen, G., Rajbhandari, S., Jaques, Y., Keizer, J.: The AGROVOC Linked Dataset. Semantic Web Journal 4(3), 341–348 (2013)

15. Caracciolo, C., Stellato, A., Rajbahndari, S., Morshed, A., Johannsen, G., Keizer, J., Jacques, Y.: Thesaurus Maintenance, Alignment and Publication as Linked Data. International Journal of Metadata, Semantics and Ontologies (IJMSO) 7(1), 65–75 (2012)

16. Morshed, A., Caracciolo, C., Johannsen, G., Keizer, J.: Thesaurus Alignment for Linked Data Publishing. In: International Conference on Dublin Core and Metadata Applications, pp. 37–46 (2011)

Retaining Consistency for Knowledge-Based Security Testing

Andreas Bernauer[1,*], Josip Bozic[2], Dimitris E. Simos[3,**],
Severin Winkler[1], and Franz Wotawa[2]

[1] Security Research, Favoritenstrasse, Vienna, 1040, Austria
[2] Techn. Univ. Graz, Inffeldgasse 16b/2, 8010 Graz, Austria
[3] SBA Research, Favoritenstrasse, Vienna, 1040, Austria
{abernauer,swinkler}@securityresearch.at,
{jbozic,wotawa}@ist.tugraz.at,
dsimos@sba-research.org

Abstract. Testing of software and systems requires a set of inputs to the system under test as well as test oracles for checking the correctness of the obtained output. In this paper we focus on test oracles within the domain of security testing, which require consistent knowledge of security policies. Unfortunately, consistency of knowledge cannot always be ensured. Therefore, we strongly require a process of retaining consistencies in order to provide a test oracle. In this paper we focus on an automated approach for consistency handling that is based on the basic concepts and ideas of model-based diagnosis. Using a brief example, we discuss the underlying method and its application in the domain of security testing. The proposed algorithm guarantees to find one root cause of an inconsistency and is based on theorem proving.

Keywords: model-based diagnosis, root cause analysis, testing oracle.

1 Introduction

Ensuring safety and security of today's software and systems is one of the big challenges of industry but also of society especially when considering infrastructure and the still increasing demand of connecting services and systems via the internet and other means of communication. Whereas safety concerns have been considered as important for a longer time, this is not always the case for security concerns. Since security threads often are caused because of bugs in software or design flaws, quality assurance is highly required.

Testing software and systems is by far the most important activity to ensure high quality in terms of a reduced number of post-release faults occurring after

---

[*] Authors are listed in alphabetical order.
[**] This work was carried out during the tenure of an ERCIM "Alain Bensoussan" Fellowship Programme. This Programme is supported by the Marie Curie Co-funding of Regional, National and International Programmes (COFUND) of the European Commission.

A. Moonis et al. (Eds.): IEA/AIE 2014, Part II, LNAI 8482, pp. 88–97, 2014.

the software deployment. Hence, there is a strong need for testing security related issues on a more regular bases. Unfortunately, testing with the aim of finding weaknesses that can be exploited for an attack is usually a manual labor and thus expensive. Therefore, automation of software testing in the domain of software security is highly demanded.

In order to automate testing we need input data for running the *system under test* (SUT) and a test oracle that tells us whether the observed system's output is correct or not. There are many techniques including combinatorial testing [2] that allows for automatically deriving test input data for SUTs based on information about the system or its inputs. Unfortunately, automating the test oracle is not that easy, which is one reason for manually carrying out testing. In order to automate the test oracle, knowledge about the SUT has to be available that can be used for judging the correctness of the received output. In the domain of security the security policies have to be available in a formal form. To solve this problem, we suggest a framework for formalizing the knowledge base, which comprises the two different parts: (1) general security knowledge, and (2) SUT specific knowledge. We distinguish these parts in order to increase reuse. The general security knowledge might be used for testing different systems while the SUT specific knowledge only comprise additional clarifications and requirements to be fulfilled by one SUT. For example, in web-based applications a login should expire at least within a certain amount of time, say 60 minutes. In a particular application the 60 minutes might be too long and a specific policy overrides this value to 30 minutes.

Unfortunately, distinguishing this two kinds of knowledge might lead to trouble. Let us discuss an example. In general no user should be allowed for adding another user to the system with the exception of a user with superuser privileges. This piece of knowledge is more or less generally valid. A specific SUT might have here a restriction, i.e., stating that some users with restricted administrator rights are allowed to add new users. When assuming that such users do not have all rights of a superuser we are able to derive a contradiction. Another example would be that a text field should never be allowed for storing Javascript in order to avoid cross-site scripting attacks (see [7] for an introduction into the area of security attacks and how to avoid them). In web application for developing web pages such a restriction does not make any sense and has to be relaxed at least for certain text fields. All these examples indicate that there is a strong need to resolve inconsistencies between general security policies and custom policies used for specific applications. In order to increase automation retaining consistency has to be performed as well in an automated way.

The challenge of handling inconsistencies in knowledge bases has been an active research question in artificial intelligence for decades. Model-based diagnosis [4,9,6] offers means for removing inconsistencies via making assumptions explicit. In this paper, we adopt the underlying ideas in order to handle inconsistencies in knowledge bases used for implementing as test oracles. Note that the presented approach is tailored to handle security testing where general and custom polices together are used for implementing a test oracle. Our approach is not restricted to a particular test suite generation method providing that the used testing method delivers test input for the SUT and that there is an oracle necessary in order to distinguish faulty from correct SUT's output.

This paper is organized as follows. In the next section we discuss the underlying problem in more detail using a running example. Afterwards, we introduce the consistency problem and provide an algorithm for solving this problem. Finally, we discuss related research, conclude the paper, and give some directions for future research.

2   Running Example

In order to illustrate the problem we want to tackle in this paper we discuss a small example in more detail. Let us consider a web application having an user authentication page where each user has to login in order to access the system. In this case the initial state is the login screen. After entering the correct user information, i.e., user name and password, the system goes to the next state, e.g., a user specific web page. The action here would be the login. The facts that hold in the next state are the user name and the information that the user logged in successfully.

Testing such a SUT requires the existence of test cases. We assume that there is a method of generating such test cases. This can be either done manually or automatically using testing methods. We call the resulting set of test cases a *test suite* for the SUT. When applying a certain test case t to the SUT there is now the question of determining whether the run is correct or not. In case a SUT behaves as expected when applying a test case t we call t a *passing test case*. Otherwise, t is a *failing test case*. But how to classify test cases as failing or passing? To distinguish test cases we introduce an application's *policy P*, which is a consistent set of first order logic rules. When using such a policy P for classifying a test case t we ask a theorem prover whether the output of the SUT after applying t is consistent with the policy P. If this is the case, the test case t is a passing test case, and a failing one, otherwise. Hence, consistency checking based on P can be considered as our test oracle.

An example for a policy P for our small example application would be: For every user access to the user specific web page is only possible if and only if the authentication is successful. The authentication is successful if and only if the user exists in the user database and the given password is equivalent to the one stored in the database. Some other examples of the rules that form a policy are the following:

1. A user (except the superuser) must not add other users to the application.
2. The database is accessed by the application only but never directly by a user.
3. A user must not be able to overwrite any configuration file.

Note that policies describe the available knowledge of the domain and a particular application. As already said we distinguish two kinds of policy knowledge:

the general domain specific knowledge and the SUT specific knowledge, i.e., a policy P comprises two parts: the general policy P_G and the custom policy P_C, i.e., $P = P_G \cup P_C$.

When separating P into two parts immediately two problems arise. First, the general policy might be in contradiction with the custom policy. In the introduction we already discussed an example. Second, there might be parts of the general policy that are not applicable for a certain SUT. For example, consider an application that uses a database to store data. A policy for this case must contain all rules that defines the database's storage functionality. In the same example, if the application does not use a database to store data, then the policy does not need to contain rules for this functionality. The latter problem is not severe because a policy that is not used in an application would never lead to a contradiction with any facts of any state of the application. Thus such a policy would never cause a test case to fail.

Let us discuss the consistency problem in more detail. For this purpose we formalize the example given in the introduction stating that ordinary users are not allowed to add new users. Only superusers might add new users. In order to formalize this general policy P_G we introduce the predicates `user(X)`, `superuser`, and `add_user`. The first order logic rules using the syntax of Prover9 [8] with the Prolog-style option for variables and constants would be:

```
all X (superuser(X) -> add_user(X)). all X ((user(X) &
-superuser(X)) -> -add_user(X)).
```

Moreover, we would also add a rule stating that superusers are also users. This is necessary in order to ensure that superusers have at least all privileges of users and sometimes additional more.

```
all X (superuser(X) -> user(X)).
```

Obviously, these three rules are consistent. We can check this by running a theorem prover like Prover9 using the rules. If there is no contradiction the theorem prover would not find any proof.

Let us now formalize a custom policy P_C. There we introduce a new kind of user that is able to add new users. We call this user, a user with administration rights and introduce a new predicate `admin` to represent such a user formally.

```
all X (admin(X) -> user(X)). all X (admin(X) -> add_user(X)).
```

In addition we add a specific user with administration rights *Julia* to the application and an ordinary user *Adam*. Hence, we add the following facts to the custom policy P_C:

```
admin(julia). -superuser(julia). user(adam).
```

Note that we explicitly have to state that *Julia* is a user with administration rights only (and not a superuser). This might be avoided when adding more knowledge to the policies. However, for illustration purposes we keep the example as simple as possible. When running Prover9 on $P_G \cup P_C$ we obtain a proof, which looks like follows:

```
3 (all A (user(A) & -superuser(A) -> -add_user(A)))
                # label(non_clause).  [assumption].
4 (all A (admin(A) -> user(A))) # label(non_clause).
                [assumption].
5 (all A (admin(A) -> add_user(A))) # label(non_clause).
                [assumption].
6 -user(A) | superuser(A) | -add_user(A).  [clausify(3)].
9 -superuser(julia).  [assumption].
10 admin(julia).  [assumption].
11 -admin(A) | user(A).  [clausify(4)].
12 -admin(A) | add_user(A).  [clausify(5)].
13 -user(julia) | -add_user(julia).  [resolve(9,a,6,b)].
15 user(julia).  [resolve(10,a,11,a)].
16 -add_user(julia).  [resolve(15,a,13,a)].
17 add_user(julia).  [resolve(10,a,12,a)].
18 $F.  [resolve(16,a,17,a)].
```

Hence, we see that there is a contradiction, which have to be resolved. In this case eliminating the contradiction is easy. This can be done by adding & -admin(X) to the rule stating that ordinary users are not allowed to add new users. However, this would change the general policy and thus limit reuse. Therefore, we need an approach that allows us to change the policy on the fly when bringing together the general and the custom policy.

3   Consistency Handling

In the previous section we discussed the underlying problem of handling consistencies in case of general and custom policies used for implementing a test oracle in the domain of security testing. In this section, we discuss how to handle cases where the general policy is in contradiction with the custom policy. Such cases require actions for eliminating the root causes of the contradiction. As already mentioned the approach introduced in this section for retaining consistency is based on the artificial intelligence technique model-based diagnosis [4,9,6]. There assumptions about the health state of components are used to eliminate inconsistencies arising when comparing the expected with the observed behavior. In the case of our application domain, i.e., eliminating inconsistencies that are caused by the rules of a policy, we have to add assumptions about the validity of the rules.

In the following, we discuss the approach using the example of Section 2. The approach is based on the following underlying assumptions:

- We assume that both the general policy and the custom policy is consistent.
- We prefer custom policy rules over general policy rules. As a consequence we search for rules of the general policy that can be removed in order to get rid of inconsistencies.
- We prefer any solutions that are smaller in terms of cardinality than others. Here the underlying rational is that we want to remove as little information as possible stored in any knowledge base.

The first step of our approach is to convert the first-order logic policy rules into a propositional form. We do this by replacing each quantified variable with a constant. The set of constants to be considered are the constants used somewhere in the policy rules. In our running example we only have the two constants julia and adam. Note that this transformation does not change the semantics in our case because all the rules are specific to a certain instantiation. The constants represent such an instantiation. Therefore, there is no negative impact on the semantics. For our example we would obtain the following set of rules after conversion to propositional logic:

```
(superuser(adam) -> user(adam)).
(superuser(julia) -> user(julia)).
(superuser(adam) -> add_user(adam)).
(superuser(julia) -> add_user(julia)).
((user(adam) & -superuser(adam)) -> -add_user(adam)).
((user(julia) & -superuser(julia)) -> -add_user(julia)).
admin(adam) -> user(adam).
admin(julia) -> user(julia).
admin(adam) -> add_user(adam).
admin(julia) -> add_user(julia).
admin(julia).
-superuser(julia).
user(adam).
```

In the next step, we introduce the predicate ab_i to indicate that a certain rule i is valid or not. If a rule is valid the predicate is *false*, and otherwise *true*. For example, (superuser(adam) -> user(adam)) might be a valid rule in the overall policy or not. In case the rule is not valid, we know that the its corresponding predicate has to be *true*. Hence, we modify the rule to:

```
        ab1 | (superuser(adam) -> user(adam)).
```

We do this for all general rules only because of our assumption to prefer custom policy rules and obtain the following set of rules:

```
ab1 | (superuser(adam) -> user(adam)).
ab2 | (superuser(julia) -> user(julia)).
ab3 | (superuser(adam) -> add_user(adam)).
ab4 | (superuser(julia) -> add_user(julia)).
ab5 | ((user(adam) & -superuser(adam)) -> -add_user(adam)).
ab6 | ((user(julia) & -superuser(julia)) -> -add_user(julia)).
admin(adam) -> user(adam).
admin(julia) -> user(julia).
admin(adam) -> add_user(adam).
admin(julia) -> add_user(julia).
admin(julia).
-superuser(julia).
user(adam).
```

The third step comprises searching for truth assignments to ab_i such that no contradiction arises. The truth assignment to all ab_i that leads to no contradiction is an explanation and all rules where ab_i is set to true can be eliminated from the general policy. Hence, the union of the rules from the general policy and the custom policy is guaranteed to be consistent by construction. Formally, we define explanations as follows:

Definition 1. *Let K be the propositional knowledge base obtained from the policy $P = P_G \cup P_C$ and AB the set of abnormal predicates ab_i used in K. A set $E \subseteq AB$ is an* explanation *if and only if setting all elements of E to true and the rest of the predicates to false causes K to be satisfiable, i.e., $K \cup \{\mathsf{ab}_i | \mathsf{ab}_i \in E\} \cup \{-\mathsf{ab}_i | \mathsf{ab}_i \in AB \setminus E\} \not\models \bot$.*

It is worth noting that this definition of explanations corresponds to the definition of diagnosis in the context of model-based diagnosis. We refer the interested reader to Reiter [9] for more information. In contrast to Reiter we do not need to define explanations as minimal explanations. Instead we say that an explanation E is minimal, if there exists no other explanation E' that is a subset of E.

For example, in our running example ab6 is an explanation, which can be easily verified by calling Prover9 with the propositional knowledge base including the logical rule -ab1 & -ab2 & -ab3 & -ab4 & -ab5 & ab6. In this case Prover9 does not return a proof. Therefore, the policy is consistent and can be used as test oracle. Similarly, we are able to check that ab5, ab6 together is also an explanation. However, this explanation is not minimal accordingly to our definitions.

Computing explanations can be done by checking all subsets of AB using a theorem prover. Such an algorithm of course would be infeasible especially for larger knowledge bases. In order to improve computing explanations we are able to use the fact that we are interested in smaller explanations. Hence, we might check smaller subsets first. We refer the interested reader to Reiter [9] and Greiner [6] for an algorithm that is based on this idea.

In the following we outline a novel algorithm **COMPEX** (Algorithm 1) that computes one single explanation for a knowledge base and the set of possible assumptions. **COMPEX** basically comprises two parts. In Line 1 to Line 8 an explanation is computed. This is done by calling the theorem prover Prover9 and extracting all ab_i predicates that are used by Prover9 to derive a contradiction. These predicates are added to the explanation E. The second part (Line 9 to 17) is for minimizing E. This is done by removing element by element from E. If we still obtain a consistent explanation when removing the element, the element is not part of the explanation. Otherwise, the element is important and, therefore, added again to the explanation (see Line 15). Although, **COMPEX** makes us of Prover9, the theorem prover can be replaced. The only requirement here is that a theorem prover returns information regarding the ab predicates used to derive a contradiction and the empty set if no contradiction can be derived.

It is worth noting that **COMPEX** does not necessarily return a minimal explanation. However, it ensures that consistency is retained. Moreover, **COMPEX** is efficient and always terminates with an explanation accordingly to Definition 1. The correctness of the output is ensured by construction because of the consistency checks performed by the theorem prover. Termination is guaranteed because in every theorem prover call more ab predicates are added. This process

---

Algorithm 1. COMPEX(K, AB)

Input: A propositional knowledge base K and a set of abnormal predicates $AB = \{\text{ab}1, \ldots, \text{ab}_n\}$
Output: An explanation

1. $E = \emptyset$
2. $CS = \{-\text{ab}_i | \text{ab}_i \in AB\}$
3. $Pr = \textbf{Prover9}(K \cup CS)$
4. **while** $Pr \neq \emptyset$ **do**
5.   Add all ab_i predicates used in Pr for deriving a contradiction to E.
6. $CS = \{-\text{ab}_i | \text{ab}_i \in AB \setminus E\} \cup \{\text{ab}_i | \text{ab}_i \in E\}$
7. $Pr = \textbf{Prover9}(K \cup CS)$
8. **end while**
9. **for** $i = 1$ **to** $|E|$ **do**
10.   Let e be the i-th element of E
11. $E = E \setminus \{e\}$
12. $CS = \{-\text{ab}_i | \text{ab}_i \in AB \setminus E\} \cup \{\text{ab}_i | \text{ab}_i \in E\}$
13. $Pr = \textbf{Prover9}(K \cup CS)$
14.   **if** $Pr \neq \emptyset$ **then**
15. $E = E \cup \{e\}$
16.   **end if**
17. **end for**
18. **return** E

---

has to stop because AB is finite and assuming that consistency can be retained using the **ab** predicates introduced.

Computational complexity of **COMPEX** is in the worst case of order $O(|AB| \cdot TPC)$ where TPC represents the worst case execution time for calling the theorem prover. Because of the fact that this time is in the worst case exponential for propositional logic, TPC is the main source of computational complexity. In order to give an indication of feasibility of **COMPEX** we carried out two case studies. In the first one, we checked the time for computing the explanations using our running example and other knowledge base comprising 39 clauses and 26 assumptions. For the running example time was less than 0.00 seconds. For the other knowledge base we required 4 theorem prover calls, each requiring 0.01 seconds at the maximum. Hence, again computing explanations was possible within a fraction of a second. The purpose of the second case study was to show the performance in one extremal case. For this purpose we borrowed the problem of computing all diagnoses of an inverter chain, where each inverter i is modeled as $\text{ab}_i |$ (i_{i-1} <-> i_i). For a chain of inverter **COMPEX** cannot reduce the number of **ab**'s in the first stage and thus has to call the theorem prover $n + 2$ times with n being the number of inverters. Note that for this example we set i_0 to be true and the final output to a value, which is not the expected one.

When using Prover9 and the inverter chain examples ranging from n equals 50 to 500 we obtained the time (in log scale) depicted in Figure 1 required for carrying out all necessary theorem prover calls. We see that the time complexity for this example is polynomial. The equation $t = 5 \cdot 10^{-7} \cdot n^{-3.487}$ fits the given curve with a coefficient of determination value R^2 of 0.99861. When assuming that we do not need to compute the reasons for inconsistency interactively, even larger knowledge bases of several hundreds of clauses can be handled within less than half an hour, which seems to be feasible for our working domain. Moreover, other theorem provers or constraint solver might lead to even better figures.

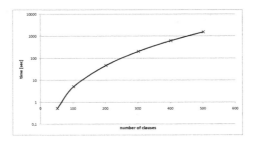

Fig. 1. Time for computing all theorem prover calls for the inverter chain example

In order to finalize this section we briefly summarize the method of constructing the overall policy to be used as test oracle. This is done by first constructing a propositional policy from the general and the custom policy where quantified variables are instantiated using constants of the policies. All propositional rules that belong to a rule of the general policies are extended with a disjunction of a corresponding abnormal predicate. After all calls to **COMPEX** those rules are eliminated where their corresponding abnormal predicate is element of the returned solution. The remaining rules are ensured to be consistent and form the knowledge-base of the test oracle.

4   Related Research and Conclusions

Retaining consistencies in knowledge bases has been considered as important research question for a long time. Shapiro [10] dealt with debugging Prolog programs in case of incorrect behavior. Later Console et al. [3] provided a model-based debugging approach with some improvements. Bond and Pagurek [1] improved the work of Console and colleagues. Based on theses previous work Felfernig et al. [5] introduced an approach for diagnosis of knowledge bases in the domain of configurations. This work is close to our work. However, there are several differences. The algorithm presented in this work is adapted towards the desired application domain, i.e., security testing, and thus does not need to provide all possible solutions. Moreover, we are not interested in finding the most general solution at the level of first order logical sentences. Therefore, finding solutions becomes decidable again. Moreover, when restricting to one solution, the computation also becomes tractable when considering the efficiency of today's theorem provers and SAT solvers.

In this paper we briefly discussed a framework for testing systems with the focus on security properties. For this purpose we assumed that we have general knowledge that captures security issues like stating that ordinary users are not allowed to add new users to a system. In order to test a specific application the general knowledge is accompanied with security properties that arise from the requirements and specification of the system itself. This separation of knowledge bears the advantage of increased reuse but at the same time causes challenges due to inconsistencies between the general and the custom knowledge.

In order to solve the inconsistency challenge we introduced a method for delivering the causes of inconsistencies. This causes have to be removed in order to retain consistency. The presented novel algorithm is specifically tailored for the application domain and is based on model-based diagnosis from which we took the basic idea. Future research include the formalization of available security knowledge in order to carry out further case studies and to provide an empirical study with the objective to prove the applicability of the approach in the security domain.

Acknowledgments. This research has received funding from the Austrian Research Promotion Agency (FFG) under grant 832185 (MOdel-Based SEcurity Testing In Practice) and the Austrian COMET Program (FFG).

References

1. Bond, G.W.: Logic Programs for Consistency-Based Diagnosis. PhD thesis, Carleton University, Faculty of Engineering, Ottawa, Canada (1994)
2. Cohen, D.M., Dalal, S.R., Fredman, M.L., Patton, G.C.: The AETG system: An approach to testing based on combinatorial design. IEEE Trans. Softw. Eng. 23(7), 437–444 (1997)
3. Console, L., Friedrich, G., Dupré, D.T.: Model-based diagnosis meets error diagnosis in logic programs. In: International Joint Conference on Artificial Intelligence (IJCAI), Chambery, pp. 1494–1499 (August 1993)
4. Davis, R.: Diagnostic reasoning based on structure and behavior. Artificial Intelligence 24, 347–410 (1984)
5. Felferning, A., Friedrich, G., Jannach, D., Stumptner, M.: Consistency based diagnosis of configuration knowledge bases. Artificial Intelligence 152(2), 213–234 (2004)
6. Greiner, R., Smith, B.A., Wilkerson, R.W.: A correction to the algorithm in Reiter's theory of diagnosis. Artificial Intelligence 41(1), 79–88 (1989)
7. Hoglund, G., McGraw, G.: Exploiting Software: How to Break Code. Addison-Wesley (2004) ISBN: 0-201-78695-8
8. McCune, W.: Prover9 and mace4, http://www.cs.unm.edu/~mccune/prover9/ (2005–2010)
9. Reiter, R.: A theory of diagnosis from first principles. Artificial Intelligence 32(1), 57–95 (1987)
10. Shapiro, E.: Algorithmic Program Debugging. MIT Press, Cambridge (1983)

Implementing a System Enabling Open Innovation by Sharing Public Goals Based on Linked Open Data

Teemu Tossavainen[1,2], Shun Shiramatsu[1], Tadachika Ozono[1], and Toramatsu Shintani[1]

[1] Graduate School of Engineering, Nagoya Institute of Technology, Japan
[2] School of Science, Aalto University, Finland
teemu.tossavainen@aalto.fi, {siramatu,ozono,tora}@nitech.ac.jp

Abstract. Social network channels are used to organise public goals, but unstructured conversation presents a challenge for identifying groups focusing on similar issues. To address the problem, we implemented a web system for creating goals from public issues and for discovery of similar goals. We designed SOCIA ontology to structure information as hierarchical goal structure. Potential collaborators can use the hierarchical information in consensus building. We proposed a method to calculate similarity between public goals on the basis of the hierarchical structure. To apply our proposed method to real-life situation, we designed and implemented an easy-to-use user interface to structure public goals by citizens. We are arranging workshops to use the system in real-life setting for gathering local issues from citizens of Ogaki city to formulate public goals to solve them. We will use the experiences gained from the workshops to improve the system for deployment for open use by communities to utilise open innovation in decision making and for facilitating collaboration between governmental agents and citizens.

Keywords: Linked open data, public collaboration, goal matching service.

1   Introduction

Societies around the world are looking ways to improve efficiency of the public sector by increasing public participation [1]. Efficiency of public decision making is important, for example, for a society's ability to recover from large-scale natural disasters, like the great earthquake in Japan 2011. In addition, many industrial countries are facing large problems with ageing population and growing dependency ratio [2]. These issues increase the pressure to improve operation of public sector.

In our research, we focus on public goals. Regional societies make various public decisions in different levels of community's internal hierarchy to address public issues. Public goals are formed in the decision making process, which leads to a large amount of different goals, some of which are conflicting or overlapping.

A. Moonis et al. (Eds.): IEA/AIE 2014, Part II, LNAI 8482, pp. 98–108, 2014.

An unsuccessfully set goal can lead to wasting of resources, for example a less important task may be given the attention, while addressing more important issues is neglected, also a goal definition that is too general or vague hinders effective participation. More importantly, lack of coordination can lead to a situation where opportunities for cooperation are missed and possible synergy benefits and resource savings are not realised. Addressing aforementioned issues could potentially be beneficial and bring resource savings.

We aim to develop technologies for facilitating open innovation through collaboration in public spheres. Open innovation and eParticipation can be seen as possible solutions for improving public sector's efficiency. In eParticipation, "the purpose is to increase citizens' abilities to participate in the digital governance" [3], participation of citizens is needed in open innovation. One of the characteristic of open innovation is increased transparency. In one hand, information is necessity for enabling the public to participate in the decision making and in other hand the public's views and opinions are seen beneficial in open decision making. Implementing these methods could be a way to improve a community's ability to generate more informed public goals and to increase public participation. Information and communication techniques (ICT), like social networks, blogs, and eParticipation systems are channels for public conversation in eParticipation. These channels are used also for discussion for organising goals and events, but the discussion is unstructured and thus it is challenging to use the accumulated information for identifying groups that are focusing on similar issues. While it is possible to programmatically access the content of the discussion for automatic information extraction, the current solutions are not able to realise high accuracy for extracting hierarchical structure of goals or issues. Hence, we need to develop a system for manual structuring of goals and for collecting training data for the automatic extraction. We implemented a public goal sharing system, for enabling open innovation. The system has functions for creating and exploring goals and issues and discovering similar goals.

The structure of this paper is as follows. In section 2 we provide an overview of related works. In section 3 we present the system architecture, the implemented method for analysing the goal similarity, and introduce the system functionality. In conclusion section we present the summary and discuss about possible future work.

2 Related Works

2.1 Goal Management Tools

In the research field of project management, goals are commonly structured as hierarchies by subdividing goals into subgoals. Instances of such structures are seen in the Thinking Process of the Theory of Constraint (TOC) [4] and the Work Breakdown Structure (WBS) in the Project Management Body of Knowledge (PMBOK) [5]. Although these models are generally used for project management within an organization, some researchers apply them to public sector problems.

Especially, TOC is tried to be applied to the recovery from disasters[6,7]. In our previous research [8], we built a linked open dataset of public goals for promoting recovery and revitalization from the Great East Japan Earthquake[1]. The dataset is based on the hierarchical structure that can be regarded as a simplification of the above models. We assumed that simplifying the models is needed for applying them to public use and for ensuring the interoperability of data.

There are many Web services for sharing tasks and managing projects, e.g., Trello[2], Cyboze Live[3], and Backlog[4]. However, they do not support functions for sharing goals and finding potential collaborators in public spheres.

2.2 Issue Sharing and Collaborator Matching

FixMyStreet[5] is a collaborative web service to enable citizens to report, view or discuss local problems such as graffiti, fly tipping, broken paving slabs or street lighting by locating them on a map [9]. Neighborland[6] is also a collaborative web service to share citizens' wishes by filling blanks in a form: "I want _____ in _____". Although the key concepts of these services are similar to that of this research, we focus on retrieval of potential collaborators and a user interface for subdividing public goals that are not dealt with the above web services.

There have been many researches about 2-sided matching problem based on the game theory [10]. We, however, do not limit the number of collaborators because dealing with public issues requires a lot of collaborators.

3   System Architecture and Implementation

In our research, we developed a system for sharing public goals. Agents use the application to input new goals and issues. Goals are structured in hierarchical order, where a goal has a set of goals as subgoals that are partial solutions for the parent goal. The goal structure is important because it allows dividing an abstract goal to more tangible and concrete subgoals, which are easier to participate in. Additionally, the agent can explore goals by filtering them by various criteria, e.g., with a keyword search, which facilitates the agent to locate relevant goals for more detailed inspection. Moreover, the agent can discover goals with a similarity search, where the system suggests similar goals. The goal discovery enables the agent to discover opportunities for collaboration over common agenda. Additionally, the agent can use the system to discover other agents, who have similar goals. The discovery function facilitates the agent to identify those parties that has same kind of aspirations, which are potential partners for cooperation. Additionally, the agent will be able to visually compare possible partners' goal

---

[1] http://data.open-opinion.org/socia/data/Goal?rdf:type=socia:Goal
[2] http://trello.com/
[3] http://cybozulive.com/
[4] http://www.backlog.jp/
[5] http://www.fixmystreet.com/
[6] https://neighborland.com/

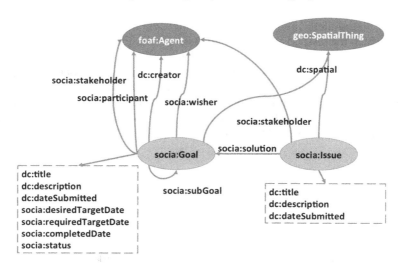

Fig. 1. SOCIA ontology to represent public goals

trees, which facilitates the agent to have better understating about the level of similarity, it also aids in a negotiation process between the agents by identifying possible conflicts of interest.

3.1 SOCIA Ontology

We use SOCIA (Social Opinions and Concerns for Ideal Argumentation) resource description framework (RDF) ontology to build linked data. Firstly, SOCIA links the background context to the goal data, secondly it describes the relations between entities and defines their structure. Figure 1 shows the relevant part of the SOCIA ontology structure. It describes goal related information and it is marked with so-cia: prefix. dc: prefix is used with Dublin metadata initiative's metadata terms[7], :foaf prefix in World Wide Web consortium's friend of a friend (FOAF) ontology[8], and geo: prefix is used with the GeoNames ontology[9]. The socia:issue class represents a public issue, which has dc:title, dc:description, socia:references, dc:spatial, dc:createdDate, and dc:creator properties. The socia:references property contains links to external data sources, for example news articles or blog posts. dc:spatial reference links an issue to a geo:SpatialThing class, which indicates the related spatial information, it also contains a socia:solution relation to a goal. The socia:Goal class represents a public goal. It contains dc:title, dc:description, dc:status, dc:spatial, dc:creator, dc:dateSubmitted, socia:desiredTargetDate, socia:requiredTargetDate, socia:status, and socia:subGoalOf properties. The socia:status property indicates the current status of the goal, which can have one of following values: "not started",

---

[7] http://dublincore.org/documents/2012/06/14/dcmi-terms/?v=terms
[8] http://xmlns.com/foaf/spec/
[9] http://www.geonames.org/ontology/documentation.html

"started", "abandoned", and "completed". The socia:subGoal property has a set of subgoals, which forms the hierarchical goal structure. The socia:participant property contains a set of agents that are participating in the goal. Figure 2 shows the classes and relations that are used to store the similarity value between a pair of goals. Socia:AnnotationInfo has a source and a target relation with the two socia:goal instances and it contains the socia:weight property that indicates the level of similarity. The socia:weight property can have values from zero to one, where greater value indicates more similar goals.

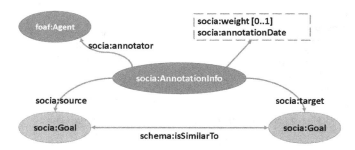

Fig. 2. SOCIA AnnotationInfo to save goal similarity

3.2   Goal Similarity

The similarity between public goals g_i and g_j can be calculated using a cosine measure between $\mathrm{bof}(g_i)$ and $\mathrm{bof}(g_j)$, a bag-of-features vectors of the goals:

$$\mathrm{sim}(g_i, g_j) = \frac{\mathrm{bof}(g_i) \cdot \mathrm{bof}(g_j)}{\|\mathrm{bof}(g_i)\|\|\mathrm{bof}(g_j)\|}. \tag{1}$$

$$\tag{2}$$

In our past research [8], we defined $\mathrm{bof}(g)$ as a weighted summation of surficial (TF-IDF) features, latent (LDA) features, and recursive (subgoal) features:

$$\mathrm{bof}(g) = \frac{\alpha}{\|\mathrm{tfidf}(g)\|}\mathrm{tfidf}(g) + \frac{\beta}{\|\mathrm{lda}(g)\|}\mathrm{lda}(g) + \frac{\gamma}{|\mathrm{sub}(g)|}\sum_{sg \in \mathrm{sub}(g)}\frac{\mathrm{bof}(sg)}{\|\mathrm{bof}(sg)\|} \tag{3}$$

$$\mathrm{tfidf}(g) = \begin{pmatrix} \mathrm{tfidf}(w_1, g) \\ \vdots \\ \mathrm{tfidf}(w_{|W|}, g) \\ 0 \\ \vdots \\ 0 \end{pmatrix} \in \mathbb{R}^{|W|+|Z|}, \quad \mathrm{lda}(g) = \begin{pmatrix} 0 \\ \vdots \\ 0 \\ \mathrm{p}(z_1|g) \\ \vdots \\ \mathrm{p}(z_{|Z|}|g) \end{pmatrix} \in \mathbb{R}^{|W|+|Z|}, \tag{4}$$

where g denotes a public goal, $\text{bof}(g)$ denotes a bag-of-features vector of g, and $\text{sub}(g)$ denotes a set of subgoals of g. Here, $w \in W$ denotes a term, $z \in Z$ denotes a latent topic derived by a latent topic model [11], and $\text{tfidf}(w, g)$ denotes the TF-IDF, i.e., the product of term frequency and inverse document frequency, of w in a title and a description of g. The $\text{p}(z|g)$ denotes the probability of z given g, $0 \leq \alpha, \beta, \gamma \leq 1$, and $\alpha + \beta + \gamma = 1$. The reason this definition incorporates a latent topic model is to enable short descriptions of goals to be dealt with because TF-IDF is insufficient for calculating similarities in short texts. Moreover, contextual information in the recursive subgoal feature is also beneficial to deal with the short goal descriptions. The parameters α, β, and γ are empirically determined on the basis of actual data.

In this paper, we redefine the bag-of-feature vector $\text{bof}(g)$, because we empirically found that contextual information is contained in not only subgoals, but also supergoals (parent goals). To incorporate features of supergoals into $\text{bof}(g)$, we newly define $\text{bof}_{\text{self}}(g)$, a bag-of-features vector extracted only from the target goal g, and $\text{bof}_{\text{cntxt}}(g)$, a contextual bag-of-features vector extacted from subgoals and supergoals.

$$\text{bof}(g) = \frac{1 - \gamma(g)}{\|\text{bof}_{\text{self}}(g)\|}\text{bof}_{\text{self}}(g) + \frac{\gamma(g)}{\|\text{bof}_{\text{cntxt}}(g)\|}\text{bof}_{\text{cntxt}}(g), \tag{5}$$

$$\text{bof}_{\text{self}}(g) = \frac{\alpha}{\|\text{tfidf}(g)\|}\text{tfidf}(g) + \frac{\beta}{\|\text{lda}(g)\|}\text{lda}(g), \tag{6}$$

$$\text{bof}_{\text{cntxt}}(g) = \sum_{subg \in \text{sub}(g)} \text{bof}_{\text{sub}}(subg) + \sum_{supg \in \text{sup}(g)} \text{bof}_{\text{sup}}(supg), \tag{7}$$

$$\text{bof}_{\text{sub}}(g) = d_{\text{sub}}\left(\text{bof}_{\text{self}}(g) + \sum_{subg \in \text{sub}(g)} \text{bof}_{\text{sub}}(subg)\right), \tag{8}$$

$$\text{bof}_{\text{sup}}(g) = d_{\text{sup}}\left(\text{bof}_{\text{self}}(g) + \sum_{supg \in \text{sup}(g)} \text{bof}_{\text{sup}}(supg)\right), \tag{9}$$

$$\gamma(g) = upper_{\text{cntxt}} \cdot \tanh(k \cdot \|\text{bof}_{\text{cntxt}}(g)\|), \tag{10}$$

where $\text{sup}(g)$ denotes a set of supergoals of g, d_{sub} and d_{sup} respectively denote decay ratios when recursively tracking subgoals and supergoals, $upper_{\text{cntxt}}$ denotes an upper limit of the weight of $\text{bof}_{\text{cntxt}}(g)$, $\alpha + \beta = 1$, and $0 \leq \alpha, \beta,$ $upper_{\text{cntxt}}, d_{\text{sub}}, d_{\text{sup}} \leq 1$. The definitions of $\text{tfidf}(g)$ and $\text{lda}(g)$ are not modified from Equation 4. The hyperbolic tangent is used for adjusting $\gamma(g)$, the weight of the contextual bag-of-features $\text{bof}_{\text{cntxt}}(g)$, according to the amount of subgoals and supergoals. Thus, $\gamma(g)$ is 0 when $\|\text{bof}_{\text{cntxt}}(g)\| = 0$ and asymptotically gets close to $upper_{\text{cntxt}}$ along to the increase of $\|\text{bof}_{\text{cntxt}}(g)\|$. The parameters α, β, $upper_{\text{cntxt}}, d_{\text{sub}}, d_{\text{sup}}$, and k are empirically determined.

In order to recommend similar goals, a pair of goals g_i and g_j satisfying $\text{sim}(g_i, g_j) > \theta_g$ is linked by the property schema:isSimilarTo that is defined by schema.org[10].

---

[10] http://schema.org/Product

In this paper, we consider to recommend not only similar goals but also potential collaborators who aim at similar goals. To recommend potential collaborators for a user u, here we formulate the similarity between users u and u_k as follows:

$$\text{sim}(u, u_k) = \frac{1}{|\text{goals}(u)|} \sum_{g \in \text{goals}(u)} \max_{g' \in \text{goals}(u_k)} \text{sim}(g, g') \tag{11}$$

where $\text{goals}(u)$ denotes a set of goals that the user u aims at. The user u is linked to users u_k who satisfy $\text{sim}(u, u_k) > \theta_u$ with the property schema:isSimilarTo in order to recommend potential collaborators u_k to the user u.

3.3 System Implementation

We implemented a web platform for linked open data for creating, exploring, and discovering public goals. The system is used by agents and decision makers in the society's government and members of the society. Figure 3 shows the structure of the application. It is divided into three parts, a website user interface, a server side API, and a linked open data storage. The website user interface is implemented with HTML and JavaScript. We utilise an open source JavaScript template library that reuses HTML document object model (DOM) elements to produce dynamic HTML content. In addition, we use a JavaScript library for localising the user interface, the client side localisation helps in producing multilingual dynamic content. Moreover, we utilise Google maps API for displaying spatial information. A requirement for the user interface is to display goal information in an easy-to-explore manner. System usability and clear visualisation are important design factors, because the system is intended to be used by broad audience with varying IT skills. The user interface design contains visual elements, e.g. agents are presented visually with an avatar image and a goal status is indicated with a presenting icon. The user is required to log in

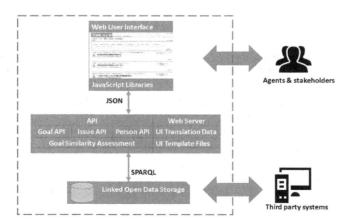

Fig. 3. Structure of the system

with Facebook authentication, to gather basic user information, e.g. name and avatar image. In future work, additional login methods will be added. The user information used to identify involved parties in the linked data. The client side application uses asynchronous HTTP requests to access the server side API, server responses are in JSON (JavaScript Object Notation) format.

The server side API contains the implementation for searching and creating goals and related entities. The API provides interfaces for fetching the data by making HTTP requests, parameters are transferred with HTTP GET and POST methods. The API verifies the parameter data validity and ensures data consistency in the data storage, for example by verifying that required entities exists before inserting triples to form a relations between them. We utilise SPARQL resource description framework query language for accessing the linked data storage. The API also provides support functions for the user interface, for example it provides data for implementing client side autocompletion feature. By handling aforementioned issues in the API, it simplifies the client side solution implementation.

The data is stored in a RDF data storage. We utilised the OpenLink Virtuoso data server[11]. The data is stored in a triple form, i.e. as subject-predicate-object sets. Predicate defines the type of relation between the linked subject and object entities, the entity keys are universal resource locators. The data storage provides a SPARQL endpoint for accessing the data with the RDF query language. The data storage and the SOCIA ontology follows open data principles, both the data and the vocabulary is available for third party applications.

The following RDF/N3 shows an example of RDF triples.

```
<http://collab.open-opinion.org/resource/goal000098>
    rdf:type <http://data.open-opinion.org/socia-ns#Goal>;
    dc:title "Charting the condition of walking ways near Mt. Ikeda"@en;
    socia:subGoal <http://collab.open-opinion.org/resource/goal000102>.
```

3.4   Issue Creation

Here we describe an example of a process of forming issues from public concerns and setting goals to address them. An agent explores internet resources in a search for public concerns, possible channels could be news networks, microblogs, social media, eParticipation and eGovernment systems. When the agent finds a resource that could constitute as an issue, he accesses the goal sharing system with a browser and logs in with a Facebook account. Afterwards, the agents navigates to the issues section and opens the issue creation dialog, seen in Figure 4, by clicking a button. The agent inputs the issue details title and description and a spatial location, he also adds the references to the relevant material found in the earlier exploration phase. We utilise the GeoNames ontology in the the spatial location input to provide a location search using human-readable place names. The issue generation phase facilitates open innovation, by using the public concerns as source of the issue. In the second phase, a decision maker logs into

---

[11] http://virtuoso.openlinksw.com

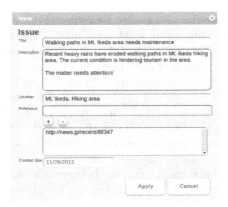

Fig. 4. User interface for adding new issues

the system, and begins exploring issues. The user can search issues by filtering the results with creation date, creator, and keyword, the resulting set is shown as a list of issues. The user opens a detail view by clicking an issue in the list. The detail view displays additional information about the issue, e.g. the description and links to the referred material. At this point, the decision makers can debate over the issue, the information provided by the issue details facilitates the decision making by improving the understanding of the issue in question. By clicking the control in the details view, user can create a goal that would be solution for the issue. Subgoals can then be added to the goal to construct a goal hierarchy to achieve more concrete goals.

3.5 Goal Discovery

Here we present an example how the system can facilitate collaboration by the process to discover similar goals. A user begins by accessing the website with a browser, after which he must login into the application. When the user explores goals, seen in Figure 5, he can first input filter options to the filter control. The user has option to filter results by creation date, desired completion date, required completion date, keyword, goal status, and creator. The search result is shown as a list of goals, which displays the basic information about the goal. The basic information contains a title, a description, a status, and a creators' avatar image. The user can open a detail view by selecting a goal by clicking the desired goal in the list. The detail view displays additional information about the goal, like lists of subgoals and participants. Clicking the detail view's "find similar goals" control initiates a new search for goals based on the similarity value. The user can then explore the goal set in a list, which is ordered by descending similarity.

Fig. 5. User interface to interact with a goal

4   Conclusion

In this paper, we present the implemented web based goal sharing system. The aim of the system is to increase transparency, advance the opportunities to utilise open innovation and eParticipation principles, and to facilitate governmental agents to collaborate. The system offers functions for creating public goals and issues, exploring and discovering similar goals, and visualising the level of similarity. We implemented a method to calculate similarity between goals that utilises the structured goal information. We present the SOCIA vocabulary that we used to structure the goal related information as linked open data. We discuss about two examples of using the system.

Currently, we are arranging workshops in Ogaki city to use the system in real-life situation, for gathering real-life issues and goals. We use the gathered data from the workshops to quantitatively measure, test, and improve the implemented system. One important concern is to determine suitable values for parameters α, β, and γ in the similarity analysis method.

We plan to deploy the web platform to complement existing solutions, like CitiSpe@k[12] and SOCIA linked data storage for facilitating communities to use open innovation in their decision making process and to enable governmental agents to collaborate with other governmental parties and citizens. Our system has a potential to get together concerned citizens and parties that can solve the concrete problem, e.g., CODE for JAPAN[13]. CODE for JAPAN advances the

---

[12] http://www.open-opinion.org/citispeak/
[13] http://code4japan.org/

cause of open innovation and eParticipation by getting people to provide solutions and tools for local communities and providing information to governmental officials.

A possible future research topic is automatic issue and goal suggestion. After citizens input their issues and goals, we will be able to construct a training corpus for automatic extraction of issues or goals from textual content. Such training data enables us to deal with the novel research topic.

Acknowledgments. This work was supported by the Grant-in-Aid for Young Scientists (B) (No. 25870321) from Japan Society for the Promotion of Science. We appreciate the Ogaki office of CCL Inc. for giving us chance to conduct a public workshop for testing our system.

References

1. IAP2: IAP2 Spectrum of public participation (2013), http://www.iap2.orgas/associations/4748/files/Kettering_FINALPreliminaryFindingsReport.pdf
2. The World Bank: Age dependency ratio, old (% of working-age population) (2013), http://data.worldbank.org/indicator/SP.POP.DPND.OL/countries
3. Sæbø, Ø., Rose, J., Flak, S.: The shape of eparticipation: Characterizing an emerging research area. Government Information Quarterly 25(3), 400–428 (2008)
4. AGI Goldratt Institute: The theory of constraints and its thinking processes - a brief introduction to toc (2009), http://www.goldratt.com/pdfs/toctpwp.pdf
5. Project Management Institute: A Guide to the Project Management Body of Knowledge. 4th edn. Project Management Institute (2008)
6. Cheng, J., Shigekawa, K., Meguro, K., Yamazaki, F., Nakagawa, I., Hayashi, H., Tamura, K.: Applying the toc logistic process to clarify the problem schemes of near-field earthquake in tokyo metropolitan area. Journal of Social Safety Science (11), 225–233 (2009) (in Japanese)
7. Ohara, M., Kondo, S., Kou, T., Numada, M., Meguro, K.: Overview of social issues after the Great East-Japan Earthquake disaster - part 3 of activity reports of 3.11net Tokyo. Seisan Kenkyu 63, 749–754 (2011)
8. Shiramatsu, S., Ozono, T., Shintani, T.: Approaches to assessing public concerns: Building linked data for public goals and criteria extracted from textual content. In: Wimmer, M.A., Tambouris, E., Macintosh, A. (eds.) ePart 2013. LNCS, vol. 8075, pp. 109–121. Springer, Heidelberg (2013)
9. King, S.F., Brown, P.: Fix my street or else: using the internet to voice local public service concerns. In: Proceedings of the 1st International Conference on Theory and Practice of Electronic Governance, pp. 72–80 (2007)
10. Roth, A.E., Sotomayor, M.A.O.: Two-Sided Matching: A Study in Game-Theoretic Modeling and Analysis. In: Econometric Society Monographs, vol. 18. Cambridge University Press (1990)
11. Blei, D.M., Ng, A.Y., Jordan, M.I.: Latent dirichlet allocation. Journal of Machine Learning Research 3, 993–1022 (2003)

Predicting Citation Counts for Academic Literature Using Graph Pattern Mining

Nataliia Pobiedina[1] and Ryutaro Ichise[2]

[1] Institute of Software Technology and Interactive Systems
Vienna University of Technology, Austria
`pobiedina@ec.tuwien.ac.at`
[2] Principles of Informatics Research Division
National Institute of Informatics, Japan
`ichise@nii.ac.jp`

Abstract. The citation count is an important factor to estimate the relevance and significance of academic publications. However, it is not possible to use this measure for papers which are too new. A solution to this problem is to estimate the future citation counts. There are existing works, which point out that graph mining techniques lead to the best results. We aim at improving the prediction of future citation counts by introducing a new feature. This feature is based on frequent graph pattern mining in the so-called citation network constructed on the basis of a dataset of scientific publications. Our new feature improves the accuracy of citation count prediction, and outperforms the state-of-the-art features in many cases which we show with experiments on two real datasets.

1 Introduction

Due to the drastic growth of the amount of scientific publications each year, it is a major challenge in academia to identify relevant literature among recent publications. The problem is not only how to navigate through a huge corpus of data, but also what search criteria to use. While the Impact Factor [1] and the h-index [2] measure the significance of publications coming from a particular venue or a particular author, the citation count aims at estimating the impact of a particular paper. Furthermore, Beel and Gipp find empirical evidence that the citation count is the highest weighted factor in Google Scholar's ranking of scientific publications [3]. The drawback about using the citation count as a search criteria is that it works only for the papers which are old enough. We will not be able to judge new papers this way. To solve this problem, we need to estimate the future citation count. An accurate estimation of the future citation count can be used to facilitate the search for relevant and promising publications.

A variety of research articles have already studied the problem of citation count prediction. In earlier work the researchers experimented on relatively small datasets and simple predictive models [4,5]. Nowadays due to the opportunity to retrieve data from the online digital libraries the research on citation behavior

A. Moonis et al. (Eds.): IEA/AIE 2014, Part II, LNAI 8482, pp. 109–119, 2014.

is conducted on much larger datasets. The predictive models have also become more sophisticated due to the advances in machine learning. The major challenge is the selection of features. Therefore, our goal is to discover features which are useful in the prediction of citation counts.

Previous work points out that graph mining techniques lead to good results [6]. This observation motivated us to formulate the citation count prediction task as a variation of the link prediction problem in the citation network. Here the citation count of a paper is equal to its in-degree in the network. Its out-degree corresponds to the number of references. Since out-degree remains the same over years, the appearance of a new link means that the citation count of the corresponding paper increases. In the link prediction problem we aim at predicting the appearance of links in the network. Our basic idea is to utilize frequent graph pattern mining in the citation network and to calculate a new feature based on the mined patterns – *GERscore* (Graph Evolution Rule score). Since we intend to predict the citation counts in the future, we want to capture the temporal evolution of the citation network with the graph patterns. That is why we mine frequent graph patterns of a special type - the so-called graph evolution rules [7].

The main contributions of this paper are the following:

- we study the citation count prediction problem as a link prediction problem;
- we adopt score calculation based on the graph evolution rules to introduce a new feature GERscore, we also propose a new score calculation;
- we design an extended evaluation framework which we apply not only to the new feature, but also to several state-of-the-art features.

The rest of the paper is structured as follows. In the next section we formulate the problem which we are solving. In the next section we formulate the problem at hand. Section 3 covers the state-of-the-art. In Section 4 we present our methodology to calculate the new feature. Section 5 describes our approach to evaluate the new feature. This section also includes the experimental results on two datasets followed by the discussion. Finally, we draw the conclusion and point out future directions for work.

2   Predicting Citation Counts

We want to predict citation counts for scientific papers. Formally, we are given a set of scientific publications $\mathcal{D}$, the *citation count* of a publication $d \in \mathcal{D}$ at time t is defined as: $Cit(d, t) = |\{d' \in \mathcal{D} : d$ *is cited by* d' *at time* $t\}|$. To achieve our goal, we need to estimate $Cit(d, t + \Delta t)$ for some $\Delta t > 0$. We can solve this task by using either classification or regression.

Classification Task: Given a vector of features $\bar{X}_d = (x_1, x_2, \ldots, x_n)$ for each scientific publication $d \in \mathcal{D}$ at time t, the task is to learn a function for predicting $CitClass(d, t + \Delta t)$ whose value corresponds to a particular range of the citation count for the publication d at the time $t + \Delta t$.

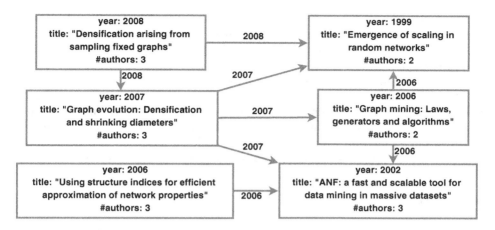

Fig. 1. Example of a citation network

Regression Task: Given a vector of features $\bar{X}_d = (x_1, x_2, \ldots, x_n)$ for a publication $d \in \mathcal{D}$ at time t, the task is to learn a function for predicting $Cit(d, t + \Delta t)$ whose value corresponds to the citation count of the publication d at the time $t + \Delta t$.

We suggest a new perspective on the citation count prediction problem. We construct a paper citation network from the set of scientific publications $\mathcal{D}$. An example of a citation network is given in Figure 1. Nodes are papers. A link from one node to another means that the first paper cites the latter. We put the year of the citation as an attribute of the corresponding link. In this setting, citation count of a paper is equal to the in-degree of the corresponding node. Its out-degree is equal to the number of references. Since node's in-degree increases if a new link appears, we can regard the citation count problem as a variation of the link prediction problem in citation networks.

3   Related Work

Yan et al. find evidence that the citation counts of previous works of the authors are the best indicators of the citation counts for their future publications [8]. However, Livne et al. observe that the citation counts accumulated by the venue and by the references are more significant [6]. Furthermore, Shi et al. discover that highly cited papers share common features in referencing other papers [9]. They find structural properties in the referencing behavior which are more typical for papers with higher citation counts. These results indicate that graph mining techniques might be better suited to capture interests of research communities. That is why we formulate the problem of the citation count prediction as a link prediction problem in the citation network. Since feature-based link prediction methods, like [10,11], can predict links only between nodes which already exist in the network, we use an approach which is based on graph pattern mining [7].

The estimation of future citations can be done with *classification* [12] or *regression* [8,6,13]. The classification task, where we predict intervals of citation counts, is in general easier, and in many applications it is enough. Furthermore, a dataset of publications from physics is used in [12], and from computer science in [8,13]. There are also two different evaluation approaches. The first one is to test the performance for the freshly published papers [6,12]. The second approach is to predict the citation counts for all available papers [8,13]. To ensure a comprehensive study of performance of our new feature and several state-of-the-art features, our evaluation framework includes both classification and regression, two evaluation approaches and two datasets of scientific publications.

4   GERscore

Our methodology to tackle the stated problem consists of several steps. First, we mine the so-called graph evolution rules in the citation network by using a special graph pattern mining procedure. Then we derive GERscore for each paper using several calculation techniques. We also calculate several state-of-the-art features. All features are obtained using data from previous years. To estimate the performance of these features, we use them in different predictive models on the testing datasets.

4.1   Mining Graph Evolution Rules

To calculate GERscore, we start with the discovery of rules which govern the temporal evolution of links and nodes. Formally, we are given a graph, in our case a citation network, $G = (V, E, \lambda, \tau)$ where λ is a function which assigns a label $l \in L_V$ to every node $n \in V$ and τ is a function which assigns a timestamp $t \in T$ to every edge $e \in E$. Though the citation network in our example is directed, we may infer the direction of links: they point from a new node towards the older one. That is why we ignore the direction and assume that the citation network is undirected.

Definition of Relative Time Pattern [7]: A graph pattern $P = (V_P, E_P, \lambda_P, \tau_P)$ is said to be a *relative time pattern* in the citation network G iff there exist $\Delta \in \mathbb{R}$ and an embedding $\varphi : V_P \to V$ such that the following three conditions hold: (1) $\forall v \in V_P \Rightarrow \lambda_P(v) = \lambda(\varphi(v))$; (2) $\forall (u, v) \in E_P \Rightarrow (\varphi(u), \varphi(v)) \in E$; (3) $\forall (u, v) \in E_P \Rightarrow \tau(\varphi(u), \varphi(v)) = \tau_P(u, v) + \Delta$.

In Figure 2a we show examples of relative time patterns. For example, the pattern in Figure 2a(1) can be embedded with $\Delta = 2007$ or $\Delta = 2006$ into the citation network in Figure 1 while the pattern in Figure 2a(3) cannot be embedded at all.

Definition of Evolution Rule [7]: An *evolution rule* is a pair of two relative time patterns called *body* and *head* which is denoted as *head* $\Leftarrow$ *body*. Given a pattern head $P_h = (V_h, E_h, \lambda, \tau)$, the body $P_b = (V_b, E_b, \lambda, \tau)$ is defined as:

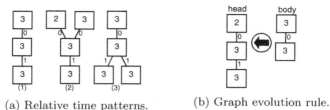

(a) Relative time patterns.   (b) Graph evolution rule.

Fig. 2. Examples of relative time patterns and graph evolution rules. Node labels correspond to the number of authors.

$E_b = \{e \in E_h : \tau(e) < max_{e* \in E_h}(\tau(e*))\}$ and $V_b = \{v \in V_h : deg(v, E_b) > 0\}$, where $deg(v, E_b)$ corresponds to the degree of node v with regard to the set of edges E_b.

An example of a graph evolution rule is given in Figure 2b. Do not get confused by the fact that body has less edges than head. The naming convention follows the one used for rules in logic. Considering the definition of the evolution rule, we can represent any evolution rule uniquely with its head. That is why relative time patterns in Figure 2a are also graph evolution rules.

To estimate frequency of the graph pattern P in the network G, we use *minimum image-based support* $sup(P) = min_{v \in V_P} |\varphi_i(v) : \varphi_i$ *is an embedding of* P *in* $G|$. The *support* of the evolution rule, $sup(r)$, is equal to the support of its head. The *confidence* of this rule, $conf(r)$, is $sup(P_h)/sup(P_b)$. Due to the anti-monotonous behavior of the support, confidence is between 0 and 1. The graph evolution rule from Figure 2b has a minimum image based support 2 in the citation network from Figure 1. The support of its body is also 2. Therefore, confidence of this rule is 1. We can interpret this rule the following way: if the body of this rule embeds into the citation network to a specific node at time t, then this node is likely to get a new citation at time $t + 1$.

Two additional constraints are used to speed up graph pattern mining. We mine only those rules which have support not less than *minSupport*, and which have number of links not more than *maxSize*. Moreover, we consider only those graph evolution rules where body and head differ in one edge. In Figure 2 all rules, except for (a3), correspond to this condition. Finally, we obtain a set $\mathcal{R}$ of graph evolution rules.

4.2   Calculating GERscore

To calculate GERscore, we modify the procedure from [7]. For each rule $r \in \mathcal{R}$ we identify nodes in the citation network to which this rule can be applied to. We obtain a set $\mathcal{R}_n \subset \mathcal{R}$ of rules applicable to the node n. Our assumption is that an evolution rule occurs in the future proportional to its confidence. That is why we put GERscore equal to $c * conf(r)$, where c measures the proportion of rule's applicability. We define three ways to calculate c. In the first case, we simply take $c = 1$. In the second case, we assume that evolution rules with higher support are more likely to happen, i.e., $c = sup(r)$. These two scores are also used for the link prediction problem in [7]. Lastly, if the evolution rule r contains more links, it provides more information relevant to the node n. We assume that such rule should be more likely to occur than the one with less edges. Since evolution rules are limited in their size by $maxSize$, we put $c = size(r)/maxSize$. Thus, we obtain three different scores: $score_1(n, r) = conf(r)$, $score_2(n, r) = sup(r)*conf(r)$, and $score_3(n, r) = conf(r)*(size(r)/maxSize)$.

Finally, we use two aggregation functions to calculate GERscore for node n:

- GERscore$_{1,i}(n) = \sum_{r \in \mathcal{R}_n} score_i(n, r)$,
- GERscore$_{2,i}(n) = \max_{r \in \mathcal{R}_n} score_i(n, r)$.

High values of GERscore can mean two things: either many rules or rules with very high confidence measures are applicable to the node. In either case, the assumption is that this node is very likely to get a high amount of citations. We may have here redundancies. For example, in Figure 2 rules (b) and (a1) are subgraphs of rule (a2). It might happen that these rules correspond to the creation of the same link. Still we consider all three rules, since we are interested to approximate the likelihood of increase in citation counts. Though the summation of individual scores is an obvious selection for the aggregation function, we also consider the maximum. It might turn out that graph evolution rules with the highest confidence are the determinants of future citations.

5   Experiment

5.1   Experimental Data

We use two real datasets to evaluate GERscore: *HepTh* and *ArnetMiner*. The first dataset covers arXiv papers from the years $1992-2003$ which are categorized as High Energy Physics Theory [12]. We mine graph evolution rules for the network up to year 1996 which has 9,151 nodes and 52,846 links. The second dataset contains papers from major Computer Science publication venues [8]. By taking papers up to year 2000, we obtain a sub-network with 90,794 nodes and 272,305 links.

We introduce two additional properties for papers: *grouped number of references* and *grouped number of authors*. For the first property the intervals are $0-1, 2-5, 6-14, 15 \leq$. The references here do not correspond to all references of the paper, but only to those which are found within the dataset. We select the intervals $1, 2, 3, 4-6, 7 \leq$ for the second property.

We construct several graphs from the described sub-networks which differ in node labels. Since we are not sure which label setting is better, we use either the grouped number of references, or the grouped number of authors, or no label. The choice of the first two label settings is motivated by [12]. With the help of the tool *GERM* [1], we obtain 230 evolution rules in the dataset HepTh, and 4,108 in the dataset ArnetMiner for the unlabeled case. We have 886 rules in HepTh,

---

[1] http://www-kdd.isti.cnr.it/GERM/

and 968 in ArnetMiner for the grouped number of authors. For the grouped number of references the numbers are 426 and 1,004 correspondingly.

In total, we obtain 18 different scores for each paper: $GERscore_{1,i}^{(j)}$ for summation and $GERscore_{2,i}^{(j)}$ for maximum, where i equals 1, 2, or 3 depending on the score calculation, and j corresponds to a specific label setting: $j = 1$ corresponds to the grouped number of authors as node labels; $j = 2$ stands for the unlabeled case; $j = 3$ is for the grouped number of references. We report results only for one score for each label setting, because the scores exhibit similar behavior. Since our new score $score_3$ provides slightly better results, we choose $GERscore_{1,3}^{(j)}$ and $GERscore_{2,3}^{(j)}$. Additionally, feature $GERscore$ is the combination of all scores.

5.2   Experimental Setting

To solve the classification task (Experiment 1), we consider three different models: multinomial Logistic Regression (mLR), Support Vector Machines (mSVM), and conditional inference trees (CIT). For the regression task (Experiment 2) we take Linear Regression (LR), Support Vector Regression (SVR), and Classification and Regression Tree (CART). We look at a variety of models because they make different assumptions about the original data. We do 1-year prediction in both tasks.

We consider two scenarios for evaluation which differ in the way we construct training and testing datasets. In Scenario 1 we predict the citation count or class label for the papers from the year t by using the data before year $t-1$, like in [6,11,12]. In Scenario 2 we take all papers from the year t and divide them into training and test datasets, e.g., as it is done in [8,13]. We also perform five times hold-out cross-validation.

To compare performance of our new feature, we calculate several state-of-the-art features: *Author Rank*, *Total Past Influence for Authors* (TPIA), *Maximum Past Influence for Authors* (MPIA), *Venue Rank*, *Total Past Influence for Venue* (TPIV), and *Maximum Past Influence for Venue* (MPIV) [8,13]. To obtain Author Rank, for every author we calculate the average citation counts in the previous years and assign a rank among the other authors based on this number. We put maximum citation count for previous papers as MPIA. TPIA is equal to the sum of citation counts for previous papers. Venue Rank, TPIV and MPIV are calculated the same way using the venue of the paper.

5.3   Experiment 1

We assign class labels in the classification task with intervals $1, 2-5, 6-14, 15 \le$ of citation counts. In Table 1 we summarize the distribution of instances according to these classes for the data which we use for the training and testing datasets.

We use *average accuracy* and *precision* to evaluate the performance of the classification. If class distribution is unbalanced, then precision is better suited for the evaluation [14]. We summarize the results of the classification task in

Table 1. Distribution of instances according to classes (% Total)

Citation Class	HepTh			ArnetMiner		
	Scenario 1		Scenario 2	Scenario 1		Scenario 2
	Year 1996	Year 1997	Year 1997	Year 2000	Year 2001	Year 2001
Class 1	42.9%	40.33%	34.09%	97.27%	96.69%	88.86%
Class 2	29.81%	26.64%	30.32%	2.51%	3.13%	7.75%
Class 3	13.70%	18.77%	19.85%	0.19%	0.18%	2.40%
Class 4	13.58%	14.27%	15.74%	0.03%	0.01%	0.99%
Total Amount	2,459	2,579	12,113	30,000	25,919	399,647

Table 2. Accuracy (%) and Precision (%) for the Classification Task

	Feature	Scenario 1						Scenario 2					
		HepTh			ArnetMiner			HepTh			ArnetMiner		
		mLR	mSVM	CIT	mLR	mSVM	CIT	mLR	mSVM	CIT	mLR	mSVM	CIT
Accuracy	GERscore	**75.56**	**75.59**	**75.37**	**98.37**	**98.37**	98.36	**76.83**	74.82	75.17	**95.65**	95.59	95.62
	Author Rank	73.62	73.56	73.57	98.35	98.36	98.36	72.61	73.19	72.57	94.31	94.47	94.41
	MPIA	73.50	73.39	72.53	98.32	98.36	98.36	70.02	70.79	69.85	94.44	94.49	94.49
	TPIA	73.62	73.58	72.99	98.36	98.36	98.36	70.56	70.97	70.88	94.49	94.48	94.49
	Venue Rank	71.59	71.46	71.55	98.36	98.36	98.36	70	70.33	70.34	94.49	94.49	94.46
	MPIV	66.34	69.73	63.85	98.36	98.36	98.36	67.89	69.08	69.45	94.29	94.49	94.49
	TPIV	70.29	69.49	67.84	98.36	98.36	98.36	69.63	69.61	70.00	94.49	94.49	94.49
	All	76.85	75.91	76.54	**98.37**	98.36	**98.36**	**81.37**	**81.11**	**79.35**	**96.11**	**96.11**	**96.05**
	w/o GERscore	74.74	73.48	74.01	98.35	98.36	98.36	74.31	74.1	74.15	94.73	94.93	94.74
Precision	GERscore	**43.71**	**36.41**	**39.15**	**39.77**	**35.15**	**26.34**	51.35	47.46	44.02	62.69	58.42	**61.2**
	Author Rank	30.92	31.04	31.05	24.17	24.17	24.17	40.87	45.57	**45.15**	35.9	29.49	36.75
	MPIA	31.55	33.03	36.91	24.17	24.17	24.17	37.32	33.37	40.31	32.08	27.17	22.17
	TPIA	30.84	31.91	36.82	24.17	24.17	24.17	37.78	38.24	41.17	22.16	24.6	24.94
	Venue Rank	27.51	24	25.54	24.17	24.17	24.17	33.42	32.62	30.34	22.17	22.17	24.96
	MPIV	24.3	13.93	16.34	24.17	24.17	24.17	21.58	28.05	24.42	25.74	22.17	22.17
	TPIV	25.49	21.99	22.58	24.17	24.17	24.17	26.97	26.85	27.54	23.76	22.17	22.17
	All	48.9	47.75	41.83	39.63	34.04	29.78	**61.47**	**61.2**	**57.56**	62.19	**63.82**	62.19
	w/o GERscore	38.81	34.72	34.46	24.17	24.17	24.17	47.55	47.09	46.63	48.79	52.66	46.82

Table 2. We mark in bold the features which lead to the highest performance measure in each column. The full model is indicated in the row *"All"*. All performance measures are average over the performance measures in 5 runs. The results indicate that the new feature is better than the baseline features and significantly improves the full model.

Due to a highly unbalanced distribution (Table 1), we observe only 1% improvement in accuracy for ArnetMiner in Scenario 2. In the case of HepTh, GERscore is at least 2% better in accuracy than the rest features. Furthermore, in Scenario 2 GERscore improves the accuracy of the full model by more than 9%. Statistical analysis shows that GERscore provides a significant improvement to the full model. If we compare precision rates, then we have that the full model with GERscore is more than 10% better than without it. Moreover, the best achieved accuracy for HepTh in Scenario 1 is 44% in previous work [12]. The accuracy of our full model mLR is 33% higher.

5.4   Experiment 2

To evaluate the performance of the regression models, we calculate the R^2 value as the *square of Pearson correlation coefficient* between the actual and predicted citation counts. In Table 3 we summarize the performance for the regression task. If a feature has "NA" as a value for R^2, it means we are not able to calculate it because the standard deviation of the predicted citation counts is zero. GERscore

Table 3. Performance measures (R^2) for the Regression Task

Feature	Scenario 1						Scenario 2					
	HepTh			ArnetMiner			HepTh			ArnetMiner		
	LR	SVR	CART	LR	SVR	CART	LR	SVR	CART	LR	SVR	CART
$GERscore_{1,3}^{(1)}$	0.011	0.028	0.07	0.02	0.009	0.021	0.063	0.069	0.137	0.138	0.13	0.154
$GERscore_{1,3}^{(2)}$	0.06	0.085	0.103	0.093	0.099	0.087	0.121	0.219	0.26	0.401	**0.431**	0.425
$GERscore_{1,3}^{(3)}$	0.009	0.053	0.091	0.157	0.14	0.169	0.009	0.011	0.065	0.188	0.211	0.209
$GERscore_{2,3}^{(1)}$	0.001	0.015	0.03	0.026	0.022	0.027	0.025	0.039	0.058	0.066	0.087	0.09
$GERscore_{2,3}^{(2)}$	0.005	0.005	NA	0.093	0.214	0.212	0.032	0.057	0.057	0.095	0.135	0.187
$GERscore_{2,3}^{(3)}$	0.069	0.094	0.088	0.097	0.125	0.162	0.001	0.009	0.002	0.094	0.102	0.108
$GERscore$	0.137	0.119	0.121	**0.233**	**0.219**	**0.213**	0.204	0.205	**0.271**	**0.483**	0.337	**0.429**
Author Rank	0.188	0.098	0.16	0.004	NA	0.004	0.204	**0.302**	0.266	0.133	0.15	0.174
MPIA	0.183	0.181	0.193	0.002	0.001	0.006	0.225	0.209	0.214	0.071	0.041	0.052
TPIA	**0.189**	**0.199**	**0.198**	0	0.001	0.005	**0.285**	0.232	0.21	0.004	0.072	0.063
Venue Rank	0.014	0.029	0.028	0.028	NA	0.014	0.051	0.061	0.05	0.037	0.058	0.054
MPIV	0.022	0.003	0.015	0.001	NA	0.014	0.039	0.048	0.035	0.024	0.023	0.037
TPIV	0.026	0.003	0.021	0	NA	0.004	0.039	0.048	0.035	0.024	0.023	0.037
All	0.245	0.192	0.161	0.235	0.184	0.175	0.371	0.357	0.395	0.513	0.317	0.544
w/o GERscore	0.203	0.120	0.164	0.01	0.004	0.013	0.312	0.289	0.274	0.157	0.149	0.19

is significantly better than the baseline features for ArnetMiner dataset. Though author related features lead to higher R^2 for HepTh, we see that GERscore still brings additional value to the best performing models (LR in Scenario 1 and CART in Scenario 2) . The analysis of variance (ANOVA) for two models, "All" and "All w/o GERscore", shows that GERscore improves significantly the full model. Our guess is that GERscore does not perform so well for HepTh due to the insufficient amount of mined evolution rules.

5.5   Discussion

Overall our new feature GERscore significantly improves citation count prediction. When classifying the future citations, GERscore is better than the baseline features in all cases. However, author-related features are still better in the regression task, but only for the dataset HepTh. HepTh provides better coverage of papers in the relevant domain, thus the citations are more complete. Another difference of HepTh from ArnetMiner is the domain: physics for the first and computer science for the latter. The last issue is the amount of mined graph evolution rules: we have only 230 unlabeled evolution rules for HepTh. We are not sure which of these differences leads to the disagreement in the best performing features. In [6] the authors argue that such disagreement arises due to the nature of the relevant scientific domains. However, additional investigation is required to draw a final conclusion.

We observe that CART performs the best for the regression task in Scenario 2 which agrees with the results in [8]. However, LR provides better results in Scenario 1. In general, the performance is poorer in Scenario 1. This means that it is much harder to predict citation counts for freshly published papers. It might be the reason why a simple linear regression with a better generalization ability performs well.

Out of all scores which constitute GERscore, the best results are gained for the scores calculated from the unlabeled graph evolution rules (see Table 3).

When aggregating separate scores, summation is a better choice compared to maximum. This is an unfortunate outcome since aggregation with maximum would allow us to speed up the graph pattern mining by setting a high support threshold. The decrease in running time is also gained through mining labeled graph evolution rules. Though $GERscore_{1,i}^{(2)}$ provides better results compared to other label settings and aggregation technique, we still receive that the other scores contribute to the combined GERscore.

Our results are coherent with Yan et al. for ArnetMiner in Scenario 2 which is the only setting that corresponds to theirs: Author Rank is better than Venue Rank [8,13]. However, we show that GERscore is even better in this case. Moreover, we arrive already at a better performance just by identifying graph evolution rules in the unlabeled citation network from the previous years.

6   Conclusion and Future Work

We have constructed a new feature - GERscore - for estimation of future citation counts for academic publications. Our experiments show that the new feature performs better than six state-of-the-art features in the classification task. Furthermore, the average accuracy of the classification is not affected much if we bring in other baseline features into the model. In the regression task the new feature outperforms the state-of-the-art features for the dataset of publications from computer science domain (ArnetMiner), though the latter still contribute to the performance of regression models. Thus, the application of graph pattern mining to the citation count prediction problem leads to better results. However, for the dataset of publications from physics (HepTh) GERscore is not as good as the author related features, i.e., author rank, MPIA and TPIA, though it does contribute to the increase of the performance. Additional investigation is required to identify the reason for the disagreement in the best performing features.

We have performed both classification and regression tasks for the prediction of citation counts in one year. It is interesting to investigate how well GERscore performs for the prediction over five and more years. Our results indicate that the performance of the model does not always improve if we include more features. Thus, an important aspect to investigate is the optimal combination of features. Ultimately, we want to include our findings into a recommender system for academic publications.

Our future work includes thorough investigation how mined evolution rules influence the predictive power of GERscore. The first issue is to study the influence of input parameters, minimum support (minSup) and maximum size (maxSize), and what is the best combination for them. We need to take into consideration that by setting maxSize high and minSupport low we will obtain more evolution rules, however the computational time will grow exponentially. Another issue is that real-world networks change considerably over time. It may lead to the fact that the evolution rules which are frequent and have high confidence at time t may become rudimentary in ten years and will not be predictive of the citation counts. Thus, we plan to investigate for how long mined evolution rules on average stay predictive. This is an important question also because mining graph evolution rules is computationally hard, and reducing the amount of re-learning GERscores is extremely important.

References

1. Garfield, E.: Impact factors, and why they won't go away. Science 411(6837), 522 (2001)
2. Hirsch, J.: An index to quantify an individual's scientific research output. Proc. the National Academy of Sciences of the United States America 102(46), 16569 (2005)
3. Beel, J., Gipp, B.: Google scholar's ranking algorithm: The impact of citation counts (an empirical study). In: Proc. RCIS, pp. 439–446 (2009)
4. Callaham, M., Wears, R., Weber, E.: Journal prestige, publication bias, and other characteristics associated with citation of published studies in peer-reviewed journals. Journal of the American Medical Association 287(21), 50–2847 (2002)
5. Kulkarni, A.V., Busse, J.W., Shams, I.: Characteristics associated with citation rate of the medical literature. PLOS one 2(5) (2007)
6. Livne, A., Adar, E., Teevan, J., Dumais, S.: Predicting citation counts using text and graph mining. In: Proc. the iConference 2013 Workshop on Computational Scientometrics: Theory and Applications (2013)
7. Bringmann, B., Berlingerio, M., Bonchi, F., Gionis, A.: Learning and predicting the evolution of social networks. IEEE Intelligent Systems 25, 26–35 (2010)
8. Yan, R., Tang, J., Liu, X., Shan, D., Li, X.: Citation count prediction: learning to estimate future citations for literature. In: Proc. CIKM, pp. 1247–1252 (2011)
9. Shi, X., Leskovec, J., McFarland, D.A.: Citing for high impact. In: Proc. JCDL, pp. 49–58 (2010)
10. Barabasi, A.-L., Albert, R.: Emergence of scaling in random networks. Science Magazine 286(5439), 509–512 (1999)
11. Munasinghe, L., Ichise, R.: Time score: A new feature for link prediction in social networks. IEICE Trans. 95-D(3), 821–828 (2012)
12. Mcgovern, A., Friedl, L., Hay, M., Gallagher, B., Fast, A., Neville, J., Jensen, D.: Exploiting relational structure to understand publication patterns in high-energy physics. SIGKDD Explorations 5 (2003)
13. Yan, R., Huang, C., Tang, J., Zhang, Y., Li, X.: To better stand on the shoulder of giants. In: Proc. JCDL, pp. 51–60 (2012)
14. Sokolova, M., Lapalme, G.: A systematic analysis of performance measures for classification tasks. Information Processing & Management 45(4), 427–437 (2009)

Co-evolutionary Learning for Cognitive Computer Generated Entities

Xander Wilcke[1,2], Mark Hoogendoorn[2], and Jan Joris Roessingh[1]

[1] National Aerospace Laboratory,
Department of Training, Simulations, and Operator Performance
Anthony Fokkerweg 2, 1059 CM Amsterdam, The Netherlands
{smartbandits,jan.joris.roessingh}@nlr.nl
[2] VU University Amsterdam, Department of Computer Science
De Boelelaan 1081, 1081 HV Amsterdam, The Netherlands
m.hoogendoorn@vu.nl

Abstract. In this paper, an approach is advocated to use a hybrid approach towards learning behavior for computer generated entities (CGEs) in a serious gaming setting. Hereby, an agent equipped with cognitive model is used but this agent is enhanced with Machine Learning (ML) capabilities. This facilitates the agent to exhibit human like behavior but avoid an expert having to define all parameters explicitly. More in particular, the ML approach utilizes co-evolution as a learning paradigm. An evaluation in the domain of one-versus-one air combat shows promising results.

1   Introduction

Serious gaming is playing a more and more prominent role to facilitate training in a variety of domains [1, 13]. The advantages of taking a serious gaming approach opposed to 'real life' training include (but are certainly not limited to) the ability to train realistic scenarios that are difficult to perform in the real world, the lower cost and the possibility to frequently repeat key learning events. In order to maximize the benefits, serious games should be populated by realistically behaving agents that for instance act as adversaries or teammates for the trainee, the so-called Computer Generated Entities (CGEs). Imagine a scenario in which an F-16 fighter pilot is trained in a virtual environment to use a specific tactic. Without having a realistic enemy to fight against, the training will not have the de-sired impact, while having to invite another fighter pilot to play the role of the enemy is inefficient in terms of training and tedious for the role-player.

One approach to obtain realistic behavior of CGEs is to distill knowledge from domain experts and build entities that incorporate the knowledge. Here, models that utilize this knowledge can be of a cognitive nature to establish human-like behavior. Another approach is to use pure learning-based techniques (e.g. reinforcement learning, evolutionary learning) and let the computer learn appropriate behavior. Both approaches however have severe disadvantages: for complex domains with limited access to domain experts the knowledge-based approach might be very difficult

A. Moonis et al. (Eds.): IEA/AIE 2014, Part II, LNAI 8482, pp. 120–129, 2014.

whereas the learning-based approaches are not guaranteed to provide realistic computer-generated entities as they might learn completely different strategies.

This paper takes a hybrid approach. It departs from a graph-based cognitive model which incorporates partial knowledge about the domain and applies evolutionary learning techniques to fine-tune the model towards a specific scenario. To be more specific, the cognitive model used is a Situation Awareness (SA) model (cf. [7]) which has been extended with a simple Decision Making (DM) model. Previously, attempts to apply learning techniques in this context have shown promising results (see [3] and [10]) but also revealed that a substantial amount of additional knowledge was needed to establish this behavior: in [3] the desired responses of the entities in all situations were needed whereas [10] has shown that learning a complex scenario can be troublesome due to characteristics of the fitness landscape. In this paper a solution to both problems is proposed. A co-evolutionary approach is used which drives two competing entities to more and more complex behavior (see e.g. [6]). It needs a limited amount of expert knowledge and circumvents the problem previously found with the relatively flat fitness landscape. The approach has been evaluated for the domain of fighter pilots.

This paper is organized as follows. In Section 2 related work is discussed as well as the domain of application. The approach is proposed in Section 3 whereas Section 4 presents the case study to investigate the effectiveness of the approach. Section 5 shows the results obtained and the paper is concluded with a discussion in Section 6.

2   Background

This section discusses the background of the approach, more in specific, it discusses cognitive modeling, the combination of cognitive models and machine learning, and the specific machine-learning approach utilized in this paper, namely evolutionary algorithms.

2.1   Cognitive Models

Cognitive models deal with the symbolic-information processing level and are thought to be largely independent of models at the physiological (neurological, milli-second) level. Some of those models are based on so-called unified theories of cognition. The goal of such a model is to show how a single control structure can handle all of the cognitive processes of which the human mind is capable. Soar [11], ACT-R [2], EPIC [9], CLARION [14], are frequently cited in connection with this class of models. In both Soar and ACT-R, cognition is largely synonymous with problem solving. They are both based on production systems (basically if .. then ..-systems) that require two types of memory: 'declarative' memory for facts and 'procedural' memory for rules. In contrast, a variety of component cognitive models have been defined that facilitate the generation of human-like behavior of role-playing agents in simulations for complex skill training. More specifically, in a previous research, a component model, a Situation Awareness (SA) model in the form of directed, a-cyclic graphs (DAGs) has been devised (cf. [7]) which is re-used in this research.

The SA model is meant to create high-level judgments of the current situation following a psychological model from Endsley [4]. The model essentially comprises of three forms of beliefs, namely simple beliefs, which can be directly derived from observations per-formed by the agent, or other simple beliefs. Furthermore, complex beliefs express combinations of simple beliefs, and describe the current situation on an abstracted level, and finally, future beliefs express expectations of the agent regarding projected (future) events in the scenario. All these beliefs are assigned an activation value between 0 and 1 and are connected via a network, in which each connection between a pair of beliefs has a particular strength (with a value between -1 and 1). Several update properties are specified that express how new knowledge obtained through observations is propagated through the net-work using the activation value of beliefs as a rank ordering of priority (the most active beliefs are updated first). For the sake of brevity, the details of the updating the algorithm have been omitted, see [7] for an in-depth treatment of the update rules. The domain knowledge that is part of the model is a so-called belief network.

Once a judgment of the situation has been created using the model above, a decision on the action given this situation should be selected. For this purpose a Decision Making (DM) model can be specified. This can be rather simple, in this case in the form of connecting the states in the SA model to actions using certain weights, i.e. again a weighted graph.

2.2   Machine Learning and Cognitive Models

As argued in the introduction, a combination of cognitive models and learning combines the benefits of both approaches whereas it also takes away some of the separate disadvantages. Hereby, cognitive models are based on coarse knowledge, and a machine-learning technique fine-tunes this knowledge. In the current effort, neuro-evolution in the form of a co-evolutionary process is applied to agents that contain a cognitive model that combines both SA and DM (that is, both models are adaptive and the DM process is contingent on the SA process), see Section 3.

In co-evolutionary learning, one can appoint coevolving species into one of the following two groups; symbiotic or parasitic. With the former variant the species cooperate and everyone benefits, while with the latter form there is usually a winning and a losing side. Irrespective, both types guide the evolution by continuously moving towards the optimum situation for all species involved. When developing an evolutionary algorithm to find a solution to a problem that involves multiple interacting agents (e.g. opposing fighter jets), one may choose to incorporate coevolution. By simultaneously evolving more than one population, and by letting the individuals from either population be each other's opponents, a situation is created in which the gain in mean fitness of the one population is directly related to a loss in mean fitness of the other population

To enable the evolution of a cognitive model (in this case, a DAG which is very much like a neural network), a neuro-evolutionary approach can be deployed. Neuro-evolution [18] concerns a group of evolutionary algorithms which specifically aim at adapting or learning a (neural) network. Most popular approaches, such as

NEAT [17], EANT [8] and EANT2 [15] grow a network from the ground up, continuously increasing complexity by adding new nodes and connections. Some do however start with a large network and try to prune it. For example, EPNet [26] starts with random overly-bloated network topologies and removes those nodes and connections that it deems redundant or irrelevant. These "pruning" algorithms have the benefit of being able to find the more optimized topologies at a lower cost than if a network-growing approach were to be used.

3   Approach

This section explains the learning approach that has been used as well as the slightly adjusted model which is subject to learning in more detail.

3.1   Cognitive Model

The cognitive model which forms the basis of the agent will be an extended version of the SA model described in [7] combined with a DM model. More specific, the default set of concepts as described in Section 2.1 will be joined by past beliefs (cf. [12]). These are beliefs on past events, which broaden the set of options to learn an appropriate model. Furthermore, the DM model used has been developed specifically for the purpose of

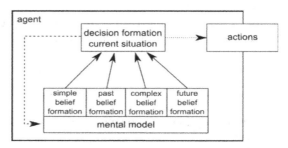

Fig. 1. Framework of the Decision Model. Decisions – sets of executable actions – are formed based on the currently active beliefs in the mental model.

this research, which also requires learning of appropriate weights. The DM model is shown in Figure 1, and consists of beliefs and decisions, with the latter being defined as a non-empty set of executable actions. A sole exception to this rule will be the decision to do nothing, which will be chosen by default when none of the decisions have a certainty c_d that exceeds its certainty threshold τ_d. For a decision to be valid, it will be required to have at least one incoming connection, whereby the source of this connection should be a belief contained within the SA model. Based on the activation value of this belief v_b and the weight $w_{db} \in [-1, 1]$ of its relation with a decision, the certainty of the latter will be calculated as formulated in Equation 1. Here, a positive outcome will denote the certainty that the selected decision is the correct one (activator), while the opposite will be true for a negative value (inhibitor).

$$c_d := \sum_{i=0}^{n-1} v_i w_{di}$$ **Equation 1**

3.2   Evolutionary Learning Algorithm

Two populations of agents, both equipped with the described cognitive model, will compete with each other, thus enabling parasitic co-evolution to occur between these populations. In addition, individuals may migrate between the populations, thus lowering the chance of a single population getting stuck in local optima.

Initially, both populations will be filled with N randomly-generated individuals, each individual representing the connections, with associated weights, of an instance of the cognitive model. To minimize bias of an 'unfair' match, each individual will be tested against Q unique opponents, randomly picked from the hostile population. Furthermore, as to minimize situational advantage, every individual-opponent pair will be evaluated R times in different (randomly-picked) scenario's. Individual fitness will then be based on the scenario's outcome O – victory (1), defeat (0), or draw (0.5) – and on the ratio of the combined size of both models g to the maximum size found amongst the population G. This ratio, in which the size of a model is defined as its total number of concepts and relations, will provide selection pressure to the more parsimonious model. Balancing this advantage will be achieved by taking the parameter for graph-size influence $I \in [0, 1]$, into account as well. Together, the two measures and the single parameter will contribute to the fitness f as formulated in Equation 2. Note that, in order to guard the bounds of the fitness range, the terms on the left and right side of the sum sign will be scaled by $(1 - I)$ and I, respectively.

$$f := O(1 - I) + I\left(1 - \frac{g}{G}\right) \qquad \textbf{Equation 2}$$

Independently in both populations, parents will be appointed by tournament selection. Depending on a probability $P_{crossover}$, this will involve either mutation or crossover. The crossover operator will primarily have an exploratory function. This will be accomplished on the chromosome level by inheriting one sub model from each of the two parents (i.e. given parent couple $SA_X DM_X$ and $SA_Y DM_Y$, two new chromosomes $SA_X DM_Y$ and $SA_Y DM_X$ will be formed).

In contrast with crossover, the more important mutation operator will perform on the genetic level, i.e. on the models. Depending on a probability $P_{topologyMutation}$, one of an individual's models will undergo either a (Gaussian) mutation of its weights, or a random mutation of its topology. The topology will be represented as weighted DAGs with the vertices and edges taking on the role of concepts and their relations, respectively (see Section 2.1). Therefore, any change in topology will encompass either the addition or removal of a vertex or edge, with several simple rules guarding the validity of the resulting models. For example, a topology mutation might involve the addition of a new *observation*. Alternatively, a weight mutation might modify the weight w_{qp} of edge e_{qp}, thereby adding a value x to said weight followed by rescaling the weights of all 'sibling' edges w_{qi} with $i \in [0, n_q)$ to distribute this change evenly.

After evaluation, the offspring will be inserted into their parents' population. Alternatively, depending on a probability $P_{migration}$, a small number of individuals may switch populations. Irrespective, any surplus will subsequently be dealt with by following an elitists approach.

4   Case Study

In order to test the suitability of the approach, a 'one versus one' (1v1) combat engagement scenario has been devised, featuring two opposing agents (simulated fighter aircraft, 'Attacker' and 'Defender'). Hence, their goal is to destroy one another. Note that this scenario is used to determine the fitness of individuals.

The primary sensor on-board the aircraft for detecting, identifying, tracking and locking the opponent is their radar. Good behavioral performance for the agent in control of the aircraft is defined as correctly detecting other aircraft on its radar, correctly identifying such aircraft, and subsequently engaging an aircraft in case it is hostile. During the engagement, the aircraft intercepts its opponent, while tracking the opponent via radar. When the opponent is within a distance that can be bridged by a missile ('weapons range'), a 'lock' can be made on the opponent (that is, focusing radar energy on its opponent), followed by the firing of the aircraft's radar-guided missiles. A scenario has been created in which this desired behavior can be exhibited. Initial positions of aircraft and initial angles between the flight paths of the aircraft are randomized at the start of each run of the scenario. However, 'Attacker' will come from the direction of the so-called FLOT (the Forward Line of Own Troops) and 'Defender', will fly towards the FLOT, such that the two aircraft will generally head towards each other and will detect each other by radar at some point (not necessarily at the same time).

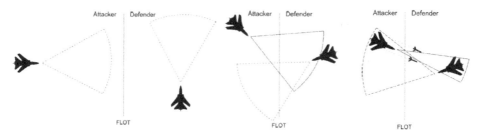

Fig. 2. Overview of 1v1 scenario

Each aircraft may perform any number of actions, among which are maneuvers that minimize the probability of detection on radar by the enemy, evasive maneuvers, tracking and intercepting an opponent, making a radar lock on the opponent, and firing missiles on the opponent. In addition, an aircraft is free in its ability to roam around, provided that the movements are limited to the horizontal plane. When the opponent succeeded to make a lock and fire a missile, the aircraft will attempt to defeat said missile, for instance by maneuvering in such fashion that the missile is unable to reach the targeted aircraft. Fig. 2 sketches the three main stages of such a scenario. First, both aircraft are on their own side of FLOT, and are unaware of each other's presence (Left).Then, 'Attacker' detects the 'Defender' on its radar (Middle). After achieving a weapons-lock on 'Defender', 'Attacker' fires two missiles. The scenario will end with the destruction of an agent, or after a pre-set maximum amount of time has passed.

5   Results

In total, four series of experiments were conducted to compare the proposed Evolutionary Algorithm (EA) with a simpler EA without coevolution and two baseline algorithms (Random Search and Random Restart Hill Climbing). Section 5.1 provides the results of the EA approach presented in Section 3.2, but without co-evolution. Section 5.2 presents the results of the full EA. Section 5.3 provides the results of experiments with the two baseline algorithms.

5.1   Simple EA (without Coevolution)

In this first series of experiments only a single population was maintained, of which the individuals (in the 'Defender' role) were evaluated against a scripted 'Attacker' (identical to the one used in [10]). Hence, these individuals were required to adapt to an opponent that demonstrated non-adaptive behavior. This provided a less dynamic fitness space, thus lowering the difficulty of the task.

The entire experiment was repeated five times with 2000 generations each, thereby starting with a fresh randomly-generated population of size $N = 1000$ on every restart. During evaluation, each individual participated in $Q = 10$ engagements, with each engagement consisting of $R = 3$ tries. The fitness in turn, was averaged each generation over these $Q \times R = 30$ evaluations, together with a graph-size influence of $I = 0.2$. In addition, a crossover probability $P_{crossover} = 0.05$ was set, as well as a topology-mutation probability $P_{topologyMutation} = 0.3$. Note that migration was omitted, as only a single population was maintained.

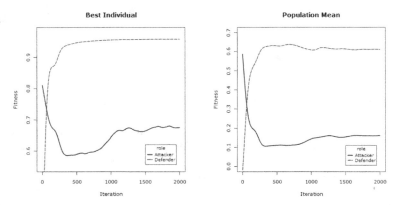

Fig. 3. Performance of EA without coevolution. Left) Fitness of both populations' best individual per generation. Right) Mean fitness of both populations.

Figure 3 shows the results. As evident by the sharp bend in the fitness curve of 'Defender', i.e. the learning agent, the algorithm found the optimal behavior fairly early: at around $t = 400$. In the case of the best individual, the curve levels out at a fitness between 0.94 and 0.96. This level, at which every engagement was won, shows a slight incline during the remaining generations which reflects the optimization of the best individual's cognitive model. In contrast, the population's mean fitness appears to show a slight decline after the curve peaked ($f \pm 0.62$) at around $t = 400$.

The fitness of the best scripted individual, i.e. 'Attacker', appears to follow an inverse pattern of that of 'Defender', showing a bend (towards a global minimum) at $t = 400$. However, from that point on, fitness increases until about t= 1200 generations. Finally, the mean fitness of the 'Attacker' population drops fast to a point between 0.1 and 0.15 after which a slight increase is observed, apparently countering the mean fitness of the 'Defender' population.

As a result of the EA, the cognitive models have lost most of their complexity through removal of redundant elements. This results in behavior that favors immediate action, with high priority for intercepting and destroying the opponent.

5.2   Full EA

The second series of EA experiments included co-evolution of the 'Attacker' and 'Defender' population, thus providing the more difficult task for the individuals of training on a dynamic fitness landscape. The parameter settings used were equal to those applied in Section 5.1. Moreover, migration was featured with a probability $P_{migration} = 0.30$ per 25 generations.

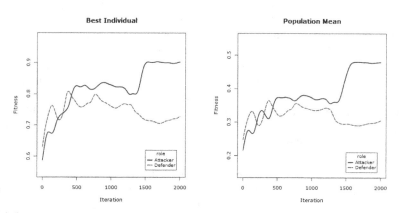

Fig. 4. Performance of EA with coevolution. Left) Fitness of both populations' best individual per generation. Right) Mean fitness of both populations.

Figure 4 shows that, with coevolution, the patterns at the individual and at the population level look remarkably similar, with only a constant different between Best Individual and Population Mean of approximately 0.4. In either case, both sides appeared to be roughly evenly matched ($\Delta f \leq 0.08$) in the first 500 generations, with 'Attacker' and 'Defender' oscillating around each other, after which 'Attacker' gains advantage with a sudden larger gap in fitness between Attacker and Defender at approximately 1400 generations. As with the EA without coevolution, the resulting models and behavior have lost most of their complexity. That is, redundant and rudimentary elements were removed, and immediate action was the preferred course of action.

5.3   Comparison with Two Baseline Algorithms

In a set of experiments with two baseline algorithms (random search and random-start hill climbing), the initial solution for each run was similar to that for an EA's randomly-generated individual. These solutions were subsequently evaluated – 5 times per solution ($R = 5$) – in the same randomly-generated scenarios as those during the simple EA. In addition, both algorithms were repeated 3 times, of which the results were averaged. Similar as with the EAs, a graph-size influence of $I = 0.2$ was set. With random search, each of the runs involved 10.000 randomly-created solutions. With random-restart hill climbing, every run was allowed 10 restarts, with each one continuing until no improvement was found for 1.000 iterations. Any improvement with random search appeared to stall after at most 4.000 iterations. At that point, fitness values between 0.31 and 0.35 were reached. No further increase was seen during the remaining iterations. During random-restart hill climbing, the better fitness values were between 0.225 and 0.27. However, most tries resulted in a lower fitness, with a minimum around 0.2.

6   Discussion

In this paper, an approach has been presented to learn parameters of cognitive models using co-evolution. Hereby, a situation-awareness model as well as a decision making model have been used which include a large number of parameters. This co-evolutionary approach has been applied to the creation of virtual opponents for fighter-pilot training. This Case Study shows that the co-evolutionary approach results in higher fitness than typically observed with evolutionary algorithms without coevolution and baseline algorithms used as benchmarks. Only with the co-evolutionary approach, CGEs can be trained against opponents that adapt themselves, which may offer additional advantages, such as better generalizability of performance against different opponents. Therefore, it is difficult to make a precise comparison between the presented approaches. When comparing the results with other work, the most obvious comparison is with the work presented in [10]. Here, a similar SA model was trained by adapting only the weights with a learning algorithm. While moderate performance was achieved, it was theorized that restricting the model's topology might have limited the solution space too severely. Therefore, the research focused on both topology and weights to be learned. Unfortunately however, building a model from scratch would result in the loss of interpretability, due to the inability of such methods to take context into account in the labeling of newly created nodes or vertices. Instead, a pruning approach was followed, such that an existing and overly-redundant model may evolve to be both slim and effective. A co-evolutionary approach has also been proposed by Smith *et al.* [16] which aims at finding new strategies in 1-v-1 air combat. However, they do not deploy a cognitive model, making the level of explainability as well as the replication of human behavior for effective training troublesome. For future work, it is envisioned to train the CGEs in the role of fighter pilots opponents for, more complex, scenarios and focus on an expert evaluation.

References

1. Abt, C.A.: Serious games. Viking Press, New York (1970)
2. Anderson, J.R.: The architecture of cognition. Harvard University Press, Cambridge (1984)
3. Gini, M.L., Hoogendoorn, M., van Lambalgen, R.: Learning Belief Connections in a Model for Situation Awareness. In: Kinny, D., Hsu, J.Y.-j., Governatori, G., Ghose, A.K. (eds.) PRIMA 2011. LNCS (LNAI), vol. 7047, pp. 373–384. Springer, Heidelberg (2011)
4. Endsley, M.R.: Toward a theory of Situation Awareness in dynamic systems. Human Factors 37(1), 32–64 (1995)
5. Floreano, D., Nolfi, S.: God Save the Red Queen! Competition in Co-Evolutionary Robotics. In: Koza, J.R., Deb, K., Dorigo, M., et al. (eds.) Proceedings of the Second Annual Conference Genetic Programming 1997, pp. 398–406. Morgan Kaufmann, San Francisco (1997)
6. Hillis, D.W.: Co-evolving parasites improve simulated evolution as an optimization procedure. In: Langton, C.G., Taylor, C., Farmer, J.D., et al. (eds.) Artificial Life II: SFI Studies in the Sciences of Complexity, vol. 10, pp. 313–324. Addison-Wesley, Redwood City (1991)
7. Hoogendoorn, M., Lambalgen, R.M., van, T.J.: Modeling Situation Awareness in Human-Like Agents using Mental Models. In: Walsh, T. (ed.) Proceedings of the Twenty-Second International Joint Conference on Artificial Intelligence, IJCAI 2011, pp. 1697–1704 (2011)
8. Kassahun, Y., Sommer, G.: Efficient Reinforcement Learning Through Evolutionary Acquisition of Neural Topologies. In: Proceedings of the Thirteenth European Symposium on Artificial Neural Networks, Bruges, Belgium, pp. 259–266. D-Side Publications (2005)
9. Kieras, D.E., Meyer, D.E.: An overview of the EPIC architecture for cognition and performance with application to human-computer interaction. In: Human-Computer Interaction, pp. 391-438 (1997)
10. Koopmanschap, R., Hoogendoorn, M., Roessingh, J.J.: Learning Parameters for a Cognitive Model on Situation Awareness. In: Ali, M., Bosse, T., Hindriks, K.V., Hoogendoorn, M., Jonker, C.M., Treur, J. (eds.) IEA/AIE 2013. LNCS, vol. 7906, pp. 22–32. Springer, Heidelberg (2013)
11. Laird, J., Rosenbloom, P., Newell, A.: Soar: An Architecture for General Intelligence. Artificial Intelligence 33, 1–64 (1987)
12. Merk, R.J.: Making Enemies: Cognitive Modeling for Opponent Agents in Fighter Pilot Simulators. Ph.D. thesis, VU University Amsterdam, Amsterdam, the Netherlands (2013)
13. Michael, D., Chen, S.: Serious games: Games that educate, train and inform. Thomson Course Technology, Boston (2006)
14. Naveh, I., Sun, R.: A cognitively based simulation of academic science. Computational and Mathematical Organization Theory, 313–337 (2006)
15. Siebel, N., Bötel, B., Sommer, G.: Efficient Neural Network Pruning during Neuro-Evolution. Neural Networks, 2920–2929 (2009)
16. Smith, R.E., Dike, B.A., Mehra, R.K., et al.: Classier Systems in Combat: Two-sided Learning of Maneuvers for Advanced Fighter Aircraft. In: Computer Methods in Applied Mechanics and Engineering, vol. 186(2-4), pp. 421–437 (2000)
17. Stanley, K., Miikkulainen, R.: Evolving Neural Networks through Augmenting Topologies. Evolutionary Computation 10(2), 99–127 (2002)
18. Yao, X.: Evolving Artificial Neural Networks. IEEE 87(9), 1423–1447 (1999)
19. Yao, X., Liu, Y.: A New Evolutionary System for Evolving Artificial Neural Networks. IEEE Transactions on Neural Networks 8(3), 694–713 (1997)

A Novel Method for Predictive Aggregate Queries over Data Streams in Road Networks Based on *STES* Methods

Jun Feng[1], Yaqing Shi[1,2], Zhixian Tang[1], Caihua Rui[1], and Xurong Min[3]

[1] College of Computer and Information, Hohai University, Nanjing 210098, China
[2] Command Information System Institute, PLA University of Science and
Technology, Nanjing 210007, China
[3] Computer Information Management Center of Nanjing Labor and Social Security
Bureau, Nanjing 210002, China
fengjun@hhu.edu.cn, yqshi_nanjing@163.com

Abstract. Effective real-time traffic flow prediction can improve the status of traffic congestion. A lot of traffic flow predictive methods focus on vehicles' specific information (such as vehicles id, position, speed, etc.). This paper proposes a novel method for predictive aggregate queries over data streams in road networks based on $STES$ methods. The novel method obtains approximate aggregate queries results by less storage space and time consuming. Experiments show that it can better do aggregate prediction compared with the ES methods based on $DynSketch$, as well as $SAES$ method based on DS.

Keywords: road networks, data streams, $STES$ methods, predictive aggregate queries.

1 Introduction

With the worsening traffic situation, intelligent transportation system (ITS) becomes a research hotspot as an effective mean to solve the traffic problems. The ITS reduces traffic congestion by implementing traffic management schemes and properly regulating traffic flow. Also it publishes travel information for trip and provides optimal path options. All above mentioned are based on a reasonable, real time, accurate prediction about road traffic state. It is the key to operate efficiently and safely in transportation system. However, with the continuous development of urban road network, intelligent transportation systems has accumulated real time, massive and complex traffic flow information, which rapidly expanded in a relatively short time. For instance, in the intelligent transportation system of a big city, supposing that there are one million GPS enabled vehicles, the vehicles emit one record every 30 seconds or 60 seconds, each record is 100 Bytes, then the total size of the data one day will be $100B \times 10^6/min \times 60/h \times 24/day \approx 144G$ [1]. The analysis and decision-making about real-time traffic flow information obtained by intelligent transportation

A. Moonis et al. (Eds.): IEA/AIE 2014, Part II, LNAI 8482, pp. 130–139, 2014.

system do not need to process specific data. The data stream approximate processing technology will be used. For example, traffic monitoring system does not need to query vehicle-specific information (such as id, position, velocity and other information) during the time period in the road. It needs data about how many cars pass under the query conditions, which used to analyze traffic congestion situation and predict the total number of car passing the same road at next time. It requires approximate aggregate query technologies.

In this paper, combined with adaptive smooth transition exponential smoothing $(STES)$ model, we propose a novel method to support predictive aggregate queries over data streams in road networks based on $DSD+$ index structure which proposed by our research group. Experiments show that the method can return high-quality predictive aggregate results.

This paper is organized as follows: Chapter 2 gives the predictive aggregate queries definition and discusses related work; Chapter 3 gives a novel method for predictive aggregate queries over data streams in road networks; Chapter 4, the experimental results are analyzed; Chapter 5 gives the conclusion.

2   Preliminaries

2.1   Predictive Aggregate Queries Definition

Predictive aggregate queries are defined as follows: Given the predicted time period $[t_0, t_1]$, the query area $q = ([x_0, x_1], [y_0, y_1])$ (where t_0 is later than the current time, $[x_0, x_1], [y_0, y_1]$ is a two-dimensional spatial coordinates description of the query area), it predicts the approximate total number of moving objects on all sections within the query region q in the period of time. Mainly discusses two cases:

(1) If $t_0 = t_1$, it expresses predictive aggregate query about moving vehicle on the query section of road networks in a single moment;

(2) If $t_0 < t_1$, it expresses the sum of prediction results obtained about predictive aggregate queries of moving object on the query section of road networks within the plurality of discrete moment in the time period $[t_0, t_1]$.

2.2   Related Work

At present, domestic and foreign dynamic traffic flow predictive theory researches are still in development stage. They have not form a more mature theoretical system, especially for short-term traffic flow prediction methods research. It is impossible to obtain satisfactory results. The traffic flow short-term predictions predict the traffic flow a few hours or even minutes later. It will be affected by many factors and cannot produce good predictor effects, such as random events interference. The uncertainty about short-term traffic flow is stronger than long-term traffic flow, and the former's regularity is more unconspicuous than the latter's. There are many solutions to predict short-term traffic flow. The main models are: time series models [2], multivariate linear regression models [3], *Kalman* filtering model [4].

Time series model has been widely used in weather forecasting, hydrological forecasting, stock market prediction, data mining and other fields. The theory is relatively mature, and it is a promising short-term traffic flow predictive method. There are ES predictive method based on the $DynSketch$ [5] index structure and $SAES$ predictive method based on the DS index structure [6,7], which are aggregate predictive techniques based on road network data stream aggregate index structures. ES predictive method is simple and requires only two values, but this method cannot adjust smoothness exponential value adaptively according to the change of data distribution. $SAES$ predictive model is the cubic exponential smoothing method of ES. The calculation is larger and more complex.

Multiple linear regression model studies interdependence between a number of explanatory variables and the dependent variable. It estimates the value of another variable by using one or more variables. Although it is easy to calculate, the factors used to create model is relatively simple. The prediction accuracy is low, and it is unsuitable for short-term traffic flow prediction. This paper studies predictive aggregate query, and considers the recent historical aggregate information, so above-mentioned method is not considered.

$Kalman$ filter model was proposed in 1960s, which has been successfully used in the field of transport demand prediction. The model is based on the model proposed in the filtering theory by $Kalman$. It is a matrix iterative parameter estimation method for linear regression model, which predictor selection is flexible and has higher precision, but this method needs to do a lot of operations about matrix and vector, resulting in relatively complex algorithm. It is difficult to apply to real-time online prediction fields. This paper studies the traffic flow in the road network which demands real-time response, therefore, it does not consider the $Kalman$ filter model.

3   Algorithm Descriptions

The data our predictive aggregation method used come from history and current aggregation query results. The recent time traffic flow place a greater impact on the prediction results of the next time, this paper uses $STES$ model in the time series models. The method inherits the advantages of ES, and mends the ES's shortcomings that exponential cannot be adjusted according to changes in the distribution of data.

3.1   Smooth Transition Exponential Smoothing

Studies have shown that the estimate about smoothing parameter must minimize the prediction error of the previous step. Some researchers believe that α parameters of the ES model [5] should be able to change over time to meet the latest features of the time series. For example, if the sequence change with the level (level shifts), the average value of exponential weighted must be adjusted such that a larger value corresponds to the latest observations. There are many adaptive methods, such as Holt and Holt-Winters methods [8], but most have been criticized because of its unstable prediction. General method for adaptive smoothing exponential formula is $f_{t+1} = \alpha_t y_t + (1 - \alpha_t)f_t$. f_{t+1} means the predictive value at time $t + 1$. α_t means the predictive smoothing exponential at time t. y_t means the true value at time t. f_t means the predictive value at time t. Adaptive smoothing methods are available as a collection species smooth transition model, the form of a simple Smooth Transition Regression Model (STR) is: $y_t = a + b_t x_t + e_t$, where $b_t = \omega/(1 + exp(\beta + \gamma V_t))$, α, ω, β, γ are constant parameters. If $\gamma < 0$ and $\omega > 0$, then b_t is a monotonically increasing function of V_t, and the value range is 0 to ω. Smooth Transition Auto Regression Model ($STAR$) is a similar form, in addition to being replaced calls lag dependencies y_{t-1}.

James W. Taylor [9,10] made a Smooth Transition Exponential Smoothing ($STES$) method in 2004, which is a new adaptive exponential smoothing method that smoothing parameter can be modeled as an user-specified variables corresponding mathematical logic function (logistic). This variable's simple choice is the size of the predictive error over past time. This method is similar to the method of modeling for the time variation parameters in smooth transition method. This paper introduces $STES$ method to traffic data stream aggregate predictive query, and designs data stream aggregate predictive model in road network. $STES$ model prediction equation is as follows:

$$f_{t+1} = \alpha_t y_t + (1 - \alpha_t)f_t, \alpha_t = \frac{1}{(1 + exp(\beta + \gamma V_t))} \tag{1}$$

The core of smooth transition model is that at least one parameter is modeled as a continuous function of converting variable V_t. As can be seen from formula 1, the function is limited α_t to the range from 0 to 1. It is different from the previous model that the historical data is used to estimate α_t, but this is used to estimate the constant parameters β and γ . If $\gamma < 0$, α_t is a monotonically increasing function of V_t . Generally, the choice of V_t is the key to the success of the model. Taking all the adaptive methods into account, the smoothing parameter depends on changes size of difference between predictive value in the last time. Clearly, the square or absolute value of the recent error can be used as conversion variables. Of course, average error, average square error, the percentage error and others can also be used as the conversion variables. Looking at $STES$ model's prediction formula, and we take $\gamma < 0$. Because when the error V_t expressed becomes large, the description should increase the weight of the true value at time t, namely the weight of y_t. It fits monotonically increasing relationship of prediction equations.

3.2   Architecture

In using adaptive smooth transition model, it usually hopes that the lag prediction error is minimal. It can be seen that when $\gamma = 0$, $STES$ model reduces to ES model. When the model $STES$ proposed in the paper chooses historical data,it orientates correct $DSD+$ structure by Chord platform. In this paper, the V_t's choice is the absolute error about prediction in the last time. It is easy to calculate and achieve real-time. β and γ are to be estimate using historical data through the recursive least two multiplications. When $\gamma = 0$, it is the ES model. When $\gamma < 0$, it can give better prediction in the case of the data having level shifts. And when $\gamma > 0$, it is used to compare the reaction about the prediction function to the abnormal value and reduce the impact about the abnormal value to the weight of predicting. Fig.1 shows the aggregation prediction flow of $STES$ model based on $DSD+$ in Chord platform.

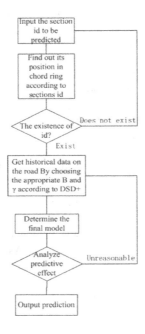

Fig. 1. $STES$ Model Predictions Flow

$STES$ model has been applied to the inspection about volatility issues in stock market. It has made more stable results. Its smoothing coefficient is not static, parameter changes can reflect the impact of data for predicting in different periods. The number of vehicles on road network is also constantly changing volatility. According to the process of Fig. 1, the aggregation predictive algorithm after finding the BPT-tree to the section's $id$is given as follows:

Assuming $qt = (t_0, t_1)$, the algorithm supports two prediction forms. one is $t_0 = t_1$, that is, predicts the next time for a query region, another is $t_1 > t_0$, which predicts the next period of the current time. The $inordernot$ algorithm is used to query approximate aggregate values in history time, root is the root node of BPT-tree, x_0, x_1, y_0, y_1, t_0, t_1, represent the x-axis range, the y-axis range, and the time range $t_0 - t_1$, respectively. And n represents the $hislength$ using by this algorithm(line: 3); $S[i]$ represents the approximate aggregate value get in $time = t - t + t_t$(line: 6); $error$ represents the prediction error absolute in $time = t - t + t_t$(line: 7); α_t represents the smoothing parameter in $time = t - t + t_t$, V_t represents the absolute value of the simple calculation error(line: 8); S_{next} represents the predictive value in $time = t + 1 - t + t_t + 1$(line: 9). The non-recursive in-order traversal BPT-tree in this method gets the approximate aggregate values based on sketch information within the query range.

Algorithm 1. agg_prediction($x_0, x_1, y_0, y_1, t_0, t_1$)

---

    input : query region: (x_0, x_1, y_0, y_1) and predictive time range: (t_0, t_1)

    output: aggregation results

1 **begin**

2      $t_t = t_1 - t_0$;

3      $S_{next} = S[t_0 - n - t_t] = inordernot(root, x_0, x_1, y_0, y_1, t_0 - n - t_t, t_0 - n)$;

4      $t = t_0 - n - t_t$;

5   **for** $i = t$ **to** $t_0 - 1$ **do**

6         $S[i] = inordernot(root, x_0, x_1, y_0, y_1, i, i + t_t)$;

7         $error = fabs(S_{next} - S[i])$;

8         $\alpha_t = 1/(1 + exp(\beta + r \cdot error))$;

9         $S_{next} = \alpha_t \cdot S[i] + (1 - \alpha_t) \cdot S_{next}$;

10   **end**

11   return S_{next};

12 **end**

---

4   Experiments

The data used in the experiments mainly come from public transport road network data in 1999, California, U.S., collected by Geographical Data Technology Inc. The map scale of road network environment used in experiments is 1:24000. The average density of the road network is 1451 road segments per km^2. First intercepted 784 road segments from the road, then simulate generated 10,000 vehicles and divided them into four kinds (car, bus, truck and autobike) with different speed and moving patterns. The data is stored as triple (tm, rid, eid), namely the time, road segment identification, vehicle identification. The information was recorded every 10 seconds. It recorded a total of 180 time vehicle information, and saved in disk with route (direction). txt.

The experiment was conducted under the Windows XP OS, with 1.5G RAM and 2.0GHZ AMD CPU. The development environment was VC++6.0. The development language was C++. The distributed Chord experiments used the online public Chord simulator [11].

The experiment mainly compared the merits of $STES$ predictive method, ES predictive method and $SAES$ predictive method. The paper mainly studies predictive aggregate from the following aspects :

(1) The influence of historical data length (*hislength*) to the accuracy of the predictive aggregate.

The experiment varied *hislength* from 1 to 5 and compared the average relative error and the maximum relative error about prediction in each hislength. In order to investigate the influence of *hislength*, first we set t_t to 1, and determine the influence of *hislength* from a single time prediction. As can be seen from Fig. 1, the relative error of approximation queries is best when the time interval is 1 and the regional interval is 5. The experiment will randomly generate 20, 30, 40, 50, 60, 70 random queries. It compared the predictive error during these queries range when *hislength* is 1 to 5. The comparison of the average relative error and the maximum relative error are in Fig. 2 and Fig. 3.

As can be seen from the two figures, when the *hislength* is 1, both relative errors are large. They will be reduced with the increment of *hislength*. When the *hislength* is 5, and it randomly generates 50 queries, the average relative error is minimal (only 11.973337), and the maximum relative error can be controlled between 30% and 42%. The two figures also show that errors are close to others' from the *hislength* = 3, the difference about the average relative errors are not more than 5%, and the maximum relative errors are controlled at about 10%. No further experiments will do to compare the predictive errors about *hislength* > 5. So the following experiment, the *hislength* is 5.

(2) Selecting the *hislength* value of best predictive accuracy, comparison is between *STES* predictive method based on *DSD*+ and *ES* predictive method based on *DynSketch*, *SAES* predictive method based on *DS*.

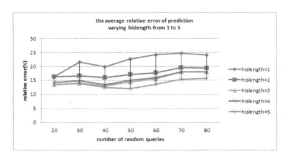

Fig. 2. The average relative error of prediction varying *hislength* from 1 to 5

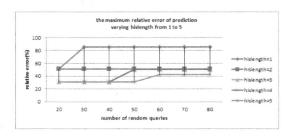

Fig. 3. The maximum relative error of prediction varying *hislength* from 1 to 5

We divides the comparison into two cases, one is for a single moment prediction, the other is the time interval prediction while $TimeInterval > 0$. The experimental changes $TimeInterval$ values from 1 to 3. When $TimeInterval = 0$, it indicates the prediction for a single time.

For a single moment prediction ($TimeInterval = 0$), we set the $hislength$ to 5, the regional interval to 5, and the time interval to 1. It generates 50 randomly queries. Because the query error of $DynSketch$ is large, the predictive error of ES is relatively larger. Therefore it is not comparable. It just compares the ES and $SAES$ method based on $DSD+$. The comparison about the relative error of $STES$, ES and $SAES$ is showed in Fig. 4.

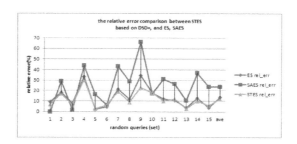

Fig. 4. The relative error comparison between $STES$ based on $DSD+$, and ES, $SAES$

AS can be seen from Fig. 4, relative error of the $STES$ predictive method based on $DSD+$ is smaller than that of the ES and $SAES$ based on $DSD+$. The performance of the $SAES$ predictive method is more unstable, and the average relative error is 23.784421.That of ES is 13.665746 and $STES$ is 11.973337. It is because $STES$ appropriately adjusted smoothness exponential α value according to the data changes in order to aggregate predict better. Next, from these random queries, observe changes of value in $hislength$ about a group which has a larger difference of predictive accuracy between ES and $STES$, that is, the ninth group of query data. The regional range and time range of data is (16.000000, 26.000000, 25.000000, 30.000000, 10.000000 10.000000), in other words, it is to predict the aggregate value in the time period of 10 using the data of first 5-9 $hislength$.

As can be seen from Fig. 5, αt changes the weight of the approximate queries value and the predictive value as the change of the previous step error to do aggregate prediction better. The last step predictive error increases, then αt will increase. That is the weight of the true approximate query value in the last moment will increase, and that of the predictive value with greater predictive error will be reduced.

The following experiment is for a time interval prediction in $TimeInterval$. The experiment will change $TimeInterval$ from 1 to 2 to compare with the value of $TimeInterval = 0$. The experiment randomly generates 50 queries. Observe the predictive relative error and the average relative error of segments which the vehicles real number is not 0. Because of the time interval variation, the query

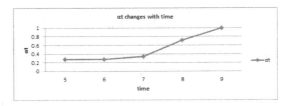

Fig. 5. αt changes with time

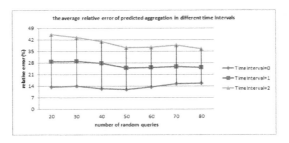

Fig. 6. The average relative error of predictive aggregate in different time intervals

regions are different in different time intervals. The experiment is generated 20-80 random queries in steps 10 to compare the average relative error in different $TimeInterval$.

From Fig. 6 we can see that the greater the predictive time interval is, the greater the relative error of predictive aggregation does. Because there is an error in approximate queries about each step. $STES$ is the improvements to ES. ES is the better predictive aggregate technology in the single time, the same to $STES$. It ensures predictive aggregate error gathered about 14%.

5   Conclusion

In recent years, with the rapid economic development, growing urban traffic congestion cause great inconvenience upon people's travel. In this paper, a novel method for predictive aggregate queries over data streams has been proposed. The experiment proved that the method can return higher quality predictive aggregate results comparing with ES predictive methods based on $DynSketch$ and $SAES$ predictive methods based on DS. It will be considered to use upstream traffic junctions to predict downstream traffic junctions in order to further improve the prediction accuracy of aggregation according to characteristics of the road network, such as STM [12] model.

Acknowledgments. This work is funded by the National Natural Science Foundation of China(No.61370091,No.61170200), Jiangsu Province science and technology support program (industry) project(No.BE2012179), Scientific Innovation program for College Graduate, Jiangsu Province (No.CXZZ12_0229).

References

1. Ma, Y., Rao, J., Hu, W., Meng, X., Han, X., Zhang, Y., Chai, Y., Liu, C.: An efficient index for massive IOT data in cloud environment. In: Proceedings of the 21st ACM International Conference on Information and Knowledge Management (CIKM 2012), pp. 2129–2133 (2012)
2. Han, C., Song, S., Wang, C.: A Real-time Short-term Traffic Flow Adaptive Forecasting Method Based on ARIMA Model. Acta Simulata Systematica Sinica 16, 146–151 (2004)
3. Li, J.Z., Guo, L.J., Zhang, D.D., Wang, W.P.: Processing Algorithms for Predictive Aggregate Queries over Data Streams. Journal of Software 16, 1252–1261 (2005)
4. Ben, M., Cascetta, E., Gunn, H., Whittaker, J.: Recent Progress in Short-Range Traffic Prediction. Compendium of Technical Papers. In: 63rd Annual Meeting, Institute of Transportation Engineers, The Hague, pp.262–265 (1993)
5. Feng, J., Lu, C.Y.: Research on Novel Method for Forecasting Aggregate Queries over Data Streams in Road Networks. Journal of Frontiers of Computer Science and Technology 4, 4–7 (2010)
6. Feng, J., Zhu, Z.H., Xu, R.W.: A Traffic Flow Prediction Approach Based on Aggregated Information of Spatio-temporal Data Streams. Intelligent Interactive Multimedia: Systems and Services Smart Innovation, Systems and Technologies 14, 53–62 (2012)
7. Feng, J., Zhu, Z.H., Shi, Y.Q., Xu, L.M.: A new spatio-temporal prediction approach based on aggregate queries. International Journal of Knowledge and Web Intelligence 4, 20–33 (2013)
8. Willaims, T.: Adaptive Holt-Winters forecasting. Journal of the Operational Research Society 38, 553–560 (1987)
9. James, W.: Smooth Transition Exponential Smoothing. Journal of Forecasting 23, 385–394 (2004)
10. James, W.: Volatility Forecasting with Smooth Transition Exponential Smoothing. International Journal of Forecasting 20, 273–286 (2004)
11. Open Chord (2012), http://open-chord.sourceforge.net/
12. Sun, J., Papadias, D., Tao, Y., Liu, B.: Querying about the past, the present, and the future in spatio-temporal databases. In: 20th International Conference on Data Engineering, pp. 202–213 (2004)

BDI Forecasting Based on Fuzzy Set Theory, Grey System and ARIMA

Hsien-Lun Wong

Minghsin University of Science and Technology
1 Hsinhsin Rd, Hsinfong, 30401 Hsinchu, Taiwan
alan@mail.mudt.edu.tw

Abstract. Baltic Dry Index (BDI) is an important response of maritime information for the trading and settlement of physical and derivative contracts. In the paper, we propose fuzzy set theory and grey system for modeling the prediction of BDI, and employ ARIMA for the calibration of data structure to depict the trend. The empirical results indict that for both short-term and long-term BDI data, fuzzy heuristic model has lowest prediction error; Structural change ARIMA fits better for the prediction in the long term, while the GM (1,1) has the greatest prediction error. Moreover, the relationship that the change between in current BDI and in previous is highly positive significance; the external interference for the current BDI index is negatively related. The conclusion of the paper would provide the bulky shipping with a beneficial reference for the market and risk assessment.

Keywords: Baltic dry index, Bulky shipping, Fuzzy set theory, Grey system, ARIMA.

1   Introduction

The freight rate of bulk shipping has been determined by both supply and demand market for a long time. The supply perspective is mainly decided by new vessel orders and overage vessel disassembly, and the demand by global economic cycle, seasons, climate and politics. The changes in freight rates are not easy to be mastered for the management decision-making. The bulk shipping companies are confronted with business uncertainties in the ocean transportation industry [1].

Baltic Freight Index (BFI) was created by London Baltic Exchange Center (BEC) in 1985, and replaced by Baltic Dry Index (BDI) in 1999. After 2006, BDI was the average of Baltic Capesize Index (BCI), Baltic Panamax Index (BPI), Baltic Supramax Index (BSI), and the multiplier for BDI was 0.99800799. Presently, Baltic Handysize Index (BHSI) is introduced, so new BDI account is equal to the same weight of the four indices multiplied by 1.192621362 to maintain BDI continuity. Thus, BDI reflects bulk shipping market sentiment as a price reference for transaction platform of bulk shipping companies and investors.

The characteristic of BDI data is that it changes with each transaction and has multi-variability. In each transaction period, there is the highest/lowest price and closing/opening price. The closing price is a fixed value that just reflects the final

A. Moonis et al. (Eds.): IEA/AIE 2014, Part II, LNAI 8482, pp. 140–149, 2014.
© Springer International Publishing Switzerland 2014

conditions of market transaction, unable to express the whole pattern. The index in nature seems accurate but has implications. Consequently, the reality of BDI data is not easily described by the traditional 0-1 logical concept.

Fuzzy set theory [2] states a fuzzy continuum logic which transits from "non-membership" to "membership". It can define a clear boundary for some uncertain data to quantify the data for calculation. In this paper, we proposed the concept of fuzzy set theory to construct a BDI prediction model for the short-term dynamic trend. Meantime, traditional time series model and the grey theory are used in order for the comparison of model capability and stability.

The remainder of this paper is organized as follows. Section 2 introduces fuzzy prediction and BDI-related literature; Section 3 presents the construction of fuzzy time series model, GM (1,1) and Grey-Markov Model and solving; Section 4 provides data source and empirical results; Section 5 gives conclusions and suggestions.

2   Literature Review

Since Zadeh [2] proposed fuzzy set theory as the tool to test uncertain membership, it has served as the theory framework in research of many fields and has solved 0-1 logic value limitation of traditional sets. Song and Chissom [3,4] established fuzzy relationship for fuzzy time series model, and introduced fuzzy theory in prediction domain.Chen [5] simplified fuzzy matrix construction procedure by Song and Chissom [3, 4, 6], used simple mathematical equation instead of complex Max-Min equation, and suggested fuzzy logical relationship group. The empirical data are superior to the results from the model [6].

Huarng [7,8] proposed two heuristic models to improve the model [5], constructed two fuzzy logical relationship groups of previous and next periods, and introduced a threshold value to construct three fuzzy logical relationship groups of previous and next periods. Hwang *et al* [9] proposed differential mode for stabilization of historical data to predict future variation not the value. The fuzzy correlation matrix is more rational, operation time is more coefficient, and predicted value is superior to the model [3,4,6]. Cheng *et al* [10] used trapezoidal membership functions to fuzzify historical data and suggested minimize entropy principle approach to find out appropriate interval median points and form unequal interval segmentation.

Yu [11] considered fuzzy logical relationship should be assigned with proper weights according to fuzzy logical relationship, reflecting the data information, and employed two interval segmentation methods [7, 8] to construct models. Huarng and Yu [12] proposed heuristic type-2 model to increase research variables to facilitate efficiency of prediction models. Huarng and Yu [13] introduced neural network nonlinear structure and employed back propagation for prediction. Huarng and Yu [14] suggested the unequal-length interval segmentation and firmly believed the same variation value has unequal validity in different intervals.

Li and Cheng [15] considered the fuzzy logical relationship group [5] is the factor that leads to prediction uncertainty and thus suggested a backtracking system to establish a sole definite fuzzy logical relationship; the findings are inconsistent with the conclusion made by Huarng [7]. Cheng et al [16] extended the concept by Wang and Hsu [17] and Yu [18] and made linear correction with linear and non-linear concepts after nonlinear fuzzy logical relationship, and assigned suitable weight to the difference between predicted value and the actual value of previous period.

Chen and Hwang [19] first introduced two-variable in fuzzy time series prediction to forecast main variables in combination with the concept by Hwang et al [9]. Cheng et al [20] applied novel fuzzy time series method to resolve the student enrollment in the University of Alabama and Taiwan stock TAIEX data. Recently, Wong *et al* [21,22,23] compared the three fuzzy prediction models in the literature and found that heuristic fuzzy time series model has the lowest prediction error [15].

After BDI study in the ocean transportation field [1,24-34], it is noted that most of them discussed the correlation analysis of shipping price variation; however, not many paid attention to the BDI prediction problem. These related studies mainly applied regression model and traditional time series model. In the paper, the fuzzy concept is applied to expressing the characteristic of BDI data for forecasting problem.

3   Research Methodology

3.1   GM Theory

In GM theory [35], one uses the generation method to pre-process the raw data which are usually incomplete and uncertain and construct a grey model through differential equation to reduce data randomness. GM $(1,1)$ model is described as follows:

(1) After Accumulated Generating Operation (AGO), the time series

$$x^{(0)}(k) = (x^{(0)}(1), x^{(0)}(2), ..., x^{(0)}(n))$$ can be transformed into:

$$x^{(1)}(k) = (x^{(1)}(1), x^{(1)}(2), ..., x^{(1)}(n))$$

where:
$$x^{(1)}(k) = \sum_{i=1}^{k} x^{(0)}(i) \quad k = 1, 2, ..., n \tag{1}$$

(2) Construct data matrix B and constant matrix Y as below:

$$B = \begin{bmatrix} -z^{(1)}(2) & 1 \\ -z^{(1)}(3) & 1 \\ .... & ... \\ -z^{(1)}(n) & 1 \end{bmatrix} \tag{2}$$

$$Y_n = \begin{bmatrix} x^{(0)}(2) \\ x^{(0)}(3) \\ ..... \\ x^{(0)}(n) \end{bmatrix} \tag{3}$$

where: $z^{(0)}(k) = \dfrac{x^{(1)}(k) + x^{(1)}(k-1)}{2}$, $k=2,3,...,n$; and let $\dfrac{dx^{(1)}}{dt} + ax^{(1)} = b$

(3) Use the method of least squares to solve coefficients a and b as follows:

$$(B^T B)^{-1} B^T Y_n = \begin{bmatrix} a \\ b \end{bmatrix} \tag{4}$$

(4) Substitute a and b into the differential equation, and the function can be obtained:

$$\hat{x}^{(1)}(t+1) = (x^{(0)}(1) - \frac{b}{a})e^{-at} + \frac{b}{a} , \quad t=1,2,...,n \tag{5}$$

where: $(x^{(0)}(1))\, e^{-at}$ is initial value term, $\dfrac{b}{a}(1-e^{-at})$ is constant term.

(5) Use the function derived from IAGO to restore $\hat{x}^{(0)}(t+1)$, and GM(1,1) can be obtained:

$$\hat{x}^{(0)}(t+1) = (1-e^a)(x^{(0)}(1) - \frac{b}{a})e^{-at} \tag{6}$$

where: $\hat{x}^{(0)}(t+1)$ denotes the predicted value, and t denotes the period.

3.2   Grey-Markov Model

Grey-Markov model is based on grey prediction model combined with Markov chain [36]. Markov chain handles a dynamic system that changes randomly and is a transition probability relying on all dimensions of state evolution. Markov transition probability is represented by $P_{ij}^{(t)}$, as shown below:

$$P_{ij}^{(t)} = \frac{m_{ij}^{(t)}}{M_i} , \quad i=1,2..., j=1,2..., t=1,2,...,n \tag{7}$$

where: $P_{ij}^{(t)}$ denotes the probability of transition from state E_i to state E_j after t step; $m_{ij}^{(t)}$ denotes the times of transition from state E_i to E_j after t step; M_i denotes the occurrence number of state E_i. The matrix $R^{(t)}$ with t step transition probability is represented by:

$$R^{(t)} = \begin{bmatrix} p_{11}^{(t)}, p_{12}^{(t)}, \dots, p_{1j}^{(t)} \\ p_{21}^{(t)}, p_{22}^{(t)}, \dots, p_{2j}^{(t)} \\ \vdots \quad \vdots \quad \vdots \\ p_{ji}^{(t)}, p_{j2}^{(t)}, \dots, p_{jj}^{(t)} \end{bmatrix}, t=1,2,\dots,n \tag{8}$$

Transition matrix $R^{(t)}$ shows that regular transposition patterns of various states in a system. Assume that the falling point state is j, the state interval range $E_i \in \left[\otimes_{1i}, \otimes_{2i} \right]$, and predicted value of j period takes middle point of between upper and lower boundaries of the state (E_i). The forecasting value is presented by:

$$\hat{X} = \frac{1}{2}(\otimes_{1j} + \otimes_{2j}) \tag{9}$$

The algorithm of Grey-Markov model can be expressed as follows: 1. Solve the residual error of GM (1,1), 2. Divide the relative state of residual error, 3. Calculate the transition probability of various states, 4. Establish transition probability matrix, 5. Predict residual error.

3.3   Fuzzy Set Theory

If U_A denotes the membership function of fuzzy set A, the membership function value $U_A(x)$ expresses the degree of membership of element x of set A. The membership function of fuzzy set $F(t)$ is $\{u_1(Y(t)), u_2(Y(t)), \dots, u_m(Y(t))\}$ where $0 \le U_i(Y(t)) \le 1$, and the fuzzy set $F(t)$ can be expressed by discrete function:

$$F(t) = \sum_{i=1}^{m} \mu_i(Y(t)) / A_i = \mu_1(Y(t)) / A_1 + \mu_2(Y(t)) / A_2 + \dots + \mu_m(Y(t)) / A_m \tag{10}$$

where: $\mu_i(Y(t)) / A_i$ denotes the membership and membership degree of original time series $Y(t)$ after experiencing linguistic variable $\{A_i\}$; "+" denotes a connecting symbol.

Fuzzy Logic Relationship
Suppose that fuzzy time series $F(t-1)$ and $F(t)$ of previous and next periods have fuzzy relation, expressed by $R(t-1,t)$. The relation between $F(t-1)$ and $F(t)$ can be

represented by $F(t-1) \rightarrow F(t)$. Assume that $F(t-1) = A_i$, $F(t) = A_j$ and two-value fuzzy relation can be represented by $A_i \rightarrow A_j$. where A_i is called LHS, and A_j is called RHS.

If the fuzzy time series $F(t)$ is affected by several previous series $F(t-1), F(t-2), ..., F(t-n)$, the fuzzy relationship can be represented by $F(t-1), F(t-2), ..., F(t-n) \rightarrow (t)$. This is called n-order fuzzy time series model. If $n=1$, it is 1-order fuzzy time series model. If the fuzzy time series $F(t)$ has multi-group fuzzy relationship, the same fuzzy relationships can be arranged into fuzzy logical relationship group.

The algorithm of fuzzy heuristic model is expressed as follows: Step 1: Define universal of data. Step 2: Define interval length, fuzzy set, and fuzzify data in linguistic manner. Step 3: Establish fuzzy logic relationship. Step 4: Integrate fuzzy logical relationships and construct a model. Step 5: Make prediction and defuzzify.

3.4   Mean Absolute Percentage Error

This study used mean absolute percentage error (MAPE) to measure prediction model accuracy. MAPE value can be represented by:

$$MAPE = \frac{1}{n}\sum_{k=1}^{n}\frac{\left|x^{(0)}(k) - \hat{x}^{(0)}(k)\right|}{x^{(0)}(k)} \times 100\ \% \tag{11}$$

When MAPE value is lower than 10%, it is treated as high prediction accuracy for model [37].

4   Empirical Results and Discussion

The BDI data were used by new formulation from Jan, 2006 to Apr, 2010 with three parts. For the 14 and 28 data, the paper applied fuzzy heuristic model and GM for test as ARIMA needs more data ($n>=30$); the 52 long data is used for all models. There are some postulates for model test: (1) The AGO=1 and 4~12 order is given for GM(1,1) test, (2) The E_i, $i=2\sim5$ and $t=1$ is given for Markov process, (3) The A_i, $i=19$ to 21 equal length and $n=1$ order is given for fuzzy model, and a triangle function used for fuzzy membership degree.

The result of BDI data from ARIMA (p,d,q) model is expressed as follows:

$$\Delta BDI_t = 0.974614\Delta BDI_{t-1} - 0.392752\varepsilon_t \tag{12}$$

where: $\Delta BDI = BDI_t - BDI_{t-1}$, R^2(adj)=0.883732, SBC=16.67470.

The ARIMA of the structural change is shown as below:

$$\Delta BDI_t = 1.97518\Delta BDI_{t-1} - 0.174640D_{1t}\ \Delta BDI_{t-1} - 0.352198\varepsilon_t \tag{13}$$

where: $D_{1t} = 0$ (period is smaller than Dec, 2007); $D_{1t} = 1$ (period is larger or equal to Dec, 2007), R^2(adj)=0.897334, SBC=16.60747.

After testing the 52 original BDI data, the GM(1,1) model with $i=4$ can be obtained:

$$\hat{x}^{(0)}(t+1) = (1 - e^{-0.0066})[x^{(0)}(1) - \frac{2967.5}{-0.0066})e^{0.0066t} \quad (14)$$

Based on the GM (1,1) result, the optimal interval for GM-Markov model is obtained by 5 error states, as shown E_1=(-140%,-70%], E_2= (-70%,-34%], E_3= (-34%,0%], E_4= (0%,34%], E_5=(34%, 70%].

According to predicted BDI of the 5 models, the following points can be found: For model fitness, the error of the GM-Markov model ($i=4,E=5$)has the lowest dispersion degree, and the predicted MAPE ranged from 0.18% to 38.70%. The ARIMA with structural change and ARIMA have larger error, and they are 0.89%~400.8% and 0.31%~415.5% receptively. For the models, largest error is similar, which occurred at the period of BDI sharp drop, i.e. between Sep, 2008 and Nov, 2008. However, the largest error produced by both GM-Markov and fuzzy heuristic models has a little lower than 84% of BDI. It indicates that the two models have good fitness for BDI data.

For prediction error distribution, the fuzzy heuristic model has the largest part of MAPE lower than 10%, ranked in high accuracy level. The fuzzy model has 39, accounting for 76%, and second GM-Markov model has 22, accounting for 46%. For the MAPE higher than 50% of irrational level, the GM-Markov model is the best, no irrational MAPE, and the fuzzy heuristic model has 1 irrational MAPE, accounting for 2%. For the MAPE ranged from 10%~50%, the fuzzy heuristic model has 11, accounting for 22%, GM(1,1) has 23, accounting for 48%.

For the overall prediction accuracy, the fuzzy heuristic model has the best accuracy, which high accuracy level accounted for the largest percentage (76%), and MAPE is the smallest 8.15%. The MAPE of GM (1,1) is the worst, 24.34%.

The 5 models may easily cause prediction error when time data of BDI has greater fluctuation. GM (1,1) model has the greatest prediction deviation. The medium/short term BDI predicted results are not compared.

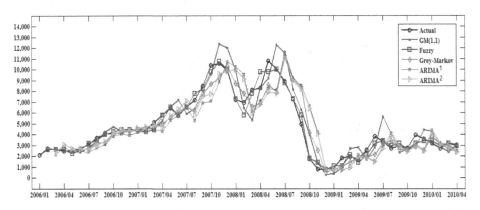

Fig. 1. The pattern of predicted BDI of fuzzy model, GM (1,1), Grey-Markov, ARIMA[1] and ARIMA[2]

Based on the prediction results, the MAPE of the former 3 models is higher with increase of the data, and the data number has the greatest impact on GM (1,1) prediction error. The fuzzy heuristic model has the lowest MAPE, irrespective of short-term or long term data, and the second is grey Markov model, and the worse is GM(1,1). For the long term or more data, ARIMA also has the high prediction accuracy. Additionally, three different periods have different BDI volatility. The fuzzy heuristic model has the smallest MAPE variation, less than 5%, followed by grey Markov and GM(1,1). The results indicate that the fuzzy time series has higher prediction stability.

The GM (1,1) model prediction performance decreases with the increase of BDI data, and its MAPE of the short/long term data prediction is 10.89% and 24.34%. The result may be related to the model's monotone increasing or monotone decreasing. GM prediction can be applied to fewer data, but if the data has greater fluctuation during the period, GM (1,1) cannot well fit the data, resulting in greater prediction error. ARIMA is superior to the GM (1,1) in long-term data fitting. If the time series data has greater fluctuation, the ARIMA considering structural change is better than ARIMA model. Thus, during the three different data periods, through the comparison of the prediction model accuracy, the prediction capability of fuzzy heuristic model is better than other prediction models.

Figure 1 illustrates BDI predicted value trend of fuzzy heuristic model, GM(1,1), GM-Markov, ARIMA and ARIMA considering structural change. During the period of Jan, 2006 and Apr, 2010, Figure 1 shows that the fuzzy time series model which fitted the data trend had rather small error in the turning place between BDI two high points (2007.10, 2008.06) and two lowest points (2008.12). The ARIMA and the ARIMA considering structural change can capture the data variation trend when the index sharply fell to low point and continuously rose from Jan, 2009 to Apr, 2010. The GM(1,1)had overestimation and after modification, GM-Markov can effectively increase prediction performance for time series data.

5   Conclusions

BDI is an important response of maritime information for the trading and settlement of physical and derivative contracts. In the paper, we propose fuzzy set theory and grey system for modeling the prediction of BDI, and employ ARIMA for the calibration of data structure to depict the trend. The data are sourced between Jan, and Apr, 2010 with monthly 52 data. MAPE is used to measure the forecasting accuracy of the models. The empirical results show that the fuzzy heuristic model has the lowest MAPE, followed by GM-Markov, ARIMA considering structural change and ARIMA model. The GM (1, 1) model is the worst.

To know economic implications of the BDI trend, this study uses ARIMA to fit the research data during the period of Jan, 2006 to Apr, 2010. The results indicate that current BDI variation is high positively related to the previous variation, and the external interference factor is negatively related to the current variation. There were two high points during the research period. This study further uses ARIMA considering structural change to fit the data. The result shows that the significant time transition occurred in Dec,2007. The explanatory ability of ARIMA model is higher than the model not considering structural change. For model usability, fuzzy heuristic model is an effective tool for business management in decision-making, especially in urgent time.

References

1. Kavussanos, M.G.: Price risk modeling of different size vessels in the tanker industry using autoregressive conditional heterskedastic models. Logistics and Transportation Review 32, 161–176 (1996)
2. Zadeh, L.A.: Fuzzy sets. Information and control 8, 338–353 (1965)
3. Song, Q., Chissom, B.S.: Forecasting enrollments with fuzzy time series-Part I. Fuzzy Sets and Systems 54, 1–9 (1993a)
4. Song, Q., Chissom, B.S.: Fuzzy time series and its models. Fuzzy Sets and Systems 54, 269–277 (1993b)
5. Chen, S.: Forecasting enrollments based on fuzzy time series. Fuzzy Sets and Systems 18, 311–319 (1996)
6. Song, Q., Chissom, B.S.: Forecasting enrollments with fuzzy time series - Part II. Fuzzy Sets and Systems 62, 1–8 (1994)
7. Huarng, K.: Heuristic models of fuzzy time series for forecasting. Fuzzy Sets and Systems 123, 369–386 (2001a)
8. Huarng, K.: Effective lengths of intervals to improve forecasting in fuzzy time series. Fuzzy Sets and Systems 123, 387–394 (2001b)
9. Hwang, J., Chen, S., Lee, C.: Handling forecasting problems using fuzzy time series. Fuzzy Sets and Systems 100, 217–228 (1998)
10. Cheng, C.H., Chang, J.R., Yeh, C.A.: Entropy-based and trapezoid fuzzification-based fuzzy time series approaches for forecasting IT project cost. Technological Forecasting and Social Change 73, 524–542 (2006)
11. Yu, H.K.: Weighted fuzzy time series model for TAIEX forecasting. Physica A 349, 609–624 (2005b)
12. Huarng, K., Yu, H.K.: A type 2 fuzzy time series model for stock index forecasting. Physica A 353, 445–462 (2005)
13. Huarng, K., Yu, H.K.: The application of neural network to forecasting fuzzy time series. Physica A 363, 481–491 (2006a)
14. Huarng, K., Yu, H.K.: Ratio-based lengths of intervals to improve fuzzy time series forecasting. IEEE Transactions on Systems, Man and Cybernetics Part B 36, 328–340 (2006b)
15. Li, S.T., Cheng, Y.C.: Deterministic fuzzy time series model for forecasting enrollments. An International Journal Computer and Mathematics with Applications 53, 1904–1920 (2007)
16. Cheng, C.H., Chen, T.L., Teoh, H.J., Chiang, C.H.: Fuzzy time-series based on adaptive expectation model for TAIEX forecasting. Expert Systems with Applications 34, 1126–1132 (2008)
17. Wang, C.H., Hsu, L.C.: Constructing and applying an improved fuzzy time series model: Taking the tourism industry for example. Expert Systems with Applications 34(4), 2732–2738 (2008)
18. Yu, H.K.: A refined fuzzy time series model for forecasting. Physica A 346, 657–681 (2005a)
19. Chen, S., Hwang, J.: Temperature prediction using fuzzy time series. IEEE Transactions on System, Man and Cybernetics,-Part B: Cybernetics 30(2), 263–375 (2000)

20. Cheng, C.H., Cheng, G.W., Wang, J.W.: Multi-attribute fuzzy time series method based on fuzzy clustering. Expert Systems with Applications 34, 1235–1242
21. Wong, H.L., Tu, Y.H., Wang, C.C.: Application of fuzzy time series models for forecasting the amount of Taiwan export. Expert Systems with Applications 37, 1465–1470 (2010)
22. Wong, H.L., Wang, C.C., Tu, Y.H.: Optimal selection of multivariate fuzzy time series models to non-stationary series data forecasting. International Journal of Innovative Computing, Information and Control 12(6), 5321–5332 (2010)
23. Wong, H.L., Shiu, J.M.: Comparisons of fuzzy time series and hybrid Grey model for non-stationary data forecasting. Applied Mathematics & Information Sciences 6(2), 409S–416S(2012)
24. Lundgren, N.G.: Bulk trade and maritime transport costs. Resources Policy 22, 5–32 (1996)
25. Cullinane, K.P.: A short-term adaptive forecasting model for BIFFEX speculation. Maritime Policy and Management 19(2), 91–114 (1992)
26. Chang, Y.T., Chang, H.B.: Predictability of the dry bulk shipping market by BIFFEX. Maritime Policy and Management 23(2), 103–114 (1996)
27. Veenstra, A.W., Franses, P.H.: A Co-integration Approach to Forecasting Freight Rates in the Dry Bulk Shipping Sector. Transportation Research A 31(6), 447–458 (1997)
28. Cullinane, K.P.: A portfolio analysis of market investments in dry bulk shipping. Transportation Research B 29(3), 181–200 (1995)
29. Akatsuka, K., Leggate, H.K.: Perceptions of foreign exchange rate risk in the shipping industry. Maritime Policy and Management 28(3), 235–24930 (2001)
30. Alizadeh-M, A.H., Nomikos, K.N.: The price-volume relationship in the sale and purchase market for dry bulk vessels. Maritime Policy and Management 30(4), 321–337 (2003)
31. Kavussanos, M.G., Nomikos, N.: Price discovery, causality and forecasting in the freight futures market. Review of Derivatives Research 28, 203–230 (2003)
32. Alizadeh-M, A.H., Nomikos, K.N.: The Price-Volume Relationship in the Sale and Purchase Market for Dry Bulk Vessels. Maritime Policy and Management 30(4), 321–337 (2003)
33. Chen, Y.S., Wang, S.T.: The empirical evidence of the leverage effect on volatility in international bulk shipping market. Maritime Policy and Management 31(2), 109–124 (2004)
34. Batchelor, R., Alizadeh, A., Visvikis, I.: Forecasting spot and forward prices in the international freight market. International Journal of Forecasting 14, 101–114 (2007)
35. Deng, J.: Control Problems of Grey System. System and Control Letters 5(3), 288–294 (1982)
36. Markov, A.A.: Extension of the limit theorems of probability theory to a sum of variables connected in a chain. In: Howard (ed.) Dynamic Probabilistic Systems, vol. 1, John Wiley and Sons, Markov Chains (1971)
37. Lewis, C.D.: Industrial and Business Forecasting Methods. Butterworths, London (1982)

Deriving Processes of Information Mining Based on Semantic Nets and Frames

Sebastian Martins, Darío Rodríguez, and Ramón García-Martínez

Information Systems Research Group. National University of Lanus. Argentina
PhD Program on Computer Science. National University of La Plata. Argentina
rgarcia@unla.edu.ar

Abstract. There are information mining methodologies that emphasize the importance of planning for requirements elicitation along the entire project in an orderly, documented, consistent and traceable manner. However, given the characteristics of this type of project, the approach proposed by the classical requirements engineering is not applicable to the process of identifying the problem of information mining, nor allows to infer from the business domain modelling, the information mining process which solves it. This paper proposes an extension of semantic nets and frames to represent knowledge of the business domain, business problem and problem of information mining; and a methodology to derive the information mining process from the proposed knowledge representations is introduced.

Keywords: information mining, requirement engineering, deriving processes of information mining, semantic nets, frames.

1   Introduction

In [1] we have proposed five processes for information mining related to the following business intelligence problems: discovery of behavior rules, discovery of groups, discovery of significant attributes, discovery of group-membership rules and weighting of behavior or group-membership rules, and the identification of information-systems technologies that can be used for the characterized processes.

The proposed processes are based on intelligent systems-based methods [2-3] such as: TDIDT algorithms (Top Down Induction Decision Trees), self-organizing maps (SOM) and Bayesian networks. TDIDT algorithms allow the development of symbolic descriptions of the data to distinguish between different classes [4]. Self-organizing maps can be applied in the construction of information cluster [5]. Bayesian networks can be applied to identify discriminative attributes in large information bases and detect behavior patterns in the analysis of temporal series [6]. Our research work lays on the hypothesis that is emerging a new discipline [7] called Information Mining Engineering (IME).

One of the early stages of IME is to understand which are the requirements that business intelligence poses to an Information Mining Project [8]. The requirement elicitation process is addressed by most commonly used IME methodologies [9-11].

A. Moonis et al. (Eds.): IEA/AIE 2014, Part II, LNAI 8482, pp. 150–159, 2014.

IME methodologies mention the necessity of business understanding as starting point for any IME project development. In general, business understanding phase may be decomposed in the following sub phases: determine business objectives (in this paper the business problem), assess situation, determine IME project goals (in this paper the problem of information mining) and produce project plan.

In this paper we address two problems: [a] how to represent: the business domain, the business problem and the problem of information mining, and [b] how to derive the processes of information mining from the selected representation.

Knowledge Engineering provides formalisms that allow capturing the business domain, the business problem and the problem of information mining. We have explored the use of frames [12] and semantic networks [13] for the capturing process.

In this context, this paper introduces knowledge representation formalisms based on frames and semantic networks to represent knowledge in an IME Project (section 2), proposes a methodology to derive the processes of information mining (section 3), presents a case study exemplifying how to derive the process of information mining from the proposed knowledge representation formalisms (section 4), and some conclusions are drawn (section 5).

2   Knowledge Representation Formalisms for Business Domain, Business Problem and Problem of Information Mining

In this section we present: proposed notation for modelling domain problem (section 2.1) and problem of information mining (section 2.2) with semantic networks; and the frames used to capture concept definitions and relations among them (section 2.3).

2.1   Proposed Notation for Modelling Domain Problem with Semantic Networks

Representation rules that apply to the modelling of the business domain, the business problem and the problem of information mining are: [a] the concepts are represented by ovals with solid lines (Fig. 1.a), [b] the attributes and values are represented by ovals with dashed lines (Fig. 1.b), [c] the identifier of an instance of the concept is defined within the oval concatenating the type of concept with a dash "-" between them (Fig. 1.c), [d] the l value or reference of the label of the arcs depend on the relationship that this represents: [d.1] if it joins an attribute with its value, it is labelled with the word "value" (Fig. 1.d), [d.2] if it denotes a relationship instance of a concept it is labelled with the word "instance" (Fig. 1.e), [d.3] if it liaises between a concept and its attribute, it is labelled with one or more words that represent the connection between the two (Fig. 1.f), [d.4] if it specifies an inheritance relationship of a concept, it is labelled with the word "subclass" (Fig. 1.g). Figure 1.d indicates a "Brand equipment" attribute, whose value is "Nokia". Figure 1.d shows that the concept of " client -1" (with 1 being the client identifier as explained in Figure 1.c) belongs to the general concept client therefore get all the features that this represents. Figure 1.f indicates a membership relation between the concept plan and the attribute length indicating that any plan has an estimated duration. Figure 1.g represents

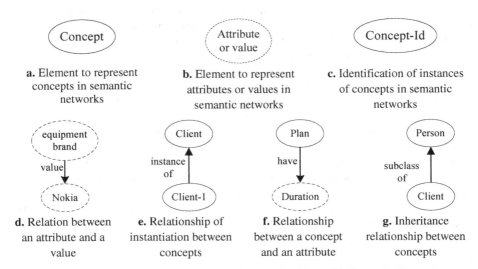

Fig. 1. Proposed Notation for Modelling Domain Problem with Semantic Networks

an inheritance relationship , in which the class inherits all the characteristics that define the parent concept, in this case the Customer concept inherits all the features of person, because a customer is a person, according to this relationship.

It is essential to emphasize that in cases presented by figures 1.d, 1.e and 1.g the possible value of the label is defined, no other value accepted for these relations, while in case presented by figure 1.f, an example is given value, because depending on the content of the concepts, attributes that relate the value of the label of the arc will vary.

2.2   Proposed Notation for Modelling Problem of Information Mining with Semantic Networks

In the Semantic Network of Problem of Information Mining, it must be identified the flow of the described problem, through the identification of elements of input and output. To provide this information additional elements are incorporated /figure 2).

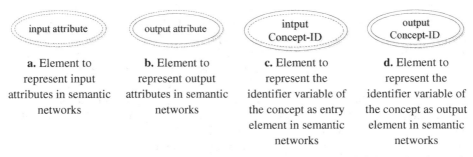

Fig. 2. Proposed Notation for Modelling Problem of Information Mining with Semantic Networks

These additional notations are: [a] the input attributes are represented with double oval with dashed lines (figure 2.a), [b] the output attributes are represented with double oval, taking the inner oval with solid line and the outer oval with intermittent line (figure 2.b), [c] the concepts instance identifiers, which are input element are represented by double oval, having its outer oval with solid line and its internal oval with dotted line (figure 2.c), and [d] the identifiers of instances of concepts, which are output element are represented by double oval whose lines are continuous (figure 2.d).

2.3   Frames Used to Capture Definitions of Concepts and Relations

The proposed frames are used to capture different aspects (concepts and relations) of business domain and information mining domain. They are:

Frame Term-Category-Definition of Business Domain: which aims to identify and classify all relevant elements within the business domain.

Frame Concept-Attribute-Relation-Domain Value of Business Domain: which aims is defining the structure of the elements of the business.

Frame Concept- Domain_Relation of Business Domain: which aims to identify the relationships between the concepts that define the business model.

Frame Term-Category-Definition of Problem of Information Mining: which aims to identify and classify the relevant elements of the problem of information mining

Extended Frame Concept-Attribute-Relation-Domain_Value of Problem of Information Mining: Which aims is to define the structures of the elements to the problem of information mining, identifying the input and output elements thereof.

Frame Concept-Problem_Relation of Problem of Information Mining: which aims to identify the relationships between the concepts that define the problem of information mining.

3   Proposed Methodology to Derive the Processes of Information Mining

We have developed a methodology (shown in figure 3) to derive the processes of information mining from frames and semantic nets. The methodology has three phases: "Analysis of Business Domain", "Analysis of the Problem of Information Mining", and "Analysis of the Process of Information Mining".

The phase "Analysis of Business Domain" develops three tasks: "Identification of the Elements and Structure of the Business Domain", "Identification of Relationships Between Concepts of Business Domain", and "Conceptualization of the Business Domain". The task "Identification of the Elements and Structure of the Business Domain" has as input the "Business Domain Description" and the "Business Domain Data"; and produces as output the "Frame Term-Category-Definition of Business Domain" and the "Frame Concept-Attribute-Relation-Domain_Value of Business Domain". The task "Identification of Relationships Between Concepts of Business Domain" has as input the "Frame Term-Category-Definition of Business Domain" and the "Frame Concept-Attribute-Relation-Domain_Value of

Business Domain"; and produces as output the "Frame Concept-Domain_Relation of Business Domain". The task "Conceptualization of the Business Domain" has as input "Frame Concept-Domain_Relation of Business Domain" and the "Frame Concept-Attribute-Relation-Domain_Value of Business Domain"; and produces as output the "Semantic Net of the Business Domain Model".

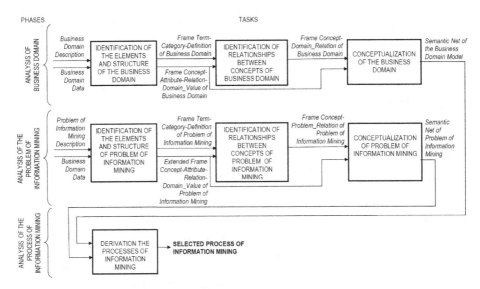

Fig. 3. Methodology to derive the processes of information mining

The phase "Analysis of the Problem of Information Mining" develops three tasks: "Identification of the Elements and Structure of Problem of Information Mining", "Identification of Relationships Between Concepts of Problem of Information Mining" and "Conceptualization of Problem of Information Mining". The task "Identification of the Elements and Structure of Problem of Information Mining" has as input the "Problem of Information Mining Description" and the "Business Domain Data"; and produces as output the "Frame Term-Category-Definition of Problem of Information Mining" and the "Extended Frame Concept-Attribute-Relation-Domain_Value of Problem of Information Mining". The task "Identification of Relationships Between Concepts of Problem of Information Mining" has as input the "Frame Term-Category-Definition of Problem of Information Mining" and the "Extended Frame Concept-Attribute-Relation-Domain_Value of Problem of Information Mining"; and produces as output the "Frame Concept- Problem_Relation of Problem of Information Mining". The task "Conceptualization of Problem of Information Mining" has as input the "Extended Frame Concept-Attribute-Relation-Domain_Value of Problem of Information Mining" and the "Frame Concept- Problem_Relation of Problem of Information Mining"; and produces as output the "Semantic Net of Problem of Information Mining".

The phase "Analysis of the Process of Information Mining" develops one task: "Derivation the Processes of Information Mining" which has as input the "Semantic

Net of the Business Domain Model" and the "Semantic Net of Problem of Information Mining"; and produces the "Selected Process of Information Mining".

4   Case Study

In this section a proof of concept on the derivation procedure of process of information mining from the business domain modelling is presented. It takes the following case [14]:

"...Description of the Business and Business Problem:
"Mobile Services Argentina SA" is a company belonging to the telecommunications sector that operates nationwide. It offers products and services in various trade and generic brands, offers a wide variety of rate plans (personal or corporate use) according to the needs of customers. It is currently promoting customized retention campaigns of clients throughout the country in order to offer new products and services according to the characteristics of each customer, to preempt competitors who are also growing nationwide. The business objective is to characterize customers in different geographical regions of coverage of the company, in order to facilitate the definition of marketing campaigns aimed at maintaining existing customers and obtaining new customers under the common preferences of each region, which will also help improve the profitability of the company. With this, it also seeks to determine the behavior of its customers and improve the knowledge we have of them. "Mobile Services Argentina SA" has the goal of customer satisfaction, which makes by offering products and quality services, dividing all mobile operations in the country in five regions: Coast, Cuyo, Pampas/Center, Patagonia and North.
Problem of Information Mining: Determine the rules, based on the variables status, type of plan and its duration, equipment brand and customer type, identifying the behavior of a set of customers according to their geographical location (region, state, city).
Data description:
The following are the relevant data that stores the company during the course of their activities:

- *Client Code: the company uses to identify him. Example: 99861598.*
- *Track Client: example: AAA (active client), BA (thin client).*
- *Date Added: belongs to the client. Example: 10/03/2003.*
- *Code of the city: to the customer's residence. Example: 379.*
- *Code of the province to which the city belongs. Example: 3.*
- *Contracted Plan: rate plan code that engages the customer. Example: U21.*
- *Duration: number of months of customer engagement. Example: 12.*
- *Customer type: example: PR / PO (prepaid customer), FF (client billing).*
- *Equipment Brand: belonging to the client. Example: NOKIA.*
- *Geographic Region: indicates the region to which each province belongs (Table 1)..."*

Table 1. Code of regions by provinces

REGION	DESCRIPTION	PROVINCES MEMBERS
1	Coast	Misiones, Corrientes, Formosa, Chaco, Santa Fe
2	Cuyo	Mendoza, San Luis, San Juan
3	Pampas Center	Buenos Aires, Federal Districl, Córdoba, La Pampa
4	Patagonia	Neuquén, Río Negro, Chubut, Santa Cruz, Tierra del Fuego
5	North	Jujuy, Salta, Tucumán, Catamarca, La Rioja, Santiago del Estero

When task "Identification of the Elements and Structure of the Business Domain" is applied, we obtain "Frame Term-Category-Definition of Business Domain" (Table 2) and the "Frame Concept-Attribute-Relation-Domain_Value of Business Domain" (Table 3). When task "Identification of Relationships Between Concepts of Business Domain" is applied, we obtain "Frame Concept-Domain_Relation of Business Domain" (Table 4).

Table 2. Frame Term-Category-Definition of Business Domain

Term	Category	Definition
belongs	Relationship	A client belongs to a region
belongs	Relationship	A client belongs to a province
belongs	Relationship	A client belongs to a city
city	Attribute	city where the client live
Client	Concept	A person who hire the service
client code	Attribute	unique client identification code
customer type	Attribute	Kind of customer by contract
date Added	Attribute	Date when the client belongs to the company
duration	Attribute	Number of months of hiring
equipment brand	Attribute	Client's equipment brand
has	Relationship	A client has a state with the company
has	Relationship	A client has a certain brand of telephone equipment
has	Relationship	A plan has a certain duration
hire	Relationship	A client hire a plan
identifies	Relationship	The client code identifies a specific client
identifies	Relationship	The plan code identifies a specific plan
is	Relationship	A customer is a certain type of customer
Plan	Concept	Kind of contract
plan code	Attribute	unique plan identification code
province	Attribute	Province where a client live
region	Attribute	Region where a client live
state	Attribute	Client's state
was registered	Relationship	A client was registered at a date

Table 3. Frame Concept-Attribute-Relation-Domain_Value of Business Domain

Concept	Attribute	Relationship	Value
Client	Code	identifies	Numeric
	state	has	alphabetical EG: AAA, BA.
	date added	was registered	dd/mm/AAAA
	province	belongs	Numeric EG: 379
	city	belongs	Numeric EG: 3
	region	belongs	1 to 5
	type	is	alphabetical EG: PR/PO, FF.
	equipment brand	has	alphabetical EG: Nokia
Plan	Code	identifies	Alphanumeric EG: U21.
	duration	has	Numeric

Table 4. Frame Term-Category-Definition of Business Domain

Concept	Concept associated	Relationship	Description
Client	Plan	hire	A client hire a plan

When task "Conceptualization of the Business Domain" is applied, we obtain the "Semantic Net of the Business Domain Model" (Figure 4). When task "Identification of the Elements and Structure of Problem of Information Mining" is applied, we obtain "Frame Term-Category-Definition of Problem of Information Mining" (Table 5) and "Extended Frame Concept-Attribute-Relation-Domain_Value of Problem of Information Mining" (Table 6). When task "Identification of Relationships Between Concepts of Problem of Information Mining" is applied, we obtain "Frame Concept- Problem_Relation of Problem of Information Mining" (Table 7). When task "Conceptualization of Problem of Information Mining" is applied, we obtain the "Semantic Net of Problem of Information Mining" (Figure 5).

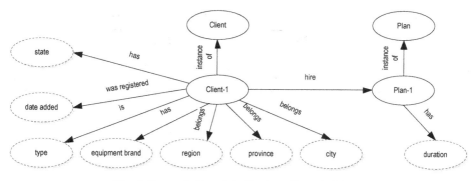

Fig. 4. Semantic Net of the Business Domain Model

Table 5. Frame Term-Category-Definition of Problem of Information Mining

Term	Category	Definition
by	Relationship	A client belongs to a group by his city
by	Relationship	A client belongs to a group by his province
by	Relationship	A client belongs to a group by his region
city	Attribute	city where the client live
Client	Concept	A person who hire the service
composed	Relationship	A group is composed by clients
Customer type	Attribute	Kind of customer by contract
define	Relationship	A group is defined by a rule
duration	Attribute	Number of months of hiring
equipment brand	Attribute	Client's equipment brand
Group	Concept	group of clients
group code	Attribute	unique group identification code
has	Relationship	A plan has a certain duration
identifies	Relationship	The plan code identifies a specific plan
identifies	Relationship	The group code identifies a specific group
identifies	Relationship	The rule code identifies a specific rule
Plan	Concept	Kind of contract
plan code	Attribute	unique plan identification code
province	Attribute	Province where a client live
region	Attribute	Region where a client live
rule	Concept	Variables that define the assignment of a client to a particular group
rule code	Attribute	unique rule identification code
state	Attribute	Client's state
subset	Relationship	The situation may be a variable that defines the mapping of the customer in a group
subset	Relationship	The plan may be a variable that defines the mapping of the customer in a group
subset	Relationship	The equipment brand may be a variable that defines the mapping of the customer in a group
subset	Relationship	The customer type brand may be a variable that defines the mapping of the customer in a group

Table 6. Extended Frame Concept-Attribute-Relation-Domain_Value of Problem of Information Mining

Concept	Attribute	Relation-ship	Input/ Output	Value
Rule	code	identifies	Output	Numeric
	state	subset		alphabetical EG: AAA, BA.
	Equip-ment brand	subset		alphabetical EG: Nokia
	type	subset		alphabetical EG: PR/PO, FF.
Plan	plan code	identifies		Alpha-numeric EG: U21.
	duration	has		Numeric
Group	group code	identifies	Output	Numeric
Client	province	by	Input	Numeric EG: 379
	city	by	Input	Numeric EG: 3
	region	by	Input	1 to 5

Table 7. Frame Concept- Problem_Relation of Problem of Information Mining

Concept	Concept associated	Relationship	Description
Plan	Rule	subset	The plan may be a variable that defines the mapping of the customer in a group
Rule	Group	define	A group is defined by a rule
Group	Client	composed	A group is composed by clients

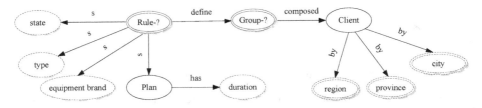

Fig. 5. Semantic Net of the Business Domain Model

When task "Derivation the Processes of Information Mining" is applied, we obtain the "Processes of Information Mining" which for the case study is "Discovery of group-membership rules" as has been proposed in [1].

5   Conclusions

This research started from the premise that the tools involved in software engineering processes do not apply to information mining projects. The authors belong to a research group that during ten years have been developing a body of knowledge for Information Mining Engineering, trying to help improve productivity of Latin-American SMEs (small and medium enterprises) in the software industry.

In this context, this paper introduced formalisms based on knowledge representation techniques (frames and semantic networks) to model the domain of business and information mining process domain. In line with the tradition of Knowledge Engineering, the process of transformation that involves the formalization allows to find explicit and implicit concepts (and relations among them) in the description of the domain and the business problem.

Based on the explanation of the concepts and relationships, a procedure to derive the exploitation process information from the proposed formalism is presented. A proof of concept that illustrates the application of the proposed procedure was developed. As future research, it is planned to develop the validation of the proposed methodology in selected information mining.

Acknowledgments. The research reported in this paper was funded by Grant-BENTR13 of the Committee on Scientific Research of the Province of Buenos Aires (CIC, Argentina), Grant-UNLa-SCyT-33A105 and partially by Project UNLa-33A167 of the Secretary of Science and Technology of National University of Lanus (Argentina).

References

1. García-Martínez, R., Britos, P., Rodríguez, D.: Information Mining Processes Based on Intelligent Systems. In: Ali, M., Bosse, T., Hindriks, K.V., Hoogendoorn, M., Jonker, C.M., Treur, J. (eds.) IEA/AIE 2013. LNCS (LNAI), vol. 7906, pp. 402–410. Springer, Heidelberg (2013)
2. Kononenko, I.Y., Cestnik, B.: Lymphography Data Set. UCI Machine Learning Repository (1986), http://archive.ics.uci.edu/ml/datasets/Lymphography (Último acceso 29 de Abril del 2008)
3. Michalski, R.: A Theory and Methodology of Inductive Learning. Artificial Intelligence 20, 111–161 (1983)
4. Quinlan, J.: Learning Logic Definitions from Relations. Machine Learning 5, 239–266 (1990)
5. Kohonen, T.: Self-Organizing Maps. Springer (1995)
6. Heckerman, D., Chickering, M., Geiger, D.: Learning bayesian networks, the combination of knowledge and statistical data. Machine Learning 20, 197–243 (1995)
7. García-Martínez, R., Britos, P., Pesado, P., Bertone, R., Pollo-Cattaneo, F., Rodríguez, D., Pytel, P., Vanrell, J.: Towards an Information Mining Engineering. In: Software Engineering, Methods, Modeling and Teaching. Edited by University of Medellín, pp. 83–99 (2011) ISBN 978-958-8692-32-6
8. Britos, P., Dieste, O., García-Martínez, R.: Requirements Elicitation in Data Mining for Business Intelligence Projects. In: Avison, D., Kasper, G.M., Pernici, B., Ramos, I., Roode, D. (eds.) Advances in Information Systems Research, Education and Practice. IFIP International Federation for Information, vol. 274, pp. 139–150. Springer, Boston (2008)
9. Chapman, P., Clinton, J., Keber, R., Khabaza, T., Reinartz, T., Shearer, C., Wirth, R.: CRISP-DM 1.0 Step by step BIguide. Edited by SPSS (2000)
10. Pyle, D.: Business Modeling and Business intelligence. Morgan Kaufmann (2003) ISBN 1-55860-653-X
11. SAS, SAS Enterprise Miner: SEMMA (2008), http://www.sas.com/technologies/analytics/datamining/miner/semma.html (last access November 24, 2013)
12. Minsky, M.: A framework for Representing Knowledge. AI-Memo 306. MIT Artificial Intelligence Laboratory (1974), http://hdl.handle.net/1721.1/6089
13. Sowa, J.: Semantic Networks. In: Shapiro, S. (ed.) Encyclopedia of Artificial intelligence, 2nd edn., Wiley & Sons (1992) ISBN 978-0471503071
14. Basso, D.Y., Vegega, C.: Case Study: Cell Phones. Business Intelligence Seminar Final Research Paper. Master Program on Information Systems Engineering. Graduate School, Regional School Buenos Aires, National Technological University, Argentine (2013)

Efficient Search of Cosine and Tanimoto Near Duplicates among Vectors with Domains Consisting of Zero, a Positive Number and a Negative Number

Marzena Kryszkiewicz and Przemyslaw Podsiadly

Institute of Computer Science, Warsaw University of Technology
Nowowiejska 15/19, 00-665 Warsaw, Poland
{mkr,P.Podsiadly}@ii.pw.edu.pl

Abstract. The cosine and Tanimoto similarity measures are widely applied in information retrieval, text and Web mining, data cleaning, chemistry and bio-informatics for searching similar objects. This paper is focused on methods making such a search efficient in the case of objects represented by vectors with domains consisting of zero, a positive number and a negative number; that is, being a generalization of weighted binary vectors. We recall the methods offered recently that use bounds on vectors' lengths and non-zero dimensions, and offer new more accurate length bounds as a means to enhance the search of similar objects considerably. We compare experimentally the efficiency of the previous methods with the efficiency of our new method. The experimental results prove that the new method is an absolute winner and is very efficient in the case of sparse data sets with even more than a hundred of thousands dimensions.

Keywords: the cosine similarity, the Tanimoto similarity, nearest neighbors, near duplicates, non-zero dimensions, high dimensional data, sparse data sets.

1 Introduction

The cosine and Tanimoto similarity measures are widely applied in many disciplines; among other, in information retrieval, text and Web mining, data cleaning, chemistry, biology and bio-informatics for finding similar vectors or near duplicate vectors representing objects of interests [4, 10, 11]. In particular, documents are often represented as term frequency vectors or its variants such as *tf_idf* vectors [9]. A number of Web mining methods was offered for approximate lossy search of similar vectors [3, 5]. Recently, however, competing approaches enabling their lossless search have also been offered [1, 2, 7, 8, 12]. In this paper, we focus on lossless search of sufficiently similar vectors. Such search is very important e.g. in data cleaning [1] as well as for plagiarism discovery, to mention a few.

In many applications, binary vectors or weighted binary vectors are of high importance. The recent lossy methods are often based on bounds on lengths of vectors (or their number of non-zero dimensions) or on the knowledge about specific subsets of non-zero dimensions of vectors. In [8], however, we examined possibilities of using bounds on lengths of vectors and subsets of non-zero dimensions in determining near

A. Moonis et al. (Eds.): IEA/AIE 2014, Part II, LNAI 8482, pp. 160–170, 2014.

duplicates of *ZPN*-vectors that are more general than weighted binary vectors. There, we defined *ZPN*-vectors as vectors each domain of which may contain at most three values: zero, a positive value and a negative value. Clearly, a weighted binary vector is a special case of a *ZPN*-vector each domain of which may contain at most two values one of which is zero.

A typical instance of a *ZPN*-vector is a vector storing results from test queries each of which may be graded in three ways: with a positive value (if an answer was positive), with a negative value (if an answer was negative) and with 0 (if no answer was provided). Such grading might be applied to discourage from guessing answers. Another possible application of *ZPN*-vectors would be more accurate evaluation of the quality of papers based on their citations. If a paper is cited as valuable, it could be graded with a positive value; if it is cited as invaluable, it could be graded with a negative value; if it is not cited, it could be graded with 0.

In this paper, we first recall the methods offered recently that use bounds on vectors' lengths and non-zero dimensions, and then offer new more accurate length bounds as a means to enhance the search of similar objects considerably. Our new method of reducing the number of candidates is presented by means of an example. We also compare experimentally the efficiency of the previous methods with the efficiency of our new method.

Our paper has the following layout. Section 2 provides basic notions used in this paper. In particular, the definitions of the cosine and Tanimoto similarities are recalled here. In Section 3, we recall bounds on lengths of cosine and Tanimoto near duplicate *ZPN*-vectors (after [8]), while in Section 4, we recall theoretical results related to using non-zero dimensions for reducing candidates for near duplicate *ZPN*-vectors (after [8]). In Section 5, we derive new more accurate bounds on lengths of *ZPN*-vectors for the cosine and Tanimoto similarity measures and show how these bounds can be used to speed up the search of near duplicates by means of an example. In Section 6, we compare the efficiency of the earlier and newly offered methods for searching near duplicates. Section 7 summarizes our contribution.

2   Basic Notions

In the paper, we consider n-dimensional vectors. A vector u will be also denoted as $[u_1, ..., u_n]$, where u_i is the value of the i-th dimension of u, $i = 1..n$.

By $NZD(u)$ we denote the set of those dimensions of vector u which have values different from 0; that is,

$$NZD(u) = \{i \in \{1, ..., n\} | u_i \neq 0\}.$$

By $ZD(u)$ we denote the set of those dimensions of vector u that have zero values; i.e.,

$$ZD(u) = \{i \in \{1, ..., n\} | u_i = 0\}.$$

A *dot product* of vectors u and v is denoted by $u \cdot v$ and is defined as $\sum_{i=1..n} u_i v_i$. One may easily observe that the dot product of vectors u and v that have no common non-zero dimension equals 0.

A *length of vector* u is denoted by $| u |$ and is defined as $\sqrt{u \cdot u}$.

In Table 1, we present an example set of *ZPN*-vectors. In Table 2, we provide its alternative sparse representation, where each vector is represented only by its non-zero dimensions and their values. More precisely, each vector is represented by a list of pairs, where the first element of a pair is a non-zero dimension of the vector and the second element – its value. For future use, in Table 2, we also place the information about lengths of the vectors.

Table 1. Dense representation of an example set of *ZPN*-vectors

Id	1	2	3	4	5	6	7	8	9
v1	-3.0	4.0			3.0	5.0	3.0	6.0	
v2	3.0	-2.0						6.0	
v3			6.0	4.0					
v4		-2.0		4.0					
v5				4.0	-3.0		3.0		
v6			-9.0	4.0		5.0			5.0
v7	3.0		6.0	4.0	-3.0		-4.0	-5.0	-2.0
v8		4.0		4.0		5.0			5.0
v9				-2.0	3.0		3.0		5.0
v10	-3.0	-2.0	-9.0						

Table 2. Sparse representation of the example set of *ZPN*-vectors from Table 1 (extended by the information about vectors' lengths)

Id	(non-zero dimension, value) pairs	length
v1	{(1,-3.0), (2, 4.0), (5, 3.0), (6, 5.0), (7, 3.0), (8, 6.0)}	10.20
v2	{(1, 3.0), (2,-2.0), (8, 6.0)}	7.00
v3	{(3, 6.0), (4, 4.0)}	7.21
v4	{(2,-2.0), (4, 4.0)}	4.47
v5	{(4, 4.0), (5,-3.0), (7, 3.0)}	5.83
v6	{(3,-9.0), (4, 4.0), (6, 5.0), (9, 5.0)}	12.12
v7	{(1, 3.0), (3, 6.0), (4, 4.0), (5,-3.0), (7,-4.0), (8,-5.0), (9,-2.0)}	10.72
v8	{(2, 4.0), (4, 4.0), (6, 5.0), (9, 5.0)}	9.06
v9	{(4,-2.0), (5, 3.0), (7, 3.0), (9, 5.0)}	6.86
v10	{(1,-3.0), (2,-2.0), (3,-9.0)}	9.70

The *cosine similarity between* vectors u and v is denoted by $cosSim(u, v)$ and is defined as the cosine of the angle between them; that is,

$$cosSim(u, v) = \frac{u \cdot v}{|u\| v|}.$$

The *Tanimoto similarity* between vectors u and v is denoted by $T(u, v)$ and is defined as follows,

$$T(u, v) = \frac{u \cdot v}{u \cdot u + v \cdot v - u \cdot v}.$$

Clearly, the Tanimoto similarity between two non-zero vectors u and v can be expressed equivalently in terms of their dot product and their lengths as follows:

$$T(u, v) = \frac{u \cdot v}{|u|^2 + |v|^2 - u \cdot v}.$$

For any non-zero vectors u and v, $T(u, v) \in \left[-\frac{1}{3}, 1 \right]$ (see [10]).

By definition, both the Tanimoto similarity and the cosine similarity can be determined based on the dot product and lengths of compared vectors. In the case of a pair of *ZPN-vectors*, their dot product and lengths are related as follows:

Proposition 1 [8]. For any *ZPN-vectors* u and v, the following holds:

a) $\forall_{i \in \{1,...,n\}} (u_i \, v_i \leq u_i^2)$.

b) $u \cdot v \leq |u|^2$.

Obviously, if $u \cdot v = 0$ for non-zero vectors u and v, then $cosSim(u, v) = T(u, v) = 0$. Hence, when looking for vectors v such that $cosSim(u, v) \geq \varepsilon$ or, respectively, $T(u, v) \geq \varepsilon$, where $\varepsilon > 0$, non-zero vectors w that do not have any common non-zero dimension with vector u can be skipped as in their case $cosSim(u, w) = T(u, w) = u \cdot w = 0$. In Example 1, we show how to skip such vectors by means of *inverted indices* [11]. In a simple form, an inverted index stores for a dimension *dim* the list $I(dim)$ of identifiers of all vectors for which *dim* is a non-zero dimension.

Example 1. Let us consider the set of vectors from Table 2 (or Table 1). Table 3 shows the inverted indices for this set of vectors. Let us assume that we are to determine vectors v similar to vector $u = v1$ such that $cosSim(u, v) \geq \varepsilon$ (or $T(u, v) \geq \varepsilon$) for some positive value of ε. To this end, it is sufficient to compare u only with vectors that have at least one non-zero dimension common with $NZD(u) = \{1, 2, 5, 6, 7, 8\}$; i.e., with the following vectors:

$$\bigcup \{I(i) | i \in NZD(u)\} = I(1) \cup I(2) \cup I(5) \cup I(6) \cup I(7) \cup I(8)$$
$$= \{v1, v2, v4, v5, v6, v7, v8, v9, v10\}.$$

Hence, the application of inverted indices allowed us to reduce the set of candidate vectors from 10 vectors to 9.   □

Table 3. Inverted indices for dimensions of vectors from Table 2 (and Table 1)

dim	I(dim)
1	<v1, v2, v7, v10>
2	<v1, v2, v4, v8, v10>
3	<v3, v6, v7, v10>
4	<v3, v4, v5, v6, v7, v8, v9>
5	<v1, v5, v7, v9>
6	<v1, v6, v8>
7	<v1, v5, v7, v9>
8	<v1, v2, v7>
9	<v6, v7, v8, v9>

Let us note that in Example 1 the value of ε has not been taken into account. To the contrary, the methods we recall in Sections 3-4 as well as our new method we present in Section 5 will use ε to speed up the search process.

3   Bounds on Lengths of Near Duplicates of ZPN-Vectors

In this section, we recall a possibility of reducing the search space of candidates for near duplicate ZPN-vectors by taking into account the bounds on lengths of ε-near duplicates of such vectors (see Theorem 1).

Theorem 1 [8]. Let u and v be non-zero ZPN-vectors and $\varepsilon \in (0,1]$. Then:

a) $\left(cosSim(u,v) \geq \varepsilon \right) \Rightarrow \left(|v| \in \left[\varepsilon |u|, \dfrac{1}{\varepsilon} |u| \right] \right)$.

b) $\left(T(u,v) \geq \varepsilon \right) \Rightarrow \left(|v| \in \left[\sqrt{\varepsilon} |u|, \dfrac{1}{\sqrt{\varepsilon}} |u| \right] \right)$.

In order to make a reasonable benefit from these bounds, it makes sense to order vectors according to their lengths. Table 4 illustrates the example set of vectors after sorting with respect to vectors' lengths. $p\#$ reflects a position of a vector in the sorted vector set (for illustration purposes in subsequent sections of the paper, we have also sorted the dimensions of vectors in a non-decreasing way with respect to the lengths of their inverted lists; that is, in the following order: 6, 8, 1, 3, 5, 7, 9, 2, 4).

Table 4. Sparse representation of the example set of ZPN-vectors from Table 2 (extended by the information about vectors' lengths and sorted w.r.t. these lengths and w.r.t. frequency of dimensions)

$p\#$	Id	(non-zero dimension, value) pairs	length
$p1$	$v4$	{(2,-2.0), (4, 4.0)}	4.47
$p2$	$v5$	{(5,-3.0), (7, 3.0), (4, 4.0)}	5.83
$p3$	$v9$	{(5, 3.0), (7, 3.0), (9, 5.0), (4,-2.0)}	6.86
$p4$	$v2$	{(8, 6.0), (1, 3.0), (2,-2.0)}	7.00
$p5$	$v3$	{(3, 6.0), (4, 4.0)}	7.21
$p6$	$v8$	{(6, 5.0), (9, 5.0), (2, 4.0), (4, 4.0)}	9.06
$p7$	$v10$	{(1,-3.0), (3,-9.0), (2,-2.0)}	9.70
$p8$	$v1$	{(6, 5.0), (8, 6.0), (1,-3.0), (5, 3.0), (7, 3.0), (2, 4.0)}	10.20
$p9$	$v7$	{(8,-5.0), (1, 3.0), (3, 6.0), (5,-3.0), (7,-4.0), (9,-2.0), (4, 4.0)}	10.72
$p10$	$v6$	{(6, 5.0), (3,-9.0), (9, 5.0), (4, 4.0)}	12.12

Example 2. Let us consider the cosine similarity measure, vector $u = v1$ (or, equivalently, $p8$) from Table 4 and let $\varepsilon = 0.75$. By Theorem 1a, only vectors the lengths of which belong to the interval $\left[\varepsilon |u|, \dfrac{|u|}{\varepsilon} \right] = \left[0.75 \times 10.20, \dfrac{10.20}{0.75} \right] \approx \left[7.65, 13.60 \right]$ have a chance to be sought near duplicates of u. Hence, only five vectors; namely, $p6$-$p10$, which fulfill these lengths condition, have such a chance according to Theorem 1a.   □

4   Employing Non-Zero Dimensions for Determining Near Duplicates of *ZPN*-Vectors

In this section, we recall a possibility of reducing the number of candidates theorems for near duplicate *ZPN*-vectors by means of non-zero dimensions (see Theorem 2).

Theorem 2 [8]. Let u and v be non-zero *ZPN*-vectors, J be a subset of non-zero dimensions of u and zero-dimensions of v and $\varepsilon \in (0,1]$. Then:

a) $(cosSim(u, v) \geq \varepsilon) \Rightarrow \left(\sum_{i \in J} u_i^2 \leq \left(1 - \varepsilon^2\right) |u|^2 \right)$.

b) $\left(T(u,v) \geq \varepsilon\right) \Rightarrow \left(\sum_{i \in J} u_i^2 \leq \left(1 - \varepsilon\right) |u|^2 \right)$.

In the beneath example, we illustrate the usefulness of Theorem 2 for reducing the number of vectors that should be considered as candidates for near duplicates.

Example 3. Let us consider again the cosine similarity measure, vector $u = v1$ (or, equivalently, $p8$) from Table 4 and let $\varepsilon = 0.75$. As follows from Table 4, $NZD(u) = \{6, 8, 1, 5, 7, 2\}$. We note that $\left(1 - \varepsilon^2\right) |u|^2 = \left(1 - 0.75^2\right) \times 10.20^2 \approx 45.50$. One may also notice that $J = \{6, 8\}$ is a minimal set of first non-zero dimensions of vector u in Table 4 such that $\sum_{i \in J} u_i^2 > \left(1 - \varepsilon^2\right) |u|^2$ (in fact, $\sum_{i \in J} u_i^2 = u_6^2 + u_8^2 = 61$). Thus, by Theorem 2a, each vector v in Table 4 for which J is a subset of its zero dimensions is not ε-near duplicate of vector u. Hence, only the remaining vectors; that is, those for which at least one dimension in J is non-zero should be considered as potential candidates for sought vectors. Thus, only vectors from the set $\bigcup \{I(i) | i \in J\} = I(6) \cup I(8) = \{p4, p6, p8, p9, p10\}$ (see Table 5 for $I(6)$ and $I(8)$ obtained for vectors from Table 4), have a chance to be sought near duplicates of vector u according to Theorem 2a. Please note that the efficiency of this method of determining candidates depends on used dimensions. A heuristic that typically leads to small number of candidates consists in using least frequent dimensions.

Table 5. Inverted indices for dimensions of sorted vectors from Table 4

dim	I(dim)
1	<p4, p7, p8, p9>
2	<p1, p4, p6, p7, p8>
3	<p5, p7, p9, p10>
4	<p1, p2, p3, p5, p6, p9, p10>
5	<p2, p3, p8, p9>
6	<p6, p8, p10>
7	<p2, p3, p8, p9>
8	<p4, p8, p9>
9	<p3, p6, p9, p10>

Application of both Theorem 1a and Theorem 2a would restrict the number of candidates to four vectors; namely, to: $\{p6, p7, p8, p9, p10\} \cap \{p4, p6, p8, p9, p10\} = \{p6, p8, p9, p10\}$.   □

5   New Bounds on Lengths of Near Duplicates of *ZPN*-Vectors

In this section, we first derive new bounds on lengths of near duplicates of *ZPN*-vectors and then, by means of an example, we introduce a method of using them to reduce the number of candidates for near duplicates of such vectors.

Theorem 3. Let u and v be non-zero *ZPN*-vectors, L be a subset of non-zero dimensions of u and zero-dimensions of v and $\varepsilon \in (0,1]$. Then:

a) $(cosSim(u, v) \geq \varepsilon) \Rightarrow |v| \leq \left(1 - \sum_{i \in L} u_i'^2\right) \dfrac{|u|}{\varepsilon}$,

b) $(T(u,v) \geq \varepsilon) \Rightarrow |v| \leq \sqrt{1 - \sum_{i \in L} u_i'^2} \dfrac{|u|}{\sqrt{\varepsilon}}$,

where $u_i' = \dfrac{u_i}{|u|}$.

Proof. Ad a) By Proposition 1a, $\forall_{i \in \{1,...,n\}} (u_i v_i \leq u_i^2)$. Hence, $cosSim(u, v) = \dfrac{u \cdot v}{|u| \|v\|} =$

$$\dfrac{\sum_{i \in \{1,..,n\} \backslash L} u_i v_i}{|u| \|v\|} \leq \dfrac{\sum_{i \in \{1,..,n\} \backslash L} u_i^2}{|u| \|v\|} = \dfrac{|u|^2 - \sum_{i \in L} u_i^2}{|u| \|v\|} = \left(1 - \sum_{i \in L} u_i'^2\right) \dfrac{|u|^2}{|u| \|v\|}$$

$$= \left(1 - \sum_{i \in L} u_i'^2\right) \dfrac{|u|}{|v|} . \text{ So, } 0 < \varepsilon \leq cos(u, v) \leq \left(1 - \sum_{i \in L} u_i'^2\right) \dfrac{|u|}{|v|} . \text{ Therefore,}$$

$$|v| \leq \left(1 - \sum_{i \in L} u_i'^2\right) \dfrac{|u|}{\varepsilon} .$$

Ad b) By Proposition 1, $|u|^2 - u \cdot v \geq 0$ and $\forall_{i \in \{1,...,n\}} (u_i v_i \leq u_i^2)$. Hence, $T(u, v) =$

$$\dfrac{u \cdot v}{|u|^2 + |v|^2 - u \cdot v} \leq \dfrac{\sum_{i \in \{1,..,n\} \backslash L} u_i v_i}{|v|^2} \leq \dfrac{\sum_{i \in \{1,..,n\} \backslash L} u_i^2}{|v|^2} = \dfrac{|u|^2 - \sum_{i \in L} u_i^2}{|v|^2}$$

$$= \left(1 - \sum_{i \in L} u_i'^2\right) \dfrac{|u|^2}{|v|^2} . \text{ So, } 0 < \varepsilon \leq T(u, v) \leq \left(1 - \sum_{i \in L} u_i'^2\right) \dfrac{|u|^2}{|v|^2} . \text{ Thus,}$$

$$|v|^2 \leq \left(1 - \sum_{i \in L} u_i'^2\right) \dfrac{|u|^2}{\varepsilon} . \qquad \square$$

Example 4. Let us consider again the cosine similarity measure, vector $u = v1$ (or, equivalently, $p8$) from Table 4 and let $\varepsilon = 0.75$. The length of u equals 10.20, its non-zero dimensions $NZD(u) = \{6, 8, 1, 5, 7, 2\}$. In Example 1, we found by Theorem 1a that only vectors $C_1 = \{p6, p7, p8, p9, p10\}$, the lengths of which belong to $[7.65, 13.60]$, can be sought vectors. In Example 2, we found by Theorem 2a that only vectors $C_2 = I(6) \cup I(8)$, which have at least one non-zero dimension in $J = \{6, 8\} \subseteq NZD(u)$, can be sought vectors. In addition, we found that J is a minimal set of non-zero dimensions of u such that each vector v for which J is a subset of $ZD(v)$ is not ε-near duplicate of vector u. Clearly, $C_1 \cap C_2$ is a superset of all sought ε-near cosine duplicates of u.

As follows from Table 5, $I(6) = \{p6, p8, p10\}$ and $I(8) = \{p4, p8, p9\}$. Let $I'(6)$ and $I'(8)$, be those vectors from $I(6)$ and $I(8)$, respectively, the lengths of which belong to the lengths' interval $[7.65, 13.60]$ (following from Theorem 1); that is,

$$I'(6) = C_1 \cap I(6) = \{p6, p8, p10\} \text{ and } I'(8) = C_1 \cap I(8) = \{p8, p9\}.$$

Then obviously, $C_1 \cap C_2 = I'(6) \cup I'(8) = \{p6, p8, p9, p10\}$.

Remembering that vectors from $I(6) \cup I(8)$ and, in consequence, vectors from $I'(6) \cup I'(8)$ have at least one non-zero dimension in J, we may split vectors in $C_1 \cap C_2$ (which equals $I'(6) \cup I'(8) = \{p6, p8, p9, p10\}$) into a number of categories (here: 3 categories) with respect to their sets of all (non-)zero dimensions in J (which consists of dimensions 6 and 8). Table 6 presents the result of this split.

Table 6. Categories of vectors in $I'(6) \cup I'(8)$ w.r.t. their non-zero dimensions (or, equivalently, w.r.t. their zero dimensions) in $J = \{6, 8\}$

p#	Id	set M of all non-zero dimensions of $p\#$ in J	set L of all zero dimensions of $p\#$ in J	length
p6	v8	{6}	{8}	9.06
p10	v6			12.12
p8	v1	{6, 8}	$\varnothing$	10.20
p9	v7	{8}	{6}	10.72

Based on Theorem 3a and the information about zero-dimensions $L \subset J$ for candidate vectors from Table 6, we reason as follows:

$L = \{8\}$: vectors $p6$ and $p10$ would have a chance to be sought near duplicates of u provided their lengths would not exceed the following tighter upper bound

$$\left(1 - \sum_{i \in L} u_i'^2\right) \frac{|u|}{\varepsilon} \approx 8.89^{1} \text{ (please note that this new upper length bound hap-}$$

pened to be less than the length 10.20 of u itself!). In fact, the lengths of $p6$ and $p10$ are greater than 8.89, so they are not are not sought near duplicates of u.

$L = \{6\}$: vector $p9$ would have a chance to be sought near duplicate of u provided its length would not exceed the following tighter upper bound

$$\left(1 - \sum_{i \in L} u_i'^2\right) \frac{|u|}{\varepsilon} \approx 10.33 . \text{ As the length of } p9 \text{ exceeds } 10.33, \text{ it is not a sought}$$

near duplicate of u.

---

[1]   The calculation of a new length bound can be determined based on non-zero dimensions in J of a candidate vector v rather than on its zero dimensions in J. Let $A = \sum_{i \in \{J\}} u_i'^2$. Then,

$\left(1 - \sum_{i \in L} u_i'^2\right) \frac{|u|}{\varepsilon}$, where L is the set of all zero dimensions of v in J, can be determined

equivalently as $\left(1 - \left(A - \sum_{i \in M} u_i'^2\right)\right) \frac{|u|}{\varepsilon}$, where M is the set of all non-zero dimensions of

v in J.

$L=\varnothing$: vector $p8$ has no non-zero dimensions in J, so the new upper length bound $\left(1 - \sum_{i \in L} u_i'^2\right)\dfrac{\lfloor u \rfloor}{\varepsilon}$ equals the general upper length bound $\dfrac{\lfloor u \rfloor}{\varepsilon}$ that follows from Theorem 1. So, the new bound will not prune $p8$ as a false candidate.

As a result of applying Theorem 3a, three out of four candidates for near duplicates of u that had remained after applying Theorems 1a and 2a, were pruned. The only candidate that remained after applying Theorem 3a was vector $u = p8 = v1$ itself.   □

6   Experiments

In Table 7, we provide a short description of used benchmark high-dimensional sparse data sets available from http://glaros.dtc.umn.edu/gkhome/cluto/cluto/download.

Table 7. Characteristics of used benchmark data sets

name	no. of vectors	no. of dimensions
cacmcisi	4 663	41 681
sports	8 580	126 373

Each vector in these data sets represents a document. A dimension represents the number of occurrences of a corresponding term in the document. The original vectors are real valued (more precisely, integer valued). Our experiments, nevertheless, were carried out on *ZPN*-vectors obtained from the original vectors by applying two kinds of transformation: *BIN* and *ZPN*, respectively. The *BIN* transformation replaced each non-zero occurrence value with 1. Thus, an original vector became a binary vector. The used *ZPN* transformation replaced each 1 with -1 and values larger than 1 with 2. Thus, an original vector was transformed to a non-binary *ZPN*-vector. The dimensions of vectors of all four resultant data sets were sorted non-decreasingly w.r.t. their frequency; that is, w.r.t. the lengths of their inverted lists.

In the experiments, we have tested the average number of candidate vectors for the cosine similarity for the methods using: 1) Theorem 1a (OldBnds), 2) Theorem 2a (NZD), 3) Theorems 1a and 2a (OldBnds&NZD), 4) Theorems 1a, 2a and 3a (OldBnds&NZD&NewBnds). The obtained results are presented in Tables 8 and 9.

Table 8. The average number of cosine candidate vectors for different methods for *cacmcisi*

	cacmcisi: BIN				cacmcisi: ZPN			
ε	OldBnds	NZD	OldBnds&NZD	OldBnds&NZD&NewBnds	OldBnds	NZD	OldBnds&NZD	OldBnds&NZD&NewBnds
0.10	4 663	1 102	1 102	476	4 660	1 096	1 096	571
0.20	4 545	1 038	1 028	291	4 282	1 034	988	354
0.30	3 821	963	864	106	3 316	975	788	118
0.40	3 102	841	658	46	2 840	862	637	52
0.50	2 771	636	464	20	2 659	670	477	28
0.60	2 486	359	229	10	2 419	404	257	18
0.70	2 182	243	131	7	2 116	286	155	12
0.80	1 517	90	34	6	1 456	117	46	7
0.90	679	23	5	3	649	30	6	3
0.95	420	13	2	2	399	15	2	2
0.99	370	12	2	2	352	12	2	2
1.00	367	12	2	2	345	12	2	2

Table 9. The average number of cosine candidate vectors for different methods for *sports*

	sports: BIN				sports: ZPN			
ε	OldBnds	NZD	OldBnds&NZD	OldBnds&NZD&NewBnds	OldBnds	NZD	OldBnds&NZD	OldBnds&NZD&NewBnds
0.10	8 580	8 375	8 374	3 640	8 578	8 365	8 364	4 746
0.20	8 569	8 249	8 243	429	8 560	8 237	8 224	1 280
0.30	8 487	8 062	7 987	140	8 364	8 070	7 882	477
0.40	8 077	7 726	7 331	121	7 811	7 797	7 162	292
0.50	7 304	7 174	6 266	101	6 979	7 342	6 133	213
0.60	6 236	6 248	4 839	79	5 883	6 575	4 793	154
0.70	4 899	4 775	3 076	53	4 540	5 297	3 142	92
0.80	3 326	2 768	1 292	25	3 038	3 349	1 419	41
0.90	1 651	817	201	7	1 491	1 122	253	10
0.95	814	217	28	2	734	316	37	3
0.99	161	18	1	1	145	26	2	1
1.00	37	7	1	1	20	7	1	1

7   Summary

In the paper, we have derived new bounds on lengths of near duplicates of
ZPN-vectors that are more strict than the bounds derived in [8]. By means of an example, we have illustrated a method of using them in determining near duplicates of
ZPN-vectors. The experiments we have carried out on benchmark high-dimensional
sparse data sets show that the new method using the new bounds is an absolute winner
in comparison with recent efficient approaches and reduces the number of candidate
vectors up to 4 orders of magnitude. The new approach has proved to be efficient also
for very challenging low values of the similarity threshold.

References

1. Arasu, A., Ganti, V., Kaushik, R.: Efficient exact set-similarity joins. In: Proc. of VLDB
   2006. ACM (2006)
2. Bayardo, R.J., Ma, Y., Srikant, R.: Scaling up all pairs similarity search. In: Proc. of
   WWW 2007, pp. 131–140. ACM (2007)
3. Broder, A.Z., Glassman, S.C., Manasse, M.S., Zweig, G.: Syntactic Clustering of the Web.
   Computer Networks 29(8-13), 1157–1166 (1997)
4. Chaudhuri, S., Ganti, V., Kaushik, R.L.: A primitive operator for similarity joins in data
   cleaning. In: Proceedings of ICDE 2006. IEEE Computer Society (2006)
5. Gionis, A., Indyk, P., Motwani, R.: Similarity Search in High Dimensions via hashing. In:
   Proc. of VLDB 1999, pp. 518–529 (1999)
6. Kryszkiewicz, M.: Efficient Determination of Binary Non-Negative Vector Neighbors
   with Regard to Cosine Similarity. In: Jiang, H., Ding, W., Ali, M., Wu, X. (eds.) IEA/AIE
   2012. LNCS (LNAI), vol. 7345, pp. 48–57. Springer, Heidelberg (2012)
7. Kryszkiewicz, M.: Bounds on Lengths of Real Valued Vectors Similar with Regard to the
   Tanimoto Similarity. In: Selamat, A., Nguyen, N.T., Haron, H. (eds.) ACIIDS 2013, Part I.
   LNCS, vol. 7802, pp. 445–454. Springer, Heidelberg (2013)
8. Kryszkiewicz, M.: On Cosine and Tanimoto Near Duplicates Search among Vectors with
   Domains Consisting of Zero, a Positive Number and a Negative Number. In: Larsen, H.L.,
   Martin-Bautista, M.J., Vila, M.A., Andreasen, T., Christiansen, H. (eds.) FQAS 2013.
   LNCS (LNAI), vol. 8132, pp. 531–542. Springer, Heidelberg (2013)

9. Salton, G., Wong, A., Yang, C.S.: A vector space model for automatic indexing. Communications of the ACM 18(11), 613–620 (1975)
10. Willett, P., Barnard, J.M., Downs, G.M.: Chemical similarity searching. J. Chem. Inf. Comput. Sci. 38(6), 983–996 (1998)
11. Witten, I.H., Moffat, A., Bell, T.C.: Managing Gigabytes: Compressing and Indexing Documents and Images. Morgan Kaufmann (1999)
12. Xiao, C., Wang, W., Lin, X., Yu, J.X.: Efficient similarity joins for near duplicate detection. In: Proc. of WWW Conference, pp. 131–140 (2008)

An Approach of Color Image Analysis by Using Fractal Signature

An-Zen Shih

Department of Information Engineering, Jin-Wen Technology University,
New Taipei City, Taiwan
anzen325@just.edu.tw

Abstract. In this paper we have tested the method to use the combination of fractal into fractal signature in image analysis. We have found that the results suggest that the fractal signature have proved itself to a very good and efficiently tool in image analysis. Also, we have discussed some concepts which are derived from the experiment. These concepts can be used in the future experiment.

Keywords: Fractal, fractal dimension, image analysis.

1   Introduction

To date there are billions of images and pictures on the network and huge numbers of visual information appear around us. Naturally, it is tedious if people have to search them for the image that he wants.

Content based image retrieval, or called CBIR, become very popular because it helps people to solve the problems mentioned above. However, to choose the proper images we need is still a very difficult task.

Feature is one of useful tool to retrieve images. Features can be color, texture, and shape. In recent year, a mathematical form called fractal was introduced. People have found that it is a very useful tool in the field of image analysis.

Fractal geometry has proved itself to be a very useful model in many fields. Fractal dimension played an important role in fractal dimension. It is an index which shows the irregularity of fractal surfaces or curves.

In this paper we used the idea of fractal and extended it in image analysis. We combine fractal dimension into a fractal signature and in the hope that this fractal signature can help us to analyze images efficiently. Our results showed that the fractal signature did suggest itself to be a very promising and efficient tool in image analysis.

The paper is organized as following. In the next section, we will present state-of-art knowledge, which includes fractal and fractal dimension concepts. Secondly, we will describe our experiment algorithm. Thirdly, we will show our experiment results. The fourth part covers the discussion about our concepts derived from our experiment and a short conclusion is followed.

A. Moonis et al. (Eds.): IEA/AIE 2014, Part II, LNAI 8482, pp. 171–176, 2014.

2   State-of-Art Knowledge

We now present some basic state-of-art knowledge in this section, which will help the readers to understand the background theory of our research

2.1   Fractal Concepts

Mandelbrot invented the word 'fractal' and described it as[1]

'Mathematical and natural fractals are shapes whose roughness and fragmentation neither tend to vanish, nor fluctuate up and down, but remain essentially unchanged as one zooms in continually and examination is refined.'

From the above words, we notice that fractal presents a strong similarity in its shape. This similarity is defined as self-similarity. In other words, the meaning of self-similarity is that each part is a linear geometric reduction of the whole, with the same reduction ratios in all directions.

Self-similarity can be deterministic self-similarity or statistical self-similarity. Deterministic self-similarity means the self-similarity object can be derived into non-overlapping parts and each non-overlapping part is exactly similar to the whole. Statistical self-similarity means that the self-similarity object can be derived into non-over-lapping parts and each non-overlapping part only looks similar to the whole but is not exactly similar.

There is also another form of similarity called self-affinity. A self-affinity object is that different ratio will be applied to different direction.

2.2   Fractal Dimension and Fractal Signature   .

A dimension of a geometrical structure is the least number of real-valued parameters which can be used to (continuously) determine the points of the structure. Fractal pattern have dimension too.[2]

However, unfortunately, to date, there is no strict definition for a fractal dimension. Mandelbrot used the Hausdoff dimension as fractal dimension but he also admitted that the Hausdoff dimension is difficult to handle so he later derived an expression of dimension which will be called 'self-similarity dimension'.

The self-similarity dimension is described below. If we suppose the fractal object is the union of N distinct (non-overlapping) copies of itself each of which is similar to the original scaled down by a ratio r, then the dimension can be derived from:

$$1 = NrD \tag{1}$$

So the dimension D is :

$$D = \log(N)/\log(1/r) \tag{2}$$

Equation (2) is used as the theoretical basis for calculating fractal dimensions.

A single Fractal dimension is often not suitable to analysis images efficiently.[][] In other words, we need to combine FD and other features to analysis images more specifically. So many material use fractal signature, which means a FD or FDs and other features, rather than a single fractal dimension to classify images.

2.3   Computing Methods

To date, there are five popular FD computing methods : difference grey, ractcal Brownian motion, difference box count, wavelet decomposition, and carpet overlay[3].

In this paper we used the popular differential boxing-counting method to calculate the fractal dimension.[4] The basic idea is that the image intensity surface is assumed to be self-similar. That is, the intensity can be divided into several non-overlapping parts with a ratio r and each divided part is similar to the entire image intensity surface.

Assumed the image contain M*N pixels and the grey level is between [0, G]. The sum of the intensity of each pixel constitutes a 3-D intensity surface. We divide the image and its highest grey level by a parameter called the scale s. That is

$$M/s = m, \; N/s = n, \; G/s = g \tag{3}$$

In this manner, we have divided the image into m*n grids and the intensity space into g grids. This can be imagined as dividing a big box of size M*N*G into many small boxes of size m*n*g.

Assume that the intensity surface is self-similar so it can be scaled down by a ratio r. This ratio can be computed by

$$r = 1/s = m/M = n/N = g/G \tag{4}$$

For each small region, m*n, in the image, assume that the highest intensity is in grid l and the lowest intensity is in grid k, then the distribution of grey levels in this region is given by

$$ar(i, j) = l - k + 1 \tag{5}$$

The total distribution on the image can be got by summing up each small region's grey level distribution

$$Ar = sigma(ar(i,j)) \tag{6}$$

Ar is counted for different value of r.
    Then by using

$$D = \log(Ar)/\log(1/r) \tag{7}$$

This We can compute fractal dimension D.

3   Experiment

Our investigation involved whether or not fractal signature could be used to analysis color images.   There had been studies showed that fractal dimension combined with other features could be very useful in image analysis and pattern recognition.[5][6] So our hypothesis was based on this assumption to develop a fractal signature which contained fractal of grey level, red color and green color for color image selection.

The propose algorithm is implemented on a PC with 2.33GHz, 1.95GB RAM, and Microsoft XP. Our image database contained 100 color scenery images which were stored in BMP forms. As the experiment proceeding, the images will be changed to grey level, red color and green color separately for our algorithm.

The algorithm can be break into several stages.   At the beginning, we computed fractal signature FS for each image in the database.   The fractal signature were composed by three parts: FDg, the fractal dimension of the image in grey level, FDr, the fractal dimension of the image in red color form, and FDG, the fractal dimension of the image in green color form.   These three fractal dimensions will compose the fractal signature of this image.   In other words, FS = f (FDg, FDr, FDG) Here we proposed a formula about FS

$$FS = \alpha FDg + \beta FDr + \gamma FDG \qquad (8)$$

Here $\alpha = 0.1$, $\beta = 0.5$, $\gamma = 0.4$, which we got these number by several tries during we tested the algorithm and found they could be a suitable choice for the experiment.

In this manner, we constituted a fractal signatures database for the images in the database.

Secondly, a query image was shown to the system.   The query image was firstly computed to get its fractal signature FSq.   Then FSq compared with FSi, i = 1, 2, ..., n, of each image i in the database.   If the comparison results were smaller than a considerable number ε, we regard the image is a possible interested image. That is, its content had the features which were similar to the query image.   We set the number ε to be 0.02.   If the algorithm found that there were two images with same FS, it will compare the FDr   to judge which one is more proper.

For each time of comparison, ten images will be chosen as the possible interested images and they will be put from highest possibility to low possibility. These images were presented to people for the final audience to confirm their correctness,

4   Results

Many studies[3][5][6] had suggested and used combination of fractal dimension and other features to analysis. According to shih[5], a single fractal dimension could not classify images very well, it can only be a good auxiliary tool in image analysis.

In this experiment we had hoped that, with the help of fractal signature, we could analyze color images more efficiently.   We used three kinds of fractal dimensions of an image, that is, the fractal dimension of the grey level of the image, the fractal dimension of the red form of the image and the fractal dimension of the green form of the image, to form the fractal signature to represent the image.

Fig.1-Fig.3 shows some results obtained in the experiment by using fractal signatures.

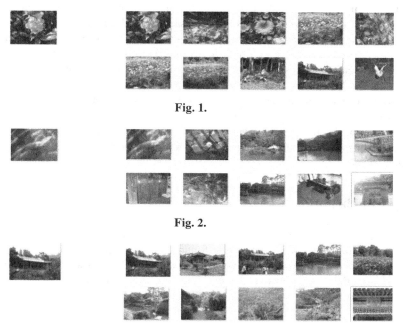

Fig. 1.

Fig. 2.

Fig. 3.

In Fig 1-3, the query image is on the left side and the selected images are on the right side. We have notice that our algorithm can include same, or similar images from the database. However, we also notice that there are still quite a few images which are just look-like but not the same content we want in the image. For this problem, we could argue that, the fractal signature is a good and efficient tool for the selection in the first stage, it help people abridge the time for search the whole image database. So we concluded that our algorithm have suggested that it is very promising and efficient.

In fig 1, fig 2 and fig 3 it can be noticed see that the query images are chosen in the ten selected images. This is a proof that the algorithm worked well. Also, we were happy to notice that the algorithm can pick up some images contented similar content. In fig 1 and fig 2 the rate are 20% and fig 3 are 30%. Here the percentage was calculated with the following equation

$$\text{Rate} = \text{images contented similar content/selected images} \qquad (9)$$

We used the fractal signature as the first stage tool for image analysis. And we left the final audience for the people.

5   Discussion

Prior work has demonstrated that fractal dimension is a very good tool for image classification. However, there is also work mentioned about that a single fractal dimension might not be able to narrow the range of the image selection. That is, there could be two different images but with same fractal dimension.

In this study we use the combination fractal dimensions to constitute a fractal signature in the hope that it can help us to improve the drawback mentioned above. Our experiment results showed that m can use fractal signature to improve this drawback and still work efficiently in image analysis. This demonstrated that it is very possible that with proper change the form of fractal dimension, we can enhance the power of fractal dimension in image selection.

To sum up, the importance of this study existed in that not only the variation of fractal dimension can analyze images but also it can improve the bad effect of fractal dimension.

6   Conclusion

In this paper, we have tested the method to use the combination of fractal into fractal signature in image analysis. We have found that the results suggest that the fractal signature have proved itself to a very good and efficiently tool in image analysis. Also, our method can improve the problem that different images but same fractal dimension. In addition, we have discussed that the significance of the experiment.

References

1. Mandelbrot, B.: Fractals-forms, chance and dimension. W. H. Freeman and xture
2. Peigen, J., Saupe: Chaos and Fractals-new frontiers of science. Springer (1992)
3. Hai-Ying, Z., et al.: A Texture Feature Extraction on Two Fractal Dimension for Content Based Image Retrieval. In: 2009 World Congress on Computer Science and Information Engineering (2009)
4. Sarkar, N., Chaudhuri, B.B., Kundu, P.: Improved fractal geometry based texture segmentation technique. In: IEE Proceedings E:Computer, vol. 140(5), pp. 233–241
5. Shih, A.-Z.: An Examination of Fractal Dimension Approach of Image Classification. In: ICMLC 2008 (2008)
6. Arun Raja, K.K., et al.: Content Based Image Retreival Suing Fractal Signature Analysis. International Journal of Advance Research in Computer Engineering and Technology 2(3) (March 2013)

A Practical Approach for Parameter Identification with Limited Information*

Lorenzo Zeni, Guangya Yang, Germán Claudio Tarnowski,
and Jacob Østergaard

Department of Electrical Engineering, Technical University of Denmark
2800 Kgs. Lyngby, Denmark
gyy@elektro.dtu.dk

Abstract. A practical parameter estimation procedure for a real excitation system is reported in this paper. The core algorithm is based on genetic algorithm (GA) which estimates the parameters of a real AC brushless excitation system with limited information about the system. Practical considerations are integrated in the estimation procedure to reduce the complexity of the problem. The effectiveness of the proposed technique is demonstrated via real measurements. Besides, it is seen that GA can converge to a satisfactory solution even when starting from large initial variation ranges of the estimated parameters. The whole methodology is described and the estimation strategy is presented in this paper.

Keywords: parameter identification, genetic algorithm, AC brushless excitation system.

1 Introduction

The need of having reliable dynamic models in power system studies calls for the development of models, and subsequently, the tuning of parameters. With regard to conventional power plants, among the components for which it is usually desirable to obtain a thorough model is the excitation system. The model of excitation systems have been extensively discussed in literature [1–4], while the IEEE standard [5] provides a wide range of models that are suitable to represent many kinds of excitation systems currently installed on modern power units.

Many techniques have been proposed and employed over the years in order to achieve reliable and computationally reliable parameter estimator. The classical identification methods can be usually in time or frequency domain [6–9]. As to the increasing complexity of the problem, heuristic methodologies have however been suggested in later studies [10, 11], where the non-linear evolutionary

---

* G. Yang and J. Østergaard is with Centre for Electric Power and Energy, Department of Electrical Engineering, Technical University of Denmark, 2800 Kgs. Lyngby, DK. L. Zeni is with Department of Wind Energy, Technical University of Denmark, 4000 Roskilde, DK. G. C. Tarnowski is with Vestas Wind Systems and Centre for Electric Technology, Department of Electrical Engineering, Technical University of Denmark, 2800 Kgs. Lyngby, DK.

A. Moonis et al. (Eds.): IEA/AIE 2014, Part II, LNAI 8482, pp. 177–188, 2014.

technique such as genetic algorithm (GA) has successfully been applied to parameter estimation of excitation systems. GA is a model-independent optimisation technique and suitable for non-linear problems. It is capable of avoiding local optima, due to its immunity to hill-climbing phenomena. Moreover, by steered search, GA performs better than purely randomly sampling algorithms [12–14].

The incomplete information regarding the excitation system and the limited number of available measurements is what distinguishes this study from literature, e.g. [10, 11]. In this investigation, the measurements are acquired during a periodical check of the excitation system and, as such, a technique based on the combination of analytical and empirical considerations, along with the application of GA, has been developed.

This paper shows a procedure based on GA to tune the parameters of a model representing an AC brushless excitation system. The system comprises of a permanent magnet synchronous machine (PMSG) and an automatic voltage regulator (AVR). The excitation system is used in a combined-heating-and-power (CHP) plant in a Danish distribution grid. The validation is requested by the distribution system operator due to the increasing need of the dynamic model for renewable integration studies. The validation is performed with limited information and measurements from the actual excitation system. The proposed validation procedure is proved to be efficient even with wide variation boundaries of parameters.

In this paper, the methodology adopted to achieve the target is presented in detail and the efficiency of the parameter estimation is illustrated and discussed. A major part of the procedure is certainly the simplification of the problem that allows for a significant reduction of the number of parameters to be optimised by GA. The paper is organised as follows. Section II describes problem to be dealt with, where the modelling of the excitation system, generator and the external grid is detailed. The identification approach is presented in section III, followed by case studies and conclusion.

2   Problem Description

In this section, the development of the models of the components for validation is presented at a sequel.

2.1   Excitation System's Layout and Representation

The targeted device is an AC brushless excitation system provided with PMSG-based pilot exciter. A static electronic AVR is employed to perform the voltage regulating action. An illustrative sketch of the system under investigation is depicted in Fig. 1.

As can be seen in Fig. 1, the system consists of,

– A terminal voltage transducer that is measuring the voltage at the generator's terminal and feeding the signal back to the voltage regulator;

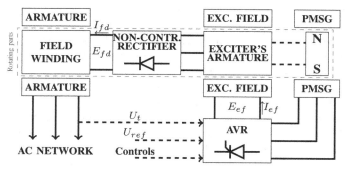

Fig. 1. Simplified physical structure of the AC brushless excitation system

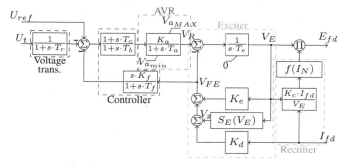

Fig. 2. Block diagram of the excitation system model - type AC1A [5]

- An AVR, which comprises of:
  - A controller that is not known in advance and may be of variegated nature.
  - An actual thyristor-based voltage regulator, that outputs the desired voltage to the main exciter.
- A main AC exciter, which field is lying on the stator and supplied by the AVR, while its armature winding is located on the rotor and supplies the generator's field.
- A non-controlled (diode-based) rectifier that serves as AC-DC converter, in order to adapt the AC voltage coming from the exciter to the DC voltage required by the generator field.

A thorough guideline towards the development of a suitable model for this kind of excitation system can be found in the IEEE standard [5], where a number of standard block diagrams are presented. The model referred to as AC1A is deemed to be appropriate to represent the system under investigation. All the components listed above can be included in this model. The system does not have under- and over-excitation limits and power system stabiliser, which are therefore ignored , as shown in Fig. 2.

In Fig. 2, the relevant parts are represented in the model, where

- The voltage tranducser is accounted for by a first order block with time constant T_r.
- The control strategy is a combination of a lead-lag feed-forward compensation (with time constants T_c and T_b) and a derivative feed-back compensation with gain K_f and time constant T_f.
- The AVR is modelled with a first order block with gain K_a, time constant T_a and output limits $V_{a_{min}}$ and $V_{a_{MAX}}$.
- The main exciter model takes into account its dynamics - gain K_e and time constant T_e -, its saturation characteristic - non-linear function $S_E(V_E)$ - and the demagnetising effect - gain K_d - of its load current, that is the generator's field current I_{fd}. A comprehensive description of this component's model can be found in [2].
- The effect of the non-controlled rectifier is represented by a non-linear function, reported in [1,5] and justified in [4], determined by the parameter K_c.

The parameters enlisted above must be set in order to bring the model to represent the actual excitation system as faithfully as possible. Besides, the *per-unit* notation of the AC exciter need to be set before assigning the parameters. This is an issue that is not exhaustively touched upon by the literature. Once the base values for such component are known, some of the parameters can be set simply by knowing the exciter's electrical characteristics. The set of base values for this important component was thus determined as:

- The AC exciter output voltage base $V_{E_{base}}$ is chosen such as the voltage that leads to having generator's field voltage $E_{fd} = 1$ pu, assuming no voltage drop over the commutating reactance and taking into account the AC-DC conversion factor $\frac{3\sqrt{2}}{\pi}$ [15].
- The corresponding exciter output base current $I_{E_{base}}$ directly derives from the equality between the AC output power of the exciter (assuming unit power factor) and the DC power entering the generator's field, i.e. $\sqrt{3}V_{E_{base}}I_{E_{base}} = E_{fd_{base}}I_{fd_{base}}$.
- The exciter field base current can be computed, by knowing the slope R_a of the no-load saturation characteristic of the AC exciter, as $I_{ef_{base}} = \frac{V_{E_{base}}}{R_a}$.
- The corresponding exciter base field voltage is calculated by multiplying the base current $I_{ef_{base}}$ by the resistance of the exciter's field winding.

The choice of such base values, combined with the knowledge of the exciter's saturation characteristics, contributed to the simplification of the problem, prior to the application of GA.

2.2   Generator and System Model

The interaction of the excitation system with the other components requires the construction of a model for the generator, transformer and external grid. Detailed and reliable representations for synchronous generator and transformer are developed, relying on standard models and the manufacturer's data. On the

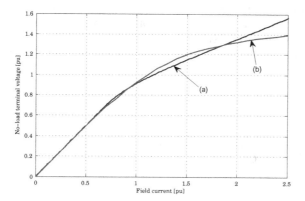

Fig. 3. Comparison, between tabular input and curve fitted generator's no-load saturation curve:
(a) blue - quadratic curve fitted (standard DIgSILENT model);
(b) red - tabular input (manufacturer's datasheet).

other hand, the external grid is modelled with its simple Thevenin equivalent, as the short circuit power at the generator's terminals is quite low and the closest voltage regulating unit is very far from the targeted power plant. In addition, the validation is based on real measurements from un-loaded conditions of the generator, therefore a detailed representation of external grids is not necessary. Nonetheless, when considering the loaded operation, it is important to achieve the appropriate level of detail of the grid.

It is important to notice that, for the sake of the parameter estimation of the excitation system model, a fundamental part of the generator's model lies in the expression of its saturation characteristic. The accuracy in the representation heavily affects both the steady-state and the dynamics of the excitation system's response, both under no-load and loaded conditions, as the generator usually functions with saturated magnetic core, if its terminal voltage is close to rated. It has been noticed that the exponential fitting equation which many dedicated tools rely upon, for example [16], does not allow for a satisfying accuracy, at least for the synchronous machine under investigation here. Thus, a tabular input has been preferred in order to fit the real curve with sufficient precision.

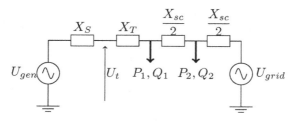

Fig. 4. System configuration for the loaded case

The apparent different between the two approaches on the targeted machine is illustrated in Fig. 3, where the terminal voltage is reported as a function of the field current for exponential fitting and tabular input respectively.

The single-line electrical diagram of the system as modelled in this study is sketched in Fig. 4.

3   Parameter Estimation Procedure

In this section, the estimation procedure on the excitation system using genetic algorithms is detailed. General considerations in parameter estimation with incomplete information are presented.

3.1   Measurement Signals

The available measurements include two different working points of the machine at no-load conditions. The measurements available are acquired through a test procedure that is periodically performed on the excitation system in order to accertain its features do not deteriorate. Hence, the acquisition of the measurements did not aim at providing a thorough set of data for the sake of parameter estimation. After searching through the available measured responses, a selection of those that are most relevant to possibly achieve a model's validation is,

- Two measured responses when machine unloaded, namely disconnected from the grid. In particular, the input reference signal U_{ref} is modulated:
    - Starting from 1 pu, it is stepped down to 0.95 pu and then back to 1 pu.
    - Starting from 1 pu, it is stepped down to 0.9 pu and then back to 1 pu.

Three signals are part of each measurement, namely the terminal voltage U_t, the exciter's field current I_{ef} - proportional to the signal V_{FE} in the model [1,5] - and the signal V_R, as shown in Fig. 2. All the recorded signals are taken on the static part of the excitation system and the device was always interacting with external elements, such as the generator. For this reason, the validation of single parts of the excitation system model was not possible. Furthermore, the problem is intrinsically non-linear with large number of variables and lack of a priori knowledge on the parameter ranges. Therefore, resorting to evolutionary algorithms such as GA is a reasonable option.

It is important to notice that the unit the measured voltages and currents are expressed with is V, that is the output unit of the instruments. No calibration factors were available for the measured voltages and the actual value of exciter field current I_{ef} and signal V_R was not known in advance, while the value of the terminal voltage could be calculated by knowing the machine ratings. Two constant correction factors - K_{corr_1} and K_{corr_2} - are therefore introduced in order to scale the value of V_{FE} and V_R during the verification. This is basically to let the optimisation procedure choose the base value of some of the physical quantities among the system.

3.2   Reduction of the Problem's Complexity

The problem is inherently complex. Referring to Fig. 2, excluding the limits $V_{a_{min}}$ and $V_{a_{MAX}}$, which cannot be set since they are never reached in the available measurements - and considering the two correction factors introduced above, one realises that GA would be required to set the values of 15 parameters. Nevertheless, some reasonable assumptions can be put forward in order to reduce the intrisic complexity of the problem. More in detail, a simple knowledge-based procedure is given below:

– The dynamics of voltage transducer and AVR should be very fast, if compared to those of the other components, due to their electronic nature. As a consequence, the determination of the time constants T_r and T_a can either be joined with the other parameters, or made by setting them to a very low value. In this case, they have been set as $T_r = T_a = 0.01\ s$ and their value may be refined in the future by acquiring more measurements.
– A proper choice of the base values in the *per-unit* notation of the exciter implies that $K_e = 1\ pu$.
– The coordinates of the two points required to represent the saturation characteristics of the exciter [1,16] - function $S_E(V_E)$ in Fig. 2 - could be extracted from the manufacturer's documentation.
– Useful indications on the analytical determination of the constant K_d were obtained from [2], where it is stated that such gain is proportional to the difference between synchronous and transient direct reactance of the exciter, i.e. $X_{S_e} - X'_{d_e}$. The value of X_{S_e} can be determined from the loaded saturation characteristic of the exciter, while the parameter X'_{d_e} was not known. However, it is usually significantly lower than X_{S_e}. Starting from this standpoint, and analysing the steady-state of the device in different operational points, a quite precise estimation of the values of K_d, K_{corr_1} and K_c could be achieved.

The simplification introduced by applying priori knowledge of the excitation system relieves the computational burden, as the solution space is dramatically reduced by dropping some of the targeted parameters, and therefore accuracy and speed of the optimisation are improved. Consequently, the values that GA is actually requested to set are:

– AVR gain K_a.
– Feed-forward control compensation loop time constants T_b and T_c.
– Feed-back control compensation loop gain K_f and its time constant T_f.
– Exciter time constant T_e.

3.3   The Application of GA to the Optimisation

Once the complexity of the problem had been reduced as much as possible, GA was applied in order to tune the values of the other parameters, starting from an initial range the values could vary within. Each population consisted of 70 chromosomes, while the maximum number of iterations was set to 500. Each chromosome contains 6 parameters, with binary coding used. The selection operator of GA uses Roulette wheel with stochasitc unversal sampling. Uniform crossover operator is used, together with Gaussian mutation operator. A Simulink block diagram including the excitation system and the generator was developed, so as to be used to obtain the response of the system with the candidate solutions.

A fitness function is written that subsequently executes what follows:

– The parameters of the block diagram are set according to the chromosomes in the current population.
– The states of the system are initialised, based on intial conditions and parameters' values.
– The simulink model is run in order to obtain the time domain response of the system and consequently extract the targeted signals.
– Using the available measurements, the value of the objective function is calculated as follows:

$$F_{obj} = \sum_{i=1}^{3} \left[W_i \cdot \int_0^{t_{end}} (x_{i_{SIM}} - x_{i_{MEAS}})^2 dt \right] \qquad (1)$$

where x_i is the i-th signal among those selected U_t, I_{ef} and V_R. The notation $x_{i_{SIM}}$ indicates simulated quantities, while $x_{i_{MEAS}}$ stands for measured i-th signals. The total simulation time is t_{end} which equals the total measurement time. The factor W_i allows for a weighting of the importance of the i-th signal in the optimisation. In this case, the priority was given to the terminal voltage and the signal V_R, since the signal V_{FE} is not to be completely representative of the exciter's field current I_{ef}. The time resolution for the comparison is directly given by the sampling time of the measurements, i.e. 1ms.

The algorithm terminates either when the progress in the objective function's value remains under a certain threshold for more than a certain number of iterations, or when the maximum number of iterations has been reached. There are totally 10 executions of the algorithm were set to achieve satisfying results. The first optimisations were performed by randomly initialising the parameters in their variation range, while the last executions were done by initialising the parameters using the previously found optimal values as a mean value and with Gaussian distribution. The initial variation range of the parameters was set to be much larger than in other studies, due to the limited information regarding the physical system.

4   Estimation Results and Discussion

At first, the optimisation procedure was run utilising as reference the first scenario, that is with machine disconnected from the grid and consequently unloaded. Initially, the machine is working in steady-state with $U_t = U_{ref} = 1\ pu$.

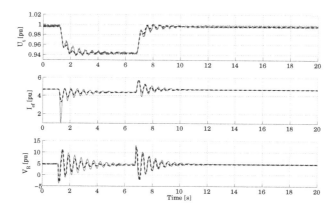

Fig. 5. Time domain response of real system and model after the optimisation procedure. The solid gray line refers to measured data, while the black dashed line represents the simulated response.

The reference voltage U_{ref} is then perturbed with a step down to 0.95 pu and, later, a step back to 1 pu.

After the application of GA, a set of optimum parameters could be obtained. As stated above, several executions were necessary in order to achieve acceptable results. The employment of the parameters deriving from the optimisation procedure in the excitation system's block diagram led to the results shown in Fig. 5, where the measured and simulated response are compared. It can be noticed that the agreement between the time series is remarkable and all the dynamics seem to be quite faithfully represented by the model, except for an initial spike in the exciter field current, which may due to the unmodeled part of the system.

The parameters tuned with the application of GA are then be used in order to check other operational conditions of the excitation system. The parameters are used to simulate the response in the second unloaded case described above, i.e. with U_{ref} going from 1 pu to 0.9 pu and then back to 1 pu. The response to such disturbances is depicted in Fig. 6. As can be noticed, the simulated curve is quite faithfully reproducing the real one even in this case, since the kind of disturbance is the same. A very good precision is achieved in the steady-state, while the error in the dynamics, mainly caused by the reasons introduced above and system nonlinearities, is increased, due to the larger magnitude of the initial spike in I_{ef}. It is noteworthy that the block diagram directly derives from IEEE standards, that were developed to be valid for few kinds of voltage regulating devices. Therefore, acquisition of more information regarding the control strategy employed on the targeted machine may suggest a modification of the control loops.

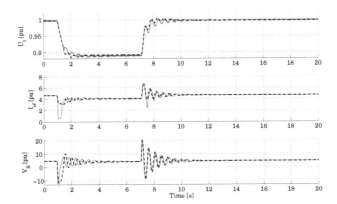

Fig. 6. Time domain response of real system and model in the second unloaded case (U_{ref} steps to 0.9 pu). The solid gray line refers to measured data, while the black dashed line represents the simulated response.

It is also worth noticing that, usually, the signals that are considered in studies of the same kind are terminal and field voltage. The latter is in this case impossible to acquire, since it would need to be measured on the rotor and no means to transmit information in its regard are present on the targeted machine. Alternatively, studies aimed at obtaining a very thorough model [8] execute more detailed measurements in every part of the machine. This is also not possible in this case.

The analysis and usage of internal signals, such as V_R and I_{ef} in this case, can lead to a more difficult parameter estimation process. On the other hand, the inclusion of such responses can help achieve a more exhaustive representation of all the dynamics of the system, improving the final result of the optimisation.

5   Discussion

In reality, establishment of models for small-sized power plants usually needs to be based on imcomplete information. The challenges may come from, eg. lack of manufacturers' data, no information regarding the control systems of the components, limited measurements, as well as inaccurate model of the external system. In such case, evolutionary algorithms, due to their modeling flexibility, make a rationale option for modelers.

Besides, available prior knowledge of the system, such as physical operation limits, possible values of parameters, should be considered during estimation to reduce the complexity and uncertainty of the problem to obtain a relatively accurate estimation.

6 Conclusion

The identification of the model for an AC brushless excitation system provided with static electronic AVR and installed on a small-size combined heat and power plant was attempted, relying on the usage of GA and a limited set of measurements acquired during a periodical test procedure performed on the voltage regulating device.

Due to the incomplete information and the limited number of available measurements, a technique that combines analytical and practical assumptions was adopted so as to fix some of the parameter values to simplify the problem. That leads to a reduction of the computational burden that GA had to withstand. Starting from 15 parameters to be set, the procedure allows for a reduction to as few as 6 values, which dramatically reduces the optimisation time.

It has be shown that the selection of the *per-unit* notation of the exciter, along with the knowledge of some electrical characteristics of the device, contributes to the reduction of the problem's complexity.

A standard IEEE model was used and the methodology permitted to obtain very good agreement between simulated and measured response with machine disconnected from the grid. GA proved to be effective at converging to the solution even when using a quite large range of variation for the paramters. A second execution of GA, initialising the parameters with a gaussian distribution centred around the first solution achieved, was effective to refine the optimum solution. Totally, ten executions of the algorithm are necessary to achieve appreciable results.

Though the acceptable results are obtained from the study, a significant discrepancy between model and measurements is noticed in the first spike of the exciter's field current. This issue must be further examined by future work and it may arise from a shortcoming of the model.

References

1. Kundur, P., Balu, N., Lauby, M.: Power System Stability and Control. McGraw-Hill Professional (1994)
2. Ferguson, R., Herbst, H., Miller, R.: Analytical studies of the brushless excitation system. AIEE Transactions on Power Apparatus and Systems Part III 79, 1815–1821 (1959)
3. Stigers, C., Hurley, J., Gorden, D., Callanan, D.: Field tests and simulation of a high initial response brushless excitation system. IEEE Transactions on Energy Conversion 1(1), 2–10 (1986)
4. Witzke, R., Kresser, J., Dillard, J.: Influence of AC reactance on voltage regulation of six-phase rectifiers. AIEE Transactions 72, 244–253 (1953)
5. Lee, D.: IEEE recommended practice for excitation system models for power system stability studies (IEEE Std. 421.5–1992). Energy Development and Power Generation Committee of the IEEE Power Engineering Society (1992)
6. Sanchez-Gasca, J., Bridenbaugh, C., Bowler, C., Edmonds, J.: Trajectory sensitivity based identification of synchronous generator and excitation system parameters. IEEE Transactions on Power Systems 3(4), 1814–1822 (1988)

7. Bujanowski, B., Pierre, J., Hietpas, S., Sharpe, T., Pierre, D.: A comparison of several system identification methods with application to power systems. In: Proceedings of the 36th Midwest Symposium on Circuits and Systems, pp. 64–67. IEEE (1993)
8. Liu, C., Hsu, Y., Jeng, L., Lin, C., Huang, C., Liu, A., Li, T.: Identification of exciter constants using a coherence function based weighted least squares approach. IEEE Transactions on Energy Conversion 8(3), 460–467 (1993)
9. Shande, S., Shouzhen, Z., Bo, H.: Identification of parameters of synchronous machine and excitation system by online test. In: International Conference on Advances in Power System Control, Operation and Management, pp. 716–719. IET (2002)
10. Feng, S., Jianbo, X., Guoping, W., Yong-hong, X.: Study of brushless excitation system parameters estimation based on improved genetic algorithm. In: 3rd International Conference on Electric Utility Deregulation and Restructuring and Power Technologies, pp. 915–919 (2008)
11. Puma, J., Colomé, D.: Parameters identification of excitation system models using genetic algorithms. IET Generation, Transmission & Distribution 2(3), 456–467 (2008)
12. Sivanandam, S., Deepa, S.: Introduction to genetic algorithms. Springer (2007)
13. Goldberg, D.: Genetic algorithms in search, optimization, and machine learning. Addison-Wesley (1989)
14. Conn, A., Gould, N., Toint, P.: A globally convergent augmented Lagrangian barrier algorithm for optimization with general inequality constraints and simple bounds. Mathematics of Computation 66(217), 261–288 (1997)
15. Krause, P., Wasynczuk, O., Sudhoff, S.: Analysis of electric machinery and drive systems. IEEE Press (2002)
16. DIgSILENT, "Synchronous generator technical documentation," DIgSILENT PowerFactory, Tech. Rep. (2007)
17. Andriollo, M., Martinelli, G., Morini, A.: Macchine elettriche rotanti. Ed. Libreria Cortina, Padova (2003)
18. Ogata, K.: Modern control engineering, 4th edn. Prentice Hall (2002)

A Novel Method for Extracting Aging Load and Analyzing Load Characteristics in Residential Buildings

Hsueh-Hsien Chang[1], Meng-Chien Lee[2], and Nanming Chen[2]

[1] Department of Electric Engineering, Jinwen University of Science and Technology, New Taipei, Taiwan
[2] Department of Electrical Engineering, National Taiwan University of Science and Technology, Taipei, Taiwan

Abstract. This study proposes a Hellinger distance algorithm for extracting the power features of aging load based on a non-intrusive load monitoring system (NILM). Hellinger distance algorithm is used to extract optimal features for load identification and the back-propagation artificial neural network (BP-ANN) is employed for the aging load detection. The proposed methods are used to analyze and identify the load characteristics and aging load in residential building. The result of aging load detection can provide the demand information for each load. The recognition result shows that the accuracy can be improved by using the proposed feature extraction method. In order to reduce the consumption of energy and send a real-time alarm of aging load to the user, the system provides the information of energy usage from the data analyses.

Keywords: Non-intrusive load monitoring system (NILM), aging load detection, Hellinger distance, back-propagation artificial neural network (BP-ANN).

1 Introduction

In traditional load monitoring methods, the sensors which connect to the main system for inner signal exchange have to be installed on each load [1]. When a load or a breaker state changes, the sensor will detect the switch state and send the message to load recorder, then the recorder transmit the data to the information center for load analysis, and the system will show the load energy information. Non-intrusive load monitoring system (NILM) does not only need no sensor on each load but also it has only one sensor to connect in the power service entry between the utility system and the residential building. It combines the energy monitoring device, which is installed on the power entrance, with household meter. The system shows the ON and OFF information of each load by analyzing the voltage and current waveforms data from the data acquisition device (DAQ). Comparing traditional load monitoring methods with NILM, the latter does not require many devices, complex installation, and high cost for every sensor [2] [3].

In the household system, the aging of electric appliances may happen because of over time or over limitation for appliances. The aging may cause the change of load characteristics in the power and harmonic components, which will increase the energy

A. Moonis et al. (Eds.): IEA/AIE 2014, Part II, LNAI 8482, pp. 189–198, 2014.

consumption and risk. Therefore, it is necessary to detect efficiently the aging of electric appliances for saving the energy and reducing the risk.

This study proposes a Hellinger distance algorithm for extracting the power features of aging load based on a NILM system. In this paper, the optimal features are
used for load identification and the back-propagation artificial neural network input
(BP-ANN). BP-ANN, which imitates animal central nervous information process
system, is employed for the aging load detection. By adjusting the weight and bias
between the neurons, the neural network can generate network outputs that are correctly corresponding to the optimal inputs [4].

There are three experiments based on Hellinger distance algorithm for detecting
and simulating the aging of loads. The experiment implements the algorithm into the
NILM system. This research measures a lot of power signatures including the root
mean square value of voltage ($Vrms$), the root mean square value of current ($Irms$),
real power (P), reactive power (Q), total harmonic distortion of voltage (V_{THD}), total
harmonic distortion of current (I_{THD}) and odd order current harmonics (I_n) for each
load by using Power Quality Monitoring Device ADX30000. Hellinger distance algorithm is used to calculate the distance between the power signatures of new and old
load and to find the optimal features. The result shows that aging load detection based
on non-intrusive load monitoring system can detect and diagnose the aging of loads in
the residential buildings.

2   Basic Theory

In the household system, the aging of appliances may cause the increasing of energy
consumption and the change of characteristics in power. Hellinger distance algorithm
is used for detecting the aging of loads by measuring the power signatures, and to find
the optimal features. The optimal features are used for load identification, especially
for serving as input in the BP-ANN. The following sections will describe the power
signatures and Hellinger distance algorithm.

2.1   Power Signatures

Power signatures include $Vrms$, $Irms$, P, Q, V_{THD}, I_{THD} and I_n. They are calculated by
the voltage and current waveforms measured from load side. The harmonic compo-
nents are calculated by Fourier series. The equation (1) and (2) is the real power and
reactive power, respectively [5].

$$P = \sum_{n=0}^{N} P_n = V_0 I_0 + \sum_{n=1}^{N} \frac{1}{2} V_n I_n \cos(\vartheta_{V_n} - \vartheta_{I_n}) \qquad (1)$$

$$Q = \sum_{n=1}^{N} Q_n = \sum_{n=1}^{N} \frac{1}{2} V_n I_n \sin(\vartheta_{V_n} - \vartheta_{I_n}) \qquad (2)$$

where n is the order of harmonic; V_0 and I_0 are the average voltage and average
current, respectively; V_n and I_n are the effective n th harmonic components of the

voltage and current, respectively; and θ_{Vn} and θ_{In} are the n th harmonic components of the voltage and current phase angles, respectively.

The total harmonic distortion of voltage and current are as shown in equation (3) and (4), respectively.

$$THD_V\% = \frac{\sqrt{|V_2|^2 + |V_3|^2 + |V_4|^2 + \dots}}{|V_F|} \times 100\% \tag{3}$$

$$THD_I\% = \frac{\sqrt{|I_2|^2 + |I_3|^2 + |I_4|^2 + \dots}}{|I_F|} \times 100\% \tag{4}$$

where V_F and I_F are the fundamental voltage and current, respectively; V_n and I_n are the n order harmonic components of voltage and current, respectively.

2.2   Hellinger Distance Algorithm

The Hellinger distance algorithm (HD) is a type of divergence which is used to quantify the similarity between two probability distributions in probability and statistics [6]. To define the Hellinger distance algorithm, let I and J denote two probability distributions, where I and J are N positive $(I_1, I_2, I_3, \dots, I_N)$ and $(J_1, J_2, J_3, \dots, J_N)$ to satisfy $\sum I = 1$ and $\sum J = 1$, respectively. The square of HD between I and J are shown in equation (5).

$$d_H^2(I, J) = \frac{1}{2}\sum_{n=1}^{N}(\sqrt{I_n} - \sqrt{J_n})^2 \tag{5}$$

The expansion of equation (5) is defined as:

$$d_H^2(I, J) = \frac{1}{2}\sum_{n=1}^{N}(I_n + J_n - 2\sqrt{I_n J_n}) \tag{6}$$

Since $\sum I = 1$ and $\sum J = 1$,

$$d_H^2(I, J) = \frac{1}{2}(2 - 2\sum_{n=1}^{N}\sqrt{I_n J_n}) = 1 - \sum_{n=1}^{N}\sqrt{I_n J_n} \tag{7}$$

where Hellinger distance satisfies $0 \leq d_H^2 \leq 1$. The Hellinger distance is minimum if $d_H^2 = 0$, or $I = J$; The Hellinger distance is maximum if $d_H^2 = 1$ or the difference between I and J is large.

While I and J are normal distributions as shown in equation (8):

$$I \sim N(\mu_I, \sigma_I^2)$$

$$J \sim N(\mu_J, \sigma_J^2) \tag{8}$$

Hellinger distance is defined as following Equ. (9).

$$d_H^2(I, J) = 1 - \sqrt{\frac{2\sigma_I \sigma_J}{\sigma_I^2 + \sigma_J^2}} e^{-\frac{1}{4}\frac{(\mu_I^2 - \mu_J^2)^2}{\sigma_I^2 + \sigma_J^2}} \tag{9}$$

where σ_I and σ_J are the standard deviation of I and J, respectively; μ_I and μ_J are the average values of I and J. Equ. (9) shows the results calculate the difference between two probability distributions based on Hellinger distance algorithm.

3   Loads Specification and Study Cases

3.1   Loads Specification

The experiment in this study uses three kinds of electric appliances. They are total six loads including new and old rice steamer, hair dryer and electrical oven to show that Hellinger distance algorithm can extract the power features of aging load. The specification of new and old appliances is the same. The modes of loads used in this experiment are the heating mode of rice steamer, the low speed mode of hair dryer, and the 4th power of electrical oven. Table 1 shows that the specification of loads.

Table 1. Specification of loads

Load	Specification
New Rice Steamer (Heating Mode)	1Φ110V, 800W, Made in 2012
New Hair Dryer (Low Speed)	1Φ110V, 355W, Made in 2013
New Electrical Oven (4th Level Power)	1Φ110V, 900W, Made in 2012
Old Rice Steamer (Heating Mode)	1Φ110V, 800W, Made in 1998
Old Hair Dryer (Low Speed)	1Φ110V, 370W, Made in 2001
Old Electrical Oven (4th Level Power)	1Φ110V, 900W, Made in 2006

3.2   Study Cases

The power signatures of each load including $Vrms$, $Irms$, P, Q, V_{THD}, I_{THD} and I_n are measured by power quality monitoring device ADX30000. The Hellinger distance is calculated between new and old load after measuring all loads, and then to extract the optimal features which have the longest distance based on the Hellinger distance algorithm. The optimal features are used to input the BP-ANN for load identification for network training and test. Fig. 1 is the process of aging load identification. The experiments implement the proposed method to test the optimal features for detecting aging load, to simulate the aging load by adjusting the voltage, and to recognize aging load in NILM system.

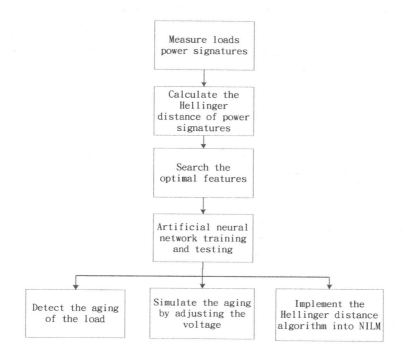

Fig. 1. The flow chart of aging load detection

4   Experimental Results

This study performs three experiments based on Hellinger distance algorithm to test the optimal features for identifying the aging of the load, to simulate the aging load by adjusting the voltage, and to implement Hellinger distance algorithm into NILM system.

4.1   Case 1 Study: Feature Extraction

In this case study, this case uses proposed method to extract optimal power signatures for indentifying a new and old rice steamer. For each rice steamer, power signatures are recorded every fifteen second. The total data is 30 in fifteen minutes. According to Hellinger distance algorithm, the average and standard deviation are calculated to find the optimal features.   Table 2 shows the results of the proposed method for a new and old rice steamer. The average and standard deviation for the P of the old rice steamer is higher than that of the new rice steamer.   Thus, the power consumption of the old load is higher than that of the new one. In Table 2, the optimal features are real power (P), reactive power (Q) and fundamental current (I_F), the HD are 1, 0.9999, and 1, respectively.

Fig. 2 is the power distributions of the new and old rice steamer in heating mode. Fig. 2 shows that the power distributions of the new and old rice steamer have a lot of difference, however P and Q. Table 3 shows the identification results of the new and old rice steamer. The average and standard deviation of optimal features P, Q and PQ are inputted into the artificial neural network. The average recognition accuracy can reach 99%

Table 2. The results of proposed method for the new and old rice steamer

Features	New Rice Steamer		Old Rice Steamer		Hellinger Distance
	μ	σ	μ	σ	
P(W)	766.955	2.729	803.186	10.573	1
Q(Var)	-2.502	0.014	-2.592	0.06	0.999
THD$_V$(%)	1.923	0.024	1.899	0.034	0.714
THD$_I$(%)	0.123	0.001	0.129	0.002	0.079
I$_F$(A)	7.033	0.021	7.403	0.084	1
I$_{3rd}$(A)	0.01	0.0007	0.009	0.0006	0.001
I$_{5th}$(A)	0.094	0.002	0.098	0.002	0.014
I$_{7th}$(A)	0.041	0.0008	0.043	0.001	0.014
I$_{9th}$(A)	0.016	0.0004	0.015	0.0007	0.042
I$_{11th}$(A)	0.056	0.002	0.061	0.001	0.053

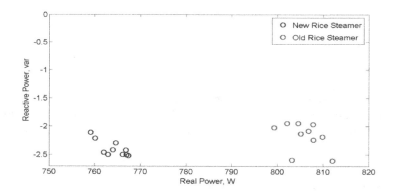

Fig. 2. Power distributions of the new and old rice steamer in heating mode

Table 3. Identification results of the Case 1 study

Features	Average Training Recognition (%)	Average Testing Recognition (%)	Average Training Time (s)
P	100	99.09	0.462
Q	100	100	0.464
PQ	100	99.09	0.458

4.2   Case 2 Study: The Simulation of the Aging Load

In this case study, the case simulates the aging of loads by adjusting the voltage. To efficiently simulate the aging of loads, an assumption is executed for aging load. The increased power of the old load is related with the increased current or voltage of the new one. Fig. 3 and Fig. 4 are the power waveform of the new rice steamer operated in 110V and 114.4V, respectively.   Fig. 5 is the power waveform of the old rice steamer operated in 110V. The figure shows that the real power waveform of the new rice steamer operated in 114.4V is similar with the real power waveform of the old rice steamer operated in 110V. As the result, the aging of loads can be simulated by adjusting the voltage.

Table 4 shows the simulation results of the new rice steamer operated in 110V and 114.4V. The average real power of the new rice steamer which simulates the old rice steamer is higher than that of the new one. The optimal features in Table 4 are real power (P), reactive power (Q), and fundamental current (I_F) while the Hellinger distance values are all 1.

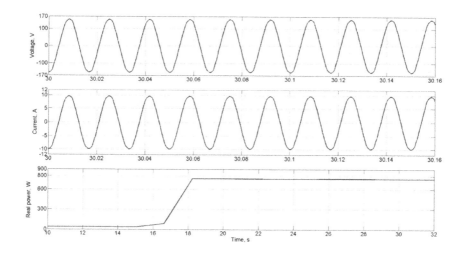

Fig. 3. Power waveform of the new rice steamer operated in 110V

In this study, the aging simulations of the new rice steamer from 110V to 110V ±20% are performed. The results show that the voltage at +4% can be used to simulate the aging of the old rice steamer. Table 5 is the identification results of the case 2 study. The input files of average of optimal features P, Q and PQ are inputted to the artificial neural network. The average recognition accuracy can reach 85%.

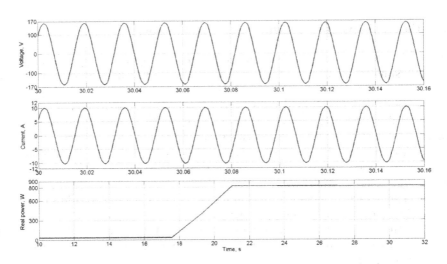

Fig. 4. Power waveform of the new rice steamer operated in 114.4V

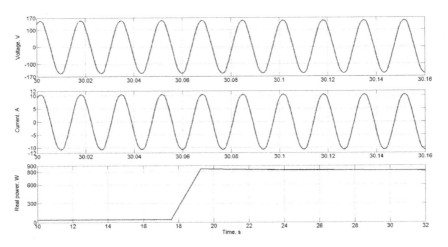

Fig. 5. Power waveform of the old rice steamer operated in 110V

Table 4. Simulation results of the new rice steamer operated in 110V and 114.4V

Features	New Rice Steamer in 110V		New Rice Steamer in 114.4V		Hellinger Distance
	μ	σ	μ	σ	
P(W)	766.955	2.729	812.633	1.502	1
Q(Var)	-2.502	0.014	-2.632	0.039	1
$THD_V(\%)$	1.923	0.024	1.969	0.025	0.998
$THD_I(\%)$	0.123	0.001	0.125	0.001	0.026

Table 4. (*Continued*)

$I_F(A)$	7.033	0.021	7.229	0.016	1
$I_{3rd}(A)$	0.01	0.0007	0.01	0.001	0.025
$I_{5th}(A)$	0.094	0.002	0.098	0.001	0.043
$I_{7th}(A)$	0.041	0.0008	0.035	0.001	0.042
$I_{9th}(A)$	0.016	0.0004	0.016	0.0006	0.012
$I_{11th}(A)$	0.056	0.002	0.057	0.0009	0.198

Table 5. Identification results of the Case 2 study

Features	Average Training Recognition (%)	Average Testing Recognition (%)	Average Training Time (s)
P	100	90.5	0.59
Q	100	92.5	0.611
PQ	100	85	0.626

4.3   Case 3 Study: Aging Load Detection in NILM System

In this case study, the results of Hellinger distance algorithm are employed to implement into NILM system. In the system, six loads are divided into the new and old of three kinds of electrical appliances. Each load is measured for voltage from 0% to ±5%. Table 6 is the identification results of this case study. The average of optimal features P, I_{THD} and I_F are inputted to the artificial neural network. The average recognition accuracy can reach 75%.

Table 6. Identification results of Case 3 study

Features	Average Training Recognition (%)	Average Testing Recognition (%)	Average Training Time (s)
$P \times I_{THD} \times I_F$	100	75	14.161

5   Conclusions

This study proposes a Hellinger distance algorithm to extract the power features of aging load based on a NILM system. The optimal features are used for load identification and the back-propagation artificial neural network training. By adjusting the weight and bias between the neurons, the neural network can produce an output that is correctly corresponding to the optimal input and to identify the aging. The results show that the proposed method of aging load detection in NILM system can be used successfully to detect and simulate the aging of loads in household system.

Acknowledgements. The authors would like to thank the National Science Council of the Republic of China, Taiwan, for financially supporting this research under Contract No. NSC 102-2221-E-228-002.

References

1. Bruce, A.G.: Reliability Analysis of Electric Utility SCADA Systems. IEEE Transactions on Power Systems 13(3), 844–849 (1998)
2. Chang, H.H.: Non-Intrusive Demand Monitoring and Load Identification for Energy Management Systems Based on Transient Feature Analyses. Energies 5, 4569–4589 (2012)
3. Chang, H.-H., Lian, K.-L., Su, Y.-C., Lee, W.-J.: Energy Spectrum-Based Wavelet Transform for Non-Intrusive Demand Monitoring and Load Identification. In: IEEE Industry Applications Society 48th Annual Meeting 2013 (IAS Annual Meeting 2013), Orlando, FL USA, October 6-11, pp. 1–9 (2013)
4. Kosko, B.: Neural networks and fuzzy systems. Prentice- Hall, New Jersey (1992)
5. Chang, H.-H., Lin, L.-S., Chen, N., Lee, W.-J.: Particle Swarm Optimization Based Non-Intrusive Demand Monitoring and Load Identification in Smart Meters. In: IEEE Industry Applications Society 47th Annual Meeting 2012 (IAS Annual Meeting 2012), Las Vegas, NV USA, October 7-11, pp. 1–8 (2012)
6. Rahul, A., Prashanth, S.K., Suresh Kumar, B., Arun, G.: Detection of Intruders and Flooding In Voip Using IDS, Jacobson Fast And Hellinger Distance Algorithms. IOSR Journal of Computer Engineering (IOSRJCE) 2, 30–36 (2012)
7. Wang, Z., Zheng, G.: Residential Appliances Identification and Monitoring by a Nonintrusive Method. IEEE Transactions on Smart Grids 3(1) (March 2012)

New Vision for Intelligent Video Virtual Reality

Wei-Ming Yeh

Department of Radio and TV,
National Taiwan University of Arts,
Taipei, Taiwan

Abstract. Since 2009, Sony DSC-HX-1 received a honor of "Best Super zoom D-camera ", in Technical Image Press Association (TIPA) Awards 2009[1]. In fact, Sony HX-1 is recognized not only having a 20x optical super zoom lens, but offering plenty intelligent video effects. It can be a new trend to judge current DSLR-like camera, such as Fujifilm FinePix HS10 received the same Award 2010[2] with same honors again. Theoretically, it is a new intelligent integration technology for video virtually reality, which provide multiple platform users for video camera, video game, and mobile phone all together. Administers from variety of fields begin to think how to integrate some hot-selling video scene effects from all possible mobile video products, developing this "dazzling" virtually reality (VR) imagination beyond limitation, to attract more potential consumers, which can be vital for small businesses, such as: photo and cell phone companies,　to survive in the future. In our experiment, we collect more than 300 cases from the telephone survey during September 2013 to December 2013. Total of 212 cases comply with the conditions. To probe mainly into the relationship between new generation video effects confidence level and 3 potential consumers: Amateur Photographer (AP), Senior Photographer (SP), and college student (CS). That is the reason what we are probe into this highly competitively market with brilliant creative design, and hope to offer an objective suggestion for both industry and education administers.

Keywords: Augmented Reality, CyberCodes, Fiduciary marker, High Dynamic Range, Smart AR, Visual Perception, Wireless Sharing Technique.

1　Introduction

Since early 2009, as soon as the digital video technology had great success in many fields, such as: video game, mobile phone, and digital video camera(DSC). Many small photo industries in Japan, realized that digital video technology could be the only way to compete with two giant photo tycoons(Canon and Nikon) in the market. In fact, Canon and Nikon have dominated the DSLR and SLR market for decades, and famous for high quality lens, CMOS and CCD sensor, processor, and accessories, not willing to follow this　trend, to develop such fancy digital scene effect for their high-level DSLR camera surprisingly. Currently, Sony, M 4/3 family, and Samsung, realized that using digital video technology may achieve a niche between high-end

A. Moonis et al. (Eds.): IEA/AIE 2014, Part II, LNAI 8482, pp. 199–206, 2014.

DSLR, and consumer-level DSC, offering shining stars, such as: Sony HX-1, HX-100V, Panasonic GF1, GF2, GF3, and many other products, recognized as prominent DSC in TIPA 2009, 2010, 2011 Awards. It is a simple fact that   DSLR-like camera and Mirrorless Interchangeable Lens Camera (MILC or EVIL)camera, with plenty of digital video technology (digital scene effects or customized scene effects), could attract many armature consumers and white-collar female consumers, offering giant marketing breakthrough, and survived in this margin-profit camera market. Actually, since 2007, the IT industries have co-operated with camera manufactures, developed many built in IC chips with brilliant customized scene effects for cameras (DSC mainly), such as: Face Detection Technology, Smile Shutter Mode, Full frame HD, CCD Anti Shake system, Live View, and many others. In fact, within few years, in early 2009, we found something new, such as: Back-illuminated CMOS image sensor[3], Sweep Panorama, Joining Multiple Exposed Patterns, Night Scene Portrait ,Motion Remover[7], that would challenge multimillion dollars business not in DC but highly profitable DSLR market, and may create a new possible attraction for photo-fans to replace their old camera.

In addition, since 2010, after the introduction of 3D movie (AVATAR), people all over the world seems to enjoy this 3D "mania"   in their life, no matter 3D monitor for hardware, and   3D Panorama video effect for photo image., and Smart AR   just another new intelligent technology wonder, which may create huge   marketing benefit ahead.

Fig. 1. Sony Smart AR
Source: http://gnn.gamer.com.tw/5/53855.html

In fact, Smart AR is one of Sony new VR technology , which is related to Augmented reality (AR) , widely adopted by sports telecasting, video   games, automobile, even jet-fighter weapon system(F-35), and ready for advanced video effect someday.   It can offer a term for a live direct or an indirect view of a physical, real-world environment whose elements are augmented by computer-generated sensory input, such as sound or graphics. It is related to a more general concept called mediated reality, in which a view of reality is modified (possibly even diminished rather than augmented) by a computer. As a result, the technology functions by enhancing one's current perception of reality. By contrast, virtual reality replaces the real world with a simulated one [5].

2   Technical Innovation

Generally, all hot-selling video scene effects are depending on the requirements from market.

In the beginning of DSC, only expensive DSLR with high quality image and cheap compact could survived, since decades ago.

As the progress of technology improvement, new categories of DSC were reveled, such as EVIL camera and DSLR-like, both equipped with many interesting video scene effects, to attract more consumers, those who enjoying private shooting, with neat size, and quality image, especially for white-collar female and low-key specialist.

Therefore, a camera with plenty of built-in video scene effects, which is designed by many experienced photographers, can help people to take good picture effortless.

Currently, there are many technical wonders of chips (Customized scene effects) for DSLR-like market, create amazing profit. We integrated some of them, which should be widely accepted, with reason price in 2009 till 2010, they were: blur-free twilight scenes shooting, Lightning-fast continuous shooting, Full HD Video, Sweep Panorama. As to another technical wonders, such as: 50× Optical Zoom, Super High Speed ISO.., they might not be as popular (accepted)as people thought, or simply limited in few models. For example, in 2008 only a few DSLRs equipped with HD video. Currently, it is hardly to find a DSC without HD video ironically (before 2009). In addition, the Fujifilm HS10 (Fall 2009), offers two special scene effects: Motion Remover and Multi Motion Capture, which won a strong attention in this market, and HS 20 (Feb,, 2011) keep the good merit of Multi Motion Capture but delete Motion Remover(too complicate to handle). In addition, new Sony HX 100V (May, 2011) offers GPS, High Dynamic Range (HDR), highly sophistic 3D Panorama, and Wireless Sharing Technique (Wi-Fi) in Alpha 7(2013), which may lead a new direction for future camera and cell phone development. Coming with the new era of 3D technology, it is easy to take 3D photo by proper camera, and manipulating 3D video image by special software, without bulky and expensive equipments. In addition, there are many brilliant 3D products available, such as: 3D TV, projector, Polarized 3D glasses, 3D game, and cell phone.

2.1   Phone Survey

In our experiment, we collect more than 300 cases from the telephone survey during September 2013 to December 2013. Total of 212 cases were effective, telephone surveys were written by two specialists, and those non-typical cases were determined after further discussion by 3 specialists.

2.2   Associated Analysis of Scene Effect Confidence Level

We took profile of contingency table proposed by Jobson (1992) to describe the preference level of scene effect associated with three user groups[7]. The study probes into preference confidence level and three specific groups, use chi-square test of independence to confirm the results.

2.3   The Different Groups in Confidence Level

The relation of preference level and three groups we present as 3 × 4 contingency table (the preference level 0, 1, 2, 3 classified low, slight, moderate and high). Seldom have preference level "3", so we amalgamate "2" and "3" to "2" (above moderate) and utilize this data to find out conditional and marginal probability, presented 3 × 3 probability table (Tab.1).

Table 1. Group × Video effects Preference: Row proportions

Video effect	Preference level		
Group	0	1	2
SP	0.2830	0.4528	0.2642
AP	0.7206	0.1324	0.1471
CS	0.5263	0.3158	0.1579
marginal prob.	0.4670	0.3255	0.2075

Table 2. Group by Video effects preference profile

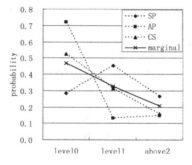

We utilize and draw profile of different groups with conditional and marginal probability of different groups preference level. It is for SP, AP and CS respectively in Fig. 3.Where solid line denotes the marginal probability of the levels, the dotted lines the conditional probability. Find levels "low"(level 0) and "slight"(level 1), there is a maximum disparity from solid line to others, nearly up to 0.2 in the AP, therefore can infer there is less preference level in the AP relatively, and greater probability of having slight confidence level in the SP, this shows it were to varying degrees influenced by the video effect preference, though with different categories of fan groups, hence the preference level associates with video effect. We use chi-squared test for independence to confirm and associated with the preference level, the χ^2 -value would be 33.7058, and P =0.

3   Proportional Odds Model

We took a test for independence to confirm the different groups preference level and three groups in section 3: The different groups' preference level is related, but this test has not used the ordinal response level to confidence level. The regular collects the classification data with order in the camera research. For example it mentioned in the article the response variables of 'low', 'slight', 'moderate' and 'high' preference, we hope to probe into the factor influencing its preference level, so we analyze utilizing proportional odds model. only one covariate, model present straight line relation to the logarithm of the accumulated odds of level p and covariate variable x, because this model supposes that there is the same slope β to all level p, this c-l straight line is parallel each other. This model is based on McCullagh and Nelder theory (1989) and Diggle, Liang and Zeger (1994). One predictable variable was included in the study, representing known or potential risk factors. They are kinds of variable status is a categorical variable with three levels. It is represented by two indicator variables x (x_1 and x_2), as follows:

Table 3. 3 groups of indicator variables

kinds	x_1	x_2
SP	1	0
AP	0	1
CS	0	0

Note the use of the indicator variables as just explained for the categorical variable. The primary purpose of the study was to assess the strength of the association between each of the predictable variables. Let

$$L_p(x_1, x_2) = \theta_p + \beta_1 x_1 + \beta_2 x_2, \quad p = 1,2 \tag{1}$$

Similar ratio tests for $b_1 = b_2 = 0$ hypothesis, the explanation of likelihood ratio tests is as follows: given L be the log likelihood of models , then G2= -2L. From hypothesis $b_1 = b_2 = 0$, we have

$$L_p(x) = \theta_p, \quad p = 1,2,\cdots,c-1 \tag{2}$$

Similar ratio tests use the difference of two deviances between model (1) and model (2) as reference value to test $H_0 : b_1 = b_2 = 0$. If reject the hypothesis, furthermore, to test if $b_1 = 0$ or $b_2 = 0$. As fact, the preference level "under moderate" to the preference level "high", the odds ratio of SP compared with CS is also $\exp(b_1)$. Now according to formula, b_1 is the logarithm of the estimated odds when the preference level "under high" to the preference level "high", the SP compared with CS. $b_1 > 0$

means the preference level presented by SP is less high than CS. Alternatively, $b_1 < 0$ means the preference level presented by SP is higher than CS.

Therefore, $b_1 - b_2$ is the logarithm of the estimated odds when the preference level "under moderate" to the preference level "high", the SP compared with AP. $b_1 - b_2 > 0$ means the preference level presented by SP is less high than AP. Alternatively, $b_1 - b_2 < 0$ means the specify preference level presented by SP is higher than AP. On the rear part in the article, we will utilize odds ratio to probe into association between fan groups and video effect preference level.

3.1   Analysis of Different Groups Confidence Level

Combining confidence level "2" and " 3" to "2"(above moderate), we utilize odds ratio to calculate, deviance equal 416.96, if $b_1 = b_2 = 0$, deviance = 444.04, the difference between two deviances is 27.087, the related χ^2-critical value will be test $H_0 : \beta_1 = \beta_2 = 0$, we find $\chi^2_{0.05} = 27.087$, and P=0.001. The null hypothesis is rejected and we conclude that β_1 and β_2 are not zero simultaneously. Furthermore, we analyze $\beta_1 = 0$ or $\beta_2 = 0$ or both not equal to zero. Table 3 is the results by maximum likelihood estimates. From Table 3 we find the hypothesis $\hat{\beta}_1 = 0$ is rejected, $\hat{\beta}_2 = 0$ is accepted, P-value is 0.017 and 0.072 respectively, $\hat{\beta}_2$ represents the logarithm of odds ratio for AP, thus the odds ratio would be estimated to be e^{β_2}. $\hat{\beta}_1 < 0$ represents the logarithm of odds ratio of preference≤ 0 rather than preference> 0 of SP is 0.423 fold than that of CS. This indicates that the logarithm of odds ratio of preference> 0 rather than preference≤ 0 of video effect preference Confidence level is about 2.36 fold for SP compared to CS customer. $\hat{\beta}_2 = 0$ means the confidence level presented by AP is the same as by CS, make sure the result is unanimous of this result.

Table 4. Analysis of the Video effect preference confidence level

Effect	B	Coefficient	Std Error	z-statistic	Sig
Intercept	$\hat{\theta}_1$	0.093	0.314	0.298	0.765
	$\hat{\theta}_2$	1.713	0.34	5.043	0.0000
Video	$\hat{\beta}_1$	-0.861	0.36	2.393	0.017
	$\hat{\beta}_2$	0.745	0.414	-1.802	0.072

3.2   Preference of Various Video Effects in Different Groups

Based on this data, AP (Amateur Photographer) may enjoy most of video effects, we assumed built in video effects could save great amount of time and budget, and pay more attention on "shooting". As to Senior Photographer(SP), usually have professional software and better equipment for years, they are famous to take quality picture by high-level DSLR with manual operation, doing post production with professional software, doing sophisticate effects alone. College student (CS) only have limited budget and time to post photos for facebook or website with limited quality frequently, .

As to the 2014, the 3D Sweep Panorama(3DP), GPS, High Dynamic Range (HDR), and Wireless Sharing Technique (Wi-Fi) can be most popular effects now. We realize that Senior Photographer (SP) is willing to accept all possible new technologies and devices, and followed by Amateur Photographer (AP), as to College student (CS) show their easy attitude all the times.

4   Conclusion and Suggrstion

Since decades ago, people have enjoyed the pleasure to take pictures or recording anything by various digital cameras (DSC/DV) and cell phone now, in replace of old film-camera, and it can take good picture easily. The future of DSC and cell phone will be full of surprised, handy in use, energy saving, affordable price, and more customized scene effects, in order to please all kinds of users, which set a new record for successors already, and push camera and cell phone manufactures to have more user-friendly innovation in entry-level model (Sony α series) especially. As to the latest wonders, Sony Smart AR combines the technology of 3D, Game, and VR, which may challenge the current hot-selling 3D panorama and VR effect in the future, and may create new 3D hardware market (DSC, Monitor/TV, Phone, NB..etc) soon, show in Fig2. As to the future, this new technology can enhance one's current perception of reality image, and to simulate a new image, can be no constraint than ever.

Fig. 2. Sony Xperia Z1 AR effect

Source: http://www.sonymobile.com/gb/products/phones/
xperia-z1/features/#camera-apps

References

1. TIPA Awards 2009 Web Site.: The best imaging products of 2009, TIPA (May 2010), http://www.tipa.com/english/XIX_tipa_awards_2009.php
2. TIPA Awards 2010 Web Site.: The best imaging products of 2009, TIPA (May 2011), http://www.tipa.com/english/XX_tipa_awards_2010.php
3. Rick User.: New Sony Bionz processor, DC View (May 2, 2010), http://forums.dpreview.com/forums/ read.asp?forum=1035&message=35217115
4. Alpha Sony.: Sony Alpha 55 and 33 Translucent Mirror. Electronista (October 18, 2010)
5. Duncan Graham-Rowe: Sony Sets Its Sights on Augmented Reality, Technology Review (MIT) (May 31 (2011), http://www.technologyreview.com/printer_friendly_article.asp x?id=37637
6. Behrman, M.: This Augmented Reality Needs No Markers to Interfere With Your (Virtual) World, Gizmodo (May 19, 2011), http://gizmodo.com/5803701/this-augmented-reality-needs-no-markers-to-interfere-with-your-virtual-world
7. Moynihan, T.: Fujifilm's Motion Remover, PC World (March 3, 2010), http://www.macworld.com/article/146896/2010/03/sonys_intelli gent_sweep_panorama_mode.html
8. Sony Web Site.: Sony Xperia Z1 AR effect, Sony (2013), http://www.sonymobile.com/gb/products/phones/xperia-z1/features/#camera-apps

An Improved Liu's Ordering Theory Based on Empirical Distributive Critical Value

Hsiang-Chuan Liu[*], Ben-Chang Shia[*], and Der-Yin Cheng

Graduate School of Business Administration, Fu Jen Catholic University
No. 510, Chung Chen Rd. Hsin Chuang Dist. New Taipei City (24205), Taiwan
lhc@asia.edu.tw, {stat1001,metiz149159}@gmail.com

Abstract. Since the Ordering Theory (OT) has only considered the item ordering relationship rather than the item non-independence, and the Item Relational Structure theory (IRS) has only focused on the item non-independence but no thought for the item ordering relationship, the first author of this paper proposed his improved theory, called Liu's Ordering Theory (LOT), which has considered both the item ordering relationship and item non-independence. However, all of the critical values of the ordering index of above-mentioned three theories are subjectively fixed numbers. In this paper, for overcoming the lack of statistical meaning, an empirical distributive critical value of the ordering index based improved LOT theory, denoted as ILOT, was proposed, this new theory is more reasonable and useful than OT, IRS and LOT. Furthermore, by using the new method ILOT, based on the theory of the ideal test proposed by the first author of this paper, we can construct the validity index of item ordering structure of any group of examinees to compare the performances of any different groups of examinees.

Keywords: Ordering relationship, non-independence, OT, IRS and LOT.

1 Introduction

For detecting the item ordering relationships or directed structures of a group of subjects, we can no more adapt the traditional symmetrical relations methods, such as the methods of correlation, distance and similarity etc, There are three well-known item ordering structure theories based on the testing performance of a group of examinees, the first is Ordering theory (OT) proposed by Bart et al in 1973 [1]-[2], and the second is Item relational structure theory (IRS) proposed by Takeya in 1991 [3]-[4], the third is Liu's Ordering Theory (LOT) proposed the first author of this paper in 2012 [8]. The OT theory has only considered the item ordering relationship rather than the item non-independence, and the IRS theory has only focused on the item non-independence but no thought for the item ordering relationship, the LOT has considered both the item ordering relationship and item non-independence.

---

[*] Corresponding Author.

A. Moonis et al. (Eds.): IEA/AIE 2014, Part II, LNAI 8482, pp. 207–217, 2014.

However, the threshold limit values of all of above-mentioned three theories are subjectively fixed values, lacking of statistical meaning, the first author of this paper, H.-C. Liu, transferred the ordering index as an approximated t value to obtained a critical t value at significant level 0.05 [5]-[6], but it is only used for nominal scale data, not for interval scale data.

In this paper, the authors consider to improve the threshold limit value by using the empirical distribution critical value of all the values of the relational structure indices between any two items, it is more valid than the traditional threshold for comparing the ordering relation of any two items.

Furthermore, in this paper, by using the new method ILOT, based on the first author Liu's Ideal Test theory, we can construct the validity index of item ordering structure of any group of examinees different groups of examinees to compare the performances of any different groups of examinees. A calculus example is also provided in this paper to illustrate the advantages of the proposed methods.

2   Well-Known Item Ordering Structure Theories

In this section, we briefly review three well-known and more important item ordering structure theories [1]–[6] as follows;

2.1   Ordering Theory

Airasian and Bart（1973）[1]-[2] thought that for any two different items in a test, if the easier item I_i is a precondition of the more difficulty item I_j, then any examinee answers item I_i wrong always so do item I_j. In other words, if item I_i is a precondition of item I_j, then the joint probability of item I_i wrong and item I_j correct is small. According to this consideration, they provided the first ordering structure theory of items, called the Ordering Theory (OT), as following:

Definition 2.1: Ordering Theory (OT)
Let $\varepsilon \in [0.02, 0.04]$, $I_T = (I_1, I_2, \cdots, I_n)$ represent a test containing n binary item scores variables with ones and zeros, and $P(I_i=0, I_j=1)$ represent the joint probability of all examinees answering item I_i wrong and item I_j correct, if $P(I_i=0, I_j=1) < \varepsilon$ then we say that item I_i is a precondition of item I_j. The relation is denoted as $I_i \rightarrow I_j$, otherwise $I_i \nrightarrow I_j$

Liu (2007) [7] rewritten this definition as follows;

Definition 2.2: Ordering Theory (OT)
Let $\delta \in [0.96, 0.98]$, $I_T = (I_1, I_2, \cdots, I_n)$ and $P(I_i=0, I_j=1)$ be defined as given above. the joint probabilities of items I_i and I_j, $P(I_i = x, I_j = y)$, $x, y \in \{0,1\}$, be listed in Table 1,

(i)   The ordering index from item I_i to item I_j, is defined as

$$\gamma_{ij}^{(OT)} = 1 - P\left(I_i = 0, I_j = 1\right) \tag{1}$$

(ii)   if $\gamma_{ij}^{(OT)} > \delta$, then we say that I_i is a precondition of I_j, denoted as $I_i \rightarrow_{OT} I_j$, otherwise $I_i \not\rightarrow_{OT} I_j$.

(iii)   If $I_i \rightarrow I_j$ and $I_j \rightarrow I_i$, then we say that I_i and I_j are equivalent, denoted as $I_i \leftrightarrow_{OT} I_j$, otherwise $I_i \not\leftrightarrow_{OT} I_j$

For convenience, let $\varepsilon = 0.03$, $\delta = 0.97$, in this paper.

Table 1. The joint probabilities of item I_i and I_j

$P(I_i=x,I_j=y)$	$I_j = 1$	$I_j = 0$	$P(I_i=x)$
$I_i = 1$	$P(I_i=1,I_j=1)$	$P(I_i=1,I_j=0)$	$P(I_i=1)$
$I_i = 0$	$P(I_i=0,I_j=1)$	$P(I_i=0,I_j=0)$	$P(I_i=0)$
$P(I_j=y)$	$P(I_j=1)$	$P(I_j=0)$	1

Examlple 1: Suppose that the joint and marginal probabilities of item I_i and I_j of a group of   subjects are listed as Table 2

Table 2. The joint probabilities of item I_i and I_j

$P(I_i=x,I_j=y)$	$I_j = 1$	$I_J = 0$	$P(I_i=x)$
$I_i = 1$	0.48	0.06	0.54
$I_j = 0$	0.02	0.44	0.46
$P(I_j=y)$	0.5	0.5	1

Table 3. The joint probabilities of item I_i and I_j

$P(I_i=x,I_j=y)$	$I_j = 1$	$I_J = 0$	$P(I_i=x)$
$I_i = 1$	0.48	0.48	0.96
$I_j = 0$	0.02	0.02	0.04
$P(I_j=y)$	0.5	0.5	1

From Table 2. we know that

(i) $\gamma_{ij}^{(OT)} = 1 - P\left(I_i = 0, I_j = 1\right) = 0.98 > \delta = 0.97 \Rightarrow I_i \rightarrow_{OT} I_j \tag{2}$

(ii) $\gamma_{ji}^{(OT)} = 1 - P\left(I_i = 1, I_j = 0\right) = 0.94 < \delta = 0.97 \Rightarrow I_j \not\rightarrow_{OT} I_i \tag{3}$

Examlple 2: Suppose that the joint and marginal probabilities of item I_i and I_j of a group of   subjects are listed as Table 3

From Table 3. we know that

(i) $\gamma_{ij}^{(OT)} = 1 - P\left(I_i = 0, I_j = 1\right) = 0.98 > \delta = 0.97 \Rightarrow I_i \to_{OT} I_j$ (4)

(ii) $\gamma_{ji}^{(OT)} = 1 - P\left(I_i = 1, I_j = 0\right) = 0.52 < \delta = 0.97 \Rightarrow I_j \not\to_{OT} I_i$ (5)

Notice that in Example 2., from (3); we know that $I_i \to I_j$
but $P\left(I_i = 0, I_j = 1\right) = P\left(I_i = 0\right)P\left(I_j = 1\right) = 0.02 \Rightarrow [I_i = 0]^{\perp\!\!\!\perp}[I_j = 1]$
Since if $I_i \to_{OT} I_j$, then I_i and I_j must not be independent, it leads to a contradiction, in other words, Ordering Theory has not consider the case of independence of two items.

2.2   Item Relational Structure Theory

Takeya (1978) [3]-[4] thought that if item I_i is a precondition of item I_j, then I_i and I_j are non-independent. Focusing on the non-independence of two items, Takeya proposed his alternative item ordering structure theory, called Item Relational Structure theory (IRS) [3]–[4] as follows;

Definition 2.3: Item Relational Structure theory (IRS) [3]-[4]
Let $I_T = (I_1, I_2, \cdots, I_n)$ and $P\left(I_i = 0, I_j = 1\right)$ be defined as given above,, $P(I_i = 0)$ represents the probability of item I_i wrong, and $P\left(I_j = 1\right)$ represents the probability of item I_j correct.
  (i) The ordering index from I_i to I_j is defined below;
$$\gamma_{ij}^{(IRS)} = 1 - \frac{P\left(I_i = 0, I_j = 1\right)}{P\left(I_i = 0\right)P\left(I_j = 1\right)} \tag{6}$$

  (ii) If $\gamma_{ij}^{(IRS)} > 0.5$, then I_i is a precondition of I_j, denoted as $I_i \to_{IRS} I_j$, otherwise $I_i \not\to_{IRS} I_j$.

  (iii) If $I_i \to_{IRS} I_j$ and $I_j \to_{IRS} I_i$, then we say that I_i and I_i are equivalent, denoted as $I_i \leftrightarrow_{IRS} I_j$, otherwise $I_i \not\leftrightarrow_{IRS} I_j$

Examlple 3: Suppose that the joint and marginal probabilities of item I_i and I_j of a group of   subjects are listed as Table 4

From Table 4, we can obtain the following two results;

$$\gamma_{ij}^{(OT)} = \gamma_{ji}^{(OT)} = 1 - P\left(I_i = 0, I_j = 1\right) = 0.98 < \delta = 0.97 \Rightarrow I_i \leftrightarrow_{OT} I_j \tag{7}$$

$$\gamma_{ij}^{(IRS)} = \gamma_{JI}^{(IRS)} = 1 - \frac{P\left(I_i = 0, I_j = 1\right)}{P\left(I_i = 0\right)P\left(I_j = 1\right)} = \frac{23}{25} > 0.5 \Rightarrow I_i \leftrightarrow_{IRS} I_j \tag{8}$$

It means that IRS and OT have the same ordering structure relationship, but IRS is without considering the case of independence of two items, it seems

Table 4. The joint probabilities of item I_i and I_j

$P(I_i=x,I_j=y)$	$I_j = 1$	$I_j = 0$	$P(I_i=x)$
$I_i = 1$	0.48	0.02	0.5
$I_j = 0$	0.02	0.48	0.5
$P(I_j=y)$	0.5	0.5	1

Table 5. $P(I_i=x,I_j=y)$ with $0.04<d<0.12$

$P(I_i=x,I_j=y)$	$I_j = 1$	$I_j = 0$	$P(I_i=x)$
$I_i = 1$	0.5-d	d	0.5
$I_j = 0$	d	0.5-d	0.5
$P(I_j=y)$	0.5	0.5	1

Examlple 4: If $0.04<d<0.12$, suppose that the joint and marginal probabilities of item I_i and I_j of a group of subjects are listed as Table 5

From Table 5, we can obtain the following two different results;

$$\gamma_{ij}^{(OT)} = \gamma_{ji}^{(OT)} = 1 - P(I_i=0,I_j=1) = 1 - d < \delta = 0.97 \Rightarrow I_i \not\nrightarrow_{OT} I_j, I_j \not\nrightarrow_{OT} I_i \qquad (9)$$

$$\gamma_{ij}^{(IRS)} = \gamma_{JI}^{(IRS)} = 1 - \frac{P(I_i=0,I_j=1)}{P(I_i=0)P(I_j=1)} > \frac{12}{25} > 0.5 \Rightarrow I_i \leftrightarrow_{IRS} I_j \qquad (10)$$

These results show that IRS and OT have opposite ordering structure relationships, in this case, the result of OT is more reasonable than which of IRS, in other words, OT is better than IRS, if two items are not independent.

2.3   Liu's Ordering Theory [8]

Example 2 shows that OT can only not be used for two independent items, and Example 4 points that OT is always better than IRS, if two items are not independent. Furthermore, Liu (2007) proposed an ordering consistence property as Theorem 2.1, which can only fit for OT theory not for IRS theory, in other words, he emphasized that OT theory is always better than IRS theory, in the meanwhile, he proposed his improved Ordering Theory based on non-independence, called Liu's OT theory, as follows [8];

Theorem 1: The ordering consistence properties of OT theory [8]

(i)
$$\gamma_{ij}^{(OT)} - \gamma_{ji}^{(OT)} = P(I_i=1) - P(I_j=1) \qquad (11)$$

(ii)
$$\gamma_{ij}^{(OT)} > \gamma_{ji}^{(OT)} \Leftrightarrow P(I_i=1) > P(I_j=1) \qquad (12)$$

Notice that Theorem 2.1 points out that the easier item always is the precondition of the more difficulty item, these properties only fit for OT theory not for IRS, in other words, OT theory is more reasonable than IRS theory.

Definition 2.4: Liu's Ordering Theory (LOT) [8]

Let $I_T = (I_1, I_2, \cdots, I_n)$ and $P\left(I_i = x, I_j = y\right)$, $x, y \in \{0,1\}$ are defined as above,

(i) The Liu's ordering index from I_i to I_j is defined below;

$$\gamma_{ij}^{(LOT)} = \left[1 - P\left(I_i = 0, I_j = 1\right)\right]^{\#} = \begin{cases} 0 & \text{if } Ind = 1 \\ 1 - P\left(I_i = 0, I_j = 1\right) & \text{if } Ind \neq 1 \end{cases} \quad (13)$$

$$\text{Where} \qquad Ind = \frac{P\left(I_i = 0, I_j = 1\right)}{P\left(I_i = 0\right) P\left(I_j = 1\right)} \quad (14)$$

(ii) If $\gamma_{ij}^{(LOT)} > 0.97$, then we say that I_i is a precondition of I_j, denoted as $I_i \rightarrow_{LOT} I_j$, otherwise $I_i \not\rightarrow_{LOT} I_j$.

(iii) If $I_i \rightarrow_{LOT} I_j$ and $I_j \rightarrow_{LOT} I_i$, then we say that I_i and I_i are equivalent, denoted as $I_i \leftrightarrow_{LOT} I_j$, otherwise $I_i \not\leftrightarrow_{LOT} I_j$

3   Improved Liu's Ordering Theory (ILOT)

From now on, we know that Liu's Item Ordering Theory is better than OT theory and IRS theory. However, all of the critical values of the ordering index of above-mentioned three theories are subjectively fixed numbers. In this paper, for overcoming the lack of statistical meaning, referring to our previous works [5]-[7], an empirical distributive critical value of the ordering index based improved LOT theory, denoted as ILOT, is proposed as follows;

Definition 3.1: Improved Liu's Ordering Theory (ILOT)

Let $I_T = (I_1, I_2, \cdots, I_n)$ represent a test containing n binary item scores variables with ones and zeros, $P\left(I_i = 0, I_j = 1\right)$, $P\left(I_i = 0\right)$, $P\left(I_j = 1\right)$ and $\gamma_{ij}^{(LOT)}$ are defined as before, The ordering index from I_i to I_j of the improved Liu's ordering theory (ILOT), denoted as $r_{ij}^{(ILOT)}$, is defined as $r_{ij}^{(ILOT)} = r_{ij}^{(LOT)}$, Let n be number of items, m be number of examinees, therefore, the \number of all ordering index $r_{ij}^{(ILOT)}$ is $n(n-1)$, then we can obtain a distribution of all ordering index $r_{ij}^{(ILOT)}$,

Let the threshold limit value of ILOT, denoted as $r_c^{(ILOT)}$, be defined as

$$r_c^{(ILOT)} = \arg \left[1 - \int_{-\infty}^{x} f\left(r_{ij}^{(ILOT)}\right) dr_{ij}^{(ILOT)} = 0.05 \right] \qquad (15)$$

where $f\left(r_{ij}^{(ILOT)}\right)$ is the probability density function of random variable $r_{ij}^{(ILOT)}$.

Test $H_0 : \rho_C \leq 0$ vs $H_1 : \rho_C > 0,$ at $\alpha = 0.05,$

using the statistic $\qquad t_c = \dfrac{r_c^{(ILOT)}}{\sqrt{\dfrac{1 - \left[r_c^{(ILOT)} \right]^2}{m-2}}} \sim t_{(m-2)} \qquad (16)$

where m is the number of sample size.

If $\quad t_c > t_{1-0.05}\left(m-2\right)$, then reject H_0, in other words, $r_c^{(ILOT)}$ can be used as a valid threshold limit value of ILOT;

(i)   If $r_{ij}^{(ILOT)} > r_C^{(ILOT)}$, then we say that I_i is a precondition of I_j, denoted as $I_i \rightarrow_{ILOT} I_j$, otherwise $I_i \not\rightarrow_{ILOT} I_j$,

(ii)   If $I_i \rightarrow_{ILOT} I_j$ and $I_j \rightarrow_{ILOT} I_i$. then we say that I_i and I_j are equivalent, denoted as $I_i \leftrightarrow_{ILOT} I_j$, otherwise $I_i \not\leftrightarrow_{ILOT} I_j$.

4   Ideal Item Ordering Matrix and Examinee Item Ordering Matrix

For any given test by using the same item ordering structure theory, how to evaluate the performances of item ordering structure of different groups of examinees is an important open problem, for solving this problem, the first author of this paper proposed his evaluating method by using matrix theory [8]-[9] as follows;

4.1   Ideal Item Ordering Matrix, Examinee Item Ordering Matrix [8]

Refer to Liu's Ideal Item Ordering Matrix and Examinee Item Ordering Matrix of LOT, we can define the Ideal Item Ordering Matrix and Examinee Item Ordering Matrix of ILOT as follows;

Definition 4.1: Ideal item ordering relation, Ideal item ordering matrix and Examinee Item Ordering Matrix of ILOT

Let $I_T = (I_1, I_2, \cdots, I_n)$ represent a test containing n binary item scores variables with ones and zeros,

(i) For any examinee taking I_T without guessing and slipping, if I_i is answered wrong, and must so is I_j, then we say that I_i is an ideal precondition of I_j, defined as $I_i \mapsto I_j$, otherwise $I_i \not\mapsto I_j$;

(ii) The ideal precondition item relation, $\mapsto$, is also called ideal item ordering relation

(iii) Let I_T be a test, if all ordering relations of its items are ideal item ordering relation, then I_T is called an ideal test.

(iv) Let the ideal item ordering matrix $T_E(I_T)$ represent the ideal item ordering

structure of I_T, where $T_E(I_T) = \left[e_{ij} \right]_{n \times n}$, $e_{ij} = \begin{cases} 1 & if \ I_i \mapsto I_j \\ 0 & if \ I_i \not\mapsto I_j \end{cases}$ (17)

(v) Let the examinee item ordering matrix $T_{E'-ILOT}(I_T)$ represent the examinee

item ordering structure of I_T by using ILOT theory,

where $T_{E'-ILOT}(I_T) = \left[e'_{ij} \right]_{n \times n}$, $e'_{ij} = \begin{cases} 1 & if \ I_i \rightarrow_{ILOT} I_j \\ 0 & if \ I_i \rightarrow_{ILOT} I_j \end{cases}$ (18)

Theorem 2: Translation rule for Ide al Item Ordering relation

$$\forall I_i, I_j, I_k \in I_T = \{I_1, I_2, ..., I_N\}, I_i \mapsto I_j, I_j \mapsto I_k \Rightarrow I_i \mapsto I_k \quad (19)$$

Notice that Ideal Item Ordering relation is a definite ordering relation satisfying the translation rule, so   do not OT, IRS, LOT and ILOT, since OT, IRS, LOT and ILOT, are probability ordering relation.

Examlple 5: illustration example for ideal item ordering relation
Let $I_T = \{I_1, I_2, I_3, I_4, I_5, I_6, I_7\}$ be a test containing 7 calculus items as follows;

$$I_1 : \frac{d}{dx} x^3 = ? \quad I_2 : \frac{d}{dx} e^x = ? \quad I_3 : \frac{d}{dx} \sin x = ? \quad I_4 : \frac{d}{dx}(x^3 + e^x) = ?$$
$$I_5 : \frac{d}{dx}(x^3 + \sin x) = ? \quad I_6 : \frac{d}{dx}(e^x + \sin x) = ? \quad I_7 : \frac{d}{dx}(x^3 + e^x + \sin x) = ?$$
   (20)

From **Definition 4.1,** we can obtain the ideal item ordering relations as follows;

$$I_1 \mapsto I_1, I_4, I_5, I_7 \qquad I_2 \mapsto I_2, I_4, I_6, I_7 \qquad I_3 \mapsto I_3, I_5, I_6, I_7$$
$$I_4 \mapsto I_4, I_5, I_6, I_7 \qquad I_5 \mapsto I_5, I_7 \qquad I_6 \mapsto I_6, I_7 \qquad I_7 \mapsto I_7 \tag{21}$$

And we can obtain Ideal item ordering matrix as follow;

$$T_E(I_T) = \begin{bmatrix} e_{ij} \end{bmatrix}_{n \times n} = \begin{bmatrix} 1 & 0 & 0 & 1 & 1 & 0 & 1 \\ 0 & 1 & 0 & 1 & 0 & 1 & 1 \\ 0 & 0 & 1 & 0 & 1 & 1 & 1 \\ 0 & 0 & 0 & 1 & 1 & 1 & 1 \\ 0 & 0 & 0 & 0 & 1 & 0 & 1 \\ 0 & 0 & 0 & 0 & 0 & 1 & 1 \\ 0 & 0 & 0 & 0 & 0 & 0 & 1 \end{bmatrix} \tag{22}$$

4.2   Validity Index of Examinee Item Ordering Matrix of ILOT

A novel validity index for evaluating the examinee item ordering structure by using OT or IRS was proposed by the first author of this paper [9], we can also define the validity index for evaluating the examinee item ordering structure by using ILOT as follows;

Definition 4.2: Validity Index of Examinee Item Ordering Matrix of ILOT

Let $T_E(I_T) = \begin{bmatrix} e_{ij} \end{bmatrix}_{n \times n}$ and $T_{E'-ILOT}(I_T) = \begin{bmatrix} e'_{ij} \end{bmatrix}_{n \times n}$ be defined as given above, (i) the Validity Index of Examinee Item Ordering Matrix of ILOT, $R\left(T_{E'-ILOT} \mid T_E\right)$, is defined below;

$$R\left(T_{E'-ILOT} \mid T_E\right) = R\left(T_{E'-ILOT}(I_T) \mid T_E(I_T)\right) = \frac{1}{2}\left[1 + \frac{\sum\limits_{i=1}^{n}\sum\limits_{j=1}^{n}\left(e_{ij} - \bar{e}\right)\left(e'_{ij} - \bar{e'}\right)}{\sqrt{\sum\limits_{i=1}^{n}\sum\limits_{j=1}^{n}\left(e_{ij} - \bar{e}\right)^2}\sqrt{\sum\limits_{i=1}^{n}\sum\limits_{j=1}^{n}\left(e'_{ij} - \bar{e'}\right)^2}} \right] \tag{23}$$

where $\qquad \bar{e} = \frac{1}{n^2}\sum\limits_{i=1}^{n}\sum\limits_{j=1}^{n} e_{ij}, \quad \bar{e'} = \frac{1}{n^2}\sum\limits_{i=1}^{n}\sum\limits_{j=1}^{n} e'_{ij} \tag{24}$

(ii) $0 \le R\left(T_{E'-ILOT} \mid T_E\right) \le 1$, the value of $R\left(T_{E'-ILOT} \mid T_E\right)$ is the larger the better

5   Steps of the Method of Improved Liu's Ordering Theory

Step 1. Deciding the Ideal Test, I_T, and its Ideal Item Ordering Matrix, $T_E(I_T)$,
According to *formula (17) and Definition 4.1*; refer to Example 5.
Step 2. Computing the probabilities $P(I_i = x, I_j = y)$, $P(I_i = x)$ and $P(I_j = y)$
Refer to Table 1. Table 4.
Step 3. Computing the item ordering indexes $\gamma_{ij}^{(ILOT)}$
According to formulas (13) and (14)
Step 4. Computing the Empirical Distributive Critical Value $\gamma_c^{(ILOT)}$
According to formula (15)
Step 5. Computing the Examinee Item Ordering Matrix: $T_{E'-ILOT}(I_T)$,
According to *formula (18)*,
Step 6. Computing the Validity Index of Examinee Item Ordering Matrix:
$R(T_{E'-ILOT} | T_E)$
According to *formulas (23) and (24)*
Step 7. Comparing the values of the Validity Index of Examinee Item Ordering Matrix for different groups of examinees $R(T_{E'-ILOT}^{(k)} | T_E)$, $k \in N$

6   Results of a Real Data Set

To illustrate the new theory, a calculus test with 23 items was administered to 212 college students in 6 classes of Chung Chou Institute of Technology in Taiwan.

For convenience, only first 7 items of the test are listed as above in formula (20), the ideal item ordering relations are listed as above in formula (21) and the ideal item ordering matrix of these 7 items of the test is listed as above in formula (22),

For convenience, to compute the examinee item ordering matrices by using the improved Liu's ordering theory, only first 2 classes of the college students were considered, two examinee item ordering matrices are listed as follows;

$$T_{E'-ilot}^{(1)}(I_T) = \begin{bmatrix} 1&0&0&1&0&0&1 \\ 0&1&0&1&0&1&1 \\ 0&0&1&0&1&0&1 \\ 0&0&0&1&0&1&1 \\ 0&0&0&0&1&0&1 \\ 0&0&0&0&0&1&0 \\ 0&0&0&0&0&0&1 \end{bmatrix}, \quad T_{E'-ilot}^{(2)}(I_T) = \begin{bmatrix} 1&0&0&1&1&0&1 \\ 0&1&0&1&0&0&1 \\ 0&0&1&0&1&1&1 \\ 0&0&0&1&1&1&1 \\ 0&0&0&0&1&0&0 \\ 0&0&0&0&0&1&1 \\ 0&0&0&0&0&0&1 \end{bmatrix} \quad (25)$$

By using formulas (23), (24) and (25), we can obtain two values of the validity indexes of Examinee Item Ordering Matrix of ILOT, $T^{(1)}_{E'-ilot}\left(I_T\right)$ and $T^{(2)}_{E'-ilot}\left(I_T\right)$, as follows;

$$R\left(T^{(1)}_{E'-ILOT} \mid T_E\right) = 0.9231, \qquad R\left(T^{(2)}_{E'-ILOT} \mid T_E\right) = 0.8835 \qquad (26)$$

Formula (26) shows that the performance of Class 1 is better than which of Class 2.

7   Conclusion

All of the critical values of the ordering index of the well-known and important ordering theories, OT, IRS and LOT, are subjectively fixed numbers. In this paper, for overcoming the lack of statistical meaning, an empirical distributive critical value of the ordering index based improved LOT theory, denoted as ILOT, was proposed, this new theory is more reasonable and useful than OT, IRS and LOT. Furthermore, by using the new method ILOT, based on Liu's ideal test theory, we can construct the validity index of item ordering structure of any group of examinees to compare the performances of any different groups of examinees. A mathematics example is also provided in this paper to illustrate the advantages of the proposed method

Acknowledgements. This work was supported in part by the National Science Council, Republic of China, under Grant NSC 100-2511-S-468 -001.

References

1. Airasian, P.W., Bart, W.M.: Ordering Theory: A new and useful measurement model. Journal of Educational Technology 5, 56–60 (1973)
2. Bart, W.M., Krus, D.J.: An ordering theoretic method to determine hierarchies among items. Educational and Psychological Measurement 33, 291–300 (1973)
3. Takeya, M.: New item structure theorem. Waseda University Press Tokyo, Japen (1991)
4. Takeya, M.: Structure analysis methods for instruction. Takushpku University Press, Hachioji (1999)
5. Liu, H.-C.: Nominal scale problem ordering theory based on information theory. Journal of Educational Measurement and Statistics 12, 183–192 (2004)
6. Liu, H.-C.: Nominal scale problem ordering theory based on certainty degree theory. Journal of Educational Measurement and Statistics 13, 110–119 (2006)
7. Liu, H.C., Wu, S.N., Chen, C.C.: Improved item relational structure theory. J. Software 6, 2106–2113 (2011)
8. Liu, H.-C.: Ideal item ordering matrix and Liu's ordering theory. Asia University (2012)
9. Liu, H.-C.: A novel validity index for evaluating the item ordering structure based on Q-matrix theory. Research Journal of Applied Sciences, Engineering and Technology 6(1), 53–56 (2013)

A Hybrid Predicting Stock Return Model Based on Bayesian Network and Decision Tree

Shou-Hsiung Cheng

Department of Information Management, Chienkuo Technology University,
Changhua 500, Taiwan
shcheng@ctu.edu.tw

Abstract. This study presents a hybrid model to predict stock returns. The following are three main steps in this study: First, we utilize Bayesian network theory to find out the core of the financial indicators affecting the ups and downs of a stock price. Second, based on the core of the financial indicators coupled with the technology of decision tree, we establish the hybrid classificatory models and the predictable rules that affect the ups and downs of a stock price. Third, by sifting the sound investing targets out, we use the established rules to set out to invest and calculate the rates of investment. These evidences reveal that the average rates of reward are far larger than the mass investment rates.

Keywords: stock returns, financial indicators, Bayesian network, decision tree.

1 Introduction

The problem of predicting stock returns has been an important issue for many years. Advancement in computer technology has allowed many recent studies to utilize machine learning techniques to predict stock returns. Generally, there are two instruments to aid investors for doing prediction activities objectively and scientifically, which are technical analysis and fundamental analysis. Technical analysis considers past financial market data, represented by indicators such as Relative Strength Indicator(RSI) and field-specific charts, to be useful in forecasting price trends and market investment decisions. In particular, technical analysis evaluates the performance of securities by analyzing statistics generated from various marketing activities such as past prices and trading volumes. Furthermore, the trends and patterns of an investment instrument's price, volume, breadth, and trading activities can be used to reflect most of the relevant market information to determine its value [1]. The fundamental analysis can be used to compare a firm's performance and financial situation over a period of time by carefully analyzing the financial statements and assessing the health of a business. Using ratio analysis, trends and indications of good and bad business practices can be easily identified. To this end, fundamental analysis is performed on both historical and present data in order to perform a company stock valuation and hence, predict its probable price evolution. Financial ratios including profitability, liquidity, coverage, and leverage can be calculated from the financial statements [2]. Thus, the focus of this study is not on

A. Moonis et al. (Eds.): IEA/AIE 2014, Part II, LNAI 8482, pp. 218–227, 2014.
© Springer International Publishing Switzerland 2014

effects but on causes that should be seeking and exploring original sources actually. Therefore, selective a good stock is the first and the most important step for intermediate- or even long-term investment planning. In order to reduce risk, in Taiwan, the public stock market observation of permit period will disclose regularly and irregularly the financial statements of all listed companies. Therefore, this study employs data of fundamental analysis by using hybrid models of classification to extract. Employing these meaningful decision rules, a useful stock selective system for intermediate- or long-term investors is proposed in this study.

In general, some related work considers a feature selection step to examine the usefulness of their chosen variables for effective stock prediction, e.g. [3]. This is because not all of features are informative or can provide high discrimination power. This can be called as the curse of dimensionality problem [4]. As a result, feature selection can be used to filter out redundant and/or irrelevant features from a chosen dataset resulting in more representative features for better prediction performances [5]. The idea of combining multiple feature selection methods is derived from classifier ensembles [6]. The aim of classifier ensembles is to obtain high highly accurate ones. They are intended to improve the classification performance of a single classifier. The idea of combining multiple feature selection methods is derived from classifier ensembles (or multiple classifiers) [7]. The aim of classifier ensembles is to obtain highly accurate classifiers by combining less accurate ones. They are intended to improve the classification performance of a single classifier. That is, the combination is able to complement the errors made by the individual classifiers on different parts of the input space. Therefore, the performance of classifier ensembles is likely better than one of the best single classifiers used in isolation.

The rest of the paper is organized as follows: In Section 2 an overview of the related works is introduced, while Section 3 presents the proposed procedure and briefly discusses its architecture. Section 4 describes analytically the experimental   results. Finally, Section 5 shows conclusions of this paper.

2   Related Works

This study proposes a new stock selective system applying the Bayesian network and decision tree algorithm to verify that whether it can be helpful on prediction of the shares rose or fell for investors. Thus, this section mainly reviews related studies of the association rules, cluster analysis and decision tree.

2.1   Bayesian Network

The Bayesian network classifier is a simple classification method, which classifies a case $d_j = \left(x_1^j, x_2^j, \cdots, x_n^j \right)$ by determining the probability. The probabilities are calculated as

$$P_r = \left(Y_i \mid X_1 = x_1^j, X_2 = x_2^j, ..., Xn = x_n^j \right) \tag{1}$$

Equation (1) also can be rewritten as

$$P_r = \frac{P_r(Y_i)P_r\left(X_1 = x_1^j, X_2 = x_2^j, ..., X_n = x_n^j \mid Y_i\right)}{P_r\left(X_1 = x_1^j, X_2 = x_2^j, ..., X_n = x_n^j\right)} \tag{2}$$

$$P_r \propto P_r(Y_i)\prod_{k=1}^{n} P_r\left(X_k = x_k^j \mid \pi_k^j, Y_i\right) \tag{3}$$

TAN Structure Learning

We use a maximum weighted spanning tree method to construct a tree Bayesian network. This method associates a weight to each edge corresponding to the mutual information between the two variables. When the weight matrix is created, the algorithm gives an undirected tree that can be oriented with the choice of a root. The mutual information of two nodes is defined as

$$I\left(X_i, X_j\right) = \sum_{x_i, x_j} P_r\left(x_i, x_j\right)\log\left(\frac{P_r\left(x_i, x_j\right)}{P_r(x_i)P_r(x_j)}\right) \tag{4}$$

We replace the mutual information between two predictors with the conditional mutualinformation between two predictors given the target. It is defined as

$$I\left(X_i, X_j \mid Y\right) = \sum_{x_i, x_j, y_k} P_r\left(x_i, x_j, y_k\right)\log\left(\frac{P_r\left(x_i, x_j y_k\right)}{P_r(x_i \mid y_k)P_r(x_j \mid y_k)}\right) \tag{5}$$

TAN Parameter Learning

Let be the cardinality of . Let denote the cardinality of the parent set of , that is, the number of different values to which the parent of can be instantiated. So it can be calculated as . Note implies . We use to denote the number of records in D for which takes its jth value. We use to denote the number of records in D for which take its jth value and for which takes its kth value.

Maximum Likelihood Estimation

The closed form solution for the parameters and that maximize the log likelihood score is

$$\hat{\theta}Y_i = \frac{N_{Y_i}}{N}$$

$$\hat{\theta}Y_{ijK} = \frac{N_{ijk}}{N_{ij}} \tag{6}$$

$$K = \sum_{i=1}^{n}(r_i - 1)\cdot q_r + |Y| - 1 \tag{7}$$

TAN Posterior Estimation

Assume that Dirichlet prior distributions are specified for the set of parameters as well as for each of the sets. Let and denote corresponding Dirichlet distribution parameters such that and. Upon observing the dataset D, we obtain Dirichlet posterior distributions with the following sets of parameters:

$$\hat{\theta}_{Y_i}^P = \frac{N_{Y_i} + N_{Y_i}^0}{N + N^0}$$

$$\hat{\theta}_{ijk}^P = \frac{N_{ijk} + N_{ijk}^0}{N_{ij} + N_{ij}^0} \tag{8}$$

The posterior estimation is always used for model updating.

2.2 Decision Tree

ID3 decision tree algorithm is one of the earliest use, whose main core is to use a recursive form to cut training data. In each time generating node, some subsets of the training input tests will be drawn out to obtain the volume of information coming as a test. After selection, it will yield the greatest amount of value of information obtained as a branch node, selecting the next branch node in accordance with its recursively moves until the training data for each part of a classification fall into one category or meet a condition of satisfaction. C4.5 is the ID3 extension of the method which improved the ID3 excessive subset that contains only a small number of data issues, with handling continuous values-based property, noise processing, and having both pruning tree ability. C4.5 decision tree in each node use information obtained on the volume to select test attribute, to select the information obtained with the highest volume (or maximum entropy compression) of the property as the current test attribute node.

Let A be an attribute with k outcomes that partition the training set S into k sub-sets S_j (j = 1,..., k). Suppose there are m classes, denoted $C = \{c_1, \cdots, c_m\}$, and $p_i = \frac{n_i}{n}$ represents the proportion of instances in S belonging to class c_i, where $n = |S|$ and n_i is the number of instances in S belonging to c_i. The selection measure relative to data set S is defined by:

$$Info(S) = \sum_{i=1}^{m} p_i \log_2 p_i \tag{9}$$

The information measure after considering the partition of S obtained by taking into account the k outcomes of an attribute A is given by:

$$Info(S, A) = \sum_{j=1}^{k} \frac{|S_j|}{|S|} Info(S_i) \tag{10}$$

The information gain for an attribute A relative to the training set S is defined as follows:

$$Gain(S, A) = Info(S) - Info(S, A) \tag{11}$$

The Gain(S, A) is called attribute selection criterion. It computes the difference between the entropies before and after the partition, the largest difference corresponds to the best attribute. Information gain has the drawback to favour attributes with a large number of attribute values over those with a small number. To avoid this drawback, the information gain is replaced by a ratio called gain ratio:

$$GR(S, A) = \frac{\text{Gain}(S, A)}{-\sum_{j=1}^{k} \frac{|S_i|}{|S|} \log_2 \frac{|S_i|}{|S|}} \tag{12}$$

Consequently, the largest gain ratio corresponds to the best attribute.

3   Methodology

In this study, using the rate of operating expenses, cash flow ratio, current ratio, quick ratio, operating costs, operating income, accounts receivable turnover ratio (times), payable accounts payable cash (days), the net rate of return (after tax)net turnover ratio, earnings per share, operating margin, net growth rate, rate of return of total assets and 14 financial ratios, the use of data mining technology on the sample of the study association rules, cluster analysis, decision tree analysis taxonomy induction a simple and easy-to-understand investment rules significantly simplify the complexity of investment rules. The goal of this paper is proposes a straightforward and efficient stock selective system to reduce the complexity of investment rules.

3.1   Flowchart of Research Procedure

The study proposes a new procedure for a stock selective system. Figure 1 illustrates the flowchart of research procedure in this study.

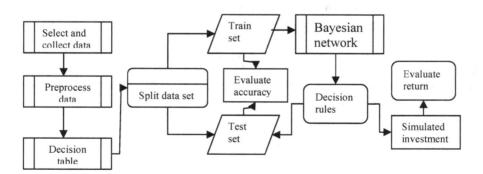

Fig. 1. Flowchart of research procedure

3.2   Flowchart of Research Procedure

This subsection further explains the proposed stock selective system and its algorithms. The proposed stock selective system can be devided into seven steps in detail, and its computing process is introduced systematically as follows:

Step 1: Select and collect the data.

Firstly, this study selects the target data that is collected from Taiwan stock trading system.

Step 2: Preprocess data.

To preprocess the dataset to make knowledge discovery easier is needed. Thus, firstly delete the records that include missing values   or inaccurate values, eliminate the clearly irrelative attributes that will be more easily and effectively pro-cessed for extracting decision rules to select stock. The main jobs of this step includes data integration, data cleaning and data transformation.

Step 3: Build decision table.

The attribute sets of decision table can be divided into a condition attribute set and a decision attribute set. Use financial indicators as a condition attribute set, and whether the stock prices up or down as a decision attribute set.

Step 4: Bayesian network.

We use a maximum weighted spanning tree method to construct a tree Bayesian network. This method associates a weight to each edge corresponding to the mutual information between the two variables. When the weight matrix is created, the algorithm gives an undirected tree that can be oriented with the choice of a root.

Step 5: Extract decision rules.

Based on condition attributes clustered in step 5 and a decision attribute (i.e. the stock prices up or down), generate decision rules by decision tree C5.0 algorithm.

Step 6: Evaluate and analyze the results and simulated investment.

For verification, the dataset is split into two sub-datasets: The 67% dataset is used as a training set, and the other 33% is used as a testing set. Furthermore, evaluate return of the simulated investment,

4   Empirical Analysis

4.1   The Computing Procedure

A practically collected dataset is used in this empirical case study to demonstrate the proposed procedure: The dataset for 993 general industrial firms listed in Taiwan stock trading system from 2008/01 to 2013/12 quarterly. The dataset contains 11916 instances which are characterized by the following 14 condition attributes: (i) operating expense ratio(A1), (ii) cash flow ratio(A2), (iii) current ratio(A3), (iv) quick ratio(A4), (v) operating costs(A5), (vi) operating profit(A6), (vii) accounts receivable turnover ratio(A7), (viii) the number of days to pay accounts(A8), (ix) return on equity (after tax) (A9), (x) net turnover(A10), (xi) earnings per share(A11), (xii) operating margin(A12), (xiii) net growth rate(A13), and (xiv) return on total assets growth rate(A14); all attributes are continuous data in this dataset.

Table 1. The Definition of various Attribute

No.	Attribute name	Notation	Definition
1	Operating expense ratio	A1	Operating expenses / net sales
2	Cash flow ratio	A2	Net cash flow from operating activities / current liabilities
3	current ratio	A3	current assets / current liabilities, to measure short-term solvency. Current assets (cash + marketable securities + funds + inventory + should be subject to prepayment), current liabilities within one year, short-term liabilities which must be spent on current assets to pay. The higher the current ratio is, the better short-term liquidity is, which is often higher than 100%.
4	Quick ratio	A4	liquid assets / current liabilities = (cash + marketable securities + should be subject to payment) divided by current liabilities = (Current assets - Inventories - Prepaid expenses) divided by current liabilities, to measure very short-term liabilities as of capacity. The higher it is, the better short-term solvency is.
5	Operating costs	A5	because of regular business activities and sales of goods or services, enterprises should pay the costs in a period of operating time, which mainly include: cost of goods sold, labor costs. Cost of goods sold can be divided into two major categories: product cost of self-made goods and purchased products. For manufacturing sector, the former usually accounts for the most majority, while other industries the latter. By the definition of accounting costs, operating costs are cost that arises throughout the manufacturing process, also known as product costs or manufacturing costs. Manufacturing costs are composed of direct materials + direct labor + manufacturing costs (including indirect materials, indirect labor and factory operations and related product manufacturing or other indirect costs).
6	Operating profit	A6	Operating profit / paid-up capital
7	Accounts receivable turnover ratio	A7	net credit (or net sales) / average receivables. Measure whether the speed of the current collection of accounts receivable and credit policy is too tight or too loose. The higher receivables turnover ratio is, the better the efficiency of collection representatives is.
8	the number of days to pay accounts	A8	Average accounts payable / operating costs * day
9	Return on equity (after tax)	A9	Shareholders 'equity is shareholders' equity for the growth rate of the year. The net income refers to the dividend earnings deducted the special stock, while equity refers to the total common equity. From the equity growth rate, we can see whether the company's business class objectives are consistent with shareholder objectives, based on shareholders' equity as the main consideration. Return on equity is acquired due to companies retain their earnings, and hence show a business can also promote the ability to grow their business even not to rely on external borrowing. It is calculated as: ROE = (Net income - dividend number) / Equity
10	net turnover	A10	net operating income / average net worth
11	Earnings per share	A11	(Net income - Preferred stock dividends) / numbers of public ordinary shares
12	Operating margin	A12	operating margin / revenue, often used to compare the competitive strength and weakness of the same industrial, showing the company's products, pricing power, the ability to control manufacturing costs and market share can also be used to compare different industries industry trends change.
13	Net growth rate	A13	the net price will fluctuate with the market increase or decrease the asset, and its upper and lower rate of increase or decrease, then is known as the net growth rate.
14	Return on total assets growth rate	A14	which represent in a certain period of time (usually one year), companies use total assets to create profits for shareholders over the previous period the growth rate.
15	Decision attribute	D1	the stock prices up or down

The computing process of the stock selective system can be expressed in detail as follows:

Step 1: Select and collect the data.

This study selects the target data that is collected from Taiwan stock trading system. Due to the different definitions of industry characteristics and accounting subjects, the general industrial stocks listed companies are considered as objects of the study. The experiment dataset contains 11916 instances which are characterized by 14 condition attributes and one decision attribute.

Step 2: Preprocess data.

Delete the 31 records (instances) that include missing values, and eliminate the 10 irrelative attributes. Accordingly, in total the data of 637 electronic firms that consist of 15 attributes and 6793 instances are included in the dataset. The attributes information of the dataset is shown in Table 1.

Step 3: Build decision table.

The attribute sets of decision table can be divided into a condition attribute set and a decision attribute set. Use financial indicators as a condition attribute set, and whether the stock prices up or down as a decision attribute set.

Step 4: Extract core attributes by Bayesian network.

The use of Bayesian network attributes reduction, the core financial attributes can be obtained. The core financial attributes are: (1) return on total assets, (2) the net rate of return after tax, (3) earnings per share, (4) net growth rate, (5) current ratio, and (6) cash flow ratio.

Step 5: Extract decision rules.

Based on core attributes extracted in step 4 and a decision attribute (i.e. the stock prices up or down), generate decision rules by decision tree C5.0 algorithm. C5.0 rule induction algorithm, is to build a decision tree recursive relationship between the interpretation of the field with the output field data divided into a subset of the and export decision tree rules, try a different part in the interpretation of the data with output field or the relationship of the results. Financial Ratios and Price Change decision tree analysis are shown in Table 2, as follows:

Table 2. The Decision rule set

No.	Decision rule set
1	If the return on total assets growth rate <= 0.070 and the return on total assets growth rate> -0.430 and earnings per share <= 0.900 and the cash flow ratio> -2.800 operating margin <= 20.120 and the business interests <= 184,323 and operating income> -97,782 price rose.
2	If the return on total assets growth rate <= 0.070 and the return on total assets growth rate> -0.430 and earnings per share <= 0.900 and the cash flow ratio> -2.800 operating margin> 20.120 price rose.
3	If the return on total assets growth rate <= 0.070 and the return on total assets growth rate> -0.430 and earnings per share of <= 0.900 and the cash flow ratio of> 36.240 and the return on total assets growth rate <= -0.240 then the share price rose.
4	If the growth rate of total assets> 0.070 and return on total assets growth rate <= 2.950 and Cash Flow Ratio <= 13.760 and earnings per share of RMB> -0.820 and net growth rate <= 16.710 price rose.
5	If the growth rate of total assets> 0.070 and return on total assets growth rate <= 2.950 and the cash flow ratio <= 13.760 per share surplus element> -0.820 and net growth rate of> 16.710 and earnings per share of $ 3.750 price rose.
6	If the return on total assets growth rate> 0.070 and return on total assets growth rate <= 2.950 and the cash flow ratio of> 13.760 and return on total assets growth rate <= 0.390 and the return on total assets growth rate <= 0.360 price rose.
6	If the return on total assets growth rate> 0.070 and return on total assets growth rate <= 2.950 and the cash flow ratio of> 13.760 and return on total assets growth rate <= 0.390 and the return on total assets growth rate <= 0.360 price rose.
7	If the return on total assets growth rate> 0.070 and return on total assets growth rate <= 2.950 and the cash flow ratio of> 13.760 and return on total assets growth rate <= 0.390 and the return on total assets growth rate <= 0.360 price rose.
8	If the return on total assets growth rate> 0.070 and the return on total assets growth rate> 2.950 share price rose.

4.2   Simulated Investment

The rules generating from Table 2 get down on stock selection from the listed companies rise in the rules in year 2011~2013. There are 24 companies in the first quarter in line with the rise in the rules, the average quarter rate of return of 11.51 %; 5 companies in second quarter 2, the average quarter rate of return of 4.64%; 2 companies in the third quarter, the average quarter rate of return of 2.12%; 11 companies in the fourth quarter, the average quarter rate of return of 4.92% in year 2011. There are 43 in the first quarter in line with the rise in the rules, the average quarter rate of return of 35.10 %; 3 companies in second quarter 2, the average quarter rate of return of 1.15%; 34 companies in the third quarter, the average quarter rate of return of 14.42%; 123 companies in the fourth quarter, the average quarter rate of return of 24.42% in year 2012. There are 50 in the first quarter in line with the rise in the rules, the average quarter rate of return of 12.20 %; 22 companies in second quarter 2, the average quarter rate of return of 9.45%; 120 companies in the third quarter, the average quarter rate of return of 10.72%; 131 companies in the fourth quarter, the average quarter rate of return of 15.52% in year 2013. The comparsion of the average quarter rate of return and the broader market quarter rate of return is shown as Table 3. These evidences reveal that the average rates of reward are far larger than the mass investment rates.

Table 3. The Average Quarter Rate of Return

		The average rates of reward	The mass investment rates
2011	First quarter	11.51 %	-5.32%
	Second quarter	4.64%	-0.55%
	Third quarter	2.12%	-14.62%
	Fourth quarter	4.92%	-7.11%
2012	First quarter	35.10 %	10.32%
	Second quarter	1.15%	-8.25%
	Third quarter	14.42%	-1.62%
	Fourth quarter	24.42%	7.51%
2013	First quarter	12.20 %	3.32%
	Second quarter	9.45%	2.25%
	Third quarter	10.72%	0.32%
	Fourth quarter	15.52%	1.51%

5   Conclusion

This paper presents a hybrid predicting stock return model based on Bayesian network and decision tree. From the results of empirical analysis obtained in this study, some conclusions can be summarized as follows:

(1.) By the dependence of each company's financial indicators and by the ups and downs of the stock, the use of Bayesian network and decision tree, we can gain a simple set of classification and prediction rules.

(2.) The average rate of return derives from the empirical results show that return on investment on stock price in the research is obvious higher than general market average.

References

1. Murphy, J.J.: Technical Analysis of the Financial Markets. Institute of Finance, New York (1999)
2. Bernstein, L., Wild, J.: Analysis of Financial Statements. McGraw-Hill (2000)
3. Abraham, A., Nath, B., Mahanti, P.K.: Hybrid intelligent systems for stock market analysis. In: Alexandrov, V.N., Dongarra, J. J., Juliano, B.A., Renner, R.S., Tan, C.J.K. (eds.) ICCS 2001. LNCS, vol. 2074, pp. 337–345. Springer, Heidelberg (2001)
4. Huang, C.L., Tsai, C.Y.: A hybrid SOFM-SVR with a filter-based feature selection for stock market forecasting. Expert System with Applications 36(2), 1529–1539 (2009)

5. Chang, P.C., Liu, C.H.: A TSK type fuzzy rule based system for stock price prediction. Expert Systems with Application 34(1), 135–144 (2008)
6. Yu, L., Wang, S.-Y., Lai, K.K.: Mining stock market tendency using GA-based support vector machines. In: Deng, X., Ye, Y. (eds.) WINE 2005. LNCS, vol. 3828, pp. 336–345. Springer, Heidelberg (2005)
7. Kim, K.J.: Financial time series forecasting using support vector machines. Neurocomputing 55, 307–319 (2003)

A Method of Bubble Removal for Computer-Assisted Diagnosis of Capsule Endoscopic Images

Masato Suenaga[1], Yusuke Fujita[1], Shinichi Hashimoto[2], Terai Shuji[2], Isao Sakaida[2], and Yoshihiko Hamamoto[1]

[1] Graduate School of Medicine, Yamaguchi University
2-16-1, Tokiwadai, Ube, Yamaguchi, Japan
{t019uh,y-fujita,hamamoto}@yamaguchi-u.ac.jp
[2] Graduate School of Medicine, Yamaguchi University
1-1-1, Minamikogushi, Ube, Yamaguchi, Japan

Abstract. Capsule endoscopy can take tens of thousands of pictures at one examination, and it is impossible for physicians to diagnose such huge number of pictures. Thus, studies of computer-assisted diagnosis of capsule endoscopic images have been conducted. Computer-assisted diagnosis includes detection of lesion areas. Besides lesion areas, capsule endoscopic images contain areas of residue, intestinal juice, and bubbles, which can be an obstacle to diagnosis of the intestinal wall. To improve the performance of diagnosis, we propose a method of removing bubbles in capsule endoscopic images. Experimental results show that the proposed method works well.

Keywords: Capsule Endoscopy, Computer-aided Region Detection, Filtering.

1   Introduction

Capsule endoscopy was developed to capture images of the small intestine, which cannot be reached by traditional upper endoscopy or colonoscopy methods [1]. Capsule endoscopy can take tens of thousands of pictures at one examination, and it is impossible for physicians to diagnose such huge number of pictures. Thus, studies of computer-assisted diagnosis of capsule endoscopic images have been conducted [2],[3]. Computer-assisted diagnosis includes detection of lesion areas. Besides lesion areas, capsule endoscopic images contain areas of residue, intestinal juice, and bubbles, as shown in Fig. 1, which can be an obstacle to diagnosis of the intestinal wall. This paper focuses especially on bubbles among residue, intestinal juice, and bubbles in the capsule endoscopic images and proposes a method of bubble removal. Finally, the efficiency of the proposed method is evaluated using 30 capsule endoscopic images.

2   Bubble Removal

Bubble area is generally defined as an area where fluid inside the small intestine becomes round through the containment of gas. Images that contain many bubble areas

A. Moonis et al. (Eds.): IEA/AIE 2014, Part II, LNAI 8482, pp. 228–233, 2014.

do not allow visualization of the intestinal wall due to reflection of light by these bubbles, and images that do not allow visualization of the intestinal wall cannot be used for diagnosis. Therefore, bubble removal in these images is required. Clusters of pixels that contain small and round edges are regarded as bubble areas. Thus, compared with areas of residue or intestinal juice, a bubble area is characterized by detected clusters of pixels indicating edges. Bubble removal in the image is conducted by focusing on this characteristic.

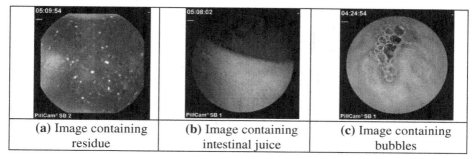

(a) Image containing residue	**(b)** Image containing intestinal juice	**(c)** Image containing bubbles

Fig. 1. Examples of images of the small intestine

2.1   Method of Bubble Removal

The method of bubble removal consists of the following three steps: edge detection, region segmentation, and detection of the bubble area. A flow chart of the process is shown in Fig. 2. Step 1 is as follows: the input color image represents an image (hereafter called a luminance image) wherein luminance Y converted from an RGB value is expressed as the pixel value [4]. Edge detection of this luminance image is performed based on the Canny algorithm [5]. Dilation is performed on the pixels containing detected edges (hereafter called edge pixels), and the edge pixels are connected. After that, noise reduction is carried out by a labeling process. In Step 2, region segmentation is performed on the luminance image based on Watersheds algorithm [6]. Step 3 is as follows: the ratio of the number of the edge pixels in the segmented region obtained from Step 1 to the number of all pixels in the segmented region obtained from Step 2 is calculated. If the ratio is higher than a threshold value (the cutoff value), the segmented region is detected as a bubble area. Then, the holes in the detected bubble area are closed by a process of expansion and contraction (closing).

2.2   Step 1: Edge Detection Based on the Canny Algorithm and Correction

In the Canny edge detection algorithm, all edges in the vertical, horizontal, and diagonal directions can be detected, and a filter is applied by which the edge curve consisting of the edge pixels is assumed to be a point of local directional maximum. After edge detection, connection of the edge pixels by dilation and noise reduction by labeling are carried out as correction processes. In this paper, after preliminary experiments, for the purpose of emphasizing edge pixels in the bubble areas, standard deviation σ at 1.0, threshold (Th_{high}) at 0.05, and threshold (Th_{low}) at 0.02 were determined as parameters of the Canny algorithm.

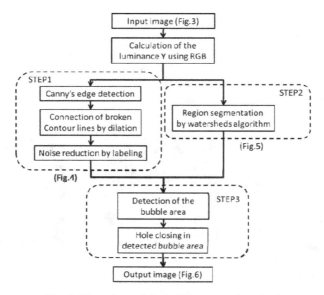

Fig. 2. Flow chart of the bubble removal method

2.3   Step 2: Region Segmentation Based on Watersheds Algorithm

We performed region segmentation using Watersheds algorithm. Watersheds algorithm is a method whereby the classic method of geoscience is applied to image processing, and luminance of an image is regarded as altitude over land, and valleys collecting water as well as watersheds of the valley are defined. Based on the definitions, the images can be divided into several regions. In Watersheds algorithm, smoothing of the input luminance image using a Gaussian filter is conducted, and a sorting step and flooding step are sequentially performed, resulting in segmentation of the regions.

2.4   Step 3: Bubble Removal

Bubble areas are removed using a combination of the results in Step 2 and the results of region segmentation in Step 3. To differentiate the bubble areas, the proportion of edge pixels contained in each segmented region to all pixels in the segmented region is calculated as the rate V, considered as a feature, with formula (1):

$$V = \frac{\text{Number of edge pixels in a segmented region}}{\text{Total number of pixels in a segmented region}} \tag{1}$$

The threshold value, which determines whether each of segmented regions is bubble area is defined as follows: 10 segmented regions visually considered as bubble areas by a human being are randomly selected from all segmented regions, and among the 10 V values, the minimum value is defined as the threshold. The segmented region in which the value of V is greater than the threshold value is selected as a region including bubbles. Next, to close the holes in the bubble region, an expansion and contraction (closing) process is performed.

3   Experimental Results

3.1   Data

An example of a capsule endoscopy image containing bubbles is shown in Fig. 3(a). This study was provided with 30 small intestine images containing bubbles from one patient by Yamaguchi University Hospital. The image size was 576 x 576 pixels, with an RGB color coordinate system (256-level gradation). To evaluate the accuracy of bubble removal, bubble areas visually defined by a human being were designated as correct and were compared with those determined by computer.

3.2   Results

Fig. 3(b) shows a resulting image of Step 1, Fig. 3(c) shows a resulting image of the segmented regions of Step 2, and Fig. 3(d) shows a resulting output image of

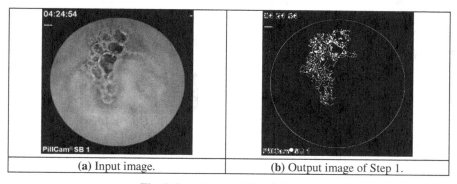

| (a) Input image. | (b) Output image of Step 1. |

Fig. 3. Input image & Result images

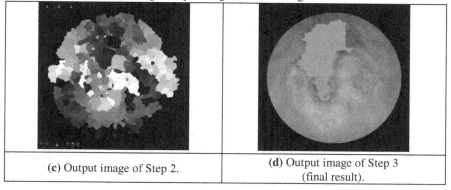

| (c) Output image of Step 2. | (d) Output image of Step 3 (final result). |

Fig. 3. Input image & Result images

Step3, i.e., the final result. The regions detected by the bubble removal method are highlighted in green.

For evaluation of bubble removal, 30 input images and their corresponding 30 output images were used. First, the number of bubbles in an input image was counted.

Second, the number of bubbles in the corresponding output image was counted. The difference between these was identified as the number of bubbles removed by computer, and the removal rate was estimated with formula (2). This procedure was performed on all 30 images, and an average removal rate of 83% was determined.

$$\text{Removal rate} = \frac{\text{Number of bubbles removed}}{\text{Number of bubbles in an input image}} \times 100 \ (\%) \tag{2}$$

4   Discussion

A successful example of bubble removal is shown in Fig. 4. The proposed method works well, particularly for small bubbles.

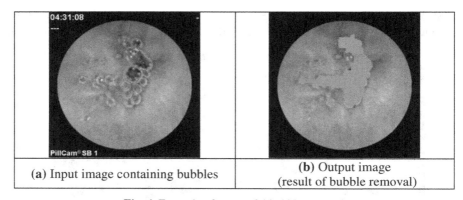

| (a) Input image containing bubbles | (b) Output image (result of bubble removal) |

Fig. 4. Example of successful bubble removal

An unsuccessful example of bubble removal is shown in Fig. 5. In this paper, the data of pixels containing edges alone were used, and the bubbles were removed. Therefore, bubble removal was successful in the regions where small bubbles had clustered. However, large bubbles as shown in Fig. 4, which did not cluster, had few pixels containing edges, and consequently bubble removal was not successful.

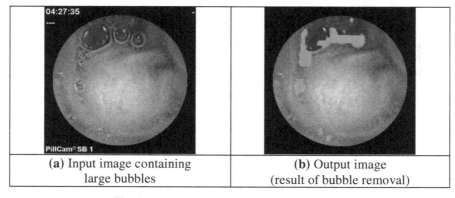

| (a) Input image containing large bubbles | (b) Output image (result of bubble removal) |

Fig. 5. Example of unsuccessful bubble removal

5   Conclusion

Bubble removal was performed on 30 images containing bubbles, resulting in an accuracy rate of successful bubble removal of 83%. However, non-clustered large bubbles failed to be removed by this method. Further studies are needed to investigate how large bubbles can be removed and whether substances other than bubbles are incorrectly removed as bubbles.

References

1. Iddan, G., Meron, G., Glukhovsky, A., et al.: Wireless capsule endoscopy. Nature 405, 417 (2000)
2. Malagelada, C., et al.: New insight into intestinal motor function via noninvasive endoluminal image analysis. Gastroenterology 135, 1155–1162 (2008)
3. Hai, V.U., et al.: Controlling the display of capsule endoscopy video for diagnostic assistance. IEICE Trans. Inf. Syst. E92-D(3), 512-528 (2009)
4. Pratt, W.K.: Digital Image Processing. John Wiley & Sons (1978)
5. Canny, J.: A computational approach to edge detection. IEEE PAMI 8(6), 679–698 (1986)
6. Vincent, L., Soille, P.: Watersheds in Digital Spaces: An Efficient Algorithm Based on Immersion Simulations. IEEE Trans. Pattern Anal. Mach. Intell. 13(6), 583 (1991)

Comparative Study of Classifiers for Prediction of Recurrence of Liver Cancer Using Binary Patterns

Hiroyuki Ogihara[1], Yusuke Fujita[1], Norio Iizuka[2],
Masaaki Oka[2], and Yoshihiko Hamamoto[1]

[1] Graduate School of Medicine, Yamaguchi University
2-16-1, Tokiwadai, Ube, Yamaguchi, Japan
{t009uh,y-fujita,hamamoto}@yamaguchi-u.ac.jp
[2] Graduate School of Medicine, Yamaguchi University
1-1-1, Minamikogushi, Ube, Yamaguchi, Japan

Abstract. Liver cancer has a high likelihood of recurrence despite complete surgical resection and is thus known as an intractable cancer. If postoperative recurrence of cancer is correctly predicted for each patient as a form of personalized medicine, effective treatment can be carried out. The purpose of this paper is to evaluate classifiers for predicting recurrence of liver cancer by use of blood test data only in patients who underwent complete surgical resection of liver cancer. We conduct comparative study of 3 classifiers with use of binary patterns, which consist of clinical data and genomic data.

Keywords: Cancer Diagnosis, Classifier, Binary Pattern, Boolean Algebra.

1 Introduction

Overcoming intractable liver cancer has become an important issue. Liver cancer has a high likelihood of recurrence, and despite complete surgical resection, the incidence of liver cancer recurrence at one year after resection is about 30%, as shown in [1]. In general, cancer varies from patient to patient. Therefore, personalized medicine, which is medicine customized for each patient, has received much attention, and easily an accessible blood test for prediction of recurrence of liver cancer is desired in the clinical setting. In this paper, we use a combination of two types of data, one is clinical data of AFP and PIVKA, which are representative blood test tumor markers, and the other is DNA methylation data of SPINT and SRD [5] as genomic data obtained from blood. Using a binary pattern consisting of these different types of data, we evaluate classifiers to predict recurrence of liver cancer in patients who underwent complete surgical resection of liver cancer.

2 Preliminaries

Blood tests are generally a mixture of non-metric data such as plus or minus, and metric data described by numbers. To cope with this, a binary pattern is used by

A. Moonis et al. (Eds.): IEA/AIE 2014, Part II, LNAI 8482, pp. 234–239, 2014.

combining binarized metric data with non-metric data. Thus, binarization of metric data has the benefit of the combined analysis of non-metric and metric data. Furthermore, another benefit of binarization is that it may absorb the variability which results in the degration of the classification performance. Therefore, it is expected that a classifier with use of such binary patterns is robust for the sample variability. In binarization of metric data, the value of cut off is determined by a medical doctor.

Prediction of recurrence of liver cancer by the combination of clinical data and genomic data is as follows: First, clinical data and genomic data obtained from the patients are binarized, and by using the binarized data, each patient is described by a binary pattern. Next, classification of this binary pattern allows for highly accurate prediction of recurrence in patients. In this paper, we discuss comparative study of classifiers.

2.1   Boolean Classifier [8]

We have n binary features. That is, let $x = (x_1, x_2, \dots, x_n)^T$, where the components x_i are either 1 or 0. Then, there are 2^n possible n-dimensional binary patterns. This means that a feature space is divided into 2^n regions. The class label for each region is assigned according to majority rule. In the Bayes rule, decision is conducted according to the a posteriori probability. Thus, the estimation of the a posteriori probability is one of critical issues in classifier design. In the Boolean classifier based on the idea of nonparametric estimation, the class ω_i most frequently represented within the region including an input pattern x is selected and x is classified to ω_i. Table 1 shows 2-dimensional binary patterns and class labeling according to majority rule. The asterisk (*) in the table indicates that both classes are equal in number of training samples or that the category does not include training samples, being classified as non-labeling.

From Table 1, a truth table of function f_1 , which assigns a binary pattern to recurrence class ω_1 , is made (shown in Table 1). In Table 1, the logical function f_1 of recurrence is defined as recurrence = 1 and non-recurrence = 0. Concretely, the principal disjunctive canonical form of f_1 from Table 1 is as follows:

$$f_1 = \bar{x}_1 \bar{x}_2 + \bar{x}_1 x_2 \tag{1}$$

In addition, the logical function f_1 is simplified to the following equation:

$$f_1 = \bar{x}_1 \tag{2}$$

On the other hand, from Table 1, the logical function f_2 of non-recurrence class ω_2 is as follows:

$$f_2 = x_1 \bar{x}_2 \tag{3}$$

Table 1. Class labeling according to majority rule

x_1	x_2	ω_1/ω_2	Class label	$f_1(x)$	$f_2(x)$
0	0	$5/3$	ω_1	1	0
0	1	$10/1$	ω_1	1	0
1	0	$2/9$	ω_2	0	1
1	1	$8/8$	$*$	0	0

As described above, in the Boolean classifier, logical functions f_1 and f_2, which are defined as discriminant functions for each class, are prepared. A logical function is given by a logical sum of minterms. Each minterm defines a class obtained from data. Note that f_1 is independent of x_2. This is considered as feature selection for class ω_1. This is one of advantages of using Boolean algebra.

2.2   Bayes Classifier for Binary Patterns [6]

We define the discriminant function:

$$g(x) = \sum_{i=1}^n [x_i \log \frac{p_i}{r_i} + (1 - x_i) \log \frac{1-p_i}{1-r_i}] \tag{4}$$

where

$$p_i = Prob(x_i = 1|\omega_1) \quad \text{and} \quad r_i = Prob(x_i = 1|\omega_2)$$

In this discriminant function, we use the following decision rule: Decide ω_1 if $g(x) > 0$; otherwise decide ω_2.

2.3   Chow Classifier [6],[7]

In general, we obtain the product expansion concerning a joint probability distribution $P(x)$:

$$P(x) = P(x_1)P\left(x_2\big|x_{j_{(2)}}\right) \dots P\left(x_n\big|x_{j_{(n)}}\right)$$

By substituting 0 or 1 for x_i and $x_{j_{(i)}}$, we get

$$P\left(x_i\big|x_{j_{(i)}}\right) = [p_i^{x_i}(1 - p_i)^{1-x_i}]^{x_{j_{(i)}}}[q_i^{x_i}(1 - q_i)^{1-x_i}]^{x_{j_{(i)}}} \tag{5}$$

where

$$p_i = Prob(x_i = 1|x_{j_{(i)}} = 1) \quad \text{and} \quad q_i = Prob(x_i = 1|x_{j_{(i)}} = 0)$$

Taking the logarithm, we obtain the Chow expansion:

$$\log P(x) = \sum_{i=1}^n \log(1 - q_i) + \sum_{i=1}^n x_i \log \frac{q_i}{1-q_i} +$$
$$\sum_{i=2}^n x_{j_{(i)}} \log \frac{1-p_i}{1-q_i} \sum_{i=2}^n x_i x_{j_{(i)}} \log \frac{p_i(1-q_i)}{(1-p_i)q_i} \tag{6}$$

In the Chow classifier, a pattern is classified to the class corresponding to the largest logarithm.

3   Experiment

3.1   Data

Data were provided by Yamaguchi University School of Medicine and obtained from patients who underwent complete surgical resection of liver cancer. They comprised 35 patients who experienced recurrence of liver cancer within 2 years after resection and 38 patients with non-recurrence. Regarding the features used for a classifier, AFP, PIVKA, SPINT and SRD were used.

Because complete surgical resection of liver cancer is performed, CT or ultrasound cannot detect liver cancer. Thus, it is quite difficult to predict recurrence of cancer non-detectable by CT or ultrasound by blood test data alone. As a reference, we describe prediction by independent use of AFP and PIVKA. In the clinical setting, the cut-off values of AFP and PIVKA are determined as 20 and 40, respectively, and recurrence of liver cancer is predicted. For example, with respect to AFP, if the measured value of AFP is ≥ 20, recurrence is predicted, and if <20, non-recurrence is predicted. We adopted the Youden index (sensitivity + specificity -1.0) as the criterion of classification performance of the classifier [2]. Youden index values of AFP and PIVKA were 0.07 and 0.02, respectively, in the 35 patients with recurrence and the 38 patients with non-recurrence, indicating that AFP and PIVKA, which are traditional tumor markers, cannot provide sufficient classification performance.

3.2   Method

Twenty samples each were randomly chosen for this experiment from the 35 recurrence samples and the 38 non-recurrence samples and were used as training samples. All the remaining samples were used as test samples. Therefore, recurrence test samples comprised 15 cases and non-recurrence test samples 18 cases. Learning of a classifier was conducted by use of training samples, and test samples were then classified by the classifier.

The higher the value of the Youden index, the higher is the classification performance of the classifier. Here, sensitivity means the rate of patients who were correctly classified as having recurrence among the patients with recurrence, whereas specificity means the rate of patients who were correctly classified as having non-recurrence among the patients with non-recurrence. Classification performance was evaluated by sensitivity, specificity, the Youden index and recognition rate for test samples obtained by the hold-out method [3]. This trial was independently conducted 30 times. The mean was estimated.

3.3   Test of Paired Data

The Youden index value of classifier A for the control was defined as x_i, and that of classifier B in comparison with classifier A was defined as y_i. When classifiers A and B classify the same test samples, x_i and y_i are considered to be paired data. Here, estimator Z used for test of paired data is defined in Eq. (7).

$$Z_1 = y_1 - x_1$$

$$Z_2 = y_2 - x_2$$

$$\vdots$$

$$Z_{30} = y_{30} - x_{30}$$

(7)

If classifier A is superior to classifier B, the result is $Z < 0$. Conversely, if classifier B is superior to classifier A, the result is $Z > 0$. $Z_1, Z_2, \ldots, Z_{30}$ described above are assumed to have been extracted from a population of unknown population mean μ and population variance σ^2. When there is no difference in classification performance between the two classifiers, the result is $\mu = 0$, being determined to be the null hypothesis.

To test the null hypothesis, estimator V is defined in the following Eq. (8) [4].

$$V = \frac{\bar{Z}}{\hat{\sigma} / \sqrt{30}}$$

(8)

where $\bar{Z}$ is the sample mean of $Z_1, Z_2, \ldots, Z_{30}$, and $\hat{\sigma}$ is the sample standard deviation of $Z_1, Z_2, \ldots, Z_{30}$. If estimate v of estimator V calculated from Eq. (8) is greater than the cut-off value, which is determined at the 1% level of significance, the null hypothesis will be rejected. Then, there is a difference in classification performance between the two classifiers.

Table 2. Comparison of classification performances

Criteria	Boolean	Chow	Bayes
Sensitivity	0.64	0.55	0.46
Specificity	0.65	0.60	0.49
Youden-index	0.29	0.15	-0.05
Recognition rate	0.64	0.58	0.48

4   Results and Discussion

The results of the classification performance are shown in Table2. Among 3 classifiers, the Boolean classifier produced the best performance, regardless of criterion.

Next, we conducted test of paired data to compare the Boolean classifier with the Chow classifier. A one-sided test was used to show the superiority of the Boolean classifier over the Chow classifier. In this test, the Youden index value of the Boolean classifier was defined as y and that of the Chow classifier was defined as x. Results show that estimate v of 3.446 was greater than the cut-off value of 2.462, which was determined at the 1% level of significance. This means that the Boolean classifier was superior to the Chow classifier.

5   Conclusion

The present paper showed that a Boolean classifier combining clinical data with genomic data was useful for   prediction of liver cancer recurrence. We intend to perform future experiments increasing the combinations of clinical data and genomic data used. In addition, medical interpretation of a minterm in a logical function obtained from data is also an important issue.

Acknowledgment. This work was supported by JSPS KAKENHI Grant Number 24500206.

References

1. Iizuka, N., Oka, M., Hamamoto, Y., et al.: Oligonucleotide microarray for prediction of early intrahepatic recurrence of hepatocellular carcinoma after curative resection. Lancet, 923–929 (2003)
2. Youden, W.J.: Index for rating diagnostic tests. Cancer 3, 32–35 (1950)
3. Jain, A.K., Duin, R.W., Mao, J.: Statistical pattern recognition: A review. IEEE Trans. Pattern Anal. Mach. Intell. 22, 4–37 (2000)
4. Guttman, I., Wilks, S.S.: Introductory Engineering Statistics. John Wiley & Sons (1965)
5. Moribe, T., et al.: Methylation of multiple genes as molecular markers for diagnosis of a small well-differentiated hepatocellular carcinoma. Int. J. Cancer 125, 388–397 (2009)
6. Duda, R.O., Hart, P.E.: Pattern Classification and Scene Analysis. John Wiley & Sons (1973)
7. Chow, C.K.: A recognition method using neighbor dependence. IRE Trans., Elec. Comp. EC-11, 683–690 (1962)
8. Ogihara, H., Fujita, Y., Iizuka, N., Oka, M., Hamamoto, Y.: Classification Based on Boolean Algebra and its Application to the Prediction of Recurrence of Liver Cancer. In: Proc. of Workshop on Recent Advances in Computer Vision and Pattern Recognition, pp. 838–841 (2013)

Using Artificial Neural Back-Propagation Network Model to Detect the Outliers in Semiconductor Manufacturing Machines

Keng-Chieh Yang[1,*], Chia-Hui Huang[2], Conna Yang[3],
Pei-Yao Chao[4], and Po-Hong Shih[4]

[1] Department of Information Management, Hwa Hsia Institute of Technology
No.111, Gongzhuan Road, Zhonghe Dist., New Taipei City 235, Taiwan
andesyoung.tw@gmail.com
[2] Department of Business Administration, National Taipei College of Business
No. 321 Jinan Road, Section 1, Taipei 100, Taiwan
leohuang@webmail.ntcb.edu.tw
[3] Institute of Business and Management, National Chiao Tung University
1001 University Road, Hsinchu 300, Taiwan
conna.yang@gmail.com
[4] Institute of Information Management, National Chiao Tung University
1001 University Road, Hsinchu 300, Taiwan
{7777682, a770401362}@gmail.com

Abstract. This study uses the artificial neural back-propagation network model to detect the outliers in semiconductor machines. The neural network model has the advantages of great precision and effectiveness. This research uses Novellus Vector Machine and its Remote Process Controller (RPC) function to collect the data. This study detects the gas transmission pressure of chamber. Our experimental results show that three-month period of network training data possesses the best results. We suggest that the prediction and model training be around 3 months.

Keywords: Semiconductor machine outliers, back-propagation network model, quality control in semiconductor manufacturing.

1 Introduction

The semiconductor manufacturing processes are usually through fault detection and classification (FDC) to collect a large number of status variable identification (SVID) as data in real-time processes. But these information are often used to adjust the machine parameters after the event. FDC can timely detect the deviations of the machine parameters when the parameters deviate from the original value and exceed the range of the specification [1,2].

Advanced semiconductor manufacturing processes are all made of very sophisticated machines. Requirements on these processes need hundreds of control parameters of the machine. For the normal operation and maintenance of

---

* Corresponding Author.

A. Moonis et al. (Eds.): IEA/AIE 2014, Part II, LNAI 8482, pp. 240–249, 2014.
© Springer International Publishing Switzerland 2014

equipment to ensure production, failures of the equipments must be diagnosed correctly and timely. In the semiconductor industry, machine monitoring to reduce the abnormal condition is necessary. According to the above conditions, the study used FDC data to analysis semiconductor machine outliers.

2   Quality Control in Semiconductor Manufacturing

High-end semiconductor manufacturing processes have more stringent quality requirements. Semiconductor manufacturers have noticed the importance of machine outliers. Hence, machine monitoring requirements have increased for these urgent situations. In this section, we will introduce the quality control method in semiconductor manufacturing company, how web-based quality control work for the factory.

2.1   Statistics Process Control (SPC) in Semiconductor Manufacturing

The semiconductor manufacturing factory is usually used SPC to detect the quality of real-time processes. These information are often used to adjust the machine parameters after the event. SPC control charts [1] as shown in Fig. 1 are important tools to diagnose abnormalities of process [2,3]. Using these charts, engineers can understand the quality of the wafer. If the output of numerical data are between the upper bound (UCL) and lower bound (LCL), it means the quality is qualified. Otherwise, the quality is faults.

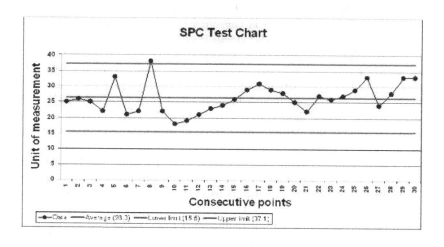

Fig. 1. SPC Control Chart

---

[1] Adopted from http://uat.qihub.scot.nhs.uk

In the semiconductor producing process, wafer manufacturing quality control methods can be classified into work-in process test and the control wafer testing machine [4]. Work-in process test is done directly on the wafer testing. However, wafer testing in the manufacturing process can be divided into the front-end processing for visual inspection, defect analysis, wafer acceptance test and posterior segment test. The control wafer testing is based on the test piece for the machine to carry out its process capability. In this test, it usually obtained information entered to statistical process control (SPC) system [2].

There are some quality control or testing methods in semiconductor manufacturing [2]. Inspection is the appearance of defections observed in the manufacturing site. Workers can view wafers in visual appearance or microscopic view. Off-line measuring machines testing is to simulate the dummy piece of result of the machine processes. Almost all of the semiconductor producing machines have this testing mode. Defect analysis uses defect analysis instruments to scan the surface of wafer, typically applying sampling method. Another method is wafer acceptance test (WAT). When designing the electronic circuit, it has been placed the test point for electrical testing. A wafer has five testing points and each point represents one-fifth of the area that must be within die quality control. Finally, Die test is run in test house. The test machine detects each die in the maximum resolution, but the feedback spends us more time. Hence, the SPC system is one of the elements in MES system. It is also used in semiconductor industry for quality control practices.

2.2 Fault Detection and Classification (FDC) in Semiconductor Manufacturing

Fault detection and classification contains two critical functions: fault detection, and the fault classification. Different fault requires different corrective actions, while fault classification function is classified based on statistics Eigen-values. So engineers can quickly refer to the machine error code and restore the machine to normal state within the least time. In other words, FDC module is built on a common platform and continuously monitors equipment parameters against pre-configured limits using statistical analysis techniques to provide proactive and rapid feedback on equipment health [3,5]. It eliminates unscheduled downtime, improves tool availability and reduces scrap. It allows engineers to analyze sensor data from manufacturing equipment, detect out-of-norm conditions and trace them back to the tool issues.

In the semiconductor wafer process, when the machine produces a certain number of wafer, some parameters will drift from original ones. So, FDC can detect deviations within a very short time. When the parameters deviate from the original value, and sometimes beyond the range of the set interval, this needs to apply the Run-to-Run adjustments to modify the parameters directly and continuously collecting the running parameters of the machine and constant feedbacks [5,6]. Based the production activities of quality control, engineers need to adjust the machine parameters to ensure that the production are within normal operations.

Production equipment engineers use the FDC monitor to ensure the information of production status correctly, including the manufacturing process, machine operating conditions, parameters, and use of the recipe. Engineers must check the machine status before the operation in issues. When the production is finished, it has caused the business loss. The FDC monitor can avoid the waste of production capacity, reduce failures and ensure the producing yield increase.

3   Methodology

In this section, we introduce artificial neural networks (ANN) as the main researching method. Second, we adopt back-propagation neural network as the researching method to analyze the semiconductor manufacturing machines outliers. Lastly, we apply gray relational analysis for justify the results.

3.1   Artificial Neural Networks (ANNs)

Artificial neural networks as shown in Fig. 2 are used in many fields. ANNs are computational models inspired from the thinking pattern of human brain [7]. ANNs can learn and generalize from experiences, and they can abstract essential information from data. ANNs are trained to "think" like humans by strengthening or weakening interconnected weights that connect processing elements of ANNs. In the literature, the terms "neuron", "cell", and "unit" are usually used interchangeably for representing processing elements of neural networks [8].

In other words, artificial neurons are computational models illuminated from the natural neurons. A natural neuron receives signals through synapses that are located on the dendrites or membrane of the neuron. When the neuron receives

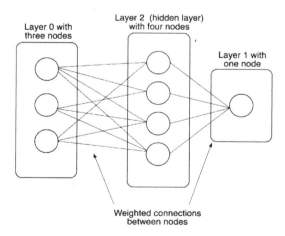

Fig. 2. The framework of artificial neural networks [11]

signals, it will be activated and emits a signal through the axon. This signal might be sent to another synapse, and/or might activate other neurons.

Biological neural network learning is to adjust the intensity of ganglion [9]. Therefore, we can say that nerve cells in the input path tree are through a lot of contact between the cell body to accept the ganglion cells around the body of the outgoing signals, and body axis of nerve cell is equivalent to the output path. We can transform information outside to the input signal into the input vector I_i and compute with the weighting value W_i. Artificial neural system can be divided into two parts. The front-end is a summation function of the input vector to be integrated, and then the rear section by a simple transferring function for the message output. Finally, the output vector Y can be the form of other neurons as input. Transferring function can normally sigmoid function [10].

3.2   Back-Propagation Neural Network (BPN)

The back-propagation neural network algorithm is a multi-layer feed-forward network trained according to error back propagation algorithm and is one of the most widely applied neural network models [12]. Back-propagation neural network model is a learning model in the neural network and the most representative one. BPN can be used to learn and store a great deal of mapping relations of input-output model, and no need to disclose in advance the mathematical equation that describes these mapping relations [12,13].

Multilayer networks for propagation algorithm is a generalized least mean squares (LMS) algorithm, and the back-propagation algorithm and LMS algorithms used mean square error (MSE) as performance indicators [13]. When each input vector are entered into the network, we can compare the gap between the network output and the target output to adjust the settings in the network variables. It is generally using minimum mean square error to measure the quality of learning.

Back-propagation neural network applies Widrow-Hoff learning rule to generalize the multi-differentiable nonlinear transferring function [8]. Back-propagation neural network has partial weight (b) and the hidden layer is hyperbolic transferring function. The output layer is a linear transferring function. Using the known input vector and its corresponding target vector, together with a sufficient number of neurons in the hidden layer, this will enable the network approximate a finite number of discontinuities in any function . When appropriately trained back-propagation neural network is given new input vector, the network will calculate a reasonable output. Using generalized characteristic in the network, the new input vector can calculate output vector. In other words, when generalized characteristic of the network achieves, we can use non-training data in the network and this can produce a satisfactory output [8].

3.3   Gray Relational Analysis (GRA)

GRA based on gray system theory was first proposed in 1982. A system containing both insufficient and sufficient information, called "black" and "white", respectively, is a gray system. Gray system theory deals with a system containing insufficient information, and the GRA is essentially believed to capture the relationship between the main factor and other factors in a system regardless whether this system has adequate information. Therefore, GRA is a method to describe the relative variance of factors during system development [14].

In the gray relational analysis, if the gaps of the reference range are excessive, certain factors will be ignored. When the direction of each factor is inconsistent, the results may cause the deviation. Hence, we must do data pre-processing for the raw data [15]. We can do the initializing, averaging or internalizing for the raw data. Each element of a sequence can satisfy two conditions: comparability and non-parallel [15].

When we develop a gray relation system, if the trend of two factors is consistent, that means a higher degree of simultaneous change. It represents a higher degree of association between the two factors. Therefore, the gray relational analysis method is based on the similarity between factors or difference between them (i.e. gray relational degree) [14].

4 Experiment Results

In this study, we use MATLAB 7.0, the neural network toolbox to analyze the data [16]. Experiment processes include data pre-processing, network variables setting, the hidden layer neurons determination, the selection of the best combination of input variables, network output results determining principles, and sensitivity analysis.

This study detects the gas transmission pressure of chamber. If the numerical data exceeds the upper bound or lower bound, this may cause the product failure. The gas may cause uneven distribution of the wafer surface, and the chemical change is likely to make the solid on the wafer surface not uniformly flat. This study uses Novellus Vector Machine and its Remote Process Controller (RPC) function to collect the data. The data collection period is between April, 2008 and December, 2011.

4.1 Model Building

In this study, the artificial neural back-propagation network model that might be applicable would be one that needs to take factors into consideration, such as the time building model and effectiveness. Also, the training data records, global/partial characteristics of the data, and number of parameters needed to include are all necessary elements that help one decide which type of artificial neural back-propagation network model. When the parameters in the neuron model are too much and the dataset is not large enough, we must go back to the data preprocessing step as mentioned earlier to see if this problem can be solved by using an algorithm with lower efficiency or a model with less neurons.

When the model training is completed, it should be compared with the parameters offered by the machine and be checked to see if it falls within the error range. If it does not, then the training requires to be processed again.

Furthermore, the MSE of the model serves as an indicator, too. When the model loses faith and requires retraining, several markers, such as the confidence index, machine parameters, recipe, environment sensor parameter and the measured figures should be served as support whether or not to go through the training again.

4.2   Model Training

There can be some difference among machines after countless times of machine maintenance and wear and tear. Thus, network training should be conducted by the different selection of network input variables. For instance, data collected for the past three months and one month should be analyzed to observe the difference in each variable.

(1) One-month network:

790 records of data were collected from the transferring pressure of the Novellus Vector December 1st, 2010 to December 31st. The tested data retrieved from January 1st 2011 to January 5th consisted of 72 records. The learning times were set to 3,000, learning rate set to 0.1, and the momentum correction coefficient set to 0. Since the training results of the artificial neural back-propagation network do not always have the same result, this study conducts the same experiment 10 times to ensure the network stability.

The results of the tests show that each round of the training and tests are slightly different. However, all the results converge within the 1,100th cycle. The overall performance is fairly well with the MSE=0.01641 and RMSE=0.59956. Table 1 part (a) shows that by comparing the Train-R data and target data and depicting prediction output figure of the network training, the default mode and outlier can be predicted (Fig. 3 (a),(b)).

(2) Three-months network:

2,036 records of data were collected from the Transfer Pressure of the Novellus Vector from October 1st, 2010 to December 31st. The test data retrieved from January 1st 2011 to January 5th consisted of 72 records. The learning times were set to 3,000, learning rate set to 0.1, and the momentum correction coefficient set to 0. Since the training results of the artificial neural back-propagation network do not always have the same result, this study conducts the same experiment 10 times to ensure the network stability.

The results of the tests show that each round of the training and tests are slightly different. However, all the results converge within the 1,400th cycle. As shown in Fig. 4, the overall performance is fairly well with the MSE=0.01406 and RMSE=0.66066. Table 1 part (b) shows that by comparing the Train-R data and target data and depicting prediction output figure of the network training, the default mode and outlier can be predicted (Fig. 4 (a), (b)).

We can see from Fig. 3 that the parameters in the training process have a fairly well learning effect. However, the extreme values of learning ability are near to perfect with no influence from the network selection. Fig. 3 shows the change in the prediction results and actual concentration values. We can see that

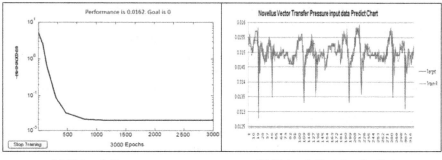

(a) Network Convergence   (b) Network Training Prediction

Fig. 3. Network convergence and Network training prediction of one month

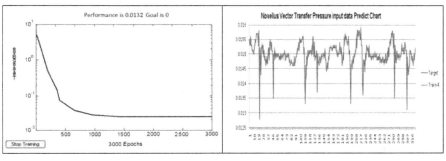

(a) Network Convergence   (b) Network Training Prediction

Fig. 4. Network convergence and Network training prediction of three months

Table 1. Network training and performance test results

Hidden neuron	1 month (a)		3 months (b)	
	MSE	Train-R	MSE	Train-R
1	0.0162	0.5873	0.0132	0.6773
2	0.0154	0.6032	0.0136	0.6832
3	0.0165	0.6244	0.0142	0.6834
4	0.0166	0.5873	0.0136	0.6583
5	0.0169	0.5972	0.0139	0.6572
6	0.0157	0.5878	0.0147	0.6978
7	0.0168	0.5983	0.0148	0.5783
8	0.0163	0.5973	0.0143	0.6373
9	0.0168	0.6036	0.0138	0.6439
10	0.0169	0.6092	0.0145	0.6899
Average	0.0164	0.5996	0.0141	0.6601

the artificial neural back-propagation network model performs better in extreme values, especially in very small values.

This study uses the mean square error (MSE) of the training output data to analyze if training results are acceptable and converge. The smaller for the MSE, the better the network training results. In the network training process, the MSE might be unstable if the network is not converged. To improve such a situation, the network learning rate and momentum correction coefficient should be readjusted to reach convergence.

This study analyzes the network training and prediction results by using the correlation coefficient R and MSE. The n in the equation signifies the total number of data inputs, μ denotes the arithmetic mean, σ denotes the standard deviation, the subscript i denotes the number of data, subscript t and a denotes the actual value and the network output value. This study analyzes the network output by calculating the correlation coefficient of the network trail output and the actual value and chooses the higher one as the optimal result.

$$MSE = \frac{1}{N} \sum_{i=1}^{n} (t_i - a_i)^2 \tag{1}$$

$$R = \frac{(\sum t_i a_i) n \mu_t \mu_a}{(n-1)\sigma_t \sigma_a} \tag{2}$$

5   Conclusion

This study detects the gas transmission pressure of chamber. If the numerical data exceeds the upper bound or lower bound, this would cause the product failure. The gas may cause uneven distribution of the wafer surface, and the chemical change is likely to make the solid on the wafer surface not uniformly flat. This would seriously affect the chip yield.

After countless times of machine maintenance and wear and tear, the machines will need to be maintained to perform well again. After long functioning periods, the machines are likely to decrease in performance. Thus, the facility engineer should be aware of such a situation. The average time of maintenance for the machine would be around 3 months. Our experimental results show that three-month period of network training data possesses the best results. Because the machine needs maintenance, the stability of machine in the first month is lower than that of three months. However, 3 months is the most stable situation in our study. Because the machine has been smoothly working, we get a better result in this experiment.

Acknowledgement. The authors are indebted to the anonymous reviewers for their careful reading and comments to enhance the quality of this article. This work is support by National Science Council Taiwan under grant no. NSC 102-2410-H-141-012-MY2 (C.H. Huang).

References

1. Chiu, C.-C., Shao, Y.E., Lee, T.-S., Lee, K.-M.: Identification of Process Disturbance using SPC/EPC and Neural Networks. Journal of Intelligent Manufacturing 14, 379–388 (2003)
2. Fan, C.-M., Guo, R.-S., Chang, S.-C., Wei, C.-S.: SHEWMA: An End-of-line SPC Scheme using Wafer Acceptance Test Data. IEEE Transactions on Semiconductor Manufacturing 13, 344–358 (2000)
3. Shao, Y.E., Lu, C.-J., Chiu, C.-C.: A Fault Detection System for an Autocorrelated Process using SPC/EPC/ANN and SPC/EPC/SVM Schemes. International Journal of Innovative Computing, Information and Control 7, 5417–5428 (2011)
4. May, G.S., Spanos, C.J.: Fundamentals of Semiconductor Manufacturing and Process Control. Wiley (2006)
5. Chen, W.-C., Lee, A.H., Deng, W.-J., Liu, K.-Y.: The Implementation of Neural Network for Semiconductor PECVD Process. Expert Systems with Applications 32, 1148–1153 (2007)
6. Card, J.P., Naimo, M., Ziminsky, W.: Run-to-run Process Control of a Plasma Etch Process with Neural Network Modeling. Quality and Reliability Engineering International 14, 247–260 (1998)
7. Dayhoff, J.E., DeLeo, J.M.: Artificial Neural Networks. Cancer 91, 1615–1635 (2001)
8. Wang, L., Fu, K.: Artificial Neural Networks. Wiley Online Library (2008)
9. Mehrotra, K., Mohan, C.K., Ranka, S.: Artificial Neural Networks. MIT Press (1997)
10. Chen, F.-L., Liu, S.-F.: A Neural-network Approach to Recognize Defect Spatial Pattern in Semiconductor Fabrication. IEEE Transactions on Semiconductor Manufacturing 13, 366–373 (2000)
11. Pu, H.-C., Hung, Y.-T.: Use of Artificial Neural Networks: Predicting Trickling Filter Performance in a Municipal Wastewater Treatment Plant. Environmental Management and Health 6, 16–27 (1995)
12. Chen, F.-C.: Back-propagation Neural Networks for Nonlinear Self-tuning Adaptive Control. IEEE Control Systems Magazine 10, 44–48 (1990)
13. Li, J., Cheng, J.-H., Shi, J.-Y., Huang, F.: Brief Introduction of Back Propagation (BP) Neural Network Algorithm and its Improvement. In: Jin, D., Lin, S. (eds.) Advances in CSIE, Vol. 2. AISC, vol. 169, pp. 553–558. Springer, Heidelberg (2012)
14. Shen, D.-H., Du, J.-C.: Application of Gray Relational Analysis to Evaluate HMA with Reclaimed Building Materials. Journal of Materials in Civil Engineering 17, 400–406 (2005)
15. Deng, J.: The Control Problems of Grey Systems. Systems & Control Letters 5, 288–294 (1982)
16. Demuth, H., Beale, M.: Neural Network Toolbox for Use with MATLAB (1993)

Transferring Vocal Expression of F0 Contour Using Singing Voice Synthesizer

Yukara Ikemiya, Katsutoshi Itoyama, and Hiroshi G. Okuno

Graduate School of Informatics, Kyoto University
606-8501 Sakyo, Kyoto, Japan
{ikemiya,itoyama,okuno}@kuis.kyoto-u.ac.jp

Abstract. A system for transferring vocal expressions separately from singing voices with accompaniment to singing voice synthesizers is described. The expressions appear as fluctuations in the fundamental frequency contour of the singing voice, such as *vibrato*, *glissando*, and *kobushi*. The fundamental frequency contour of the singing voice is estimated using the subharmonic summation in a limited frequency range and aligned temporally to chromatic pitch sequence. Each expression is transcribed and parameterized in accordance with designed rules. Finally, the expressions are transferred to given scores on the singing voice synthesizer. Experiments demonstrated that the proposed system can transfer the vocal expressions while retaining singer's individuality on two singing voice synthesizers: the *Vocaloid* and the *CeVIO*.

1 Introduction

Every singer has unique vocal expressions and singing style, which characterize his or her singing. The goal of our study is to create a library of vocal expressions and styles that can be applied to consumer-generated media and music information retrieval [1]. Such a library would enable the vocal expressions of favorite singers to be transferred other songs by using a singing voice synthesis system, such as the Vocaloid [2], and retrieval of songs based on singing style. A demonstration of our vocal expression transfer is available on-line[1].

This paper describes a system that transfers vocal expressions involving variation and fluctuation of the fundamental frequency (F0), such as *vibrato*, *kobushi*, and *glissando*, which are extracted from singing voices with instrumental accompaniment. Changes in F0 characteristics affect singing voice quality and individuality more than changes in spectral characteristics [3,4]. Fig. 1 shows a typical template for each expression. Vibrato is a deliberate, periodic fluctuation in the F0 contour. Kobushi is short tremolo that appears in Japanese folk songs such as *enka* and *min-yo*. Glissando is generally separated into two types: *glissdown*, which is a glide down in pitch for an offset note, and *glissup*, which is a glide up in pitch for an onset note.

The proposed transfer system consists of three steps as follows:

---

[1] winnie.kuis.kyoto-u.ac.jp/members/ikemiya/demo/sst2013.html

A. Moonis et al. (Eds.): IEA/AIE 2014, Part II, LNAI 8482, pp. 250–259, 2014.

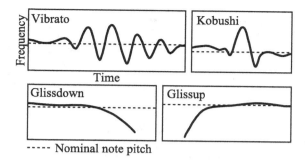

Fig. 1. Vocal expressions

1. Estimation of singing voice F0 from polyphonic music.
2. Transcription of vocal expressions in F0 contour.
3. Transfer of extracted expressions into a new song.

We use a sequence of symbolized, chromatic pitches of the song, such as (G4, A#4, C5, ...), to achieve F0 estimation with high accuracy. Note that the sequence does not contain note values. The F0 contour is searched for in a limited frequency range by considering the smoothness.

About steps 2 and 3, we discuss problems with a number of existing studies. While some studies have been aimed at making synthesized singing voices more human-like by adding pitch fluctuations [5, 6], they use manually-tuned vibrato expressions and they do not represent singers' individuality. The VocaListener2 [7] transfers pitch, volume, and timbre from the user's singing to the Vocaloid system directly. The problem of the VocaListener2 that the transfer can be applied to the same song because the VocaListener2 simply extracts moment-to-moment fluctuations, not vocal expressions such as vibrato. Although a statistical model of vocal F0 fluctuation by the second-order transfer function has been proposed [8], this model also cannot represent separately each vocal expression. Singing voice synthesis systems based on the hidden Markov model (HMM) [9–11] learn singing styles as the distributions of the feature vectors and reconstruct them in other songs. A problem of the HMM-based systems is that they require many sets of singing voices without accompaniments and effects and corressponding scores for learning. Yasuraoka *et al.* have proposed a musical instrument sound synthesizer which learns the pitch, volume, and timbre of the instrument sounds such as guitar and reconstructs them in other melodies [12]. The objective of their system is very close to ours except for the target sounds. In addition, their system focuses on musical instrumental sounds, not vocal with accompanying sounds, as a target musical signal.

2   Estimating Fundamental Frequency

Our proposed transfer system requires a method for estimating the F0 contour of a singing voice with accompaniments with both high accuracy and high

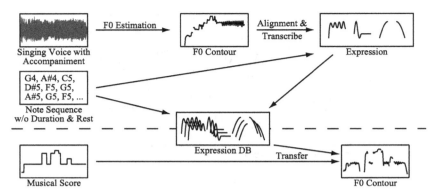

Fig. 2. Overview of proposed vocal expression transfer system

frequency-resolution. The search range for F0 is limited, from $L - 400$ to $H + 400$ cent[2], where L and H are the lowest and highest pitches in the given pitch sequence.

We assume that the F0 of singing voices has the following properties:

1. A singing voice usually performs the predominant part.
2. A large movement in the F0 takes a long time.
3. It has inertia: a moving F0 continues to move and sometimes overshoots the desired pitch.

We developed an objective function that satisfies these properties to obtain an optimal F0 contour.

The objective function to be maximized is defined as

$$
\hat{F} = \underset{F \in \{f_1, \dots, f_T\}}{\arg\max} \left(\sum_{t=1}^{T} \log P_M(f_t) + \sum_{t=2}^{T} \log P_{\Delta F0}(f_t - f_{t-1}) \right.
$$
$$
\left. + \sum_{t=3}^{T} \log P_{\Delta\Delta F0}(f_t - 2f_{t-1} + f_{t-2}) \right).
$$

An optimal F0 contour is computed by using the Viterbi algorithm. Each term on the right-hand side corresponds to a property described above.

The first term is for property 1 on predominancy:

$$
P_M(f_t) = \frac{\text{SHS}(t, f)}{\int_{L-400}^{H+400} \text{SHS}(t, f') \, df'},
$$
$$
\text{SHS}(t, f) = \sum_{n=1}^{N} 0.84^{n-1} \, \text{CQ}(t, f + 1200 \log_2 n).
$$

---

[2] Cent is the logarithmic unit of frequency, and a harftone is equal to 100 cent.

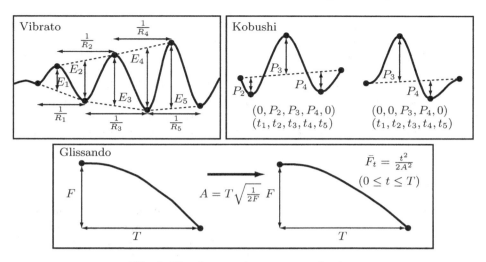

Fig. 3. Vocal expression parameterization

We use the subharmonic summation (SHS) [13] to calculate the likeliness of the predominant F0 at time t and the N overtones to calculate frequency f. Here we set $N = 7$. $CQ(t, f)$ is the spectrogram obtained the constant-Q transform [14].

The second term is for property 2 on F0 change:

$$P_{\Delta F0}(f) = \begin{cases} 1/200 & \text{if } |f| < 100 \\ 0 & \text{otherwise} \end{cases}.$$

This term takes a uniform positive value when the F0 adjacent frame difference is less than a certain level or zero.

The third term is for property 3 on inertia. It is defined as a truncated Gaussian function:

$$P_{\Delta\Delta F0}(f) \propto \begin{cases} \exp(-f^2/(2 \times 50^2)) & \text{if } |f| < 50 \\ 0 & \text{otherwise} \end{cases}.$$

This term takes a large value when the second-order derivative of F0 is close to zero.

3   Transcribing Vocal Expressions

This section describes a method for transcribing vocal expressions from the estimated F0 contour and storing them as parameters. We assume that the expressions are strongly connected to the notes on the score. In other words, the expressions do not exceed the note boundaries. We thus temporally align the

F0 contour and the pitch sequence. This alignment is regarded as a problem of hidden state estimation. Let the observation and the hidden state sequence be the F0 contour and pitch sequence, respectively. The alignment is regarded as a problem of state estimation for each observation. We thus define the state likeliness for the given F0 as the squared difference of the F0 and the pitch so that this problem can be solved using the Viterbi algorithm. When the F0 contour does not have values for a certain span, the span is detected as a rest, and the state changes to the next one. An overview of expression transcription is shown in Fig. 3.

3.1   Vibrato

We detect vibrato sections by using a previously proposed method [15] that uses short-time Fourier transform to find a sharp peak that corresponds to the vibrato rate (the number of vibrations per second). We uniquely restrict the range of the extent (the amplitude of vibration) to $30 - \infty$ cent and that of the rate to 3 – 8 Hz because enka and min-yo songs have a vibrato with a much larger extent and a lower rate than the restrictions proposed previously [15].

A vibrato section is represented as a sequence of pairs of two parameters, rate R_i and extent E_i, such as $((E_1, R_1), (E_2, R_2), (E_3, R_3), \ldots)$. Let I be the number of peak points of the vibrato, and let f_i and t_i be the logarithmic frequency and time of the i-th peak point in the F0 contour ($i = 1, \ldots, I$). The extent and rate parameters $((E_1, R_1), \ldots, (E_{I-2}, R_{I-2}))$ are calculated as

$$R_i = \frac{1}{t_{i+2} - t_i} \quad \text{and}$$
$$E_i = |(f_{i+2} - f_i)(t_{i+1} - t_i)R_i + (f_i - f_{i+1})|.$$

3.2   Glissando

Glissdown (glissup) sections are extracted by detecting a monotonic decrease (increase) of more than F_{least} cent from a phrase end (beginning). On the basis of the results of our preliminary experiments, F_{least} set to 200 cent.

A glissdown (glissup) is modeled as a parabola, and stored as the parameters of the parabola curve and its duration. Since they have bilateral symmetry, we describe only glissdown here. Let T s and F cent be the duration of the detected glissdown and the frequency decrease. Coefficient A of a parabola is calculated as

$$A = T\sqrt{\frac{1}{2F}}.$$

3.3   Kobushi

Before detecting kobushi sections, we extract all peak, valley, and point which cross the pitch of the corresponding note from the F0 contour. Although the pattern of kobushi is not well defined among professional singers, we have found by observing the F0 contour of enka and min-yo songs that kobushi follows three rules.

1. Kobushi sections do not overlap vibrato sections.
2. A kobushi section has only one peak greater than 150 cent (main peak).
3. In front of and behind the main peak, one or no small valley(s) (sub-peak) appears.

We define that a kobushi section contains the main peak and sub-peaks, and that the gradient between the peaks is more than V cent/s. Here we set $V = 1000$.

A kobushi section is stored as a quintuple: a starting point, a left sub-peak, a main peak, a right sub-peak, and an end point. Each element of the quintuple is a pair of an extent of the peak and its time. If a sub-peak does not exist, the extent of the corresponding element is set to zero. The extent P_i of the i-th peak is calculated as

$$P_i = f_i - \left(\frac{f_5 - f_1}{t_5 - t_1}(t_i - t_1) + f_1 \right),$$

where t_i and f_i denote the time and log-frequency of the i-th peak, respectively.

4 Transferring Vocal Expressions

This section describes the process of transferring vocal expressions to a vocal synthesizer by using a vocal expression library. The synthesizer is assumed to provide musical score information for synthesized vocal expressions and a mechanism for handling pitch.

4.1 Vocal Expression Library

A vocal expression library consists of sets of vocal expression parameters and note information for each vocal expression (vocal expression set). Note information includes four elements: pitch, duration, musical intervals, and label. Musical intervals are the differences in pitch from one note to the next. Note label represents whether a note is at the beginning, the middle, or the end of a musical phrase. If there is an unvoiced section of over 200 ms between two notes, the first note is labeled as the end of the phrase, and the second note is labeled as the beginning.

4.2 Preprocessing

The vocal pitch range of the input score may be completely different from that of the library. This makes it difficult to transfer vocal expressions on the basis of a simple rule. Thus, the pitch range in the library is adjusted by shifting the lowest pitch in the library to that of the input score. Furthermore, the note information for all notes in the score is acquired.

4.3   Transfer Rule

The following process is performed to each note in the input score. First, a set of vocal expressions matching four conditions are extracted from the library.

for all expressions Note label is the same as that of the target note.

for all expressions Difference between note pitch (note number) of vocal expression set and that of target note is smaller than M.

for kobushi and glissando Full length of vocal expression is shorter than the length of target note.

for kobushi Signs of note transitions are the same as that of the target note.

The smaller the M, the stricter the rule for transfer.

Second, from the extracted set, the set of notes nearest to the target note are chosen for each vocal expression. When no vocal expressions are extracted, no vocal expressions are applied to the target note. The nearness of notes is determined using two indices and priority.

1. Difference in note pitch
2. Difference in note length

If the full length of the selected vibrato is larger than the length of the target note, the vibrato is trimmed to end at the end of the note or the beginning of the glissdown, and if smaller, it is extended with the rate to the end of the note. Vocal expressions are transferred by resynthesizing the expression in accordance with the selected parameters and pasting it on the F0 contour of the target note.

5   Evaluation

5.1   Experimental Settings

All musical pieces for our experiments were converted to a 16-kHz sampling rate with 16 bits per sample. A constant-Q spectrogram was calculated with a time resolution of 10 ms, a frequency resolution of 6 cent, a frequency range of 60 to 6000 Hz, and a Q value of $(1/(2^{0.01} - 1))/5$. Additionally, we postulated that voiced sections were detected in advance.

5.2   Transcription with Commercial Recordings

We applied our method to two commercial recordings, a verse part of "Jinsei Ichiro (by a Japanese famous singer, Misora Hibari)" and a chorus part of "Crispy (by a Japanese famous singer, Spitz)". The former is an enka song, while the latter is a Japanese pop song.

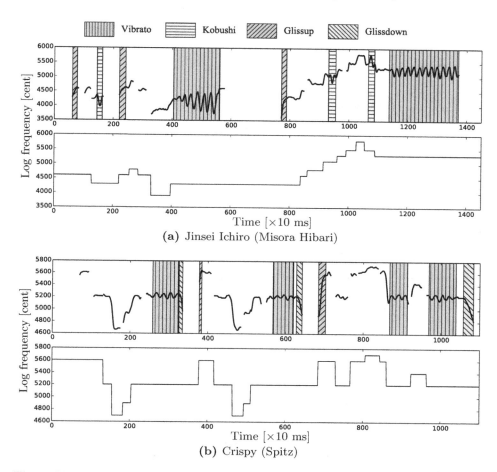

Fig. 4. Vocal expression identification. Upper figures show estimated F0 contour and identified expressions; bottom figures show alignment of note sequence with F0 contour.

Fig. 4 shows the results of vocal expression identification. On the left (Fig. 4(a)), we can see that both long and short vibrato, kobushi characteristic of enka and glissup attached to strained singing, were identified. On the right (Fig. 4(b)), we can see that frequent glissdown best characterizes the singing style of the "Spitz" vocal. Fig. 5 shows the result of resynthesis of the vocal expressions. Figs. 5(a)-(b) correspond to the second and third glissdowns in Fig. 4(b), Figs. 5(c)-(d) correspond to the second and third kobushi expressions in Fig. 4(a). The root mean square errors for glissdown and kobushi were 22.3 and 16.0 cent, respectively. Considering that a semitone is 100 cent, we can say that each expression was precisely resynthesized despite differences in scale and shape.

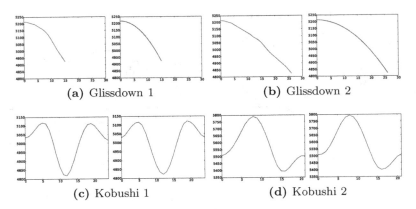

(a) Glissdown 1   **(b)** Glissdown 2

(c) Kobushi 1   **(d)** Kobushi 2

Fig. 5. Resynthesized vocal expressions. Left figures (green) show original contour; right figures (blue) show synthesized contour.

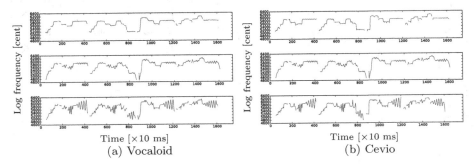

(a) Vocaloid   **(b)** Cevio

Fig. 6. F0 contour of synthesized singing voice with vocal expression. Top, center, and bottom figures are without expressions, with Spitz expressions, and with Misora Hibari expressions, respectively.

5.3   Transferring Expressions with Singing Voice Synthesizers

For transferring vocal expressions, we used two singing voice synthesizers: the *Vocaloid* and the *CeVIO*[3]. In Vocaloid, F0 can be controlled with pitch bend parameters, which change the pitch up and down from the chromatic regular pitch. In CeVIO, F0 can be controlled by directly setting the logarithmic values of the pitch.

Figure 6 shows the results of vocal expression transfer to the singing voice synthesizers. From the top, the figures of F0 contour with no vocal expression, with vocal expression of "Spitz", and with vocal expression of "Misora Hibari" are shown. We can confirm that vocal expression such as vibrato, kobushi, and glissando are transferred similarly in both synthesizers.

---

[3] http://cevio.jp/

6   Conclusion

Our developed system for transferring vocal expressions achiveved to resynthesize expressions that reflect singer's individuality extracted from singing voices with accompaniments. Vocal expressions are detected from the F0 contour and parametrized based on the designed rules, and then transferred to the input score using the Vocaloid and CeVIO singing voice synthesizers. Experimental results demonstrated that our method can transcribe vocal expressions from commercial songs and resynthesize them precisely. In future work, we intend to expand our method to other types of expressions. We also intend to apply the library of the expressions for retrieving musical pieces on the basis of the singing styles.

References

1. Downie, J.S.: Music information retrieval. Annu. Rev. Inf. Sci. Technol. 37, 295–340 (2003)
2. Kenmochi, H., Ohshita, H.: Vocaloid - commercial singing synthesizer based on sample concatenation. In: INTERSPEECH 2007, pp. 4009–4010 (2007)
3. Saito, T., Goto, M.: Acoustic and perceptual effects of vocal training in amateur male singing. In: INTERSPEECH 2009, pp. 832–835 (September 2009)
4. Guzman, M.A., Dowdall, J., Rubin, A.D., Maki, A., Levin, S., Mayerhoff, R., Jackson-Menaldi, M.C.: Influence of emotional expression, loudness, and gender on the acoustic parameters of vibrato in classical singers. Journal of Voice 26(5), 675–681 (2012)
5. Stables, R., Athwal, C., Bullock, J.: Fundamental frequency modulation in singing voice synthesis. In: International Conference on Speech, Sound and Music Processing: Embracing Research in India, pp. 104–119 (2012)
6. Umbert, M., Bonada, J., Blaauw, M.: Generating singing voice expression contours based on unit selection. In: SMAC (July 2013)
7. Nakano, T., Goto, M.: VocaListener2: A singing synthesis system able to mimic a user's singing in terms of voice timbre changes as well as pitch and dynamics. In: ICASSP 2011, pp. 453–456 (2011)
8. Ohishi, Y., Kameoka, H., Mochihashi, D., Kashino, K.: A stochastic model of singing voice F0 contours for characterizing expressive dynamic components. In: Proc. INTERSPEECH (September 2012)
9. Oura, K., Mase, A., Yamada, T., Muto, S., Nankaku, Y., Tokuda, K.: Recent development of the HMM-based singing voice synthesis system - Sinsy. In: Proc. ISCA Tutorial and Research Workshop on Speech Synthesis, pp. 211–216 (September 2010)
10. Saino, K., Tachibana, M., Kenmochi, H.: A singing style modeling system for singing voice synthesizers. In: Proc. INTERSPEECH, pp. 2894–2897 (September 2010)
11. Lee, S.W., Ang, S.T., Dong, M., Li, H.: Generalized F0 modelling with absolute and relative pitch features for singing voice synthesis. In: Proc. ICASSP, pp. 429–432 (March 2012)
12. Yasuraoka, N., Abe, T., Itoyama, K., Takahashi, T., Ogata, T., Okuno, H.G.: Changing timbre and phrase in existing musical performances as you like. In: ACM Multimedia 2009, p. 10 (2009)
13. Hermes, D.J.: Measurement of pitch by subharmonic summation. J. Acoust. Soc. Am. 83(1), 257–264 (1988)
14. Brown, J.C.: Calculation of a constant q spectral transform. J. Acoust. Soc. Am. 89(1), 425–434 (1991)
15. Nakano, T., Goto, M.: An automatic singing skill evaluation method for unknown melodies using pitch interval accuracy and vibrato features. In: Proc. INTERSPEECH (September 2006)

Minimum Similarity Sampling Scheme for Nyström Based Spectral Clustering on Large Scale High-Dimensional Data

Zhicheng Zeng, Ming Zhu*, Hong Yu, and Honglian Ma

School of Software, Dalian University of Technology,
Tuqiang street. 321, 116620 Dalian, China
z.alex.zeng@gmail.com,
{zhuming,hongyu,mhl}@dlut.edu.cn

Abstract. Large-scale spectral clustering in high-dimensional space is among the most popular unsupervised problems. Existed sampling schemes have different limitations on high-dimensional data. This paper proposes an improved Nyström extension based spectral clustering algorithm with a designed sampling scheme for high-dimensional data. We first take insight into some existed sampling schemes. We illustrate their defects especially in high dimension scene. Furthermore we provide theoretical analysis on how the similarity between the sample set and non-sampling set influences the approximation error, and propose an improved sampling scheme, the minimum similarity sampling (MSS) for high-dimensional space clustering. Experiments on both synthetic and real datasets show that the proposed sampling scheme outperforms other algorithms when applied in Nyström based spectral clustering with higher accuracy, and lowers the time consumption for sampling.

Keywords: Large-scale, High dimensionality, Spectral Clustering, Nyström extension, Sampling.

1 Introduction

Spectral clustering [10,7,9,8,12] is a widely used unsupervised methods due to its adaptability and flexibility. Spectral clustering combines Laplacian Eigenmaps (LE) and Kmeans together. Rather than some other algorithms, spectral clustering is less limited by geometries. Clusters are not required to be convex.

In spite of these virtues, the high time complexity $O(n^3)$ makes it unadaptable to large-scale data. Among many solutions, Nyström based spectral clustering [4] is Nyström low-rank approximation and spectral clustering combined. By Nyström method, the similarity matrix and its eigenvectors are estimated by a small mount of samples. However, a bad approximation will lower the accuracy, and the approximation is susceptible to the samples. Therefore researchers have attempted to improve the sampling scheme.

---

* Corresponding Author.

A. Moonis et al. (Eds.): IEA/AIE 2014, Part II, LNAI 8482, pp. 260–269, 2014.

This paper concerns four existed sampling schemes, random sampling (RS), weighted sampling (WS), Kmeans based sampling (KS), and incremental sampling based on variance (IS). WS [1] is designed for Nyström approximation, yet it is seriously time consuming. KS [14] is proposed to lower the Nyström approximation error. Huang [5] gives an error bound of spectral clustering, and presents an algorithm named KASP [13]. KASP uses the Kmeans centroids as the representations of other points. On the contrary, Shinnou [11] removes the points near the Kmeans centroids. However, KS suffers the limitation of Kmeans. IS [15] is based on the similarity variance. The influence of clusterability to the accuracy is quantified to show that a sample set with a small similarity variance is good for clustering. However, it works badly on high-dimensional data.

High dimensionality is not just concerned recently. Beyer [2] figures out that the Euclidean distance of high-dimensional points would be meaningless. Yet no existed sampling scheme has taken the high dimensionality into consideration.

To yield an improvement on sampling, approximation and clustering, we design a new sampling scheme. Given a dataset X and a sample set S, we first discuss how the similarity between S and $X - S$ effects the approximation error. Inspired by our analysis, we present minimum similarity sampling (MSS), mainly to adapt into the high-dimensional cases. Experiments show that MSS outperforms the others on clustering accuracy and computational efficiency.

Section 2 introduces Nyström method tersely. The relationship between sampling and approximation is presented in Section 3. In Section 4, four existed sampling schemes are discussed. Especially we figure out the shortages of KS and show when IS performs badly based on the analysis in Section 2. Then we present the new sampling scheme. Section 5 gives convictive experimental results. Finally we summarize up our work.

2   Brief of Nyström Extension Based Spectral Clustering

Given the dataset X and a sample set S ($|S| \ll |X|$), W_S denotes the similarity matrix of S, and W_{NS} denotes the similarity matrix between S and $X - S$. What is about to be approximated is the similarity matrix of $X - S$, W_N. See the function below.

$$\int_0^1 W(x,y)u(y)dy = \lambda_i u(x), \ i = 1, 2, \ldots, n \ . \tag{1}$$

where $u(x)$ denotes the eigenfunction. Discretize the integration by the rectangular formula. According to [4], we have

$$WU = m\overline{U}\Lambda \ . \tag{2}$$

Note that $w_{ij} = W(\rho_i, \rho_j)$, $U = \{u_1', u_2', \ldots, u_m'\}$, $\Lambda = diag\,(\lambda_1, \lambda_2, \ldots, \lambda_n)$, $\overline{U} = \{u_1', u_2', \ldots, u_n'\}$, $u_j' = u'(\rho_j)$ is the approximated $u(\rho_j)$, $\rho_j \in [0, 1]$.

Substitute W above with $W_{NS}{}^T$, and Λ with $\Lambda_S = 1/m \times diag\,(\lambda_1, \lambda_2, \ldots, \lambda_n)$. Where λ_i is the ith eigenvalue of W_S. Let U_S be the eigen matrix of W_S, we have $W_{NS}{}^T U_S = U_{NS}\Lambda_S$ or $U_{NS} = W_{NS}{}^T U_S \Lambda_S{}^{-1}$.

The approximated eigenvectors of W_X is given by

$$U' = [U_S; U_{NS}] = [U_S; W_{NS}{}^T U_S \Lambda_S{}^{-1}] . \tag{3}$$

Note that U_S is orthogonal. According to the eigen decomposition $W'_X = U' \Lambda_S U'^T$, we have $W_N = W_{NS}{}^T W_S{}^{-1} W_{NS}$.

Approximating the eigenvectors is introduced in [4], [14], and [15]. The time complexity of Nyström approximation algorithm is $O(m^3 + nm^2)$.

3   Error Reanalysis of Nyström Low-Rank Approximation

3.1   Theoretical Analysis from the Probability View

Given a dataset X and a sample set S, $|X| = n$ and $|S| = m$. We discuss when the error $err = \left\| W_N - W_{NS}{}^T W_S{}^{-1} W_{NS} \right\|_F$ might be large. The item $W_{NS}{}^T W_S{}^{-1} W_{NS}$ is known as Schur complement.

Empirically, the more samples concentrate, the denser W_S and sparser W_{NS} we have, because a non-sampling points is more likely outside the clusters covered by S. Suppose $x_i \in S$ and $x_j \in X - S$, $k(x_i, x_j)$ indicates the Gaussian similarity, $E[k(x_i, x_j)]$ is more likely to be very small and w_{NSij}, the corresponding component in W_{NS}, tends to be zero. Then $W_{NS}{}^T W_S{}^{-1} W_{NS}$ tends to be sparse, and err tends to euqal with $\|W_N\|_F$.

Consider the bipartition case ($k = 2$), when the sample set $S = \{x_1, x_2\}$. Suppose that the similarities between points of different clusters are zero. And the real cluster is C_1, $C_2 \subset X$. Let w_{NSj} be the jth column vector of W_{NS}, $W_{NS} = \{w_{NS1}, w_{NS2}, \ldots, w_{NSn-m}\}$.

Suppose that x_1, $x_2 \in C_k$. Given another point $x_j \in X - S$, the corresponding column vector in W_{NS} is w_{NSj}. We have

$$P(w_{NSj} \neq 0) = P(x_j \in C_k) = \frac{|C_k| - m}{n - m} . \tag{4}$$

Then the expectation of w_{NSj} is

$$E\left[w_{NSj}\right] = \begin{bmatrix} k(x_1, x_j) \\ k(x_2, x_j) \end{bmatrix} \times \frac{|C_k| - m}{n - m} . \tag{5}$$

where, $k(x_i, x_j)$ is the Gaussian similarity of x_i and x_j.

In the other case when x_1 and x_2 belong to C_1 and C_2 respectively, $\forall x_j \in X$, $P(w_{NSj} = 0) = 1 - P(w_{NSj} \neq 0) = 1 - \frac{|C_1| + |C_2| - |S|}{n - |S|} = 0$, which means no zero vector in W_{NS} and a small err.

For the general case ($k \geq 2$), by the same way we have the probability that for each $x \in X - S$, $s \in S$, $k(x, s) = 0$.

$$P(w_{NSj} = \vec{0}) = \frac{\sum_{l=1}^{K} \mathbf{1}\left\{|S \cap C_l| = 0\right\} \times |C_l|}{n - |S|} . \tag{6}$$

The expectation of the number of nonzero column vectors in W_{NS} is

$$\mathrm{E}\left[\sum_i^{n-|S|} 1\left\{w_{NSi} \neq \vec{0}\right\}\right] = n - |S| - \sum_{l=1}^{k} 1\left\{|S \cap C_l| = 0\right\}. \tag{7}$$

Evidently, the more clusters covered by S, the larger $\mathrm{E}\left[\sum_i^{n-|S|} 1\left\{w_{NSi} \neq \mathbf{0}\right\}\right]$ is, and the more favorable the sample set is to Nyström approximation. And a sample set such that the average similarity between the samples is smaller is likely to cover more clusters. If we find an S such that the inside similarity is minimum, then the error of Nyström low-rank approximation can be efficiently lowered and thus the clustering quality is improved.

3.2   An Example

The following instance illustrates how the zero vectors in W_{NS} work. Given $X = \{x_1, x_2, \ldots, x_7\}$ and its similarity matrix

$$W = \begin{bmatrix} E_3 & & \\ & E_2 & \\ & & E_2 \end{bmatrix}. \tag{8}$$

The similarities of points from different clusters are set to be zero and the similarities inside one cluster are always 1. There are three clusters in X, $\{x_1, x_2, x_3\}$, $\{x_4, x_5\}$, and $\{x_6, x_7\}$. Suppose we have an $S = \{x_1, x_4, x_6\}$. By Nyström low-rank matrix approximation technique we have

$$W_N' = W_{NS}{}^T W_S{}^{-1} W_{NS} = \begin{bmatrix} E_2 & & \\ & E_1 & \\ & & E_1 \end{bmatrix}. \tag{9}$$

After sorting the columns by the original index, we have the approximated matrix $W' = W$ such that $\|W_N - W_N'\|_F = 0$. This owes to that S covers all the clusters and contains enough cluster characteristics.

Now consider another $S = \{x_1, x_2, x_3\}$. And we have $A = E_3$, $B = [0]$, $W_N' = W_{NS}{}^T W_S{}^{-1} W_{NS} = [0]$. And the approximation error is $\|W_N\|_F = 8$. Such a large error will make the clustering result undesired.

4   Minimum Similarity Sampling Algorithm

Despite its simpleness and low complexity, RS is limited by the equiprobable distribution assumption. WS lacks practicability as it's time consuming.

KS is considerable in most cases but suffers many limitations. First, Kmeans assumes the potential clusters to be convex and with close sizes and densities. Second, the samples are "created", which causes instability. Third, it reruns when empty clusters emerge; experiments reveal that this problem seriously raises the time cost. Moreover, Kmeans is susceptible to acnodes and noises.

IS is based on the variance of similarity. The storage complexity is $O(nm)$ and the time complexity is $O(tnm)$ as it chooses t random points when computes the variance [15]. It works badly on high-dimensional data, as the variance of Euclidean distance tends to be zero [2], so that the inappropriate point will be chosen. Note that the Gaussian kernel just magnifies the high-dimensional feature in limited degree.

As is mentioned above, a sample set such that the inside similarity is the minimum can efficiently lower the Nyström low-rank approximation error.

Algorithm 1. Minimum Similarity Sampling on High-dimensional Data

---

Require:
  X as the dataset
  m as the mount of samples
Ensure: Sample set S
  $NS \leftarrow X$
  Randomly choose $s \in X$
  $S \leftarrow s$
  $NS \leftarrow NS - s$
  $\boldsymbol{a} \leftarrow Similarity(s, X)$
  repeat
    $t \leftarrow PointOfMin(\boldsymbol{a})$
    $S \leftarrow S \cup t$
    $\boldsymbol{a} \leftarrow \boldsymbol{a} + Similarity(t, X)$
  until $|S| \geq m$

The function $Similarity(s, X)$ denotes a vector $\boldsymbol{a}$ where a_i is the similarity between $x_i \in X$ and $s \in S$. The function $PointOfMin(\boldsymbol{a})$ returns the point corresponding to the minimum component in $\boldsymbol{a}$.

The extra storage complexity is $O(n)$ for $\boldsymbol{a}$, or $O(nm)$ if we meanwhile want the final W_{NS}. The time complexity is $O(mn)$. For each point in an iteration the algorithm just calculates the similarity with the new sample instead of every sample. When choosing the new sample, MSS simply search for the minimum similarity instead of calculating the variances. Compared with IS, the computation complexity is improved.

5   Experiments

All the datasets used are chosen from UCI datasets. Gaussian kernel function is used to compute similarity. Parameters are chosen empirically, because this paper does not focus on parameters. The clustering quality is measured with NMI [6] and accurate rate. Considering that sampling schemes are sensitive to dataset, and that every existed schemes contains randomness, we repeat the algorithms for 50 times to get the average results.

5.1   A Special Dataset

Consider the case in Fig. 1. Given a two-dimensional dataset $X = x_{iN}$ with 48 points in three clusters. Note that $x_1 = (0,\ 0)$, $x_2 = (0,\ 1)$, ..., $x_5 = (1,\ 0)$, ..., $x_N = (13,\ 3)$. The dataset X is linearly separable.

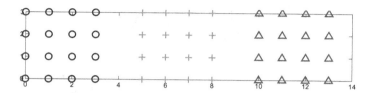

Fig. 1. A special dataset with 48 points and 3 clusters

When $|S| = 3$, IS firstly chooses $x_{20}(5,3)$ and $x_{28}(7,3)$ randomly then adds $x_{21}(6,0)$ as the next sample. Note that $S = \{x_{20}, x_{28}, x_{21}\}$ concentrates upon the second cluster. The matrix W_{NS} has 32 zero columns, which leads to a sparse Schur complement and a large error. While the situation of MSS is much better.

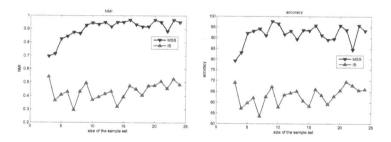

Fig. 2. Comparing IS and MSS on the special dataset

The sample sequences of different sizes are shown in Tab. 1. Samples by MSS cover the clusters well. The experimental results conform and support the theoretical analysis we present above.

5.2   Comparison on Synthetic Data

We choose three classical UCI datasets, Smile, Three Circles and Normal7. Fig. 3 plots the points in different clusters with different colors and shapes.

   Tab. 2 shows the size, clusters, and the used parameter δ of each synthetic dataset. The sizes range from 200 to 7000. All of the datasets are not linearly separable except Normal7.

Table 1. The sampling sequences of MSS and IS

$|S|$	MSS and IS	Sampling series
3	IS	20, 28, 21
	MSS	32, 1, 45
4	IS	11, 1, 9, 6
	MSS	21, 1, 48, 16
5	IS	21, 39, 31, 30, 29
	MSS	46, 1, 24, 29, 12

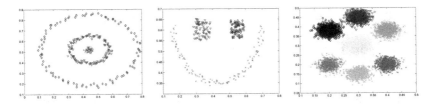

Fig. 3. Three synthetic datasets

Table 2. Details of the synthetic datasets

Dataset	Size	Number of clusters	The kernel parameter δ
three circles	266	3	0.06
smile	299	3	0.1
Normal7	7000	7	0.5

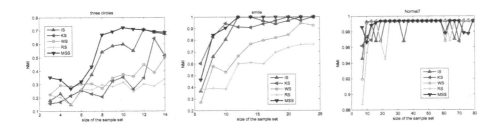

Fig. 4. NMI of the schemes on the synthetic dataset

The performance of MSS is better than others on Three Circles and Smile. On Normal7, the stability of MSS is superior. Note that when $|S|$ ranges from 7 to 22 and 61 to 76, all of the algorithms suffer degradation.

The result shows WS is significantly costy. Thus Fig. 7 bellow plots the cpu run time of them without WS. MSS is faster than all of the other schemes except for RS. IS calculates the similarity variance in its iterations, which costs much more time than calculating the sum. KS is unstable when empty clusters emerge. One example is the run time when $|S| = 5$, 8, 12, 13 in Fig. 7(a).

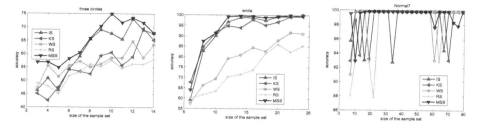

Fig. 5. Accuracy of the schemes on the synthetic dataset

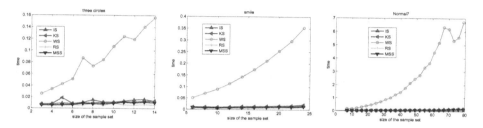

Fig. 6. Runtime of the schemes on the synthetic dataset

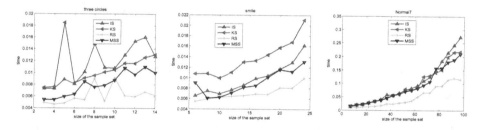

Fig. 7. Runtime of the schemes except WS on the synthetic dataset

5.3 Comparison on Real Data

We choose three UCI real datasets with dimensionality ranging from 14 to 64, to test the performance of our sampling scheme in high-dimensional cases. Tab. 3 shows the size, dimensionality, and the used parameter δ of each dataset.

Since the tendency of the NMI value is similar with that of the accuracy, for the length of the paper only the accuracy figures are given in the following. Note that the accuracy of MSS on Wine increase with the sample set grows and higher than those of IS, while the other algorithms are significantly unstable. When $|S| \geq 9$, MSS outperforms any other sampling schemes and the accuracy still increases slowly and stably.

On real datasets, WS is still time consuming. Again we remove the run time line of WS and clarify the superior efficiency of MSS.

Table 3. Details of the real datasets

Dataset	Size	Dimensinality	Number of clusters	The kernel parameter δ
wine	178	14	8	0.055
zoo	101	17	7	0.4
air	359	64	3	0.1

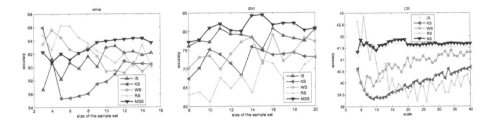

Fig. 8. Accuracy of the schemes on the real dataset

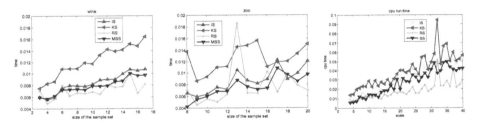

Fig. 9. Runtime of the schemes except WS on the real dataset

6   Summary

We focus on the improvement of Nyström based spectral clustering on high-dimensional data. We take insight into the existed sampling schemes and figure out their drawbacks especially in high-dimensional space. We find that a sample set with small inner similarity lowers the Nyström approximation error. A new sampling scheme call minimum similarity sampling (MSS) is proposed to improve the accuracy and the computational efficiency. Experiments on both synthetic and real UCI datasets reveal the defects of WS, RS, KS, and IS. And MSS shows its stability, computational efficiency and advanced improvement on clustering.

The following is some future works. First, Nyström method is not the only approach that depends on sampling. Other algorithms like LSC [3] also can be improved by choosing the right sampling scheme. Second, thought estimating the comentropy of a single sample from a large-scale dataset is tough, a sampling scheme based on the maximum comentropy might be potential.

References

1. Belabbas, M.A., Wolfe, P.J.: Spectral methods in machine learning and new strategies for very large datasets. Proceedings of the National Academy of Sciences 106(2), 369–374 (2009)
2. Beyer, K., Goldstein, J., Ramakrishnan, R., Shaft, U.: When is "nearest neighbor" meaningful? In: Beeri, C., Bruneman, P. (eds.) ICDT 1999. LNCS, vol. 1540, pp. 217–235. Springer, Heidelberg (1998)
3. Chen, X., Cai, D.: Large scale spectral clustering with landmark-based representation. In: AAAI (2011)
4. Fowlkes, C., Belongie, S., Chung, F., Malik, J.: Spectral grouping using the nystrom method. IEEE Transactions on Pattern Analysis and Machine Intelligence 26(2), 214–225 (2004)
5. Huang, L., Yan, D., Taft, N., Jordan, M.I.: Spectral clustering with perturbed data. In: Advances in Neural Information Processing Systems, pp. 705–712 (2008)
6. Hunter, B., Strohmer, T.: Performance analysis of spectral clustering on compressed, incomplete and inaccurate measurements. arXiv preprint arXiv:1011.0997 (2010)
7. Kannan, R., Vempala, S., Vetta, A.: On clusterings: Good, bad and spectral. Journal of the ACM (JACM) 51(3), 497–515 (2004)
8. MeilPa, M., Shi, J.: Learning segmentation by random walks (2000)
9. Ng, A.Y., Jordan, M.I., Weiss, Y., et al.: On spectral clustering: Analysis and an algorithm. Advances in Neural Information Processing Systems 2, 849–856 (2002)
10. Shi, J., Malik, J.: Normalized cuts and image segmentation. IEEE Transactions on Pattern Analysis and Machine Intelligence 22(8), 888–905 (2000)
11. Shinnou, H., Sasaki, M.: Spectral clustering for a large data set by reducing the similarity matrix size. In: Preeedings of the Sixth International Language Resouces and Evaluation, LREC (2008)
12. Von Luxburg, U.: A tutorial on spectral clustering. Statistics and Computing 17(4), 395–416 (2007)
13. Yan, D., Huang, L., Jordan, M.I.: Fast approximate spectral clustering. In: Proceedings of the 15th ACM SIGKDD International Conference on Knowledge Discovery and Data Mining, pp. 907–916. ACM (2009)
14. Zhang, K., Tsang, I.W., Kwok, J.T.: Improved nyström low-rank approximation and error analysis. In: Proceedings of the 25th International Conference on Machine Learning, pp. 1232–1239. ACM (2008)
15. Zhang, X., You, Q.: Clusterability analysis and incremental sampling for nyström extension based spectral clustering. In: 2011 IEEE 11th International Conference on Data Mining (ICDM), pp. 942–951. IEEE (2011)

Performance Improvement of Set Partitioning Embedded Block Algorithm for Still Image Compression

Shu-Mei Guo[1], Yun-Wei Lee[1], Chih-Yuan Hsu[1], and Shy-Jen Guo[2]

[1] Department of Computer Science and Information Engineering,
National Cheng Kung University,
Tainan, 701, Taiwan, R.O.C.
guosm@mail.ncku.edu.tw
[2] Department of International Trade,
National Taichung University of Science and Technology,
Taichung, 404, Taiwan, R.O.C.

Abstract. The set partitioning embedded block (SPECK) algorithm is a fast and efficient technique for still image compression. In this paper, we propose a novel wavelet-based coding scheme, called prepartition SPECK (PSPECK), on the extension of SPECK. In order to improve the peak signal-to-noise ratio (PSNR) performance, we predict the significance of each set in the list of insignificant sets (LIS) by exploiting inter-subband correlation for reducing the bit budget. Furthermore, the proposed method can be combined with other quadtree-based coding techniques. Experimental results show that the proposed method outperforms SPECK, especially at high bit rates.

Keywords: Discrete wavelet transform, Image compression, SPECK.

1 Introduction

In the past decade, the discrete wavelet transform (DWT) has been widely employed for image compression due to its excellent energy compaction nature in both space and frequency. Thus, several wavelet-based image compression schemes, including the JPEG2000 standard codec [1], have been proposed. Those wavelet-based coding schemes can be roughly categorized into two main groups: zerotree-based and zeroblock-based algorithms.

In 1993, the first zerotree-based algorithm, called embedded zerotree wavelet (EZW), was introduced by Shapiro [2]. According to Shapiro's work, Said and Pearlman introduced an efficient coding scheme called set partitioning in hierarchical trees (SPIHT) [3] becoming the benchmark in image compression. On the other hand, zeroblock-based coding is also an efficient coding technique. Instead of using inter-subband correlations, the zeroblock-based algorithms exploit intra-subband correlations for image compression. The well-known zeroblock-based algorithms include embedded block coding with optimized truncation (EBCOT) algorithm [4], the set partitioned embedded block coder (SPECK) [5, 6], subband hierarchical block partitioning (SBHP) [7], and embedded zero block coding (EZBC) [8]. The SPECK

A. Moonis et al. (Eds.): IEA/AIE 2014, Part II, LNAI 8482, pp. 270–278, 2014.

recursively partitions the transformed image into contiguous subblocks and performs the significance test on individual subblocks. However, the SPECK does not exploit spatial correlations across subbands at different levels of a wavelet pyramid, which means that the compression efficiency of SPECK can be further improved by removing some redundancy from the self-similarity of coefficients at different subbands. Thus, zeroblock-based and zerotree-based coding algorithms which exploit both inter-subband and intra-subband correlation have been developed. Previous works exploiting both zerotree and zeroblock coding algorithm include the Hybrid Block Coder (HBC) [9], the Embedded Zero Block Tree Coder (EZBTC) [10], the hybrid Block-Tree (BT) [11], and the Wavelet Block Tree Coding (WBTC) [12, 13].

To achieve better performance of the SPECK algorithm by exploiting inter-subband correlation, we propose a novel algorithm termed prepartition SPECK (PSPECK) in this paper. The proposed method predicts the significance of each block in the LIS by the significance of its parent, child, and neighbors, so bit savings are made at prepartition stage. Moreover, the static Huffman code is used to improve coding efficiency. The experimental results show that the coding performance of PSPECK is better than SPECK at various bit rates and is comparable to other state-of-the-art algorithms.

This paper is organized as follows. In Section 2, we review the zerotree-based and zeroblock-based coders. Section 3 fully describes the quadtree-based coding techniques which are the most commonly adopted technique among zeroblock-based algorithm. Then, the proposed PSPECK is analyzed and presented in Section 4. The performance and comparison with other state-of-the-art codecs of PSEPCK are given in Section 5. In the last section, we summarize the results of our proposed method.

2   Background

Due to different set partition strategies, those wavelet-based algorithms can be broadly classified into zerotree-based and zeroblock-based algorithms using intra-subband and inter-subband correlation, respectively.

2.1   Zerotree-Based Coding

The zerotree structure is an efficient expression of zero coefficients due to the fact that there are strong parent-child dependencies across scales. The zerotree structure is defined as the spatial relationships among wavelet coefficients on the hierarchical wavelet pyramid. If a wavelet coefficient at a coarse scale is insignificant with respect to a given threshold T, then all wavelet coefficients of the same orientation in the same spatial location at finer scales are likely to be insignificant with respect to T [2].

Among the zerotree-based coding algorithms, the SPIHT is a well-recognized algorithm due to its excellent coding performance. It substantially improves the EZW algorithm by exploiting a new zerotree structure and using lists to store the significance of coefficients. The SPIHT exploits three lists, a list of insignificant sets (LIS), a list of insignificant pixels (LIP), and a list of significant pixels (LSP), to store the significance information. It efficiently partitions wavelet coefficients into these three lists; unlike that in EZW, particular symbols are used to indicate the significance of wavelet coefficients. Specifically, EZW is defined as a degree-0 zerotree coder and SPIHT is defined as a degree-2 zerotree coder. More discussion of the reason that SPIHT is better than EZW is described in [14].

2.2   Zeroblock-Based Coding

The zeroblock structure is an alternative expression of zero coefficients. The SPECK, rooted in AGP [15], SWEEP [16], and SPIHT, is a successful algorithm with high coding performance and low complexity among the zeroblock-based algorithms. In SPECK, rectangle sets (blocks) are exploited to group the wavelet coefficients, and quadtree decomposition strategy is applied to the significant sets recursively as the significance test works with a specific threshold. Similar to the SPIHT, the LSP and LIS store the significant pixels and insignificant sets, respectively.

Due to high coding efficiency of SPECK, the two well-known algorithms, SBHP and EZBC, are developed. The SBHP, incorporated into the JPEG2000 framework as a low complexity option, is based on SPECK. As the trade-off between coding performance and complexity, SBHP does not use arithmetic coding. As a result, SBHP encoder runs about 4 times faster, and the decoder is about 6 to 8 times faster than EBCOT, depending on the platform [7]. On the other hand, The EZBC [8] is a higher complexity variant of SPECK. It integrates more complicated arithmetic coding with sophisticated context modeling into SPECK and its compression efficiency is comparable to EBCOT.

2.3   Discussion

It is clear that there are inter-subband and intra-subband correlations within wavelet subbands. However, the zeroblock-based algorithms do not exploit self-similarity across different scales. And, the zerotree-based algorithms do not use intra-subband correlations. To improve the coding efficiency, zerotree-based and zeroblock-based algorithms have been developed. Among previous attempts of using both correlations, the hybrid Block-Tree (BT) algorithm [11], is a successful algorithm. To further improve coding efficiency, an adaptive scanning order [17] is applied to BT instead of conventional raster scanning order. In [12], another zerotree-based and zeroblock-based algorithm called the Wavelet Block Tree Coding (WBTC) is proposed, it partitions a transformed image into coefficient blocks and then blocktrees are formed. In a blocktree, significant blocks are found using tree partitioning concept of SPIHT, whereas significant coefficients within each block are found using the quadtree partitioning of SPECK. In this paper, we propose a different approach to improve the coding efficiency of SPECK by taking advantage of self-similarity across scales. In addition, simple fixed Huffman code is exploited in the PSPECK under consideration to both coding performance and complexity.

3   Quadtree-Based Coding Technique

There are three kinds of bits produced by SPECK: the significance map, magnitude refinement and sign bits. The significance map bits are chosen only to entropy-code. The significance maps are the paths which record the decomposition decisions of a set in the quadtree. The existing methods [5, 17, 18] which reduce the bit budget for coding the significance map are discussed in the following subsections.

3.1   Simple Coding

If a set S is significant with respect to a quantization threshold, the set is equally divided into fours subsets $O(S) = \{S_0, S_1, S_2, S_3\}$. Then, the four-subset group $O(S)$ is tested for their significances against the same threshold in the raster scanning order, i.e. $S_0 \rightarrow S_1 \rightarrow S_2 \rightarrow S_3$. If S_0, S_1, and S_2 are all insignificant, S_3 will be significant because there are at least one significant subset in $O(S)$. Therefore, 1 bit which indicates the significance of S_3 can be saved.

3.2   Complex Coding

The complex coding, proposed in [17], is a sophisticated coding method. If a set S is significant, extra one bit is required to identify whether the set contains only one significant subset or more than one significant subset. If there is only one significant subset in the set S, two bits are needed to imply the location of the significant subset. Otherwise, 3.5 bits on average are required to record the significance information of the four-subset group $\{S_0, S_1, S_2, S_3\}$. The actual number of bits needed to represent more than one significant subset.

3.3   Huffman Coding

The Huffman coding is an efficient entropy coding algorithm for data compression. In [7] and [18], static Huffman coding is chosen to improve coding efficiency without substantially increasing the complexity. During the encoding process, a significant set S is split into four subsets (or pixels) $O(S) = \{S_0, S_1, S_2, S_3\}$. It is clear that at least one subset is significant. Hence, there are 15 possible results of their significances (0001, 0010, 0011, ..., 1111) and they can be entropy-coded by static Huffman coding with 15 symbols. In [18], the Huffman code table is established based on average probability distribution as illustrated.

Thus, the average code length of the three above-mentioned coding techniques can be estimated. The direct coding is the simplest method that needs 4 bits to record the significance of subsets. And, the simple coding, complex coding, and Huffman coding employ 3.86, 3.73, and 3.68 bits on average, respectively. As a result, the Huffman coding is chosen to encode the significance of subsets in the proposed method.

4   Improvement of SPECK Algorithm

To improve the coding performance of the SPECK algorithm, the proposed PSPECK exploits inter-subband correlation. We analyze the bit budget for coding the roots of quadtrees and establish a prediction model. Before encoding, each set in the LIS is predicted for its significance. If the prediction is accurate, the coding performance is improved. Moreover, the weighting values used in our prediction model can be optimized by learning-based algorithms.

4.1   Analysis

As mentioned in Section 2, during the bit plane coding of SPECK, each set in the LIS should be tested for their significance against the current threshold. If there is a significant coefficient in a set S, the set is quadrisected recursively until the bottom quadtree level is reached.

During the quadtree partitioning process, each set corresponds to a node in Fig. 1. As can be seen in Fig. 1, the black nodes are significant, and the scanning order is $N_0 \rightarrow N_1 \rightarrow N_2 \rightarrow N_3 \rightarrow N_4 \rightarrow N_5 \rightarrow ...$, and so on. The coding techniques that we mentioned in the Section 3 are only applied to $N_1 \sim N_8$. It is apparent that the root of the quadtree N_0 is not entropy-coded in the previous approaches, i.e., the bit budget for coding the roots of the quadtrees is not reduced. It is clear that it still exists some coding redundancy which can be removed. Thus, we propose an efficient method to reduce the bit budget by taking advantage of self-similarity across scales. The proposed method increases the coding efficiency by reducing the bits which indicate the significance of the quadtree roots. Especially, it improves much more coding performance at high bit rate due to there are more sets added to the LIS while the bit plane encoder works on lower bit planes. The proposed method is introduced in detail in the next subsection.

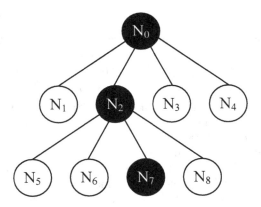

Fig. 1. Tree representation of quadtree splitting process

4.2   Proposed Algorithm – PSPECK

This subsection describes the underlying ideas in the proposed PSPECK. In the SPECK, the sets in the LIS should be tested for their significance during the encoding process. There are strong parent-child dependencies across subbands [2] so that conditional probability is used to predict the significance of those sets. In the proposed PSPECK, each set in the LIS should be predicted for their significance before coding its significance. If a set S is predicted to be significant, the set will be prepartitioned into four subsets and added to the LIS. First, if we predict that a set S is significant and the set is actually significant, then the bit, which indicates the set is significant, can be saved due to the fact that we have partitioned the set in the prepartition stage. In other words, we need 0 bit to code the significance of the set during the encoding process. Second, if we predict that a set S is significant but the set is actually insignificant, 4 bits are needed to indicate that four-subset group $\{S_0, S_1, S_2, S_3\}$ is insignificant. The reason is that we have partitioned the insignificant set S into four insignificant subsets in the prepartition stage. Finally, once we predict that a set is insignificant, we will do nothing in the prepartition stage. As a result, we still need 1 bit to record the significance of the set during the encoding process.

Let n be the number of sets which lie in the LIS, t_p represent the number of true positive predictions, and f_p be the number of false positive predictions. If the bit budget can be reduced by our prepartition scheme, the following equation is satisfied

$$(n \times 1) \geq (t_p \times 0 + f_p \times 4 + (n - t_p - f_p) \times 1) . \tag{1}$$

The left-hand side of Eq. (1) is the bit budget for coding the roots of quadtrees without prepartition scheme. If there are n quadtrees (sets) in the LIS, we need n bits to encode the roots of quadtrees since one bit is essential for each tree root to indicate its significance. On the other hand, the right-hand side of Eq. (1) is bit budget for coding the roots of quadtrees with prepartition scheme. Solving Eq. (1), we have

$$t_p \geq 3f_p . \tag{2}$$

If the bit savings can be made by prepartition scheme, the number of true positive predictions must be greater than the number of false positive predictions 3 times.

Assume that a set S in the LIS has a parent, a child and four neighbors. To predict the significance of the set S, one parent set P, one child set C, four neighbor sets $NE_0 \sim NE_3$, and its height (or width) L are considered. Then, heuristic weighting values are given to P, C, $NE_0 \sim NE_3$, and L for prediction. If P has been significant in the previous bit planes, it is weighted by 5. Otherwise, it is weighted by 0. And, If C has been significant in the previous bit planes, it is weighted by 13 since the significance of a child set is more important for prediction. Then, a neighbor set is weighted by 3, if it has been significant in the previous bit planes. Finally, the height (or width) of S is weighted by 1. When the significance of a set is predicted in the prepartition stage, the sum of all weighting values is calculated against a heuristic threshold T_w ($T_w = 12$). If the sum of the weighting values is below T_w, the set is predicted to be insignificant. Otherwise, it is predicted to be significant.

In addition, the heuristic weighting values proposed in PSPECK can be optimized by learning-based algorithms. In this paper, we use chaos evolutionary programming (CEP) [19] to find the optimal weighting values. The optimization of weighting value is discussed in the next subsection.

4.3   Weighting Value Optimization

The weighting values exploited in PSPECK are optimized by CEP which is an optimization technique to find an optimal solution. The CEP is integrated by Evolutionary Programming (EP) and Chaos Optimal Algorithm (COA) [20]. In CEP process, an optimal solution is approximated by EP which uses mutation and selection processes, then COA is applied to search a better solution than the approximated one. During the EP process, each individual represents a solution (a vector of weighting values) for PSPECK. It yields offspring by mutation and then fitness of all individuals are evaluated based on PSNR. The best-fit individuals are selected to the next generation. After the EP is terminated, the COA is employed in searching the optimal solution. It can go non-repeatedly through every state. The details of CEP can be found in [19].

5   Experimental Results

Experimental results of PSPECK are presented in this section. Five-level wavelet decomposition with the 9/7-tap filters of [21] and a "reflection" extension at the image edges are used. The rate/distortion (R/D) performances measured at various bit rates are compared with four state-of-the-art algorithms SPECK, SPIHT, WBTC, and BT-A. Both coding performance and computational complexity are presented in this section.

The evaluation results are shown in Table 1 with the two standard monochrome, 8 bpp, 512×512 images, i.e., Lena and Goldhill. All results are obtained without arithmetic coding. The BT-A and WBTC are the zerotree-based and zeroblock-based codec. Especially, BT-A uses the complex coding and adaptive scanning order to improve coding performance. Table 5 shows that the PSPECK outperforms other coders especially at high bit rates. Since there are more sets added to the LIS at high bit rates, more bits can be saved during the prepartition stage of PSPECK.

The results of the second image set are exhibited with six monochrome, 8 bpp, 512×512, images namely Baboon, Boat, Barbara, Peppers, Remote, and F-16. The images are tested with the proposed method, SPECK, and BT-A at various bit rates. Similarly, the PSPECK is slightly better than BT-A in terms of coding performance in most cases. The proposed method gives 0.01 to 0.18 dB gain over BT-A.

Then, the run time is performed by PSPECK and BT-A at the rates 0.25, 0.5, and 1.0 bpp for the two standard images. The PSPECK is faster than BT-A because BT-A employs a more complicated coding method than the static Huffman code. In addition, the adaptive scanning exploited in BT-A works on each node (except the root) of a quadtree, but our approach is only applied to the tree root. It can be observed that the PSPECK is about 30-40% faster than BT-A in both encoding and decoding.

Table 1. Comparison of lossy coding algorithms for two standard images

Coding method	PSNR (dB)			
	0.25 bbp	0.50 bbp	1.00 bbp	2.00 bbp
Lena image (512×512)				
SPIHT	33.70	36.85	39.99	44.35
SPECK	33.37	36.49	39.65	---
WBTC	33.67	36.86	40.03	44.50
BT-A	33.89	37.02	40.20	44.74
PSPECK	33.89	37.03	40.24	44.84
Goldhill image (512×512)				
SPIHT	30.22	32.71	36.00	41.12
SPECK	30.21	32.58	35.67	---
WBTC	30.24	32.63	36.03	41.09
BT-A	30.41	32.93	36.29	41.56
PSPECK	30.43	32.98	36.36	41.72

6   Conclusions

In this paper, a novel wavelet-based coding scheme PSPECK for still image compression is proposed. On the extension of SPECK, we introduce an efficient method to reduce the bit budget for coding the roots of the quadtrees by exploiting the self-similarity across subbands, and the proposed method can be combined with existing quadtree coding techniques. Experiment results show that the proposed method leads to gain in PSNR performance over the original SPECK especially at high bit rates. Therefore, the PSPECK is attractive for medical image compression when the quality of images is the major concern. However, our approach increases complexity slightly.

Acknowledgements. This work was supported by the National Science council of Republic of China under contacts NSC 102-2221-E-006-199 and NSC 102-2221-E-006-208-MY3.

References

1. Acharya, T., Tsai, P.S.: JPEG2000 Standard for Image Compression: Concepts, Algorithms and VLSI Architecture. John Wiley & Sons, Inc., Hoboken (2004)
2. Shapiro, J.M.: Embedded image coding using zerotrees of wavelets coefficients. IEEE Trans. Signal Processing 41, 3445–3462 (1993)
3. Said, A., Pearlman, W.A.: A new, fast, and efficient image codec based on set partitioning in hierarchical trees. IEEE Transactions on Circuits and Systems for Video Technology 6, 243–250 (1996)
4. Taubman, D.: High performance scalable image compression with EBCOT. IEEE Transactions on Image Processing 9, 1158–1170 (2000)

5. Pearlman, W.A., Islam, A., Nagaraj, N., Said, A.: Efficient, low-complexity image coding with a set-partitioning embedded block coder. IEEE Transactions on Circuits and Systems for Video Technology 14, 1219–1235 (2004)
6. Pearlman, W.A.: Wavelet image compression. In: Synthesis Lectures on Image, Video, and Multimedia Processing (January 2013)
7. Chrysafis, C., Said, A., Drukarev, A., Islam, A., Pearlman, W.A.: SBHP—A low complexity wavelet coder. In: IEEE Int. Conf. Acoust., Speech Signal Processing (ICASSP), vol. 4, pp. 2035–2038 (June 2000)
8. Hsiang, S.T., Woods, J.W.: Embedded image coding using zeroblocks of subband/wavelet coefficients and context modeling. In: IEEE Int. Conf. Circuits and Systems (ISCAS), vol. 3, pp. 662–665 (May 2000)
9. Wheeler, F.W., Pearlman, W.A.: Combined spatial and subband block coding of images. In: IEEE Int. Conf. Image Processing (ICIP 2000), vol. 3, pp. 861–864 (September 2000)
10. Arora, H., Singh, P., Khan, E., Ghani, F.: Memory Efficient Image Coding with Embedded Zero Block-Tree Coder. In: Proc. IEEE Int. Conf. Multimedia and Expo 2004 (ICME 2004), vol. 1, pp. 679–682 (May 2004)
11. Yin, X.W., Fleury, M., Downton, A.C.: A structure for tree and block encoders. In: Proc. Int. Conf. on Pattern Recognition, ICPR 2004 (2004)
12. Moinuddin, A.A., Khan, E., Ghanbari, M.: Efficient algorithm for very low bit rate embedded image coding. IET Image Process. 2, 59–71 (2008)
13. Chandandeep, K., Sumit, B.: Listless Block Tree Coding with Discrete Wavelet Transform for Embedded Image Compression at Low Bit Rate. International Journal of Computer Applications 70, 32–36 (2013)
14. Cho, Y., Pearlman, W.A.: Quantifying the coding performance of zerotrees of wavelet coefficients: Degree-k zerotree. IEEE Transactions on Signal Processing 55, 2425–2431 (2007)
15. Said, A., Pearlman, W.A.: Low-complexity waveform coding via alphabet and sample-set partitioning. Proc. SPIE Visual Communications and Image Processing 3024, 25–37 (1997)
16. Andrew, J.: A simple and efficient hierarchical image coder. In: Proc. IEEE Int. Conf. Image Processing (ICIP), vol. 3, pp. 658–661 (October 1997)
17. Yin, X.W., Fleury, M., Downton, A.C.: Prediction and adaptive scanning in block-set wavelet image coder. IEE Proc.-Vision. Image Signal Process. 153 (April 2006)
18. Guo, S.M., Wu, B.H., Tsai, J.S.H.: Improved set-partitioning embedded block scheme for still image compression. Journal of Electronic Imaging 17 (September 2008)
19. Guo, S.M., Chang, W.H., Tsai, J.S.H., Zhuang, B.L., Chen, L.C.: JPEG 2000 wavelet filter design framework with chaos evolutionary programming. Signal Processing 88, 2542–2553 (2008)
20. Yan, D.Z.C.X.F., Hu, S.X.: Chaos-genetic algorithms for optimizing the operating conditions based on RBF-PLS model. Computers and Chemical Engineering 27, 1393–1404 (2003)
21. Antonini, M., Barlaud, M., Mathieu, P., Daubechies, I.: Image coding using wavelet transform. IEEE Transactions on Imam Processing 1, 205–220 (1992)

Similarity-Based Classification of 2-D Shape Using Centroid-Based Tree-Structured Descriptor

Wahyono, Laksono Kurnianggoro, Joko Hariyono, and Kang-Hyun Jo

Graduate School of Electrical Engineering
University of Ulsan
Ulsan, 680-749, Korea
{wahyono,laksono,joko}@islab.ulsan.ac.kr,
acejo@ulsan.ac.kr

Abstract. This paper introduces a novel shape descriptor invariant to rotation and scale, namely centroid-based tree-structured (CbTs), for measuring shape similarity. For obtaining CbTs descriptor, first, the central of mass of a binary shape is computed. It will be regarded as the root node of tree. The shape is divided into b sub-shapes by voting each foreground pixel point based on angle between point and major principal axis. In the same way, the central of masses of the sub-shapes are calculated and these locations are considered as level-1 nodes. These processes are repeated for a predetermined number of levels. For each node corresponding to sub-shapes, five parameters invariant to translation, rotation and scale are extracted. Thus, a vector of all parameters is carried out as descriptor. To measure dissimilarity between shapes, we employ vector-based template matching with X^2 distance measurement. Results are presented for MPEG-7 dataset.

Keywords: Shape descriptor, shape classification, centroid, tree-structured.

1 Introduction

In real life, when people are faced with a problem to recognize an object, they will naturally identify the object based on its shape rather than other characteristics. This behavior is then adopted for computer as well as robot in order to recognize objects in surrounding environment. However, implementing this task is quite difficult. Point of view difference of object will cause different shape captured by robot. The object usually appears in 3D form, but the robot will capture 2D shape of object. Thus, shape recognition plays an important part in object recognition.

In state of the art of shape matching and recognition, generally, it is assumed that the various shapes have been known previously regarded as templates. If given a new shape, our objective is to identify whether it is belonging to one of the template shapes. It can be done by matching their descriptor, namely shape matching. Shape matching deals with transforming a shape, and measuring the resemblance with another one, using some similarity measure [1] such as Euclidean distance, city block distance, X^2 distance, and others.

A. Moonis et al. (Eds.): IEA/AIE 2014, Part II, LNAI 8482, pp. 279–288, 2014.

In this paper, we introduce a new shape descriptor, namely Centroid-based Tree-structured (CbTs) for shape matching as well as object recognition. The CbTs is an effective, simple and invariant to rotation, translation and scale. The CbTs is constructed based on the centroid of shape and its sub-shapes. The detail of this process will be discussed in Section 3. To prove the robustness of our descriptor, some experiments are conducted using well-known public dataset in shape matching such as MPEG-7 [2]. Then the experimental results are compared with other descriptors. The summary of our experimental results will be presented in Section 4. Section 5 will conclude our work and discuss the future development of our works. However, prior to all above, the related works of shape matching will be discussed in Section 2.

2 Related Works

In the past few decades, researchers in many areas have considerable given attention for shape matching problem. One of the earliest researches in this field is based on global features extracted from whole image shape [3], such as invariant moments [4], Zernike moments [5], and combination of invariant moments and edge oriented histogram [6]. Although the global features have the advantages of being compact and efficient for matching, local information and spatial configuration of image shape are lost [7]. To overcome this problem, another class of methods extracts features for shape based on part of shape i.e. edge. In [8], the authors utilized edge pixel of shape to construct shape context. The shape context describes the coarse arrangement of the shape with respect to a point inside or on boundary of the shape. For matching cost, it was combined with a conventional appearance-based descriptor, such as local orientation. Since this descriptor is based on sample of edge position and the distance was computed by Euclidean distance, it causes less discriminability for complex shape with articulations. Thus, reference [9] suggested using inner-distance to capture better shape structure. However, the inner-distance is sensitive to shape topology which sometimes causes problems. Authors in [10] utilized a Hidden Markov Model for shape curvature as descriptor for individual classes and the weighted likelihood function for discriminating classes. Nevertheless, occlusion remains unsolved problem in this method.

Another approach was suggested in [11], rather than extract feature from edge shape that unsolved occlusion problem, the authors use all points of the shape. First, the shape is divided into four sub-regions by two principal axes corresponding to the two eigenvectors at centroid of shape. Each sub-region is subdivided into four sub-regions in the same way. Several parameters then are extracted for the all sub-regions representing the shape descriptor. These parameters are invariant to translation, rotation and scale. However, it did not work well when the pixels of a shape are circularly distributed due to high sensitivity of eigenvector direction. Thus, our descriptor utilizes the advantages of parameters in [11], and expands the number of regions and level of sub-shapes in order to reduce this sensitivity impact.

3   Centroid-Based Tree-Structured

In this section, we describe in detail our shape descriptor. The process is started with tree-structured constructing from given shape based on its centroid. Once all nodes are obtained, the five parameters which are scale and rotation invariance are calculated. The last, we define the matching cost as well as matching process for determining the similarity between shapes.

3.1   Tree-Structured Generating

Let $p_i=[x_i\ y_i]^T$ be the location of a pixel i, L be the number of levels of tree, and b be the number of split regions, where $i=1...N$, and N be the total number of foreground pixels belonging to the shape. Tree-structured from a shape is constructed by the following procedure:

Step 1.   Compute the centroid, Ce, and orientation, θ, of the shape by:

$$Ce(\bar{x}, \bar{y}) = \frac{1}{N} \sum_{i=1}^{N} p_i \tag{1}$$

$$\theta = \frac{1}{2} \arctan \left(\frac{2\mu_{11}'}{\mu_{20}' - \mu_{02}'} \right) \tag{2}$$

where

$$\mu_{qr}' = \frac{1}{N} \left(\sum_{i=1}^{N} \sum_{j=1}^{N} (x_i - \bar{x})^q (y_j - \bar{y})^r \right), \tag{3}$$

$\bar{x}$ and $\bar{y}$ are centroid in x and y axis respectively, and μ_{qr}' is second order central moment. The orientation is needed to obtain rotation invariant.

Step 2.   For each point p_i, determine the unsigned angle, α_i, between vector p_iCe and x-positive axis.

$$\alpha_i = \arctan \left(\frac{x_i - \bar{x}}{y_i - \bar{y}} \right) \tag{4}$$

Step 3.   Classify point p_i into b bins based on $\alpha_i + \theta$ voting. By performing this step in current level, the b sub-shapes will be obtained.

Step 4.   For sub-shape j $(j=1...b)$, repeat step 1 until step 3.

Step 5.   The process will be terminated if the Lth level is reached.

Step 6.   Finally, the $Lb+1$ centroid, Ce_{ij} defined as jth centroid in ith level $(i=1...L$ and $j=1...b)$, will be acquired where each centroid represents a corresponding node in the tree-structured. For each node of the tree, five parameters are then extracted. A shape descriptor is represented as a vector of all parameters obtained. Fig. 1 illustrates the tree generating.

(a) Original shape taken from MPEG-7 database

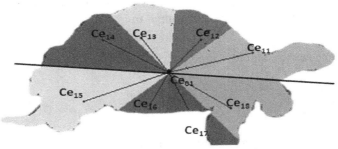

(b) Shape after performing tree generating

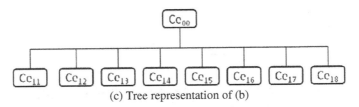

(c) Tree representation of (b)

Fig. 1. Illustration of tree generating ($L=1$ and $b=8$). The centroid location of each sub-shape is denoted with text. Bold line represents the major principal axes.

3.2   Parameters Extracting

For jth node at ith level, except root node, five parameters such as *normal_angle, axes_ratio, distance, occupation_ratio* and *compactness*, are extracted. Let *parent(i, j)* be the parent of node j at level i.

normal_angle(i, j): the unsigned angle between $Ce_{i,j}Ce_{parent(i,j)}$ and major principal axes corresponding to parent region of node j, where $Ce_{i,j}Ce_{parent(i,j)}$ is vector formed between $Ce_{i,j}$, the centroid of the node j, and $Ce_{parent(i,j)}$, the centroid of the parent node of the node j. Fig. 2 shows how to define the *normal_angle* of first node at first level.

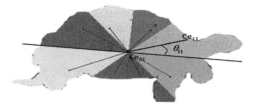

Fig. 2. Example of the normal angle at first node in first level, θ_{11}

axes_ratio(i, j): the ratio of $ma_{i,j}$ to $Ma_{i,j}$, where $ma_{i,j}$ and $Ma_{i,j}$ are denoted as length of minor and major principal axes of node j at level i, respectively. This value is formulated by (5). Note that length of two principles axes equal the eigenvalues λ_1 and λ_2 of covariance matrix $C_{i,j}$ of node j at level i, defined by (6).

$$axes\_ratio\,(i,\,j) = \frac{ma_{i,j}}{Ma_{i,j}} \tag{5}$$

$$C_{i,j} = \frac{1}{N}\sum_{k=1}^{N} p_k\,p_k^T - Ce_{i,j}Ce_{i,j}^T \tag{6}$$

distance(i, j): the distance between $Ce_{i,j}$ and its parent's centroid location $Ce_{i-1,j}$. In order to make this value be scale invariant, it is divided Ma_{i-1} by the length of major principal axes of parent region. Hence, it can be expressed as

$$dist\,(i,\,j) = \frac{\left| Ce_{i,j} - Ce_{parent\,(i,j)} \right|}{Ma_{parent\,(i,j)}} \tag{7}$$

occupation_ratio(i, j): the ratio of $R_{i,j}$, the area of the region corresponding to node j at level i, to $B_{(i,j)}$, the area of the bounding box corresponding to node j at level i,. This value can be expressed as follow formula

$$occupation\_ratio\,(i,\,j) = \frac{R_{i,j}}{B_{i,j}} \tag{8}$$

compactness(i, j): the ratio of $R_{i,j}$, the area of the region corresponding to node j at level i, to $R_{parent(i,j)}$, the area of the associated region of its parent. This value can be expressed as follow formula

$$compactness\,(i,\,j) = \frac{R_{i,j}}{R_{parent\,(i,j)}} \tag{9}$$

3.3   Transformation Invariances

To provide the robust results, a shape descriptor is required to be invariant to rotation, translation and scale. Thus, in this part, we discuss how these invariants can be addressed by our shape descriptor.

1. **Translation Invariant:** The translation invariant can be achieved by descriptor, if an absolute reference frame with a fixed origin is used [8]. Our descriptor is translation invariant since everything is measured with respect to location of centroid of shape.
2. **Scale Invariant:** Our descriptor invariant to scale due to it calculated the normalized parameter *distance* by divided it with the length of major principal axes.
3. **Rotation Invariant:** The simplest method to make descriptor invariance to rotation is measure all angles for an object relative to its principal axes. Thus, since our normal_angle parameter computation is relative to the major principal axes, the CbTs invariance to rotation.

3.4   Matching

In determining the similarity between two shapes, we utilize descriptor-based template matching with X^2 distance as matching cost. Defining $CbTs_i$ and $CbTs_j$ are the vector descriptor of ith and jth shapes. The similarity between these shapes, S_{ij} is determined by X^2 distance formulated as:

$$S_{ij} = \frac{1}{2} \sum_{k=1}^{M} \frac{\left[CbTs_i(k) - CbTs_j(k) \right]^2}{CbTs_i(k) + CbTs_j(k)} \tag{10}$$

where $CbTs_i(k)$ and $CbTs_j(k)$ denote the kth element of vectors descriptor of ith and jth shapes, respectively, and M represents the vector size. The value of S_{ij} ranges from 0 to 1. Our objective is to minimize the cost of matching. Hence, the ith shape can be considered to be similar to jth shape each other, if the similarity value S_{ij} is minimal among others.

4   Experimental Results

In experiment, the proposed descriptor is tested using MPEG-7 CE Shape-1 Part-B [2]. Our descriptor is programmed in C++ and is executed on PC 3GHz Pentium 4 with 2GB Ram under Windows 32 bits. Furthermore, we set the value of b varying, i.e. 4 and 8, and the value of node level as 1 and 2. Note that $CbTs$-b-l is denoted as the centroid-based tree-structured with b sub-shapes and l level division.

MPEG-7 CE Shape-1 Part-B dataset consists of 1400 shape samples, 20 for each class [2] as shown in Fig. 3. According to data set, we divide evaluation test into two cases. First, we include all samples for matching. Second, we follow [10] and [12] that used a subset of this data set in shape classification experiments. The subset contains following seven classes: Bone, Hearth, Glass, Fountain, Key, Fork, and Hammer, where each class includes 12 samples. As seen in Fig. 4, the shape classes are very distinct, but the data set shows substantial within-class variations [10]. We compare our method with seven invariant Hu moments (HM), Zernike moments (ZM), and method from [10]. Furthermore, in some classes, the shapes appear rotated

Fig. 3. Some of the shapes in the MPEG-7 dataset. One image per class for the first 40 classes (the database has 70 classes).

Fig. 4. Part of MPEG-7 CE Shape-1 Part-B shape classes. It was used in [10] and [12].

and flipped, which we address using a modified distance function. The distance dist(A, B) between a reference shape B and a query shape A is defined as

$$dist(A,B) = \min\{dist(A,B^u), dist(A,B^v), dist(A,B^h)\} \quad (11)$$

where B^u, B^v, and B^h denote three versions of B: unchanged, vertical flipped, and horizontal flipped. The evaluation protocol is conducted into two protocols:

(1) **Nearest Neighbor Score.** Query shape is assigned as the same class of the nearest similarity from reference shapes.
(2) **Bull's Eye Score.** The retrieval rate is measured by the so-called bull's eye score. Every shape in the database is compared to all other shapes, and the number of shapes from the same class among the 40 most similar shapes is reported. The bull's eye retrieval rate is the ratio of the total number of shapes from the same class to the highest possible number (which is 20 × 1400). Thus, the best possible rate is 100%.

Table 1. Results for MPEG-7 CE Shape-1 Part-B Shapes Using NN Score

Method	NN Score
CbTs-4-1	73.53%
CbTs-8-1	77.64%
CbTs-4-2	74.23%
CbTs-8-2	85.12%
HM [4]	69.94%
ZM [5]	75.04%
Method [11]	67.89%

Table 2. Results for MPEG-7 CE Shape-1 Part-B Shapes Using Bull's eye score

Method	Bulls eye Score
GT [13]	91.61%
Inner Dist. [9]	85.40%
CbTs-8-2 (ours)	80.73%
SC [14]	79.92%
Distance Set [15]	78.38%
Shape Cont. [8]	76.51%
CbTs-4-2 (ours)	76.46%

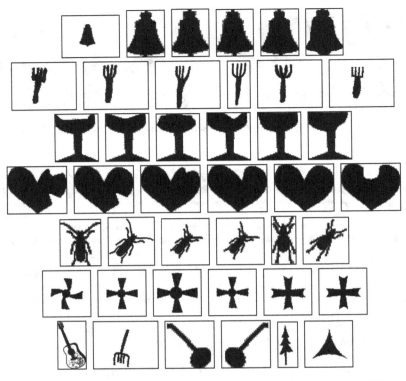

Fig. 5. Retrieval results using our descriptor: Each row shows instances of a different shape class. The first image of each row denotes as query, the second image is 1st ranked closest matches, the third image is 2nd ranked closest matches, and so on. Note that several of the shapes require rotation, translation, scale, and occlusion invariant. However, our descriptor failed to retrieve the last row (guitar) due to unfilled shape.

Table 3. Results in Percentage for Subset of MPEG-7 CE Shape-1 Part-B Shapes

Method Class	CbTs 4-1	CbTs 8-1	CbTs 4-2	CbTs 8-2	HM [4]	ZM [5]	HMM [10]
Bone	95	95	100	100	85	100	83.3
Hearth	90	100	100	100	90	85	100
Glass	100	100	100	100	100	100	100
Fountain	100	100	100	100	95	100	100
Key	90	85	90	96.67	85	100	100
Fork	75	75	80	100	65	60	100
Hammer	75	75	80	96.67	100	100	100
Overall	89.29	90	92.86	99.04	88.57	94.29	97.6

Table 1 shows the classification rate of our descriptor using nearest neighbor score for all 70 classes in the dataset. The CbTs with 4 sub-shapes outperforms Hu moment [4] and method in [11] except Zernike moment that gives better accuracy. Hence, we increased the number of sub-shapes to be 8. It contributes on a significant improvement of NN score as much as 4.11%. The CbTs descriptor with 8 sub-shapes and 2 level divisions achieves 85.12% accuracy which is the best result among other methods. Regarded the results, it can be concluded that increasing the number of sub-shapes as well as the number of levels will obtain a significant improvement in classification rate. This is because a large number of sub-shapes can record local information and spatial configuration of image shape well.

Table 2 evinces the classification rate of the proposed descriptor using bull's eye score for all 70 classes. When we set the number of sub-shapes equal to 4 and level equal to 2, the CbTs gains the lowest result among others that is 76.46%. In other hand, if we set the number of sub-shapes equal to 8 with the same level, the CbTs classification rate of 80.73% is better than [14], [15] and [8], with rate value 79.92%, 78.38, and 76.51%, respectively. Yet, the best performance is achieved by Bai et al. [13], with a retrieval rate of 91.61%, followed by Ling et al [9] at 85.50%.

Hereafter, our proposed descriptor is evaluated by subset of MPEG-7 which was used by [10] and [12]. The comparison of classification rate between the CbTs and other descriptors can be seen in Table 3. Bicego et al. [12] reported that the best classification accuracy in their experiment is 98.8% while the worst is 92.9%. Thakoor et al [10] achieved the highest accuracy at 97.62% and the worst one is 80%. Meanwhile, in our experiment, CbTs achieves the best rate at 99.04% for $b=8$ and $L=2$, while the lowest one at 89.29% for $b=4$ and $L=1$.

Selected examples of retrieval MPEG-7 dataset by our descriptor are shown in Fig. 5. The first image of each row represents the query image, and the remaining images are the retrieval result start from the closest matches. However, our descriptor fails to retrieve the unfilled image shape as seen in the last row of Fig. 5, where guitar shape retrieved fork, spoon, tree, and device shape, respectively.

5   Conclusions

In this paper, a new shape descriptor for measuring similarity in 2-D image shapes has been addressed. Our descriptor, centroid-based tree-structured (CbTs), is based on the centroid of the shape. For extracting CbTs, first, the central of mass of a binary shape is determined considered as the root node of tree. The shape is divided into b sub-shapes by major principal axis. In the same way, the central of masses of the sub-shapes are computed as level-1 nodes. These processes are repeated for a certain level. Five parameters invariant to translation, rotation and scale are extracted at each node. Thus, a vector of all parameters is combined as descriptor. The X^2 distance function is employed in order to measure similarity between shapes.

As shown before at the experimental results, the performance measurement values verify the robustness and effectiveness of our descriptor for matching and classifying 2-D shapes. Based on our experiment, it can be concluded that increasing the number of sub-shapes as well as the number of levels will obtain a significant improvement in classification rate. A large number of sub-shapes can record local information and spatial configuration of image shape well. However, the proposed descriptor still has some limitations to be rectified in our future works. Although our experiments have obtained good result, occlusion remains unsolved problem in our descriptor. Furthermore, since CbTs descriptor is based on shape centroid, it may be sensitive to shape topology which sometimes causes problems.

Acknowledgements. This work was supported by the National Research Foundation of Korea (NRF) Grant funded by the Korean Government (MOE) (2013R1A1A2009984).

References

1. Veltkamp, R.C.: Shape matching: similarity measures and algorithms. In: Proceeding of International Conference on Shape Modeling and Applications, pp. 188–197 (May 2001)
2. Jeannin, S., Bober, M.: Description of core experiments for MPEG-7 motions/shape. Technical report ISO/IEC JTC 1/SC/29/WG 11 MPEG99/N2690, MPEG-7, Seoul (March 1999)
3. Sharvit, D., Chan, J., Tek, H., Kimia, B.: Symmetry-based indexing of image databases. J. Visual Communication and Image Representation (1998)
4. Hu, M.K.: Visual Pattern Recognition by Moment Invariants. IRE Trans. Information Theory 8, 179–197 (1962)
5. Kim, Y.S., Kim, W.Y.: Content-Based Trademark Retrieval System Using a Visually Salient Feature. Image and Vision Computing 16(12-13), 931–939 (1998)
6. Jain, A.K., Vailaya, A.: Shape-Based Retrieval: A Case Study with Trademark Image Databases. Pattern Recognition 31(9), 1369–1390 (1998)
7. Alajlan, N., Kamel, M., Freeman, G.: Geometry-based image retrieval in binary image databases. IEEE Trans. on PAMI 30(6), 1003–1013 (2008)
8. Belongie, S., Malik, J., Puzicha, J.: Shape matching and object recognition using shape context. IEEE Transaction on Pattern Analysis and Machine Intelligence 24(24), 509–522 (2002)
9. Ling, H., Jacobs, D.: Shape classification using the inner-distance. IEEE Transaction on Pattern Analysis and Machine Intelligence 29(2), 286–299 (2007)
10. Thakoor, N., Gao, J., Jung, S.: Hidden Markov Model-Based Weighted Likelihood Discriminant for 2-D Shape Classification. IEEE Transactions on Image Processing 16(11), 2707–2719 (2007)
11. Kim, H.-K., Kim, J.-D.: Region-based shape descriptor invariant to rotation, scale and translation. Signal Processing: Image Communication 16, 87–93 (2000)
12. Bicego, M., Murino, V.: Investigating hidden Markov models' capabilities in 2-D shape classification. IEEE Transaction on Pattern Analysis and Machine Intelligence 26(2), 281–286 (2004)
13. Bai, X., Yang, X., Latecki, L.J., Liu, W., Tu, Z.: Learning Context Sensitive Shape Similarity by Graph Transduction. IEEE Trans. Pattern Analysis and Machine Intelligence, PAMI (2009)
14. Xie, J., Heng, P., Shah, M.: Shape matching and modeling using skeletal context. Pattern Recognition 41(5), 1756–1767 (2008)
15. Grigorescu, C., Petkov, N.: Distance sets for shape filters and shape recognition. IEEE Trans. on Image Processing 12(7), 729–739 (2003)

Ego-Motion Compensated for Moving Object Detection in a Mobile Robot

Joko Hariyono, Laksono Kurnianggoro, Wahyono, Danilo Caceres Hernandez, and Kang-Hyun Jo

Graduate School of Electrical Engineering, University of Ulsan, Ulsan 680–749 Korea
{joko,laksono,wahyono,danilo}@islab.ulsan.ac.kr,
acejo@ulsan.ac.kr

Abstract. This paper presents a moving object detection method using optical flow in an image obtained from an omnidirectional camera mounted in a mobile robot. The moving object is extracted from the relative motion by segmenting the region representing the same optical flows after compensating the ego-motion of the camera. To obtain the optical flow, image is divided into grid windows and affine transformation is performed according to each window so that conformed optical flows are extracted. Moving objects are detected as transformed objects are different from the previously registered background. In omnidirectional and panoramic images, the optical flow seems to be emerging on focus of expansion (FOE), on the contrary, it to be vanishing on focus of contraction (FOC). FOE and FOC vectors are defined from the estimated optical flow and used as reference vectors for the relative evaluation of optical flow. In order to localize the moving objects, histogram vertical projection is applied with specific threshold. The algorithm was tested in a mobile robot and the proposed method achieved comparable results with 92.37% in detection rate.

Keywords: Moving object detection, Omnidirectional camera, Mobile robot, Ego-motion compensated.

1 Introduction

Vision-based environment detection methods have been actively developed in robot vision [1]. Detecting moving object is one of the essential tasks for understanding environment. It is important to segment out and detecting moving objects in order to avoid an obstacle and control locomotion of the mobile robot in real-world environment. However, the vision system can provide not only a huge amount of information but also intensity and feature information in the populated environment. The omnidirectional vision system supplies a wide view of 360 degree, so they have been popularly used in many applications such as the motion estimation, environment recognition, localization and navigation of a mobile robot.

In the past few years, moving object detection and motion estimation methods for a mobile robot using the optical flow have been actively studied and developed [2].

A. Moonis et al. (Eds.): IEA/AIE 2014, Part II, LNAI 8482, pp. 289–297, 2014.
© Springer International Publishing Switzerland 2014

Fig. 1. An omnidirectional camera mounted on a mobile robot

A qualitative obstacle detection method was proposed using the directional divergence of the motion field. The optical flow pattern was investigated in perspective camera and this pattern was used for moving object detection. Also real-time moving object detection method was presented during translational robot motion. The optical flow pattern in a perspective camera is different from the pattern in an omnidirectional camera because of the distortion of an omnidirectional mirror.

Several researchers have been also developed for ego-motion estimation and navigation of a mobile robot with an omnidirectional image [3], [4], [5] and [6]. [3], [4] tried to measure camera ego-motion itself using omnidirectional vision, and [5] gave analysis related to translation and rotation motion using optical flow. They used Lucas Kanade optical flow tracker and obtained corresponding features of background in the consecutive two omnidirectional images. Use analyzing the motion of feature points, camera ego-motion was calculated, but it was not used for moving object detection. They set up an omnidirectional camera on a mobile robot and obtained panoramic image transformed from omnidirectional image. They obtained camera ego-motion compensated frame difference based on an affine transformation of two consecutive frames where corner features were tracked by Kanade-Lucas-Tomasi (KLT) optical flow tracker [6]. But detecting moving objects resulted in a problem that only one affine transformation model could not represent the whole background changes since the panoramic image has many local changes of scaling, translation and rotation of pixel groups. For this problem, each affine transformation of local pixel groups should be tracked by KLT tracker. The local pixel groups are not a type of image features such as corner or edge. We use grid windows-based KL T tracker by tracking each local sector of panoramic image while other methods use sparse features-based KLT tracker. Therefore we can segment moving objects in panoramic image by overcoming the nonlinear background transformation of panoramic image.

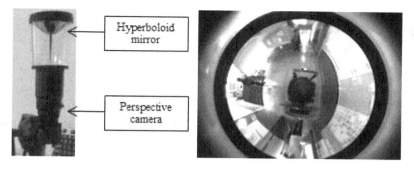

Fig. 2. The structure of omnidirectional vision and its image

In this paper, a method for detection moving object using ego-motion compensated was proposed in an omnidirectional camera mounted on a mobile robot. We focus on the optical flow in omnidirectional camera. First, an omnidirectional image converted into a panoramic image. The moving object is detected in panoramic image. In omnidirectional image, the length of optical flow becomes large according as the radial distance goes away from the center point. Otherwise, in panoramic image, the length of optical flow becomes is not affected from radial distance of omnidirectional image. Then, optical flow pattern is analyzed in panoramic images. The moving object is segment out through the relative evaluation of optical flows. The image divides as grid windows then compute each affine transform for each window. Moving objects can be detected from the background transformation-compensated using every local affine transformation for each local window. In order to localize the moving objects, we applied histogram vertical projection with specific threshold. The proposed algorithm was tested in mobile robot motions straight forward and rotation.

2   Mobile Robot with Omnidirectional Camera

This section presents the omnidirectional camera system which used in this work and how to detect moving object from an omnidirectional camera mounted on the mobile robot. The mobile robot shows in Fig. 1. The omnidirectional camera consists of perspective camera and hyperboloid mirror as shown in Fig. 2. It captures an image reflecting from the mirror so that the image obtains reflective scene and not perspective. It is easier to recognize whether image contains moving object or not, so it is necessary to transform the obtained image into panoramic image [7].

In order to perform the omnidirectional camera in mobile robot, before we apply in higher level task, it need to calibrate and investigated the accuracy. When it applied in structure from motion, we need to recover the metric information from environment [9]. In this work the omnidirectional camera was calibrated using checker board as a pattern with control points [8]. We used a flexible calibration method for omnidirectional single viewpoint sensors from planar grids. However this method was based on an exact theoretical projection function and some parameters as distortion were added to consider real-world errors.

Fig. 3. In omnidirectional and panoramic images, the optical flow seems to be emerging on focus of expansion and to be vanishing on focus of contraction

The sphere model was used by [8] and didn't consider the image flip. This approach adds to this model distortion parameters to consider real world errors. This method is multi view, which means that it requires several images of the same pattern containing as many points as possible. This method needs the user to provide prior information to initialize the principal point and the focal length of the catadioptric system. The principal point is computed from the mirror center and the mirror inner border. The focal length is computed from three or more collinear non-radial points. Once all the intrinsic and extrinsic parameters are initialized a non-linear process is performed. From this step we got intrinsic and extrinsic camera parameter that is useful to apply this omnidirectional camera system in real application for mobile robot system [9].

3   Ego-Motion Compensated

In omnidirectional and panoramic images, the optical flow seems to be emerging on focus of expansion (FOE), on the contrary, the optical flow seems to be vanishing on focus of contraction (FOC) [5]. If mobile robot moves then the panoramic image changes like Fig. 3. The translational motion of mobile robot makes two scaling points, and pixels move from FOE to FOC making curve trajectories. The rotational motion of mobile robot makes all pixels move to left or right. These movements of pixels can clearly happen when there is no moving object. If both translation and rotational motions happens together, these pixel movements look quite nonlinear that is why only one affine transformation model cannot represents these motions.

3.1   Moving Object Segmentation

In order to obtain moving object from omnidirectional image in mobile robot, it is not easy to segment out only moving object areas when the camera also moving caused by camera ego-motion [10]. Using [6], we apply KLT Optical Flow Tracker in order to deal with several conditions. Brightness constancy which projection of the same point looks the same in every frame, small motion that points do not move very far and spatial coherence that points move like their neighbors.

The frame difference represents all motions caused by camera ego-motion and moving object in the scene. It needs to compensate this effect from frame difference to segment out only moving object motion, so how much the background image has

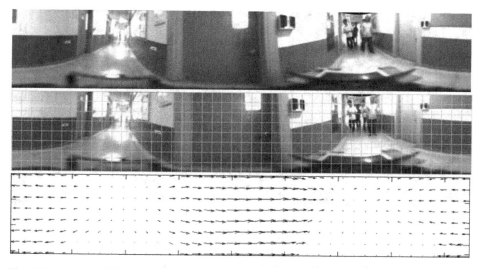

Fig. 4. From an omnidirectional image transformed to panoramic image (top image), we decide grid windows (middle image) and track each windows in the next image (bottom image)

been transformed in two sequence images. Affine transformation represents the pixel movement between two sequence images as in (1),

$$P' = AP + t \qquad (1)$$

where P' is pixel location in second image and P is pixel location in first image. A is transformation matrix and t is translation vector. Affine parameters can be calculated by least square method using at least three corresponding features in two images.

From the input omnidirectional image then transform to panoramic image, we decide grid windows and then compare and track each window in the next image, as shown in Fig. 4. Using method from [6], find the motion $d(i,j)$ of each group $g_{t-1}(i,j)$ by finding most similar group $g_t(i,j)$ in the next image.

$$g_{t-1}(i,j) = g_t(i + d_x(i,j), j + d_y(i,j)) \qquad (2)$$

It represented as affine transformation of each group as (3)

$$g_t(i,j) = Ig_{t-1}(i,j) + d_x(i,j) \qquad (3)$$

The camera motion compensated frame difference I_d is calculated based on the tracked corresponding pixel groups using (4)

$$I_d(i,j) = |g_{t-1}(i,j) - g_t(i,j)| \qquad (4)$$

where $I_d(i,j)$ is a pixel group located at (i,j) in the grid.

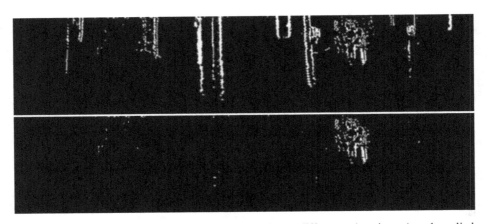

Fig. 5. From two consecutive images we applied frame difference (top image) and applied frame difference with ego-motion compensated (bottom image)

Suppose two consecutive sequence panoramic images shown in Fig. 4. It is not easy to segment out moving object using frame difference, then we apply frame difference with ego-motion compensate could obtain moving objects area shown in Fig. 5.

3.2   Object Localization

Each pixel output from frame difference with ego-motion compensated could not show clearly as silhouette. It just gives information of motion area from moving object. To obtain detected object, it is important to localize moving object area from the image. In this work, we define detected moving objects are represented by the position and width in $x - axis$. Using projection histogram h_x by vertically project image intensities into $x - coordinate$.

$$h_x = P_x I_d = [I, ..., I] I_d \tag{5}$$

where P_x is a projection vector which size is same as the height of panoramic image. An obtained h, is shown in Fig. 5.

We detect moving object based on the constraint of moving object existence that the bins of histogram in moving object area must be higher than a threshold and the width of these bins should be higher than a threshold as below

$$h_x(i \pm 10) > A \, max(h_x) \tag{6}$$

where A is a control constant and the threshold of bin value is dependent on the maximum bin's value. In order to get threshold of bin value, in this work using 10 omnidirectional images for training. Each image contains one or more moving objects shape with different shape and high, it is related to the distance from camera to object.

From Fig. 6 shows localization results. Top image show image result from frame difference with ego-motion compensated. Middle image shows histogram vertical

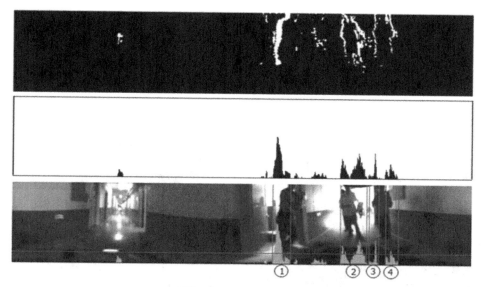

Fig. 6. Detection result

projection from above image, and bottom image shows there are four detected moving object, obtain from the region where have the number of bin above of the horizontal line as threshold value.

4   Experimental Results

In this work, our robot system is run in corridor with constant speed and detected moving object surround its path. Then evaluate our method from those image sequences. Proposed algorithm was programmed in MATLAB and executed on a Pentium 3.40 GHz, 32-bit operating system with 8 GB Random Access Memory

Mobile robot moves in constant speed at 20 centimeters per second. From omnidirectional camera it captured image sequences with frame rate 4 hertz in indoor environment. We perform two kinds of evaluation with difference of the number of image frames. The first we apply the system for around 800 image frames, for which is the number of moving objects are 2,335. The second, we conduct for 30 minutes of image sequence consist of 15,673 objects. In this case, all of moving object captured by omnidirectional camera is human walking in corridor. Evaluation process obtain from calculate the true positive detection and false positive detection.

In table 1, when the robot moving at constant speed for the first evaluation the accuracy of moving object detection result shown the system could detect 2,157 (92.37%) objects and 93 false positive detections. When we apply the system for long image sequences, the detection rate shown almost consistence with 92.33% and less than 4% in false positives detection rate. Fig. 7 shows detection results several images taken from omnidirectional camera.

Fig. 7. Successful moving objects detection results

Table 1. Detection comparison

Number of Image Frames	The number of human	True positives	False positive	Detection rate
800	2,335	2,157	93	92.37%
7,000	15,673	14,472	572	92.33%

5   Conclusions

This paper presents moving object detection applied in mobile robot which mounted by an omnidirectional camera. The moving object is segment out through the relative evaluation of optical flows to compensate ego-motion of camera. The image is divided as grid windows then compute each affine transform for each window. Moving objects can be detected from the background transformation-compensated using every local affine transformation for each local window. In order to localize the moving objects, we applied histogram vertical projection with specific threshold. The algorithm was tested in mobile robot motions straight forward and rotation. The proposed method achieved comparable results with 92.37% in detection rate and less than 4% in false positive detection.

In order to improve detection rate of the system, in the future work it need consider to combines object detection based on moving object detection with geometrical approach to calculate object position or kinematic model of robot movement relative to static objects environment.

Acknowledgement. This work was supported by the National Research Foundation of Korea (NRF) Grant funded by the Korean Government (MOE) (2013R1A1A2009984).

References

1. Wolf, J., Burgard, W., Burkhardt, H.: Robust vision-based localization by combining an image-retrieval system with Monte Carlo localization. IEEE Trans. on Robotcis 21(2), 208–216 (2005)
2. Talukder, Goldberg, S., Matthies, L., Ansar, A.: Real-time detection of moving objects in a dynamic scene from moving robotic vehicles. In: Proc. of Int. Conf. Intelligent Robotics and Systems, pp. 1308–1313 (2003)
3. Vassallo, R.F., Santos-Victor, Schneebeli, H.: A General Approach for Egomotion Estimation with Omnidirectional Images. In: Proceedings of the Third Workshop on Omnidirectional Vision, pp. 97–103. Copenhagen (2002)
4. Liu, H., Dong, N., Zha, H.: Omni-directional Vision based Human Motion Detection for Autonomous Mobile Robots. Systems Man and Cybernetics 3, 2236–2241 (2005)
5. Kim, J., Suga, Y.: An Omnidirectional Vision-Based Moving Obstacle Detection in Mobile Robot. International Journal of Control, Automation, and Systems 5, 663–673 (2007)
6. Tomasi, C., Kanade, T.: Detection and Tracking of Point Features. International Journal of Computer Vision 9, 137–154 (1991)
7. Hoang, V.D., Vavilin, A., Jo, K.H.: Fast Human Detection Based on Parallelogram Haar-Like Feature. In: The 38th Annual Conference of the IEEE Industrial Electronics Society, Montreal, pp. 4220–4225 (2012)
8. Mei, C., Rives, P.: Single view point omnidirectional camera calibration from planar grids. In: International Conference on Robotics and Automation, Roma, pp. 3945–3950 (2007)
9. Hariyono, J., Wahyono, Jo, K.H.: Accuracy Enhancement of Omnidirectional Camera Calibration for Structure from Motion. In: International Conference on Control, Automation and Systems, Gwangju (2013)
10. Hariyono, J., Hoang, V.-D., Jo, K.-H.: Human Detection from Mobile Omnidirectional Camera Using Ego-Motion Compensated. In: Nguyen, N.T., Attachoo, B., Trawiński, B., Somboonviwat, K. (eds.) ACIIDS 2014, Part I. LNCS (LNAI), vol. 8397, pp. 553–560. Springer, Heidelberg (2014)

A Flexible Network Management Framework Based on OpenFlow Technology

Chun-Chih Lo[1], Hsuan-Hsuan Chin[1], Mong-Fong Horng[2],
Yau-Hwang Kuo[1,3], and Jang-Pong Hsu[4]

[1] Department of Computer Science and Information Engineering,
Center for Research of E-life DIgital Technology (CREDIT)
National Cheng Kung University, Tainan, Taiwan
[2] Department of Electronic Engineering,
National Kaohsiung University of Applied Sciences, Taiwan
[3] Department of Computer Science, National Chengchi University, Taipei, Taiwan
[4] Advance Multimedia Internet Technology Inc., Tainan, Taiwan
{cobrageo,bashell,kuoyh}@cad.csie.ncku.edu.tw,
mfhorng@cc.kuas.edu.tw

Abstract. In this paper, we present a network management framework based on OpenFlow technology, to enhance flexibility in network management to conform more service requirements from diverse user identities. To this aim, user-level based QoS differentiation is introduced to provide fine-grained handling of network flows under different user and service constraints. With user identity and service requirement in place, we are able to provide dynamic bandwidth adjustment with the support of traffic statistics and even allocate and share unutilized network resources to satisfy user service requirement and improve resource utilization in networks. Furthermore, we also present a unified but simple network management interface to help network administrators to manage a network as a single entity. With this interface, the agents in switches are developed to detect and response to anomalies of flows. In summary, the proposed framework improves the original OpenFlow framework to provide a flexible and agile way to manage networks with dynamic resource allocation and to meet the diverse QoS requirements in future cloud services.

Keywords: Software Defined Networking, OpenFlow, Network Management, QoS, Flexibility.

1 Introduction

Advanced technologies have evolved tremendously over the past few decades as network administrators must take heterogeneous systems, different networking technologies, and a wide variety of applications into consideration. As network applications grows and quality of user experience becomes important, network management plays a significant role in providing efficient management services for large and complex networks to ensure good quality of network services and performance. In this respect, it is not

A. Moonis et al. (Eds.): IEA/AIE 2014, Part II, LNAI 8482, pp. 298–307, 2014.

surprising that the conventional networking approaches are not sufficient enough to cope with increasingly complex challenges.

Recently, the rise of Software Defined Networking (SDN) has created a great opportunities for network innovation in network control and management. SDN separates the network's control plane from the data plane [1][2], in which the data plane keeps the data forwarding decisions locally at each OpenFlow switch in the network and delegate the control functionality to a logically centralized software-based controller. In doing so, the centralized approach have the global knowledge of the network state and make decisions and facilitates dynamic configuration to all network devices in a logically single location to overcoming the need for complicated distributed control protocols. This means that SDN provides flexible programmability to the network, in which network configuration methods are more agile and flexible to enhance a better management over the network.

OpenFlow [3] is a popular example of SDN that standardizes the way that a controller communicates with the switches in SDN architecture [4]. It provides a rich northbound API and a global view of the network, which make it feasible to analyze traffic statistics using software and create an abstraction of the networking environment to network administrators. In this regard, policy-based flow management can be enforced within a network, forwarding information can be dynamically updated, and different types of traffic can be identified and managed by network administrators in any way desired. Although OpenFlow offers great opportunities for network innovation, but it also faces challenges such as security, compatibility, reliability and ease of management issues.

This paper provides a network management framework called Flexible Network Management Framework based on OpenFlow Technology (FNMF-OF). The proposed framework defines user-level based QoS differentiation to provide fine-grained handling of network flows that require to be handled under different performance constraints. That is, it employs user identity to allow the provision of dynamic policy management and fine-grained policy enforcement according to user identification then automates the network configurations across the entire network. By doing so, we are able to provide flexible network management and dynamic bandwidth adjustment. In addition, the network administrators are able to enforce access control to the network and provide an automated way to leverage and maximize the utilization of network bandwidth. In this way, the framework may not only provide a flexible way of configure and manage networks, but also gives more agility to the establishment and reallocation of unutilized bandwidth to meet the diverse bandwidth and QoS requirements to ensure the performance of a service. Moreover, this paper also introduces a unified but simple network management interface to enable the network administrators to manage the network as a single entity and eliminates the management burden by using the global view of the network to control and allocate network resources.

The rest of this paper is structured as follows. In Section 2, the related work is outlined. In Section 3, basic concepts and the operation of the framework are discussed. In Section 4, a proof of concept evaluation on the proposed framework is conducted. Finally, we conclude this work in Section 5.

2   Related Work

In the OpenFlow network architecture, a centralized controller uses a secure channel to interconnect and manages a distributed collection of OpenFlow switches. Packet forwarding is executed in switches on the basis of a flow table. Flow entries in the flow table contain policies that can be used to match incoming packets and number of actions is associated with it. If a packet arrives to a switch that does not match an existing policy in the flow entries, it is encapsulated and sent to the controller over the secure channel. The controller can then investigate the packet further and update flow entries in the network, and send the packet back to the switch. In this way, the forwarding decisions are migrated to a controller, and allowing the controller to add and remove forwarding entries in the switch. By removing network control plane functionality from network devices and centralize it, it gives network administrators the flexibility to configure, manage and manipulate the flow tables on the switch and keep close watch over the behavior of the network, so network administrators can respond rapidly to user demands and traffic fluctuations.

OpenFlow was initially deployed in campus networks where separation of research and production traffics is crucial for achieving coexistence of experimental and production traffics without disrupting the operation of the campus network. A flow table is used to govern packet processing in an OpenFlow switch and flow entries are designed to manage flows. A typical flow entry is shown in Fig. 1. There are three fields, including rule, action and statistics, to describe a flow entry. Rules are used to identify packets. Actions are the procedure to process the packets matching the rule. Statistics present a measure related to packets matching the rule.

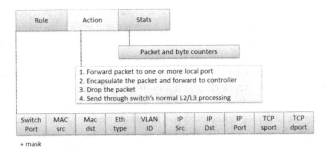

Fig. 1. A typical structure of flow entry

Several large testbeds based on OpenFlow technology such as GENI [5] in the U.S., OFELIA [6] and FIRE [7] in E.U., JGN-X [8] in Japan and FIRST [9] in Korea have been conducted. The success of OpenFlow can also be seen from the large number of recent researches such as network management in data center and enterprise networks [10][11] that aims to simplify network management through OpenFlow. Ethane [10] is an architecture proposed for managing the network of an enterprise that defines and creates centralized policies managed by the controller. Another example is an event-driven network control framework based on SDN called Procera has been implemented in [11]. It translates high-level policies into a set of forwarding rules that are used to enforce the policy on the underlying switches using OpenFlow. Inspired by these researches, it has motivated us to build a network management framework based on OpenFlow.

3   A Flexible Network Management Framework Based on OpenFlow Technology

Now days, many enterprises have own private network that provide connectivity to personal computers, laptops and other related devices across departments and workgroup networks. To manage such type of network is extraordinarily complex; to efficiently manage and properly allocate resources plays a key role in improving network performance in this kind of environment.

3.1   Network Scenario

In this paper, we consider a Small and Medium Enterprise (SME) networking environment that consists of a RADIUS server, OpenFlow controller and OpenFlow-enabled switches as shown in Fig. 2.

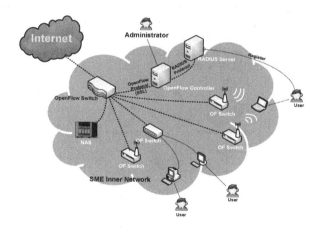

Fig. 2. Operation scenario

In this environment, users may be connected to the network via LAN or WLAN through different switches. The RADIUS server is responsible to enforce user-level based QoS differentiation where different level of QoS control can be administrated through user identity information and provide flexible network management and dynamic bandwidth adjustment. The controller is responsible for monitoring and managing the configuration of switches to ensure that QoS requirements are consistently satisfied in an efficient manner. It collects the current status of each switch to maintain complete visibility of the network and generate a set of policies for the switches in the network and configures them accordingly through a secure channel. In order to monitor and manage the network efficiently, the controller needs timely access to bandwidth usage and network load statistics in the network to maximize the performance in the network. In this respect, a unified but simple network management interface is introduced to provide a unified view of the network and simplifies the interaction required to control and manage the network.

3.2   Flexible Network Management Framework

The proposed network management framework consists of three main functions. That is, 3.2.1) User-level based QoS differentiation; 3.2.2) Flexible bandwidth management and 3.2.3) Unified but simple network management interface. Detailed descriptions of these functions are described as follows:

3.2.1   User-Level Based QoS Differentiation

Providing per user per flow control capability in enterprise network may potentially enable flexible policy management, where network administrators can specify and deliver differentiated QoS requirement for various departments to meet business and operational objectives in the enterprise based on user identity. To achieve this, the proposed network management framework incorporate a RADIUS server in the network environment to provide the controller with the ability to identify each user based on user access credentials and define and enforce user policies based on user identity. Therefore, we are not only able to facilitate fine-grained user-level policy enforcement, but also automate the process of policy configurations and reduce administrative overhead during the authentication process. A ladder diagram illustrating the process of user-level based authentication of FNMF-OF is shown in Fig. 3.

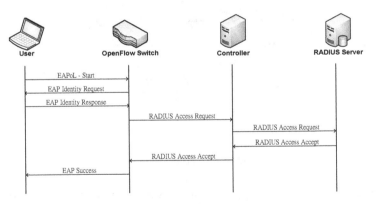

Fig. 3. Ladder diagram of the authentication process

The authentication process works as follows. A user sends an authentication request to OpenFlow switch to gain access to the network using access credentials. If the switch receives such request, it passes them to the designated RADIUS server through the controller to request for authorization to grant access via the RADIUS protocol. If credential provided by the user is correct, the designated RADIUS server authenticates the user by sending a success message to the user. At the same time, controller is notified that the user is authorized to use the network as requested and the controller initializes the policy entries for that specific user and returns the user's configuration information to the switch through the controller. This is done by providing the controller with user and service profiles that indicates the identity of an authorized user and what type of service and bandwidth is specified to the user as well as what is needed to fulfill user's QoS requirements. In order to maintain a scalable flow management, controller only sets up policy entries for authorized users and other policy entries are temporary suspended until user logon again.

By combining user and service profile with the switch location where the user is connected to, we are able to dynamically map the user to a set of policy that need to be enforced as user roams from a switch to another switch. By doing so, users are ensured that the user-level QoS requirements are consistently satisfied among different locations in an efficient manner. In addition, users may be group together based on their credentials and given a set of policy according to the service requirement for that group. Thus, policies can be enforced to support different service requirement per user group and it also makes it possible for policy configuration to be applied across entire group without requiring changing the service requirement attached to individual users. An example of user-level based QoS differentiation is illustrated in Fig. 4. For each user in the user profile, the controller specifies policies to the user according to its service profile to configure and applies them on the switches. As a result, more flexible and effective user-level QoS requirements can be applied to the network no matter where or how users are connected to the enterprise network.

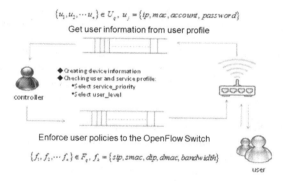

Fig. 4. User-level based QoS differentiation

3.2.2   Flexible Bandwidth Management

With OpenFlow controller, network administrators are able to obtain a global view of the entire network along with its configuration. This makes it feasible to analyze traffic statistics and create an abstraction of the networking environment that enables network administrators to facilitate dynamic configurations. With the traffic statistics, it is possible for administrators to keep track of bandwidth usage per user per flow and ensures resources are available according to the user and service profile. However, current bandwidth provisioning relies on user-defined bandwidth allocation and policy rules to manage the available bandwidth. In this respect, sometimes network resources are not fully utilized due to bandwidth reservation policies that assigns and guarantee certain amount of bandwidth to a flow. As a result, if the bandwidth reserved is not used or underutilized, such network resources are wasted.

To overcome this problem, we introduces a flexible bandwidth management scheme based on a combination of user and service-level hierarchy that takes various criteria, such as traffic statistics, user and service profiles into consideration to maximum bandwidth usage in the network. The goal is defined and set for a user based on the type of service that is required. Based on this idea, if excess bandwidth is available, including bandwidth that are not used, underutilized or not allocated, the excess bandwidth is divided among users that demand extra bandwidth to fulfill their service requirements. This division is done in a way that the excess bandwidth is divided in proportion to their configured bandwidth. By doing so, the controller is able to self-reconfigures the flow tables of the switches according to this hierarchy in order to provide more efficient bandwidth usage.

For example, an enterprise CEO is guaranteed with the highest level of network service quality. Research and Development department receive a higher level of network service quality than the Human Resource department and VoIP traffic is always given a higher priority over HTTP. In this case, a set of policy is assigned to differentiate different levels of bandwidth according to above description. With traffic statistics, we are able to determine which user and application is using the network resources and how much excess bandwidth is available. In the case where Human Resource department needs to make a conference call, but the allocated bandwidth is unable to maintain such service. In the other hand, the bandwidth allocated to the CEO is underutilized and it is considered as excess bandwidth. With the flexible network management scheme, it identifies the amount of bandwidth needed by the service and how much excess bandwidth is available. Based on such information, an appropriate amount of the extra bandwidth is allocated to Human Resource department to ensure user service requirement is achieved. In doing so, we not only provide dynamic and flexible resource allocation that meet the user and service requirements, but the acquisition of excess bandwidth also improves resource utilization in networks.

3.2.3   Unified but Simple Network Management Interface

In this section, we present a unified but simple network management interface as illustrated in Fig. 5. The interface enables the network administrators to manage the network as a single control entity and access all management features by menus. Topology discovery scheme is used to obtain the physical network topology and collect traffic statistics of flows. A graphical viewer called topology viewer is used to display the discovered network topology which enables the network administrator to deploy polices and monitor traffic statistics by selecting a user or a switch from the viewer. In addition, the topology viewer enables the network administrator to use traffic statistics to dynamically manage flows and transfer operations such as creation, modification and deletion of a flow into requests to the controller.

The network management interface also provides user and service profile of a user that states a set of policy that is assigned to a user according to its service requirement. This information can be used by the network administrator as a reference when performing reconfiguration on specific user. Moreover, traffic statistics also present information such as the description of each flow, the number of sent packets, the number of lost packets and the duration of each flow. By using these information together with user and service profile of a user, flow anomalies can be detected and create a report in the log viewer. An agent is used to monitor any anomalies presented in the log to detect and response to an event such as a user service requirement is not achieved due to lack of network resources and congestion. Based on logged historical information, the agent can react to such situation automatically without the intervention of a network administrator. However, if the agent is unable to handle an anomaly, a report message is send to the network administrator to alert such event and request for assistance. Therefore, the network management interface simplifies the interaction needed by the network administrator and the time required to maintain the network is also reduced.

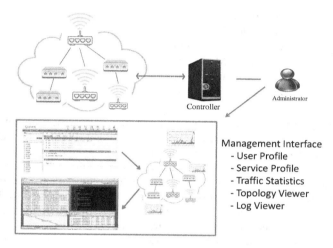

Fig. 5. Network management interface

4   Evaluation

In this section, a functional validation is conducted to verify the correct operation of FNMF-OF and to provide a proof of concept over a real but small, controllable topology. We deployed an experimental network using two 1.7 GHz Intel Duo-Core Embedded Platforms, with 8 GB of memory. An Ethernet extension module is also added to run OpenFlow Software Switch [12]. The controller runs on an Intel Pentium 4 3GHz computer with 8 GB of memory. First, users gained access to the network using access credentials through a web portal. The controller then identifies each user based on user access credentials and enforces user policies based on user identity. At this stage, policies for each individual user is deployed according to user and service profiles that indicate what type of service and bandwidth is specified to the user as well as what policies is needed fulfill user's QoS requirements.

Our first experiment uses user identity and service profile to evaluate the flexibility and functionality of FNMF-OF in flow control for different type of services by implementing several different network functions and policies. Therefore, a set of policy is deployed to enables users to access certain type of services according to user identity and these services include FTP, mail services, streaming, web browsing and Dropbox. As shown in Table 1, the controller uses user identity to validate whether or not a service is valid for a specific user. Once the service is permitted, it automatically allocates bandwidth according to the users' service profiles and traffic analysis. As a result, FNMF-OF works well in services that is currently covered by OpenFlow, and services such as web-based streaming and web-based mail service does not work in the proposed framework due to the lack of higher layer support in OpenFlow.

Table 1. Service validation

Service Type	Flow Control
FTP	✓
Mail Server	✓
RTP Video Streaming	✓
Web-based Streaming	✗
Web-based Mail Service	✗
Web browsing	✓
Dropbox	✓

The second experiment is conducted to exam the consistency of policies in the network. In order to maintain a scalable flow management, FNMF-OF sets up policy entries for authorized users and other policy entries are temporary suspended if the user is not logon. Therefore, achieving consistent policy enforcement throughout the network is critical to ensure that the user-level QoS requirements are consistently satisfied among roaming users. In this experiment, a user is connected to the network via WLAN and roams between two switches. When a user roams to a switch, controller updates the network topology using topology discovery scheme and discovers the topological changes between two switches. When controller realized such changes, it enables a set of policy that was defined for that specific user to the switch where the user has roamed to. As a result, user policies are enforced in the switch where user is connected to and policy consistency is achieved. Fig. 6 shows an example of topology discovery that that discovers the topological changes in a switch.

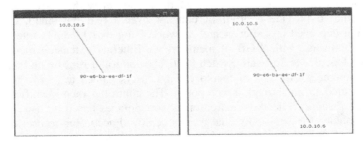

Fig. 6. Topology discovery

5   Conclusions

In this paper, we presented a flexible network management framework that provides a more agile and flexible method to simplify management and enhance a better management of switches on networks. First, we define user-level based QoS differentiation that provides fine-grained handling of network flows based on user identity and service requirement. This allows the provision of dynamic policy management and fine-grained policy enforcement according to user identification then automates the network configurations across the entire network. Furthermore, the management framework combines user-level based QoS differentiation and traffic statistics provide a more flexible bandwidth management scheme that enables the sharing of any excess network resources to improve resource utilization in the network. Moreover, we also presented a unified but simple network management interface that enables the network administrators to manage the network as a single entity of control. This interface provide a graphical interface that allowed the network administrator to select user or a switch in the topology viewer to manage flows and uses user and service profile as a reference to manipulate and reconfigures policies for a specific user. Moreover, an agent is used to detect and respond to anomalies with a flow, which simplifies and reduce the time needed to maintain the network.

Acknowledgement. The authors would like to express their sincere appreciation for the financial support from National Science Council and Southern Taiwan Science Park (STSP) under the reseach grants of 102-2218-E-151 -005 - , 101-2221-E-006 -259 -MY3 and 102CC02.

References

1. Gude, N., Koponen, T., Pettit, J., Pfaff, B., Casado, M., McKeown, N., Shenker, S.: NOX: Towards an operating system for networks. ACM SIGCOMM Computer Communication Review 38(3), 105–110 (2008)
2. Open Networking Foundation, https://www.opennetworking.org
3. McKeown, N., Anderson, T., Balakrishnan, H., Parulkar, G., Peterson, L., Rexford, J., Shenker, S., Turner, J.: Openflow: Enabling innovation in campus networks. ACM SIGCOMM Computer Communication Review 38(2), 69–74 (2008)
4. Sherwood, R., et al.: Carving research slices out of your production networks with OpenFlow. ACM SIGCOMM Computer Communication Review 40(1), 129–130 (2010)
5. GENI, http://www.geni.net
6. OpenFlow in Europe: Linking Infrastructure and Application (OFELIA), http://www.fp7-ofelia.eu
7. Future Internet Research and Experimentation (FIRE), http://cordis.europa.eu/fp7/ict/fire
8. Japan Giga Network 2 (JGN2), http://www.jgn.nict.go.jp/english
9. Hahm, J., Kim, B., Jeon, K.: The study of Future Internet platfom in ETRI. The Magazine of the IEEK 36(2), no. 3 (March 2009)
10. Casado, M., Freedman, M.J., Pettit, J., Luo, J., McKeown, N., Shenker, S.: Ethane: taking control of the enterprise. ACM SIGCOMM Computer Communication Review 37(4), 1–12 (2007)
11. Kim, H., Feamster, N.: Improving network management with software defined networking. IEEE Communications Magazine 51(2), 114–119 (2013)
12. Openflow, http://archive.openflow.org/wk/index.php/Ubuntu_Install

A Data Driven Approach for Smart Lighting

Hongyu Guo, Sylvain Letourneau, and Chunsheng Yang

National Research Council of Canada
1200 Montreal Road, Ottawa, Ontario
{hongyu.guo,sylvain.letourneau,chunsheng.yang}@nrc-cnrc.gc.ca

Abstract. Smart lighting for commercial buildings should consider both the overall energy usage and the occupants' individual lighting preferences. This paper describes a study of using data mining techniques to attain this goal. The lighting application embraces the concept of Office Hotelling, where employees are not assigned permanent office spaces, but instead a temporary workplace is selected for each check-in staff. Specifically, taking check-in workers' light requirements as inputs, a collective classification strategy was deployed, aiming at simultaneously predicting the dimming levels of the shared luminaries in an open office sharing light. This classification information, together with the energy usages for possible office plans, provides us with lighting scenarios that can both meet users' lighting comfort and save energy consumption. We compare our approach with four other commonly used lighting control strategies. Our experimental study shows that the developed learning model can generate lighting policies that not only maximize the occupants' lighting satisfaction, but also substantially improve energy savings. Importantly, our data driven method is able to create an optimal lighting scenario with execution time that is suitable for a real-time responding system.

1 Introduction

Smart lighting, which aims to improve on both the overall energy usage and the occupants' individual lighting comfort, has been identified as a potential market of 4.5 billion dollars in revenue by 2016[1]. Such smart lighting is of importance, not only for the "green" concept in terms of energy efficiency, but also for "personalized" office space.

Recent research has shown that buildings consume one-third of the total primary energy in the U.S., and of which, lighting, in particular, accounts for about 30% [9,10]. To cope with this increasing operational expenditure, modern lighting systems aim to be designed to minimize the energy consumption. Equally important, modern lighting also needs to take into account the occupants' lighting preferences. Studies have indicated that lighting comfort, for example, can dramatically impact workers' moods and thus productivity [8,13,14]. This is especially true under the context of Office Hotelling, where a company does not assign permanent office spaces for employees; instead it selects a temporary

---

[1] http://www.nanomarkets.net/

A. Moonis et al. (Eds.): IEA/AIE 2014, Part II, LNAI 8482, pp. 308–317, 2014.

workspace for each check-in staff. As a result, in addition to minimizing energy saving, introducing personalized lighting for occupants in commercial buildings is also of great importance [4].

Fig. 1. The mock-up smart lighting office with six cubicles

This paper discusses a study of using a data driven approach to attain the above goals. Specifically, we apply recent data mining techniques to generate lighting scenarios for an open office sharing light, within the context of Office Hotelling. Figure 1 depicts the demonstration laboratory being set up for this application. This laboratory includes six cubicles, and sensors were installed in various positions of each cubicle in order to measure the environmental data such as temperature and light level. The sensor positions are shown at the bottom-right corner of Figure 1. The nine (9) shared lights are on the ceiling, and can be adjusted by either the computer in each cubicle or the center control system installed. The lighting policy generating unit here takes aim at creating lighting scenarios that not only minimize energy consumption but also satisfy users' light requirements, based on occupants' lighting preferences.

To generate a lighting scenario, the light requirements for the six desks are first obtained and used as inputs for the smart lighting system. Next, the dimming levels for the nine lights on the ceiling are determined by a machine learning classification model. By doing so, such classification information will be able to provide us with lighting scenarios that can both match users' preferred lighting and save energy consumption, provided that we have the energy usages for possible office plans. To this end, to obtain the various energy usages of potential office arrangements, we shuffle the workplaces of the employees, which is a practical approach within the office hotelling context where workers typically have different offices each time they check in. In this way, an energy saving lighting scenario, for instance, could be assigning closer offices to workers with similar lighting preferences. When compared with four other alternative lighting control strategies, our study shows that the developed data driven learning model can generate, within reasonable responding time, lighting policies that not only max-imize the occupants' lighting satisfaction, but also substantially improve energy savings.

The rest of the paper is organized as follows. Section 2 outlines our prediction task and challenges. Next, in Section 3, we discuss our modeling approach. This is followed by an empirical study in Section 4. Finally, Section 5 concludes the paper and outlines our future research directions.

2   Data Mining Task and Challenge

2.1   Task Description

Constrained by the lighting preferences of the check-in employees in the office, the aim of the smart lighting system is to generate an energy saving lighting scenario. Also, the system has to be able to create the optimal lighting scenario in a reasonable execution time which is suitable for a real-time responding system.

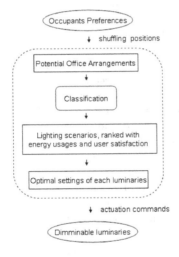

Fig. 2. Framework of the smart light control system

Figure 2 depicts the framework of our smart lighting control system. In detail, the input of this lighting system is the check-in occupants' light preferences. With such input, we can have 720 different positioning scenarios for the system. That is, we can assign an occupant to any one of the six desks in the room. In our approach, before initiating the classification system, we shuffle the positions of the individuals, and then use the shuffled scenarios as inputs. Subsequently, the classification system generates the output for each of the 720 positioning arrangements. In this way, we then can rank them in a specific order. Hence, the automatic system is able to choose the one that both saves energy and satisfies users' preferences. By doing so, for the checked-in occupants in the room, the classification system is able to generate multiple position and lighting arrangements, each with different preference-matched level and energy usage. As highlighted in Figure 2, the optimal one is then chosen to actuate the luminaries.

2.2   Modeling Setup

Practically, it is a common approach to use a Synthetic Imaging System for light design [1]. For instance, the RADIANCE system is often used by domain experts to simulate different lighting scenarios and to foresee the effects of tailored configurations. In our studies, the laboratory as depicted in Figure 1 was simulated using the RADIANCE system, where each of the configurable light sources has three dimming levels, i.e. Low, Median, and High. By doing so, the light on the ceiling of the room can be controlled by the RADIANCE system, and the illuminance values on each of the desks in the room can be accurately measured in Lux. Consequently, the data for the machine learning task was produced through this simulated environment. To this end, the collected data include all the combinations of the lighting levels on the ceiling, and the resulting measurements on the desks. In total, 19683 instances were collected, each is composed of six attributes, corresponding to the illuminances on the desks, as well as nine label sets, which reflect the nine luminaries' dimming levels.

2.3   Modeling Challenge

From a data mining perspective, this task can be mapped into a multi-target classification problem. That is, using six inputs X (i.e., occupants' light requirements on the desks) to predict the nine outputs Y (i.e., the dimming levels of the nine light sources on the ceiling): $Y = f(X)$. It is worth noting that, in such applications, correctly classifying all target variables of an instance is required. As a result, one needs to consider a classifier has classified an instance correctly only if all target variables of that instance are correctly determined (i.e., "exact match"). In other words, the overall accuracy here refers to the "exact match" accuracy. Consequently, the main aim for such classification algorithm is to take aim at achieving higher "exact match" accuracy through learning a function f that maps X to Y.

To deal with such multi-target tasks, one straightforward approach is to learn a binary or multi-class classifier for each set of labels, and then each trained learner independently assigns a corresponding label for the test object. However, such an approach tends to result in poor predictive accuracy in terms of correctly classifying all labels simultaneously. This is because there is a large number of possible labels for each object to be classified, as discussed previously. For example, as observed in our experiments, a decision tree learning method [12] can achieve an average accuracy of 85% over nine (9) independent classifiers against this lighting application. Nevertheless, the predictive accuracy in terms of simultaneously predicting all correct labels for the nine label sets was only 22.57%. Unfortunately, as mentioned earlier, in such problems, simultaneously predicting the correct labels for all label sets is of importance. For example, imagine that we correctly predict eight out of the nine light sources, but misclassify one of them.

In such scenario, the resulting luminance values of the six desks will be very different from that of correctly classifying all nine light sources. This is because the misdetermined light source contributes its light to all the six desks. That type of incorrect determination will result in the dissatisfaction of all occupants in the open offices.

3   Modeling Methodology

To address the above mentioned modeling challenge, we deployed a state-of-the-art multi-target learning strategy, as presented by Guo and Létourneau in [5]. As reported in [5], when compared with several popular multi-target classification algorithms, the so-called Iterative Multi-target Classification (IAMC) approach can meaningfully enhance the "exact match" accuracy. Instinctively, the IAMC method benefits from being able to not only employ many accurate, mature single-target learning approaches to model each of the target attributes, but also utilize an iterative learning strategy to exploit the relationships among multiple related target attributes, thus achieving higher accuracy, when compared with other popular learning strategies for multiple targets problems [5].

Due to the IAMC method's superior predictive performance in terms of "exact match", we adapt this strategy for our smart lighting application. In particular, we significantly improve the IAMC method's predictive accuracy in our lighting application through integrating an ensemble strategy, namely the AdaBoost approach. Next, we will discuss the IAMC method and our extension in detail.

Algorithm 1. The Training of the IAMC Algorithm

Input: Object set with X attributes and Y labels, and a single target method f.
Output: Classification model $Y = f(X)$

1: **Training begins**
2: **for** each $y_i \in Y$ **do**
3:     Build a model f_i^s using X only;
4:     Apply AdaBoost strategy to model f_i^s; obtaining predictive accuracy $\epsilon^s(i)$;
5: **end for**
6: **for** each $y_i \in Y$ **do**
7:     Build a model f_i^r using $X \cup (Y \backslash Y_i)$
8:     Apply AdaBoost strategy to model y_i^r; obtaining predictive accuracy $\epsilon^r(i)$
9: **end for**

3.1   Iterative Approach For Multi-target Classification

The IAMC method includes two phases: training and inference. The training stage constructs two collections of single-target classifiers, while the inference stage aims at exploiting the relationships among target attributes through these constructed classifiers. Specifically, as depicted in Algorithm 1, the IAMC method firstly constructs two collections of classifiers: one utilizes the descriptive attributes only while the other is augmented with provided target attributes in the training data. Next, as described in Algorithm 2, these two collections of classifiers are used for iterative inference, as follows. The first collection is used to initiate the iterative process, where all values of the target attributes in the test data set are unknown. The second one is then deployed to continue the inference procedure until the process stops. In each iteration, the current target attribute estimates, resulting from the previous iteration, are used to enhance the learning models. The above iterative process repeats until all of the labels have stabilized or a pre-set number of iterations have been reached. As stated in Algorithm 2, the IAMC outputs the labels of the last iteration.

Note that a detailed description of the IAMC algorithm falls beyond the scope of this paper. Interested readers are referred to [5] for more discussions on this strategy.

Algorithm 2. The Joint Inference of the IAMC Algorithm

1: generate descending ordering O based on prediction improvement $\epsilon^r(i) - \epsilon^s(i)$
2: **for** each object t in the test set **do**
3:     obtain y_i using f_i^s;update y_i in the test set
4: **end for**
5: **repeat**
6:     **for** each object t in the test set **do**
7:         **for** each $y_i \in O$ **do**
8:             compute y_i using f_i^r; update y_i in the test set
9:         **end for**
10:     **end for**
11: **until** pre-set threshold number of iterations have elapsed or all labels have stabilized

---

3.2   Improving the IAMC Strategy

Recall from Algorithm 1 that, the IAMC method requires a single-target learning method as input. Our studies show that the accuracy of this single-target learning method has a significant impact on the overall "exact match" accuracy of the IAMC approach. Basing on this observation, we meaningfully improve the predictive accuracy of the IAMC method in our application with a boosting strategy, as will be discussed next.

In a nutshell, the IAMC method falls in the learning framework of collective classification. As pointed out by Neville and Jensen in [7], one of the necessary conditions for the success of collective classification is that the system must be able to make some initial inferences accurately. Following this thought, we intend to improve the individual classifiers' prediction before initiating the collective inference procedure. Particularly, we look into the AdaBoost ensemble method, which have been proven to be able to meaningfully improve predictive accuracy of a high error classifier, namely the so-called weak classifier [3]. The underlying principle of the Boosting strategy states that by using a weak learning algorithm several times on a sequence of carefully constructed training examples, the weak learning algorithm can be converted into an algorithm with a predictive performance that surpasses the original weak algorithm [11]. While learning, the Boosting algorithm first focuses on the production of a series of dependent classifiers, in which each classifier is better able to predict hard examples for which the previous classifier performance was poor [11]. The outputs of these classifiers are then combined using weighted voting in the final prediction of the model.

In the implementation of our lighting application, the single-target learning method in the IAMC strategy as depicted in Algorithms 1 and 2 are replaced by a Adaboost ensemble. That is, the f function there is replaced by a H function as defined as following.

$$H(x) = sign(\sum_{m=0}^{M} \alpha_m h_m(x))$$

Here, h_m is the m-th classifier of the series of M dependent classifiers in the boosting ensemble, and α_m represents the corresponding weight of the classifier h_m.

Promisingly, our experimental studies, as will be presented in the next section, suggest that boosting the performance of individual learners before the collective inferences can significantly improve the collective classification's prediction as measured by the "exact match" metric. Therefore, in our lighting application, we apply the AdaBoost algorithm [3] to boost the single-target learners' accuracy before deploying them for the IAMC strategy.

4 Experiments

4.1 Predictive Accuracy Achieved

In this experiment, we present our evaluations on using the C4.5 decision trees [12] and the Artificial Neural Networks [2] as the single-label learning methods of the IAMC approach. The C4.5 decision tree learner was used due to its de facto standard for empirical comparisons. Also, Artificial Neural Networks were chosen because they have proven to be surprisingly successful in many real-world knowledge discovery applications [6]. Each of these experiments produces results using 10-fold cross validation. In addition, the number of iterations for the collective inferences was heuristically set to 20 for each experiment.

We compared the predictive accuracy, in terms of simultaneously predicting all correct labels, obtained by the three approaches, namely 1) the straightforward approach that learns a classifier for each set of labels, and then each trained learner independently assigns a corresponding label for the test object (noted as the Intrinsic model), 2) the collective classification strategy IAMC, and 3) the collective classification strategy with the AdaBoost method applied (noted as BoostIAMC). We presented the accuracies obtained by the three approaches with decision trees and neural networks as single-label learning methods in Tables 2 and 1, respectively. In these two tables, we also described the performance improvement of both the IAMC algorithm and the BoostIAMC method over that of the Intrinsic strategy. The statistic significance of these results was examined using a paired t-test.

Table 1. Accuracy obtained by the Intrinsic, IAMC, and BoostIAMC methods using decision trees as the single-label learning method, along with the prediction improvement of both the IAMC and BoostIAMC strategies over that of the Intrinsic approach ($p < 0.001$ in the paired t-test)

	Predict all lights simultaneously	Accuracy Improvement
Intrinsic Model	22.57%	
IAMC	41.37%	18.80%
BoostIAMC	68.33%	45.76%

Table 2. Accuracy obtained by the Intrinsic, IAMC, and BoostIAMC methods using neural networks as the single-label learning method, along with the prediction improvement of both the IAMC and BoostIAMC strategies over that of the Intrinsic approach ($p < 0.001$ in the paired t-test)

	Predict all lights simultaneously	Accuracy Improvement
Intrinsic Model	29.37%	
IAMC	64.23%	34.86%
BoostIAMC	84.86%	55.49 %

Results as shown in Tables 2 and 1 indicate that the BoostIAMC approach can statistically and significantly increase the predictive accuracy of the Intrinsic models in terms of simultaneously predicting all of the correct labels. That is, the experimental results suggest that the AdaBoost strategy and the collective inference technique were successfully employed in this lighting application. For example, when decision trees were applied as single-label learners, the accuracy obtained by the Intrinsic method was very low in terms of simultaneously predicting all of the correct labels. The accuracy was only 22.57%. In this case, the collective inference process increased its accuracy to 41.37%. Furthermore, this prediction was improved by applying the AdaBoost approach before the collective classification. As a result, the final predictive accuracy achieved by the BoostIAMC algorithm was 68.33%.

When considering deploying the neural network algorithm, the accuracy for the Intrinsic and IAMC methods were 29.37% and 64.23%, respectively. Significantly, this accuracy was improved to reach as high as 84.86% when the AdaBoost algorithm was employed before the collective inferences, as achieved by the BoostIAMC approach. Results from these two tables have also shown that these prediction improvements were statistically significant. The results indicate that the p-values achieved by the paired t-test were less than 0.001.

These results indicate that the BoostIAMC model with neural networks applied was very promising. The final accuracy of this model against all luminaries was 84.86%, which was more than 16% higher than that of applying decision trees as the single-label learning methods. Thus, in our smart lighting application as described in Section 2.3, we deployed a BoostIAMC strategy with neural network algorithms as its single-target learning methods for our classification task.

4.2   Comparison with Alternative Control Strategies

In this section, we compare our classification system with four other lighting control strategies which are commonly used in the lighting domain, namely all-on, all-off, half-on, and exhaustive search. The all-on model refers to a configuration of the office where all of the luminaries are wired to be turned on all together. On the contrary, the all-off model will turn off all of the lights in the room at the same time. In contrast, the half-on model sets each light to its Median dimming level. When the exhaustive search model is applied, the system will exhaustively search all of the possible lighting scenarios in the light data set and then choose the best-matched one as its output.

We present the comparison results in Table 3, where the energy consumption, average discrepancy, and response time are described for each strategy. We calculated the discrepancy of a lighting scenario by comparing each desk's illuminance value, generated by the lighting strategy, with its true value obtained. The measurement of the discrepancy was computed using the following Euclidean Distance function:

$$Discrepancy = \sqrt{\sum_{i=1}^{n}(p_i - q_i)^2}$$

Here p_i is the illuminance value of the ith desk in the room, resulting from the lighting scenario generated by the lighting strategy; q_i is the illuminance value of the ith desk measured, given the current lighting scenario.

Table 3. Energy consumption required and satisfaction discrepancy measured, along with the execution time needed, to generate a lighting scenario

	Energy Consump.	Average Discrep.	Response Time (sec.)
All ON Mod.	100%	1454.44	0
ALL OFF Mod.	0%	1460.26	0
Half ON Mod.	50%	490.13	0.0
Exhaustive Search	50%	0	20
Data Mining Mod.	50.04%	4.93	2

The results, as shown in Table 3, indicate that the all-on, all-off, and half-on models produced a much larger gap between the occupants' light preferences and the ones they would receive if such a lighting system was employed than that of the exhaustive search and data mining models. For example, the average discrepancy was less than 5 for the two latter strategies, compared to that of over 490 for the former two approaches.

When comparing the exhaustive search with the data mining models, the results suggest that the response time required for the exhaustive search model was large. It required 20 seconds to generate a lighting scenario, compared to that of only 2 seconds needed from the data mining strategy. These observations imply that the exhaustive search approach is not suitable for a real-time responding environment.

5   Conclusions

Smart lighting needs to take into consideration both the overall energy usage and the occupants' lighting comfort. This paper describes a data-mining approach to attain this goal. Specifically, taking check-in workers' light requirements as inputs, a collective classification strategy was deployed to simultaneously predict the dimming levels of the shared luminaries. This classification information, together with the energy usage for potential office plans, provides us with lighting scenarios that take into account both the users' preferred lighting and energy consumption. We evaluate our method against four other lighting control strategies. Our study shows that our method can generate lighting policies that both maximize the occupants' satisfaction and improve the overall energy savings.

Our future work will include the dynamic lighting settings, such as people entering and leaving the room.

References

1. RADIANCE SYNTHETIC IMAGING SYSTEM, Lawrence berkeley national laboratory, Berkeley (2006), http://radsite.lbl.gov/radiance/ (retrived March 2009)
2. Duda, R., Hart, P., Stork, D.: Pattern Classification. Wiley-Interscience (2000)
3. Freund, Y., Schapire, R.E.: Experiments with a new boosting algorithm. In: ICML 1996, pp. 148–156 (1996)
4. Gu, Y.: The Impacts of Real-time Knowledge Based Personal Lighting Control on Energy Consumption, User Satisfaction and Task Performance in Offices. PhD thesis, Carnegie Mellon University (2011)
5. Guo, H., Létourneau, S.: Iterative classification for multiple target attributes. J. Intell. Inf. Syst. 40(2), 283–305 (2013)
6. Mitchell, M.T.: Machine Learning. McGraw Hill, New York (1996)
7. Neville, J., Jensen, D.: Iterative classification in relational data. In: AAAI Workshop on Learning Statistical Models from Relational Data, 13-20 (2000)
8. Newsham, G., Veitch, J.: Individual control over office lighting: Perceptions, choices and energy savings. Construction Technology Updates (1998)
9. Online. THE INTERLABORATORY WORKING GROUP ON ENERGY-EFFICIENT AND CLEAN-ENERGY. In: Scenarios for a Clean Energy Future: Interlaboratory Working Group on Energy-Efficient and Clean-Energy Technologies (2000), http://www.nrel.gov/docs/fy01osti/29379.pdf
10. Online. U.S DEPARTMENT OF ENERGY, BUILDING TECHNOLOGY PROGRAM. ENERGY SOLUTION FOR YOUR BUILDING (2000), http://www.eere.energy.gov/buildings/info/office/index.html
11. Opitz, D., Maclin, R.: Popular ensemble methods: An empirical study. Journal of Artificial Intelligence Research 11, 169–198 (1999)
12. Quinlan, J.R.: C4.5: programs for machine learning. Morgan Kaufmann Publishers Inc., USA (1993)
13. Wen, Y.-J., Agogino, A.: Control of wireless-networked lighting in open-plan offices. Lighting Research and Technology 43, 235–248 (2011)
14. Wen, Y.-J., Bonnell, J., Agogino, A.M.: Energy conservation utilizing wireless dimmable lighting control in a shared-space office. In: Proceedings of the 2008 Annual Conference of the Illuminating Engineering Society (2008)

On-Board Applications Development via Symbolic User Interfaces

Bora İ. Kumova

İzmir Institute of Technology, Department of Computer Engineering, Turkey
borakumova@iyte.edu.tr, iyte.edu.tr/~borakumova

Abstract. becerik is a functional language consisting of symbolic commands for managing and composing applications. Application commands consist of symbols that are associated with reading sensor values, computing those values and executing actuator values. It is the result of a co-design of mechatronic functionality and robotic behaviour. The requirements given for mechatronic functionality were those of simple robotics kits that are used in school education or as toys. The requirements given for the behaviour were to provide a reflexive one, consisting of triggering simple computations and actuations from simple sensor values. becerik currently lives as a leJOS application on NXT robots and enables developing simple applications using the standard display and buttons of the NXT brick. In this paper we introduce the symbolic user interfaces of becerik.

Keywords: Human-computer interaction, on-board application development, robot operating systems, embedded applications.

1   Introduction

State-of-the-art in robot programming is that robot applications are first developed on a computer, thereafter loaded onto the robot, where they cannot be changed any more. If an application needs to be changed, then it is developed again on a computer. However, as robot utilisation areas continue expanding and their usage as consumer products become widespread, one of the most important properties of consumer products, which is the requirement of consumers to adapt a product to own needs, directly on the product, will become inevitable for robots as well. In other words, consumers will require to adapt the hardware, as well as the software of their robots according their needs. We can observe this trend already today with the robotics kits that become increasingly popular. Since most middle-sized and large-sized robots have a built-in computer, all application development can be performed there. However, consumers would require simpler interaction interfaces.

Robotics kits fascinate people from pre-school to late adult ages. It has been shown that robotics kits were successful in developing cognitive skills and manual abilities in active learning [4], [22]. In the process of developing the hardware of a robotics kit, manual abilities are improved, while cognitive skills are improved, when developing the software. The kit approach can push the child to develop its robot even further, by inspiring the child further with every modification on the robot.

A. Moonis et al. (Eds.): IEA/AIE 2014, Part II, LNAI 8482, pp. 318–327, 2014.

Developing applications on-board a robot, without separately using a computer, including modifying, saving versions and saving different applications are all required for facilitating an interactive robot development.

A further advanced requirement for the robot is to autonomously develop its behaviour. That can be achieved by the robot, by inductively learning from its environment. For instance, collecting statistical data, as it perceives from and actuates in the environment. Hence, the robot can increase its autonomy. As soon as a child starts discovering that the robot can learn, it will wonder about what and how the robot can learn and then will try establishing stronger cognitive relationships with the robot. As a result of this process, while a child tries developing even more complex robots, it will develop own cognitive skills as well [20].

Most toy robots belong to the class of small-sized robots, because of their hardware and software capabilities. Small-sized robots have relative restricted processors and memories, since their design objective is to use the sensors and actuators in real-time, mostly reactively, without any cognitive processes. Mostly controllers that also operate like processors are used, instead of general purpose processors. Their operating systems (OS) are relatively shrank-down versions of common computer OSs [1]. Therefore, processing high-level cognitive processes that require more memory and processing power and multi-tasking, is restricted. On-board application development environments are rare and existing ones have many restrictions, such as no support for all sensors or actuators, restricted number of application steps, loading only a single application, no multi-tasking, no learning capabilities. In order to remedy these drawbacks, we have developed an on-board application development environment that can operate even on such restricted small-sized robots in real-time [6], [12].

In this contribution, we present the symbolic version of becerik and the command language becerik. Both constitute the becerik tool for on-board application development.

First, small-sized robot operating systems and application development environments are reviewed, the capabilities of the NXT brick is summarised, the operation environment of the becerik tool presented, the becerik user interfaces discussed and finally concluding remarks summarised.

2   Small-Sized Robot Operating Systems and Their Application Development Environments

In an earlier study on small-sized robots we have found out that on-board application development environments are rare and those few existing ones are very restricted [12]. In this section, we briefly discuss OSs and on-board application development environments for small-sized robots.

2.1   Operating System Properties

We will briefly discuss some properties of widely used small-sized OSs. For a detailed discussion and comparison of the below OSs, the reader is referred to [12].

FreeRTOS is an open source software, written in C [5]. Since only the kernel is distributed, any device driver needs to be adapted to the kernel explicitly.

Applications are first compiled and linked with the OS on a computer, thereafter loaded onto the robot. Similar properties apply to RobotC [19] and pbLua [11].

NXT OS [7] is developed in a C derivative, interpreted language and is adapted solely to NXT robots. Not like the above OSs, applications are not linked with the OS. After the OS is loaded to the robot, multiple applications can be transferred to the file area at any time.

leJOS NXJ [2] is an OS with a built-in Java Virtual Machine (JVM), but itself is written in C. It was primarily developed for NXT robots. Like NXT OS, it is separated from applications and multiple applications may be transferred at any time. As the JVM supports multi-threading, applications may be executed in multi-tasking mode.

2.2   Application Development Environments

Since most OSs do not provide an on-board application development environment or the provided one is too restricted, a separate computer software is required, for developing applications.

Computer software distributions for developing robot applications have following common capabilities:

- Development: develop applications and transfer them to the robot OS on the robot
- Updating: flashing the robot RAM with an OS
- Simulation: virtual and visual execution of applications, before transfer to the robot

Most of this software support further loading the OS separately from applications or first linking OS and applications to a single executable image.

For instance, the NXT application development software for the computer, NXT-G [9], [10], provides symbolic building blocks that can be related to each other, in order to specify a programme [8], which is simpler for children to learn. However, the symbols of the NXT-G programming blocks are visually very complex graphics, even for adults. ROBOLAB [18] is basically a redesign of NXT-G, but with much easier graspable graphical symbols. The NXT robot is discussed more detailed below. Urbi [3] supports Java, C++ and Matlab, which is even more difficult for children, as programming becomes the more difficult the younger the age. Similarly, Webots [15] supports C and C++. These development environment provide a simulation environment, for testing applications, before loading onto the robot. The simulator can visually show the selected robot, within a specified environment and execute the application virtually, for verification purposes.

Scratch [16] falls into the same category with the above environments and languages. The only remarkable difference is that, instead of pure graphical symbols like the above, Scratch symbols consist of graphics with build-in alpha-numerical words. Therefore, it is less suitable for illiterates. Scratch programmes read just like any imperative programming language, but with a block-look. Consecutive commands are presented as geometrically and smoothly attached symbols.

Besides the NXT on-board development environment, ICON [21] seams to provide some capabilities similar to becerik. Out of the poorly documentation of ICON, one may conclude following capabilities: on-board application development by choosing motor movements, recording of motor movements, managing multiple applications and generation of NXT-G compatible code, which enables for optionally editing the code further in NXT-G on the computer.

3   Operation Environment of Becerik

Becerik is an application development software that runs itself as an ordinary application under leJOS. The symbolic user interfaces have the same functionality like the becerik versions with English or Turkish user interfaces [12].

3.1   The Becerik Interpreter

It has been shown experimentally that pre-school children can perceive a word symbolically, if the association of that symbol to a known object is exemplified. They can grasp the relationship and, to some extend, can apply it [14]. In robotics courses that we give in our laboratory, we observe that children, who do not know English, are able to use the English menus on the robot properly, even though the English words are explained to them only once experimentally.

The first becerik command language was designed according to these experiences [13]. It consists of the following 6 English commands: new, set, var, get, compute and jump. These commands are no more used by the symbolic becerik interpreter that we introduce here. The symbolic becerik language consists of symbols for linguistic values that are associates with simple sensing, computing and actuating primitives. The symbolic becerik interpreter associates every symbol with a different execution code.

The symbolic becerik language consists of a collection of sensor, computer and actuator symbols (Table 3; Table 4; Table 5). Every symbol is associated with a linguistic value that covers the value range of the related sensor, computer command or actuator.

A becerik application consists of a random sequence of symbols. Since any symbol sequence is allowed, no syntax checking is performed. However, there is only one exception, the loop command. It is allowed to optionally use it at most ones in any slot of a particular application. This constraint is managed by the becerik user interfaces.

The becerik language semantics consists of only a few rules:

- Unconditional loop: the loop command is executed unconditionally; the command sequence after the loop command is executed in an endless loop
- Conditional actuation: if a sensor command is placed right before an actuator command, then the actuator command is executed only, if the actually read sensor value matched the sensor command
- Continuous motor: ones a forward, backward or turn command is executing, this execution will continue, until the stop command or until the next forward, backward or turn command
- Cumulative motor: while a forward, backward or turn command is executing, the speed value of any following motor command that is related to these motors, will be added to the current speed value, for the duration of the WaitVeryShort command

The symbolic becerik interpreter executes every symbol by directly calling the associated sensor or actuator driver of leJOS.

3.2   Compound Application

In order to facilitate the reuse of frequently concurrently executed applications, several already saved applications may be added to a compound application.

A compound application is not allowed to contain any command symbols on its own or other compound applications. It may include solely applications. One or more applications are allowed.

3.3   Multi-tasking

If more than one application is loaded, then the run operation will schedule all loaded applications for execution. If compound applications are loaded, then all their containing applications are scheduled. However, out of the 16 possible distinct applications each one is scheduled at most ones, even if a particular application is contained in several compound applications. Thus up to 16 distinct applications may be scheduled for concurrent execution.

The relative simple multi-tasking implementation follows these policies:
• Scheduling: all applications are scheduled in the order they where loaded and all are executed with equal priority in round-robin fashion
• Context switching: at most one command of the active application is executed, thereafter the next scheduled application becomes active
• Critical region: the execution of a command is always finalised, before switching the context; an exception is made for the wait command, whose time continues passing in wall clock time

As a result of these policies, all scheduled applications appear the user to be executed quasi simultaneously.

4   Becerik User Interfaces

The objective is to provide clear user interfaces with clear functionalities, where a new user should be able to capture the principle capabilities of becerik in a few trials. With the same objective in mind, hardware required to use the interfaces is kept minimal. An LCD display and 4 button are perfectly sufficient, just that what is available on an NXT brick.

4.1   Display Organisation

The NXT LCD has 100*64 pixel display. becerik symbols have a standard size of 16*16 bits (Figure 1). The entire screen is divided into 4 rows. The first row has 8 pixel reserved as status bar for the operating system leJOS. Then there is a gap of 4 pixel between the becerik symbols. The next 3 levels have each 16 pixels and a 2 pixels gap. becerik symbols appear in these 3 rows, separated by 5 pixels each. Therefore, at most 5 symbols can be displayed at a time in every row.

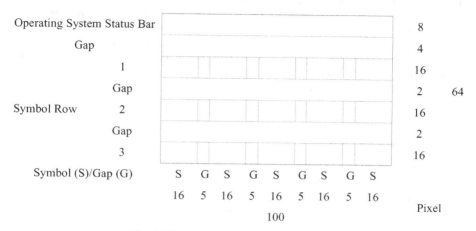

Fig. 1. Display Design for the becerik menus

The 3 physical rows are shared by 3 menu levels of the becerik menu hierarchy, where every becerik level constitutes a menu. The first level displays all becerik applications and compound applications that are available under leJOS and that have not been loaded into becerik yet. The second level displays all loaded applications and compound applications. The third level displays all commands of the current application or compound application (Appendix B).

The levels may be scrolled up or down the physical rows, depending on user choices. A level may be scrolled horizontally over one or more physical rows, if the number of symbols do not fit into a single row.

A pop-up window is available for every symbol, at every level. The operations provided are context-sensitive on every symbol. That is the only method used to provide sub-menus. Following generic operations are available in pop-up menus (Table 1):

- Run: start executing current application
- New: create a new application of command
- Delete: delete current application or command
- Load: load current application or all commands of current application into next menu level
- Save: save current application outside of becerik

The operations available in a pop-up window are the same for all symbols on a particular level, but they differ from level to level:

- Unloaded applications level: delete, load
- Loaded applications/compound applications level: new, delete, load, save, run
- Application commands level: new, delete

If the load operation is applied to an already loaded compound application, then all containing applications are displayed in the next physical display row. This level is handled as an intermediate one between the second and third menu levels.

Selecting operation new at the application command level displays a new pop-up menu that provides the available application commands. Since there are more than 60 such commands available (Table 3; Table 4; Table 5), the size of this pop-up menu is reduced to less than 20 commands, by providing here only commands with the smallest linguistic values. Choosing anyone of these commands will pop-up a further window with all linguistic values. At this pop-up level, a hierarchy of three pop-up menus is reaches. However, only the current pop-up menu is displayed. Pop-up menus are not displayed cascaded, previous pop-menus are simply overwritten on the display. Pressing the return button always exits directly to the application command level.

4.2   Keyboard Organisation

Only 4 buttons are required to use the becerik user interface:
- Enter: display a pop-up menu for current symbol or select current choice in pop-up menu
- Return: exit pop-up menu or move to next upper display row or stop all currently running applications
- Left: move cursor to left menu choice; may initiate scrolling
- Right: move cursor to right menu choice; may initiate scrolling

Scrolling down is not directly possible using any button. This is achieved by first calling the pop-up menu of the current symbol and selecting the load symbol. Scrolling up is achieved by pressing the return button.

5   Conclusion

We have presented the on-board application development environment becerik and the symbolic becerik command language, in which becerik applications are coded. The language consists solely of simple symbols that are associated with easy graspable sensing, computing and actuating primitives. Symbolic becerik is designed to facilitate fast learning of basic robotics concepts by using these primitives, even by pre-school children.

Acknowledgement. Thanks are due to Dipak Kumar Singh for contributing to the design of the symbols and to Vural Can Yenel for partially implementing an initial version of becerik.

References

[1]   Andersson, K., Andersson, R.: A comparison between FreeRTOS and RTLinux in embedded real-time systems. Linköping University (2005)
[2]   Bagnall, B.: Core LEGO MINDSTORMS Programming. Prentice Hall, lejos.sourceforge.net (2002)

[3] Baillie, J.C., Akim, D., Hocquet, Q., Nottale, M., Tardieu, S.: The Urbi Universal Platform for Robotics. In: Simulation, Modeling and Programming for Autonomous Robots (SIMPAR), Venice, Italy (2008)

[4] Barker, B.S., Ansorge, J.: Robotics as Means to Increase Achievement Scores in an Informal Learning Environment. Journal of Research on Technology in Education, International Society for Technology in Education (2007)

[5] Barry, R.: Using the FreeRTOS Real Time Kernel - A Practical Guide. Real Time Engineers Ltd. (2010)

[6] becerik; sourceforge.net/projects/becerik

[7] Ferrari, M., Ferrari, G., Astolfo, D.: Building Robots with LEGO Mindstorms NXT. Syngress Publishing (2007)

[8] Gasperi, M.: LabVIEW for LEGO MINDSTORMS NXT. NTS Press (2008)

[9] Grega, W., Pilat, A.: Real-time control teaching using LEGO® MINDSTORMS® NXT robot. In: International Multiconference on Computer Science and Information Technology, IMCSIT (2008)

[10] Griffin, T.: The Art of Lego Mindstorms NXT-G Programming. No Starch Press (2010)

[11] Ierusalimschy,   R.:   Programming   in   Lua,   lua.org   (2006),
     http://www.hempeldesigngroup.com/lego/pbLua

[12] Kumova, B.İ., Takan, Ş., Tos, U., Geçer, E.C., Aytar, A.: Türkçeleştirilmiş bir Robot İşletim Yazılımı İle Robot Üzerinde Uygulama Geliştirme (On-Board Robot Application Development Using A Turkish Localised Robot Operating System) Ulusal Otomatik Kontrol Toplantısı (TOK); GYTE (2010)

[13] Kumova, B.İ., Takan, Ş.: Developing Applications On-Board of Robots with becerik. Advanced Materials Research. Journal at scientific.net, MEMS, NANO and Smart Systems (2011)

[14] Mason, J.M.: When Do Children Begin to Read: An Exploration of Four Year Old Children's Letter and Word Reading Competencies. Reading Research Quarterly (1980)

[15] Michel, O.: Webots: A powerful realistic mobile robots simulator. In: Workshop on RoboCup. Springer (1998)

[16] Olabe, J.C., Olabe, M.A., Basogain, X., Maiz, I., Castaño, C.: Programming and Robotics with Scratch in Primary Education. In: Mendez-Vilas, A. (ed.) Education in a Technological World: Communicating Current and Emerging Research and Technological Efforts. Formatex (2011)

[17] nxtturkish. SourceForge, sourceforge.net/projects/nxtturkish

[18] Portsmore, M.: ROBOLAB: Intuitive Robotic Programming Software to Support Life Long Learning. APPLE Learning Technology Review (1999)

[19] RobotC, http://www.robotc.net

[20] Silk, E.M., Schunn, C.D., Shoop, R.: Synchronized robot dancing: Motivating efficiency and meaning in problem solving with robotics. Robot Magazin (2009)

[21] http://www.teamhassenplug.org/NXT/ICON

[22] Wendell, K.B., Connolly, K.G., Wright, C.G., Jarvin, L., Rogers, C., Barnett, M., Marulcu, I.: Incorporating engineering design into elementary school science curricula. In: American Society for Engineering Education Annual Conference & Exposition, Louisville, USA (2010)

Appendix A. Symbol Set of the Command Language Becerik

The becerik symbol set for a minimal robotics kit configuration is listed in the below tables. This configuration covers the most used sensors and actuators.

Table 1. Generic Operation Symbols

▶	New object
✖	Delete object
▲	Load object
▼	Save object
⁙	Run all loaded applications and compound applications

Generic operations (Table 1) are displayed only in a pop-up window.

Application management symbols (Table 2) are available at the two levels for unloaded applications and for loaded applications.

Table 2. Application Management Symbols

▣▣▣▣ ... ▣	Application 0, 1, 2, 3, …, 15
▣▣▣▣ ... ▣	Compound application 0, 1, 2, 3, …, 15

Computer command symbols are (Table 4) available at the application command level along with all sensor command symbols (Table 3) and actuator command symbols (Table 5).

Table 3. Sensor Command Symbols

✍	Touch
◁ ◁) (◁) (◁))	UltrasoundVeryLow, Low, High, VeryHigh
☼ ☼ ☼ ☼	LightVeryDim, LightDim, LightBright, LightVeryBright
♪ ⟨♪ «♪ «♪«	SoundVeryLow, SoundLow, SoundHigh, SoundVeryHigh

Table 4. Computer Command Symbols

⟳	Loop
⧗ ⧗ ⧗ ⧗	WaitVeryShort, WaitShort, WaitLong, WaitVeryLong

Table 5. Actuator Command Symbols

↑ ↑ ↑ ↑	ForwardVeryFast, Fast, Slow, VerySlow
↓ ↓ ↓ ↓	BackwardVeryFast, Fast, Slow, VerySlow
⌐ ↗ ⌐ ⌐ ⌐	TurnRightSlightly, Half, Full, BackHalf, BackFull
↑ ↖ ↰ ↰ ↰	TurnLeftSlightly, Half, Full, BackHalf, BackFull
✋	Stop forward, backward, turn motors
	MotorARightVerySlow, Slow, Fast, VeryFast
	MotorBRightVerySlow, Slow, Fast, VeryFast
	MotorCRightVerySlow, Slow, Fast, VeryFast
	MotorALeftVerySlow, MotorALeftSlow, MotorALeftFast, MotorALeftVeryFast
	MotorBLeftVerySlow, Slow, Fast, VeryFast
	MotorCLeftVerySlow, Slow, Fast, VeryFast
	BulbDim, BulbBright
🔊 🔊 🔊 🔊	VolumeVeryLow, VolumeLow, VolumeHigh, VolumeVeryHigh

Appendix B. Sample User Interface Scenario

After starting becerik, initially all applications and compound applications that are available for becerik, will be displayed at the application menu level of becerik, starting in the first display row (Figure 2 (a)). Pressing the enter button on the current selection will pop up a window containing the operations new, delete, edit, save and run, which are the available operations at this level (Figure 2 (b)). Selecting edit, will display the current application at the editing level in the next available display row. At the editing level, first the chosen application is displayed, followed by the content of that application, up to 4 commands (Figure 2 (c)). The application can now be edited, by moving the cursor stepwise left or right within the application content.

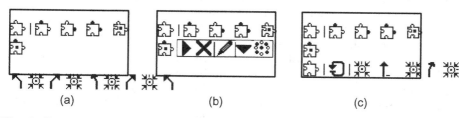

(a)   (b)   (c)

Fig. 2. Editing application 0 from (a) available applications 0, 1, 2, 3 and compound applications 0, 1; (b) pop-up menu for application 0; (c) first 4 commands of application 0 loaded in row 3 of display

Implementation of an Event Information Sharing Support System Utilizing OpenStreetMap

Kohei Hayakawa, Tomomichi Hayakawa, and Takayuki Ito

Nagoya Institute of Technology,
Gokiso-cho, Showa-ku, Nagoya City, Aichi Pref., Japan
hayakawa.kohei@itolab.nitech.ac.jp,
{hayakawa.tomomichi,ito.takayuki}@nitech.ac.jp
http://www.itolab.nitech.ac.jp/

Abstract. Information systems that enable adding and sharing of event information on maps continue to attract more and more attention. In this paper, we propose a mapping information system that focuses on adding and sharing event information. We also introduce a new cache mechanism to more efficiently process OpenStreetMap API and implement a recommender function based on collaborative filtering as an event information filtering method. Our system was employed at Aichi Triennale 2013, a performing-arts event in Japan. We evaluated the effectiveness of our system by analyzing web accesses and interviews with NPO volunteers who served as joint researchers.

Keywords: Information System, OpenStreetMap, Collaborative filtering.

1 Introduction

Information systems that enable users to add and share event information on maps are attracting more and more attention. Such web map systems, as Google Maps and Foursquare, are spreading rapidly. However, no current system focuses on adding and sharing event information. In this paper, we define general information as that pertaining to parks or convenience stores. General information is necessary information for map systems because it is frequently demanded by users. Event information is specific information, for example, the history or the culture of a place. Event map systems require both general and event information.

In this paper, we developed a map system that adds and shares both general and event information utilizing OpenStreetMap (OSM) [1][2][3], which is a collaborative project to create an open editable map of the world. Our system, which is digitalizing event maps based on the usual paper medium, was used at Aichi Triennale 2013 [9], an performing-arts event held in Aichi, Japan from August 10 to October 27, 2013. Many new kinds of valuable information was added to our system and shared by users on the web during the event's duration. With our system, users got much new information that is not available on other existing systems.

A. Moonis et al. (Eds.): IEA/AIE 2014, Part II, LNAI 8482, pp. 328–337, 2014.

Our system is also aiming to vitalize local communities by adding and sharing city information. Using OSM, the practical use of Information and Communication Technology (ICT) might promote local community activation. During Aichi Triennial 2013, much event information was city information because it was held in Nagoya and Okazaki, both of which are cities in Aichi prefecture.

The rest of the paper is organized as follows. In section 2, we describe the existing map systems and the related studies. In section 3, we define event information and describe ways to add and share it. In section 4, we describe our system in detail. In section 5, we assess it by analyzing access logs, questionnaires, and interviews. Finally, in section 6, we conclude with a discussion of possible avenues for future work.

2   Related Work

Google Maps is a popular application to get general information [6]. But it does not focus on adding and sharing event information. It does has a custom map service called My Maps, but it is only available for personal use and not for information sharing by many users.

Foursquare is a location-based social networking service where users "check in" at venues by mobile websites, text messaging, or device-specific applications by selecting from a list of venues of applications located nearby [7]. We can add new information with Foursquare, but it doesn't focus on sharing information.

Finally, we describe location-based social networking services [8]. By focusing on locations, users can learn about the activities of other users, but the information is dependent on the time or user preferences because it focuses on social networking.

In this paper, we develop a map system that enables event information to be added and shared that is not considered by other existing systems. Event information is beneficial for many users. Our proposed event information sharing support system is a custom map system that focuses on adding and sharing event information. Its demand is high for event or hazard maps. Our system also aims to vitalize local communities.

3   Event Information Sharing

We introduced our system at Aichi Triennale 2013, a performing-arts event held at Nagoya and Okazaki cities from August 10 to October 27, 2013 that attracted almost 630,000 visitors. Our system was used for information maps at event sites and to introduce the two cities. The Aichi Triennale 2013 City Vitalization Project [10] was our system's administrator and uploaded much event information, including the area's history, culture, and arts.

Event information has the following properties: IDs, titles, tags, photos, comments, locations, and dates. An ID is a unique number assigned to a piece of event information. Titles are the names of the event information. Tags are sets of string literals attached to the event information. There are ten types of tags:

History/Culture, Resting Areas, Souvenir Shops, Art Information, Food, Stories, Knowledge, Popular City Spots, Art, and Children. At least one tag must be attached to a piece of event information. Photos are image data related to the event information. Comments are detailed explanations of event information. Locations are composed of latitude and longitude. The date is the date and the time that the event information was contributed. Event information also has impressions and clicks. Impressions are the number of times that it was viewed by users. The impressions and the number of clicks are used for quantitative valuation basis during system assessment.

For this event, only administrators added event information, even though our system can enable general users to upload information. However, we decided to reduce the risks of malicious users who might upload unrelated information. Administrators assign titles, tags, photos, comments, and locations. IDs, dates, impressions, and clicks are automatically assigned by the system. Assigning locations is done by clicking on maps. Titles, tags, photos, and comments are editable. Users can show event information in board and map styles.

4 Implementation of Our Event Information Sharing Support System

4.1 Implementation Outline

We implemented our Event Information Sharing Support System as a web application that uses a web browser as a client. Our system can be accessed at `http://mappingo.com/machitori`. Fig. 1 shows its interface. We built client-side programs using HTML 5, CSS3, and JavaScript (jQuery, Leaflet, and Twitter Bootstrap as a library) and server-side programs using Ruby on Rails as a framework. We confirmed system operation on such browsers as Safari, Google Chrome, and Internet Explorer. Our system is composed of three functions: sharing maps, sharing general information, and adding/sharing event information.

We use OSM as a map system and Leaflet, which is an open-source JavaScript library for mobile-friendly interactive maps. The event places are the cities of Nagoya and Okazaki, and users can select a place on our system's top menu. The basic map functions are "moving," "scaling," and "getting current position."

4.2 Sharing General Information

For general information, we use OSM data, which are available using an OSM API called Overpass API. OSM data include modern architecture, artwork, museums, toilets, convenience stores, and parks. Users can select general information on the top menu. It is shown on maps as markers with balloon icons. Users click on markers and place names popup.

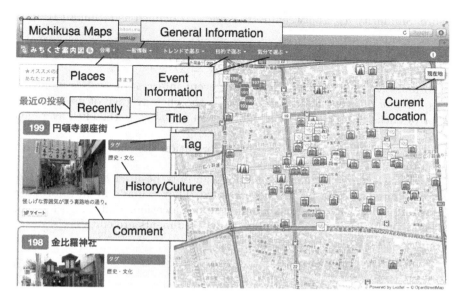

Fig. 1. System interface

4.3   Sharing and Adding Event Information

Event information is shown on the left side of our system and can be viewed by scrolling. Event information is also on the map as markers whose icons are event information IDs.

Our system filters event information based on four criteria: (1) date, (2) popularity, (3) recommendation function and (4) tags. We subscribe them below.

(1) Date criterion filtering arranges event information in a new date order using its date.

(2) The popularity criterion is based on all users' interests. As an indicator of interest, we use the click-through rate (CTR) [4], which is defined below:

$$CTR = \frac{Clicks}{Impressions}. \tag{1}$$

In formula (1), *Clicks* is the number of times that the event information was clicked on by users, and *Impressions* is the number of times that the event information was viewed by users. CTR filtering arranges the event information in a high CTR order.

(3) The recommendation function is based on the personal interests of users. We use collaborative filtering [5]. Our system matches users whose browsing history similarities are high. User valuations of the event information are the number of clicks. A click is a positive valuation to the event information; there are no

negative valuations. We use the Pearson correlation coefficient as a similarity measure. Similarity $w_{u,v}$ between users u and v is defined as formula (2) below:

$$w_{u,v} = \frac{\sum_{i \in I}(c_{u,i} - \overline{c_u})(c_{v,i} - \overline{c_v})}{\sqrt{\sum_{i \in I}(c_{u,i} - \overline{c_u})^2}\sqrt{\sum_{i \in I}(c_{v,i} - \overline{c_v})^2}}. \tag{2}$$

In formula (2), I is the total amount of event information, $c_{u,i}$ is number of clicks of user u of event information i, $\overline{c_u}$ is the click average of user u, $c_{v,i}$ is number of clicks of user v of event information i, and $\overline{c_v}$ is the click average of user v.

(4) Tag-based filtering provides event information that has been attached to selected tags.

The adding event information function can only be used by administrators who are required to login by ID and password. They can add event information by assigning titles, tags, photos, comments, and locations and assign locations by clicking on maps. Titles, tags, photos, and comments are editable.

4.4   System Architecture

Figure 2 shows our system architecture. Our system was implemented as a web application that can be accessed from web browsers. We used OSM for the maps and Leaflet (an open-source JavaScript library) for the OSM maps. We built a client-side program using HTML 5, CSS3, and JavaScript (jQuery, Leaflet and Twitter Bootstrap as a library) and a server-side program using Ruby on Rails as a framework. We implemented an API that handles event information on the server-side. The database is SQLite3, and the server OS is CentOS.

4.5   OpenStreetMap Data Cache Mechanism

Due to the access concentration on the OSM server, the OSM API often loads data slowly or not at all. So we implemented a cache function that caches OSM data on our system's server to solve the above problem. Fig. 3 compares ways of getting OpenStreetMap data. In the conventional method, servers return data when they are requested by clients. The OSM server returns XML data when clients request them. Fig. 4 shows a mechanism for getting OSM data by caches. In our cache mechanism, OSM data are automatically obtained by a curl command from ShellScript. The XML data obtained by the curl are saved on a public directory on the system server. Clients read the cached XML files when they use the OSM data. By this cache mechanism, we successfully got stabler and faster access.

5   System Evaluation

In this section, we explain how we evaluated our system. We analyzed 65 days of access logs from August 10 to October 6, 2013 and interviewed members from the Aichi Triennale 2013 City Vitalization Project.

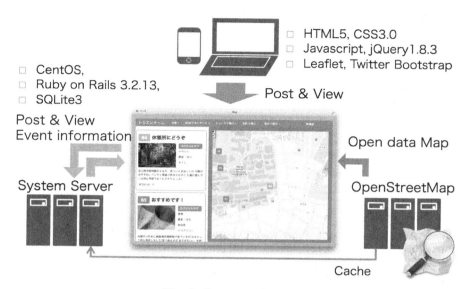

Fig. 2. System architecture

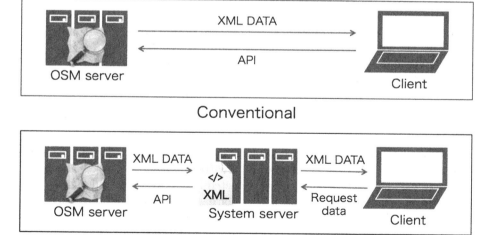

Fig. 3. Comparing ways to get OpenStreetMap data

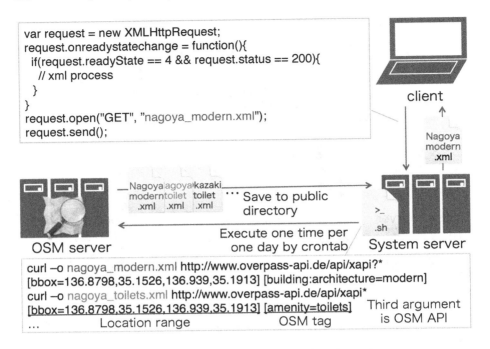

```
var request = new XMLHttpRequest;
request.onreadystatechange = function(){
  if(request.readyState == 4 && request.status == 200){
  // xml process
  }
}
request.open("GET", "nagoya_modern.xml");
request.send();
```

client

Nagoya
modern
.xml

Nagoya agoya kazaki
modern toilet toilet ··· Save to public
  .xml   .xml .xml   directory

Execute one time per
one day by crontab

OSM server   System server

curl –o nagoya_modern.xml http://www.overpass-api.de/api/xapi?*
[bbox=136.8798,35.1526,136.939,35.1913] [building:architecture=modern]
curl –o nagoya_toilets.xml http://www.overpass-api.de/api/xapi*
[bbox=136.8798,35.1526,136.939,35.1913] [amenity=toilets] Third argument
...   Location range   OSM tag   is OSM API

Fig. 4. Mechanism of getting OSM data by cache

The access logs were collected by Google Analytics and our system itself. Google Analytics collected page views, user visits, etc. Our system collected impressions, clicks, etc.

First, we analyzed the access logs collected by Google Analytics. Fig. 5 shows a total of 2,691 page views over the 65 days. The daily average was 41. Our system's demand was constant, and its largest amount of access was on August 10 and September 29.

Figure 6 shows new visits, which received a high value, because users only accessed our system when they visited the event site the first time. This shows that our system is useful for first time visitors who want to geographically learn about the event information. Increases in new visits equal an increase in system recognition. This is the result of advertising at event sites and on the web, especially on such SNSs as Facebook and Twitter.

Second, we analyzed the access logs collected by our system itself. There were 222 pieces of event information as of October 6. 156 were related to Nagoya and 66 to Okazaki. Fig. 7-10 shows the event information by tags, the impressions by tags, the clicks by tags, and compares the event information, the impressions, and the click rates.

The event information consists of all of the event information added by the administrators. Deleted event information was not included. The impressions equal the number of times filtering by done by tags. High impressions mean that users have much interest in the tags. Clicks, which are the number of times that

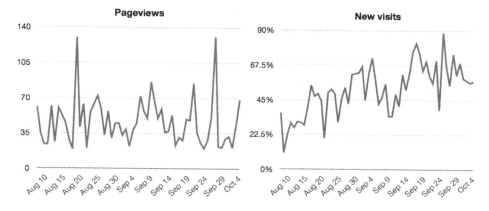

Fig. 5. Page views   **Fig. 6.** New visits

the event information was clicked by users, reflect the valuable event information in the tag. If the clicks are high the tag has much valuable event information. In this case, users need resting places and food information because these impressions are high. Clicks are high in Art because Aichi Triennale 2013 was an artistic event. Resting places and food also received many clicks.

Figure 10 shows that the art and food information numbers are low in spite of high interest in them. Adding more art and food information might increase the user satisfaction of our system.

Finally, we describe the results of our interview surveys with the administrators.

- "The system got good valuations from the event participants, our staff, and the local residents."
- "Your system emphasized some of the hidden charms of our cities."
- "We want to use this system for community planning based on its performance at Aichi Triennale 2013."

From the results of interviews, we confirmed that our system contributed to the vitalization of local communities.

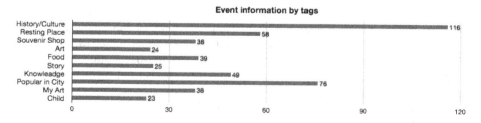

Fig. 7. Event information by tag

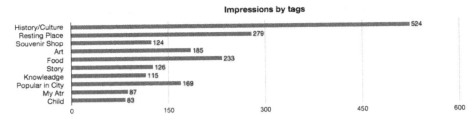

Fig. 8. Impressions by tags

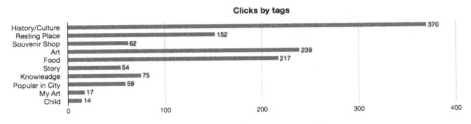

Fig. 9. Clicks by tag

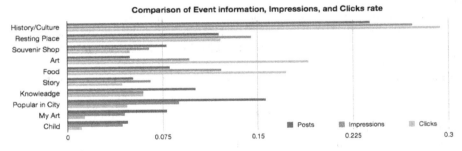

Fig. 10. Comparison of event information, impressions, and click rates

6   Conclusion and Future Work

We proposed and developed a mapping information system that focuses on adding and sharing event information because such information systems are attracting more and more attention. We used OSM as a base map to satisfy the required high flexibility that existing mapping applications lack. Our system was introduced at Aichi Triennale 2013 that was held in the cities of Nagoya and Okazaki in Aichi prefecture from August 10 to October 27, 2013. 222 pieces of event information were added and shared as of October 6, and many were completely new information that could not be discovered in many other existing mapping applications.

Future works must verify access speed by cache mechanism. We will show verification by measured values. Next are filtering evaluations by popularity and a recommender system. This evaluation will be done by analyzing questionnaires.

Finally, we must also search for a way to use our system to vitalize local communities. Our system was developed for a specific event (Aichi Triennale 2013). But since it can provide much other valuable city information, we are searching for ways to use our system as city information maps to share event information.

Acknowledgement. This work is partially supported by the Funding Program for Next Generation World-Leading Researchers (NEXT Program) of the Japan Cabinet Office.

References

1. OpenStreetMap: http://www.openstreetmap.org
2. Hayakawa, T., Imi, Y., Ito, T.: An Analysis on Community Activity based on Geo Information Data of OpenStreetMap in Japan. In: The Fifth International Workshop on Emergent Intelligence on Networked Agents (WEIN 2013) (May 6, 2013)
3. Hayakawa, T.: An Introduction about a Use Case of the OpenStreetMap at Shinshiroshitara Promotion Office, Aichi Prof., Japan (2011),
   http://www.ospn.jp/osc2011-fall/pdf/osc2011fall_OSM_MikawaSankan.pdf
4. Joachims, T., Granka, L., Pan, B., Hembrooke, H., Gay, G.: Accurately interpreting click-through data as implicit feedback. In: The 28th Annual International ACM SIGIR Conference on Research and Development in Information Retrieval (SIGIR 2005), August 15-19 (2005)
5. Segaran, T.: Programming Collective Intelligence. O'Reilly Media (August 2007)
6. Miller, C.C.: A Beast in the Field: The Google Maps Mashup as GIS/2. Cartographica: The International Journal for Geographic Information and Geovisualization 41(3) (Fall 2006)
7. Cramer, H., Rost, M., Holmquist, L.E.: Performing a check-in: Emerging practices, norms and 'conflicts' in location-sharing using foursquare. In: MobileHCI 2011 Proceedings of the 13th International Conference on Human Computer Interaction with Mobile Devices and Services, pp. 57–66 (2011)
8. Suzuki, R., Kawaguchi, S., Otsuka, T., Ito, T.: An Implementation to Smartphone of A Geolocation Information-sharing System. In: The 74th National Convention of IPSJ (2012)
9. Aichi Triennale 2013 Official Web Site: http://aichitriennale.jp/english/
10. NPO: Aichi Triennale 2013 City Vitalization Project: http://machitori2013.com

A Non-invasive Approach
to Detect and Monitor Acute Mental Fatigue

André Pimenta, Davide Carneiro, José Neves, and Paulo Novais

CCTC/DI - Universidade do Minho
Braga, Portugal
apimenta@di.uminho.pt,
{dcarneiro,jneves,pjon}di.uminho.pt

Abstract. In our day to day, we often experience a sense of being tired due to mental or physical workload. Along with that, there is also a feeling of degrading performance, even after the completion of simple tasks. These mental states however, are often not felt consciously or are ignored. This is an attitude that may result in human error, failure, and may lead to potential health problems together with a decrease in quality of life. States of acute mental fatigue may be detected with the close monitoring of certain indicators, such as productivity, performance and health indicators. In this paper, a model and prototype are proposed to detect and monitor acute acute fatigue, based on non-invasive Human-computer Interaction (HCI). This approach will enable the development of better working environments, with an impact on the quality of life and the work produced.

Keywords: Acute Mental Fatigue, Human Computer Interaction, Behavior Biometrics, Monitoring, Pattern Recognition.

1 Introduction

Fatigue is regarded as one of the main causes of human error. Its symptoms are frequently ignored, as well as its importance for a good mental and physical condition, key for human performance and health. Mental fatigue is usually characterized by a lack of mental energy, a feeling of tiredness and drowsiness [1], mental exhaustion, loss of initiative and difficulties concentrating. This lack of focus is also often the source of errors that would easily be avoided in a normal situation [2]. In more objective terms, a person affected by mental fatigue also faces a performance loss as well as an increase in the number of errors and increased difficulty processing information. Thus, mental fatigue must be taken into serious consideration, especially in critical scenarios such as the ones of operating vehicles and machines, or in jobs of high risk/responsibility.

Mental Fatigue can occur at any moment and its effects may persist from only a few hours to several consecutive days. Depending on its duration and intensity, fatigue can make the carrying out of daily tasks increasingly difficult or even impossible [3]. In severe or prolonged cases it can cause illnesses such as depression or chronic fatigue syndrome.

A. Moonis et al. (Eds.): IEA/AIE 2014, Part II, LNAI 8482, pp. 338–347, 2014.

There are many possible causes for mental fatigue: excessive and prolonged brain activity, poor nutrition, significant changes in one's environment and irregular or inadequate sleep patterns have the power to cause fatigue of the mind and the body. Other possible causes for fatigue include recreational drugs, alcohol, and specific medications. However, aspects such as the accumulation of physical exercise, monotony, prolonged discomfort or stress peaks should also not be despised as chief causes of acute mental fatigue [4].

Very often fatigue is simply ignored and seen as yet another effect of our busy and active lifestyle. It is not recognized by people as a medical issue. However, people who suffer from fatigue should remember that it may be a sign that a more serious underlying health problem is occurring. People who experience continued or chronic fatigue should discuss their symptoms with an expert: medical evaluation is often recommended for these patients, as fatigue could be an indication of chronic fatigue syndrome or other serious health problems [5,1].

Given this problem and its context, an approach to detect and monitor fatigue is proposed. It aims to detect different degrees of mental fatigue of an individual in a non-invasive and transparent way, focusing especially on acute mental fatigue in which the loss of performance and increase of errors are more significant. Specifically, a set of features extracted from the individual's interaction with the computer that can be affected by fatigue are studied. These features describe the way the individual uses the computer's peripherals and include the velocity and acceleration of the mouse, the typing rhythm, the distance travelled by the mouse, among others. A similar approach has already been employed successfully in previous work to study the influence of stress on such behaviours [6,7].

Undeniably, the detection and classification of fatigue will be based on notion of Behavioural Biometrics, depicted further ahead. Specifically, it relies on keystroke and mouse dynamics. A simple logger application was developed that acquires information about each mouse and keyboard event (e.g. mouse button down, mouse button up, key down, key up, mouse movement), with the detail needed to generate the features pointed out below. Following this approach it is possible to collect data that will allow to learn the behavioural patterns of interaction with the keyboard and mouse of each user, in a non-intrusive way [8].

This non-invasive and transparent approach on the analysis of fatigue will open the door to the development of better and intelligent working environments, that are sensitive to their users' mental states. This may improve the decision-making of team managers and increase the productivity of the team as well as the quality of its work. This will also have a positive effect on the quality of life of the members of the working community which should not be despised.

1.1   Human-Computer Interaction

Personal computers are increasingly present in our lives. They are used daily in our workplace, in our homes, for entertainment or to get informed. As they gradually take over the space of other objects such as the television, the radio or books, they start to become an irreplaceable part of our lives. In this transformation, intuitive and natural interaction mechanisms between humans and computers are fundamental. This is the main goal of Human-Computer interaction (HCI): to improve the interaction between users and computers by making computer interaction mechanisms adapt to the users and the tasks instead of the other way around. In a more long-term perspective, HCI aims to design systems that minimize the barrier between the Human's cognitive model of what they want to accomplish and the computer's understanding of the user's task, as well as help in solving real world problems [9].

An especially interesting topic in HCI is affective computing: the branch of artificial intelligence that aims to design computer systems that are capable of recognizing, interpreting and processing human emotions and behaviours. Under this approach, interaction can be driven by (among other issues) the emotional state of the user, resulting in computer systems that are sensitive to stress, tension, satisfaction or fatigue and are able to deliver better and contextualized services.

Affective computing essentially aims to make human-computer interaction closer to human-human interaction [10]. In human-human interactions factors such as emotions, body language or signs of stress are taken into consideration. Often, these factors influence the decisions and the course of action of an person. The inclusion of these issues in computer models will allow to develop systems that care for the user's problems, that understand the motivation of the users and that communicate in a more appropriate manner. While computers will never learn to care or feel *for real* [11], the modelling and simulation of some of these traits could result advantageous.

1.2 Behavioural Biometrics

Behavioural Biometrics is the science that seeks to identify individuals based on their behaviour while carrying out daily tasks. Its underlying assumption is that each individual behaves differently and that these behaviours are unique enough to identify us with a satisfiable degree of certainty.

Just like physical biometrics look at the size of the hand, the fingerprint or the iris, behavioural biometrics look at features like the way we talk, the way we look at a screen or the way we walk. Moreover, such behaviours tend to vary with factors such as mood, fatigue or stress [12]. Thus, behavioural biometrics allow to, not only identify the individual, but also identify specific states of an individual. This analysis of behavioural patterns may be conducted in a non-intrusive and non-invasive way, by the mere observation of the user.

2   A Framework for Monitoring the Effects of Mental Fatigue

The main aim of the proposed framework is to detect symptoms of mental fatigue and, through notifications or other actions on the environment, prevent eventual errors or accidents that frequently result from fatigue. To assess the level of fatigue, the framework looks at the interaction patterns of the user with the keyboard and the mouse. This process is implemented using a background application that captures the events fired by the keyboard and mouse in an entirely transparent way. This way, individuals are able to carry out their tasks as usual, without being influenced by the monitoring. The features considered are used by or inspired in behavioral biometrics. In what concerns the keyboard, the features considered in this work are:

- Keydown time: the time during which a key is pressed down while typing; Units: Milliseconds (ms)
- Time between keys: the time between the release of a key and the pressing of the following one; Units: Milliseconds (ms)
- Errors per key: the number of times that the backspace or delete keys are used in comparison with the remaining keys;

Concerning the mouse, the features considered in this work are:

- Mouse Velocity: the velocity at which the cursor of the mouse travels in the screen; Units: Pixel/Milliseconds (px/ms)
- Mouse Acceleration: the acceleration of the cursor of the mouse at a given time; Units: Pixel/Milliseconds^2 (px/ms^2)

These features were selected from a wider group in a previous study [13], given the statistically significant effects that fatigue produced on them.

2.1   The Architecture

The architecture of the proposed system includes not only the simple acquisition and classification of the data, but also a perception of the environment. Users may interact with devices, that are integrated into the environment and provide information about users, their interaction patterns and their surroundings and context.

This system is organized into several layers that separate the acquisition and the data processing tasks:

- Data Sensing - The Data Sensing layer is responsible for acquiring data that characterizes the user in terms of the features studied, that describe their behavioural patterns;
- Data Processing - This layer processes and transforms the data to be sent to the next layer, synchronizing data from different sources and constructing the appropriate software objects. It also filters outlier values that would have a negative impact on the analysis (e.g. a key pressed for more than a certain amount of time);
- Classification - This layer is responsible for interpreting data from the mental fatigue indicators and build the meta-data that will support decision-making. To do it, this layer uses the machine learning mechanisms detailed below;

— Data access - This layer is responsible for providing structured access to the data of each user and managing their complete information. It provides, not only, access to this data in real-time but also provides access to a behavioural historic and profile of each user, allowing studies within longer time frames;
— Presentation - This layer includes the mechanisms to build intuitive and visual representations of the users' state.

The classification of the mental state of a user is achieved through the use of the k-Nearest Neighbour algorithm (k-NN). It is a method of classification based on closest training samples in the feature space. The data used to train the model used by the k-NN were acquired from the study detailed in the Case Study section. The classification model uses the following features: Mouse Acceleration, Mouse Velocity, Keydown Time, Time Between Keys and Error per Key. These are the features that showed more statistically significant differences due to fatigue in the studies conducted.

The process of analysing and classifying the behaviours is detailed in the work flow depicted in Figure 1. It starts with the acquisition of data from the mouse and the keyboard and continues with its processing and filtering in real time. The results of this classification are depicted graphically in real-time through the interfaces developed and are also stored in a database for future use.

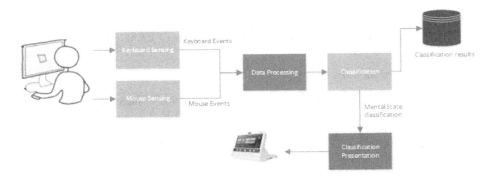

Fig. 1. The flow of data in the Mental Fatigue Monitoring Framework

It was also an objective to make these interfaces intuitive and simple to read. Indeed, the analysis of the information compiled should not constitute, in itself, a drawback. Despite the complexness of the information compiled, the it is shown, in a first instance, in a very minimal and intuitive interface. There is also no explicit or conscious interaction with the system: the events fired by the use of the mouse and the keyboard are stored and analysed entirely in background. The whole system runs in background, and only the output (classified mental state) is shown, through an icon depicting the mental state, as detailed in Figure 2.

It is possible to observe the current mental state of the user as well as the historic. For this purpose, it was developed a second graphical interface that shows the history of the user's fatigue, its current value, as well as the intensity of previous values. Both the mental states and their intensity are the result of the classification process. An example of a monitoring session is shown in Figure 3, where it is possible to observe the current mental state and respective intensity of the user, as well as its historic.

Fig. 2. Icon located on the taskbar of the Operating System, showing the current state of the user with a minimal interface. The arrow points to all the icons used to characterize the state.

3   Case Study

For the work detailed in this paper, the system was installed and used in a real working environment with the objective of monitoring in acute mental fatigue. This type of fatigue is common in nowadays working environments, in which the pressure of timelines, competition and challenging objectives results on short peaks of mental fatigue, with real-time visible effects on the behaviour of the individuals.

To study these behavioural differences due to acute mental fatigue, a study was conducted with collection of data in two different moments, for each one of the twenty participants. The described monitoring framework was installed in the computers of the participants and kept running in the background, having no effect on the normal behaviour and carrying out of the work of the participants. The participants were volunteer researchers from the University of Minho, performing their regular tasks at their laboratories, aged between eighteen to fifty. All these individuals were familiar and had previously used the computers.

The first moment of data collection took place in the morning, when participants were fully rested and starting their regular day of work. The second moment took place at the end of the day of work. Both collection moments took place in the same day for each participant. At the end of the day the user feels the effects of fatigue due to a full day's work [14]. In this work we argue that these effects are also felt and measurable in the user's interaction patterns, as described in the Results section.

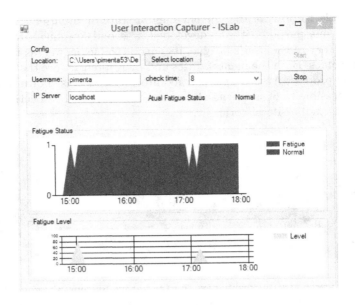

Fig. 3. The monitoring system displaying the current mental state and respective intensity, as well as the historic of the current monitoring session. It is possible to see peaks of mental fatigue around 15:00 and 17:00 hours.

4   Results

To check if there really is an effect on the behaviour of the participants between the two moments of data collection, a statistical analysis of the data was carried out.

First, it was determined, using the Pearson's chi-squared test, that most of the distributions of the data collected are not normal. In that sense, the Mann-Whitney test is used to test the hypothesis described ahead. This test is a non-parametric statistical hypothesis test for assessing whether one of two samples of independent observations tends to have larger values than the other, and thus prove the existence of distinct behaviours. The null hypothesis considered is: H_0 represents the equal medians of the two distributions. For each two distributions compared, the test returns a $p-value$, with a small $p-value$ suggesting that it is unlikely that H_0 is true. Thus, for every Mann-Whitney test whose $p-value < \alpha$, the difference is considered to be statistically significant, i.e., H_0 is rejected. In this work, a value of $\alpha = 0.05$ is considered.

The results obtained through this approach show that the handling of the keyboard and mouse in a normal state (without fatigue) and a state of mental fatigue are different. A significant difference between the data collected in the two different moments was observed which proves the existence of different behaviours. Figures 4 and 5 show that the use of mouse and keyboard is in generally slower. In these images, data from the first moment of data collection is depicted in green and the from the second moment is depicted in red. Figure 4 depicts an increase in the *keydown time* feature, which quantifies the time, in milliseconds, that a key is pressed down while typing. Figure 5 shows a decrease in the velocity of the mouse, measured in pixels/milliseconds. This velocity is computed between each two consecutive clicks. Both features show a statistically significant decrease in performance.

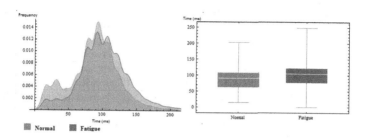

Fig. 4. Histograms and Box Plots comparing the data of the two distributions for the feature Keydown Time: fatigued individuals tend to write slower

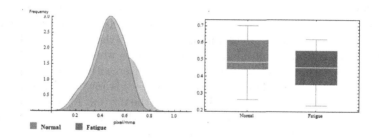

Fig. 5. Histograms and Box Plots comparing the distributions of the data collected in the two moments for the feature Mouse Velocity: fatigued individuals move the mouse slower

Table 1 depicts the results of the Mann-Whitney test for each of the features, when comparing the distributions of the data in the two collection moments. It also shows the trends observed in the participants (e.g. 100% of the participants evidence an increase in the keydown time due to fatigue) and the values of the mean and medians for each of the features.

In a first instance these results prove, once more, that the presence of mental fatigue is accompanied by a loss of performance and an increase in errors. This is not surprising nor was it the main objective of this work. More important in this scope is the fact that it is indeed possible to detect mental fatigue through the way people use the keyboard and the mouse. Moreover, this detection can be carried out using non-invasive tools, in real-time.

Table 1. Results of the statistical analysis of the data for the 20 participants. Only the features that have shown significant differences are included. The "trend" column depicts the percentage of participants that have a given trend. For example, the mean value of the velocity of the mouse decreases for 90% of the students, when fatigued.

Metric		Normal	Fatigued	Trend	p-value
Keydown Time	Mean:	79.827	87. 119	Increases in 100%	$0.7 * 10^{-4}$
	Median:	77.601	81.502	Increases in 60%	
Time between keys	Mean:	469. 193	1040.26	Increases in 100%	$1.23 * 10^{-144}$
	Median:	215.75	386.55	Increases in 90%	
Mouse Acceleration	Mean:	0.4238	0.3829	Decreases in 90%	$3.01 * 10^{-11}$
	Median:	0.2202	0.2010	Decreases in 100%	
Mouse Velocity	Mean:	0.5002	0.4401	Decreases in 90%	$5.03 * 10^{-15}$
	Median:	0.2680	0.2537	Decreases in 100%	
Time between Clicks	Mean:	3081.35	3257.61	Increases in 50%	$5.8 * 10^{-4}$
	Median:	1733.30	1863. 15	Increases in 50%	
Error per key	Mean:	7.643	9.002	Increases in 90%	$2 * 10^{-2}$
	Median:	7.444	8.598	Increases in 90%	

5 Conclusion

This paper detailed a framework for the detection and monitoring of fatigue based on pure Human-computer Interaction. It configures a process that is non-invasive and transparent to the users, relying on the sheer observation of the behaviour. Through the use of mouse and keyboard and the respective behavioural biometrics, keystroke dynamics and mouse dynamics were observed and used to analyse the different patterns of interaction of a user with the computer.

The results obtained evidence not only an already known and expected effect of fatigue on the user's performance throughout the day but also, and more interestingly, that it is possible to measure and classify these effects, in real time. This work opens the door to the development of leisure and work environments that are sensible to their user's level of fatigue. From this point on we envision the creation of decision-support systems that will improve the performance and quality of life of the individuals by suggesting better time-management strategies that optimize the work schedule. Additionally, within the context of the CAMCoF project, the long-term goal is to develop environments that are autonomous and take actions concerning the management of the environment towards the minimization of fatigue and the increased performance of a group of individuals, by adjusting aspects such as work schedules, ambient sound, pauses or individual musical selection.

Acknowledgments. This work was developed in the context of the project CAMCoF - Context-aware Multimodal Communication Framework funded by ERDF - European Regional Development Fund through the COMPETE Programme (operational programme for competitiveness) and by National Funds through the FCT - Fundação para a Ciência e a Tecnologia (Portuguese Foundation for Science and Technology) within project FCOMP-01-0124-FEDER-028980.

References

1. Hossain, J.L., Ahmad, P., Reinish, L.W., Kayumov, L., Hossain, N.K., Shapiro, C.M.: Subjective fatigue and subjective sleepiness: Two independent consequences of sleep disorders? Journal of Sleep Research 14(3), 245–253 (2005)
2. van der Linden, D., Eling, P.: Mental fatigue disturbs local processing more than global processing. Psychological Research 70(5), 395–402 (2006)
3. Lorist, M.M., Klein, M., Nieuwenhuis, S., De Jong, R., Mulder, G., Meijman, T.F.: Mental fatigue and task control: Planning and preparation. Psychophysiology 37(5), 614–625 (2000)
4. Kobayashi, H., Demura, S.: Relationships between Chronic Fatigue, Subjective Symptoms of Fatigue, Life Stressors and Lifestyle in Japanese High School Students. School Health 2, 5 (2006)
5. Joyce, J., Rabe-Hesketh, S., Wessely, S.: Reviewing the reviews: The example of chronic fatigue syndrome. JAMA: The Journal of the American Medical Association 280(3), 264–266 (1998)
6. Carneiro, D., Castillo, J.C., Novais, P., Fernández-Caballero, A., Neves, J.: Multimodal Behavioural Analysis for Non-invasive Stress Detection (July 2012)
7. Novais, P., Carneiro, D., Gomes, M., Neves, J.: Non-invasive Estimation of Stress in Conflict Resolution Environments. In: Demazeau, Y., Müller, J.P., Rodríguez, J.M.C., Pérez, J.B. (eds.) Advances on PAAMS. AISC, vol. 155, pp. 153–159. Springer, Heidelberg (2012)
8. Yampolskiy, R.V., Govindaraju, V.: Behavioural biometrics: A survey and classification. International Journal of Biometrics 1(1), 81–113 (2008)
9. Picard, R.W., Wexelblat, A., Clifford, I., Nass, C.I.N.I.: Future interfaces: social and emotional. In: CHI 2002 Extended Abstracts on Human Factors in Computing Systems, pp. 698–699. ACM (2002)
10. Picard, R.W.: Affective computing: Challenges. International Journal of Human-Computer Studies 59(1), 55–64 (2003)
11. Picard, R.W., Klein, J.: Computers that recognise and respond to user emotion: Theoretical and practical implications. Interacting with Computers 14(2), 141–169 (2002)
12. Jain, A., Bolle, R., Pankanti, S.: Introduction to biometrics. In: Biometrics, pp. 1–41. Springer (2002)
13. Pimenta, A., Carneiro, D., Novais, P., Neves, J.: Monitoring mental fatigue through the analysis of keyboard and mouse interaction patterns. In: Pan, J.-S., Polycarpou, M.M., Woźniak, M., de Carvalho, A.C.P.L.F., Quintián, H., Corchado, E. (eds.) HAIS 2013. LNCS, vol. 8073, pp. 222–231. Springer, Heidelberg (2013)
14. Rogers, A., Spencer, M., Stone, B., Britain, G.: Validation and development of a method for assessing the risks arising from mental fatigue. HSE Books Sheffield (1999)

Vitae and Map Display System for People on the Web

Harumi Murakami[1], Chunliang Tang[1], Suang Wang[1], and Hiroshi Ueda[2]

[1] Graduate School for Creative Cities, Osaka City University
3-3-138, Sugimoto, Sumiyoshi, Osaka 558-8585 Japan
harumi@media.osaka-cu.ac.jp
http://murakami.media.osaka-cu.ac.jp/
[2] ATR Creative Inc.
2-2-2, Hikaridai, Seika, Soraku, Kyoto 619-0288 Japan

Abstract. We present a system that displays a curriculum vitae with a map to understand people. Our method is based on the following processes: (1) creating curriculum vitae using related work [1], (2) extracting the names of places where the person studied and worked from the vitae, (3) getting such location information as latitudes, longitudes, and addresses from the place names using Google Maps API, and (4) displaying a vitae along with a map using Google Maps JavaScript API. We developed a prototype and evaluated our algorithms that extract place names and convert them into location information from web search results for 56 person names.

Keywords: web people search, curriculum vitae, resume, map display, location information.

1 Introduction

Due to an increase in the number of people about whom the web can provide information, the popularity of web searches that identify people continues to rise. Such detailed and organized information might provide valuable insight into search targets.

In this paper, we develop an interface that extracts such detailed information as personal histories and organizes it into a vitae with a map to understand people. We obtain important, relatively long-term location information for the person and focus on such life stages as schools and workplaces.

Extracting schools and workplaces from web search results is difficult. We must solve the following main problems: (a) finding an "event-time-place" tuple for the person, (b) removing redundant information, and (c) obtaining location information that matches the extracted place names.

For (a) and (b), we use related work [1], which creates a curriculum vitae when a person's name is input and gives web search results. We create a vitae and extract place names (schools and workplaces) from it. For (c), we use Google Maps API to obtain location information from the extracted place names.

A. Moonis et al. (Eds.): IEA/AIE 2014, Part II, LNAI 8482, pp. 348–359, 2014.

Below, we explain our approach and give examples of our implemented prototype in Section 2 and describe our experiments in Section 3. We discuss the significance of our research in Section 4. The examples in this paper were translated from Japanese into English for publication.

2 Approach

2.1 Overview

Our approach is based on the following processes: (1) creating curriculum vitae using related work [1], (2) extracting the names of places where the person studied and worked from the vitae, (3) getting location information from the place names using Google Maps API, and (4) displaying a vitae that includes location information along with a map using Google Maps JavaScript API.

Figure 1 shows an overview of our approach.

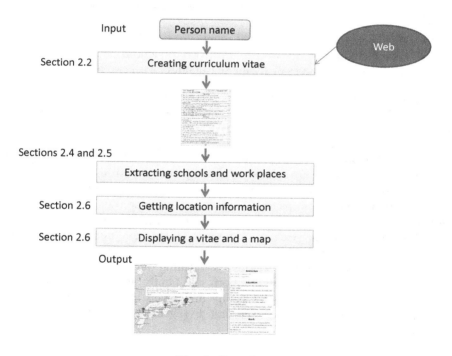

Fig. 1. Overview

2.2 Creating Vitae

The related work [1] is based on the following: (1) extracting event sentences (a character string that includes at least one time and one event) using heuristics and filtering them, (2) judging whether the event sentences are related to the person by SVM mainly using the patterns of HTML tags, (3) classifying these sentences into four categories (background, education, history, and awards) by SVM, and (4) clustering the event sentences including both identical times and events. The history categories contain event sentences except for the background, education, and award categories. Finally, (5) such background information as date and place of birth is generated.

Extracting Event Sentences. First, sentences are extracted from web pages by cutting-sentence heuristics and event sentences are selected from those sentences using sentence-extraction heuristics that include the time (or judging year) listed below.

- The following cutting-sentence heuristics segment character strings:
  - those ending with period
  - those included in *tr*, *li*, *h1-h5*, *title*, *p*, or *div* tags without periods
  - those ending with *br* tags without periods
- The following sentence-extraction heuristics extract sentences that include time (judging year):
  - those including four sequential numbers (e.g., 2000)
  - those including an apostrophe with two sequential numbers (e.g. '00)
  - those including names of Japanese calendars with two numbers with year

Next, 21 stop words (e.g., copyright, post) and 20 stop patterns (e.g., expressing such detailed times as 2008 12-31 12:00:00) were used to filter unnecessary event sentences.

Judging Relation to a Designated Person. Event sentences are judged on the degree to which they are related to a designated person using SVM.

Five person names including politicians and researchers were used as queries and 200 search results (web pages) for each person name (i.e., 5 * 200 = 1,000) were obtained from Google Web APIs. 7,211 event sentences were obtained through the event-sentence-extraction process. They were judged manually and 2,266 correct and 4,474 incorrect training data were obtained.

We learned 31 patterns for SVM as the training data. Example patterns include: "The event sentence includes both the family and given names," "the nearest *h1* to *h5* tags include the family and given names," "the number of nouns inside the *title* tag," and "the first *tr* tag in the *table* tag includes the family name."

Classifying Event Sentences into Four Categories. Using the one-versus-rest method of SVM, event sentences, which are judged to be related to a designated person, are classified into four categories: background, education, history, and awards.

We obtained 500 web pages that were expected to include profile information for the training data: 200 for the query "personal history site:ja.wikipedia.org", 200 for the query "personal history inurl:profile", and 100 for the query "alma mater site:read.jst.go.jp". After the event-sentence-extraction process, 14,974 event sentences were obtained and classified into the four categories. We morphologically analyzed the event sentences and calculated the tf-idf values. The idf values were the number of event sentences. The tf-idf values and the number of terms were used to form SVM patterns. The non-linear SVM was used for each category.

Clustering Event Sentences Including Both Identical Times and Events. Next we group event sentences with the same meanings. For example, we must combine the following sentences: "A entered X university in 1982" and "In April 1982, A enrolled at Univ X." We must also distinguish the combined sentences from such examples as, "A graduated from X university in 1986." Since these sentences share many words, standard clustering algorithms may not be able to distinguish them.

We focus on time and cluster event sentences. The tf-idf, cosine, and single-path methods were used for the clustering. The following is the algorithm.

Step 1 Event sentences are obtained that include "year, month, and day." Each year-month-day cluster is generated, and event sentences are clustered in year-month-day clusters.

Step 2 Event sentences, which include "year and month" and years/months that are identical with the existing clusters, are clustered by the single-path method.

Step 3 Event sentences, which include "year" and years identical with the existing clusters, are clustered by the single-path method.

Step 4 Each event sentence, which includes a "year and month" and does not belong to any existing clusters, becomes a year-month cluster. The clusters with identical years/months are clustered by the single-path method. Event sentences that include "year" and years identical with the generated clusters are clustered by the single-path method.

Step 5 Each event sentence, which includes a "year" and does not belong to any existing clusters, becomes a year cluster. The clusters with identical years are clustered by the single-path method.

Generating Background Information. We extracted the dates of birth and death from the background category by simple heuristics.

- date of birth: select a cluster that includes the most frequent "date of birth," "birthday," or "being born" and extract a date as "date of birth" from the cluster.
- date of death: select a cluster that includes the most frequent "date of death," or "being dead" and extract a date as "date of death" from the cluster.
- place of birth: extract the most frequent prefecture name in the cluster from which the date of birth was extracted.

Example of Created Vitae. Figure 2 shows the process for creating a vitae for *Naoto Kan*, a former Japanese prime minster. On the left is shown a created vitae from the implemented related work [1]. The background, education, and history categories are displayed. The education and history categories include event sentence clusters.

When a user selects [+], the clustered event sentences are displayed. When a user selects "L", the original web page containing the event sentence is displayed.

Since the amount of event sentence clusters in a history category is large, more than 11 clusters are hidden. When a user selects "more," the hidden clusters will be displayed.

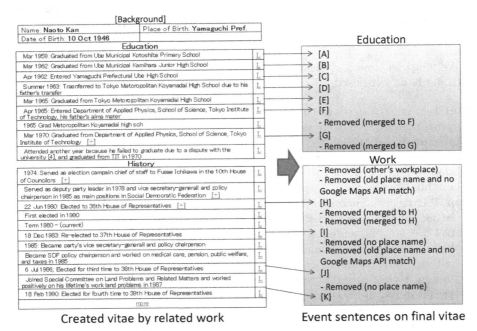

Created vitae by related work   Event sentences on final vitae

Fig. 2. Creating vitae for Naoto Kan

2.3 Issues of Using Vitae Created by Related Work

There are three main issues for extracting place names from the created vitae: extracting (1) schools from the education category, (2) schools from the history category, and (3) workplaces from the history category.

(1) Extracting schools from the education category is the easiest among the three since the schools themselves are relatively easy to extract because character strings for them are typical and such events are also typical (e.g., enter, graduate, transfer). The main problem here is filtering the same event sentences as those that weren't filtered by the previous step (creating vitae). Identifying locations from ambiguous place names (e.g., universities with multiple campuses) is another important problem; however, this paper ignores it.

(2) Extracting schools from the history category (as workplaces) is slightly more difficult, since such events include various positions (e.g., RA, researchers, professors), degrees, and institutes.

(3) Extracting workplaces from the history category is the most difficult step in our research. This is why the related work used the "history" category instead of the "work" category. Working history is ambiguous, and workplaces are ambiguous as well. For example, when a politician joins a party, is this treated as working history? If he is elected, is this working history? If so, what are appropriate working places, party offices, private offices, or parliament, etc.?

Extracting schools and workplaces from event sentences are described in Sections 2.4 and 2.5.

2.4   Extracting Schools

The process for extracting schools is listed below.

Step 1 Event sentences are extracted that contain the following: "(a piece of) school," or "(a piece of) graduation."
Step 2 Morphological analysis was performed on the event sentences.
Step 3 School candidates were extracted from the event sentences.

> **3-a** When nouns are judged to be organizations or locations in the event sentences:
> The following nouns are connected: those starting from being judged to be organizations or locations and those ending at these noun phrases: (a) "graduate school" when it contains "graduate school;" (b) "university" or "college" when it contains "university" or "college" but not "school;" (c) otherwise, "school."
> e.g.) "Ube Municipal Kotohira Primary School"
> **3-b** When no noun exists that is judged to be an organization or a location in the event sentences:
> The following nouns are connected (with backtracking) starting at these noun phrases: "graduate school," or "school."
> e.g.) "Azabu High School"

Step 4 Identical school candidates were filtered.

> **4-a** The school candidates are grouped into three categories if they contain the following terms: (1) "enter," (2) "transfer," or (3) "graduation," or "graduate."
> **4-b** The character strings of school candidates in each category are compared, and when all the characters of a school candidate are included in other school candidates, the shorter candidate is removed.
> e.g.) "Tokyo Metoropolitan Koyamadai High School" and "Metoropolitan Koyamadai High School" are compared, and the latter is removed.

2.5 Extracting Workplaces

The process for extracting workplaces is listed below.

Step 1 Event sentences are extracted that contain the following keywords: "elected," "position," "resignation" (for politicians), "university," "school," "research" (for researchers), "join," "traded," "retirement" (for athletes), "role," "being in charge," (for entertainers), "employed," "joining a company," "leaving a company" (for company workers).

Step 2 Morphological analysis was performed on the event sentences.

Step 3 Event sentences are removed that do not belong to the designated person. When a noun is judged to be a "family name" that is different from the designated person with Japanese particles that link a subject with nouns, the event sentence is removed.

Step 4 When one or more nouns are judged to be organizations in the event sentences:

    4-a When there is only one noun, it is extracted.

    e.g.) "House of Representatives"

    4-b With more than one noun, the nearest one to the keyword is extracted.

Step 5 When no noun is judged to be an organization in the event sentences:

    5-a When "company limited" or "Co. Ltd." are found, the nouns before and after the character strings are connected.

    e.g.) "Mikasakosan Co. Ltd." and "Kurehaboseki Company Limited"

    5-b When "minister" is found, a minister's name is extracted from a minister dictionary.

    e.g.) "Transportation Minister"

    5-c When "mayor" or "governor" and nouns judged to be locations are found, the nearest noun and "city" (for mayor) or "prefecture" (for governor) are connected.

    e.g.) "Akune City"

Step 6 When "university," "school," or "research" (for researchers) are found in the event sentences:

    6-a The same process as in Section 2.4 was performed. For 3-b, the noun phrases of the starting point of the backtracking were "graduate school" and "research center."

    e.g.) "Graduate School of Information Science and Technology, Nara Institute of Science and Technology" and "National Institute of Informatics"

2.6 Getting Location Information and Displaying Vitae and Maps

Each place name obtained in Sections 2.4 and 2.5 was converted to location information (latitudes, longitudes and addresses) using the Google Maps API v3.

Next, the vitae created in Section 2.2 was modified. We extracted the event sentence clusters that include the above place names and location information

(Fig. 2 (right)). Seven out of the nine event sentence clusters in the education category are extracted as A to G. The seventh event sentence clusters are filtered since they are combined with the previous one. For the history category, many event sentence clusters were removed. For example, the first-event sentence cluster was removed since there was no workplace for the designated person. The second-event sentence cluster was filtered since there is no workplace, or exactly speaking, there was an old work place name without a hit in the Google Maps API.

Although related work [1] failed to create work categories, this research successfully identified important work history by extracting workplaces.

Finally, a vitae that includes location information along with a map is displayed using Google Maps JavaScript API v3.

2.7 Prototype

Based on the person name input, we developed a prototype that displays (1) a vitae with background information, education, and work information that includes location information in the event sentences, and (2) a map on which icons express places included in the event sentences.

When the user selects an event sentence in the vitae or an icon on the map, the event sentence, the place name, and the address hit in the Google Maps API are displayed.

Figure 3 shows an example for *Naoto Kan*. Almost all (7/9, 78%) the event sentence clusters in the vitae's education category are displayed in the education category, and the selected (19/41, 46%) event sentences in the vitae's history category are displayed in the work category.

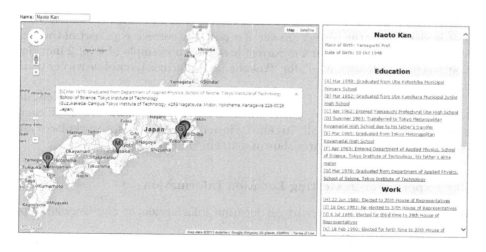

Fig. 3. Prototype

3 Experiment

3.1 Dataset

We used the names (mostly famous) of 56 Japanese people: 17 politicians, 14 athletes, 12 entertainers, 10 researchers, 1 entrepreneur, 1 manga artist, and 1 historical figure. We obtained 50 web search results for each person name and removed unrelated web pages. Our implemented [1] method generated 56 vitae.

We identified the correct data (place names) included in the vitae.

3.2 Experiment 1: Extracting Place Names

The following are the evaluation measures for extracting place names (schools and workplaces):

$$Precision = \frac{\text{correct extracted place names}}{\text{extracted place names}} \tag{1}$$

$$Recall = \frac{\text{correct extracted place names}}{\text{correct place names to be extracted}} \tag{2}$$

Note that correct place names are duplicated. For example, when person A entered and graduated from X university, it counts as two in the education category. If person B also entered X university, it counts as three in the education category. If person A worked at X university, it counts as three in the education category and one in the work category.

As stated before, since workplaces are ambiguous, we defined some guidelines to judge them in this research.

In addition, we classified their levels of correctness. 1 is exact: extracted place name is identical as the correct answer. 2 is partially correct: e.g., part of the correct answer, abbreviation for the correct answer. An example of level 2 includes "Marinos (soccer team name)" for "Yokohama F Marinos (complete soccer team name)."

Table 1 shows the result. The results of extracting schools are fine (Precision: 91-97% and Recall: 84-90%), however, those of the workplaces are insufficient (Precision: 65-73% and Recall: 70-90%). Since our guidelines for workplaces are not strict enough, these figures show maximum values.

3.3 Experiment 2: Getting Location Information

We used the correct answers to get location information by the Google Maps API v3.

We judged the top ranked results. For schools, 0.90 (53/59), and for workplaces, 0.53 (81/154) were correct.

The most common error for workplaces was the name of political parties, e.g., LDP. Party offices were sorted from north to south in the Google Maps API

Table 1. Experiment 1 results

	Precision		Recall	
	p1	p2	r1	r2
Schools	0.91	0.97	0.84	0.90
	(59/65)	(63/65)	(59/70)	(63/70)
Work places	0.65	0.73	0.70	0.90
	(154/241)	(166/241)	(154/233)	(166/233)

and the northernmost offices were top ranked. However, most correct offices are located in Tokyo, in central Japan.

If school names were correctly extracted, Google Maps API returned good results. For workplace names, the result was poor. We need to try different geocoding services or consider better algorithms and create dictionaries using Google Maps API.

4   Related Work and Discussion

WePS-3 conducted a competitive evaluation on person attribute extraction on web pages [2]. Schools were included in the extraction candidates, but not workplaces since they are ambiguous.

Murakami et al. [3] only extracted "one representative" piece of location information, which is typically the most current places from the web search results. Our research extracts schools and workplaces about the designated person.

Other research assigns information to learn about people. Wan et al. assigned titles (similar to vocations) [4]. Ueda et al. assigned vocation-related information including vocations, organizations, and works [5], Mori et al. assigned keywords [6], and Murakami et al. assigned library classification numbers [7] to person clusters. This research is part of a project to develop interfaces to select and understand people on the web [8].

Some research extracted event sentences from web pages. Kimura et al. [9] extracted event sentences to create personal histories and listed them. Our research creates vitae and displays schools and workplaces on a map.

Although related work [1] failed to create work categories, this research successfully identified important work history by extracting workplaces. Therefore, we not only used the related work but also improved it.

Some geocoding services (e.g. [10]) convert place names to location information. In our survey, Google Maps API is currently the best service among them.

We believe that our work's main contribution is that it displays study and work places on a map with curriculum vitae for people on the web. To the best of our knowledge, this is the first research to do so.

Although our research is limited to Japanese, the idea is easily applicable to other languages.

Future work must improve the extraction of workplaces. We also need to consider algorithms to find exact workplaces for geocoding services. The algorithms should be evaluated using different datasets (e.g., ordinary people).

5   Conclusions

We presented a system that displays curriculum vitae on a map to understand people. Our method is based on the following processes: (1) creating curriculum vitae using related work [1], (2) extracting the names of places where the person studied and worked from the vitae, (3) getting location information (latitudes, longitudes and addresses) from the place names using Google Maps API, and (4) displaying a vitae along with a map using Google Maps JavaScript API. We evaluated our algorithms that extract place names and converted them into location information from 56 person name search results.

Acknowledgements. This work was supported by JSPS KAKENHI Grant Number 22500219, 25330385.

References

1. Ueda, H., Murakami, H., Tatsumi, S.: Creating Curriculum Vitae for Understanding People on the Web. Transactions of the Japanese Society for Artificial Intelligence 25(1), 144–156 (2010)
2. Artiles, J., Borthwick, A., Gonzalo, J., Sekine, S., Amigo, E.: WePS-3 Evaluation Campaign: Overview of the Web People Search Clustering and Attribute Extraction Tasks. In: CLEF 2010 (2010)
3. Murakami, H., Takamori, Y., Ueda, H., Tatsumi, S.: Assigning Location Information to Display Individuals on a Map for Web People Search Results. In: Lee, G.G., Song, D., Lin, C.-Y., Aizawa, A., Kuriyama, K., Yoshioka, M., Sakai, T. (eds.) AIRS 2009. LNCS, vol. 5839, pp. 26–37. Springer, Heidelberg (2009)
4. Wan, X., Gao, J., Li, M., Ding, B.: Person Resolution in Person Search Results: WebHawk. In: Proceedings of the Fourteenth ACM Conference on Information and Knowledge Management (CIKM 2005), pp. 163–170. ACM Press, New York (2005)
5. Ueda, H., Murakami, H., Tatsumi, S.: Assigning Vocation-Related Information to Person Clusters for Web People Search Results. In: Proceedings of the 2009 Global Congress on Intelligent Systems (GCIS 2009), vol. 4, pp. 248–253. IEEE Press, New York (2009)
6. Mori, J., Matsuo, Y., Ishizuka, M.: Personal Keyword Extraction from the Web. Journal of Japanese Society for Artificial Intelligence 20, 337–345 (2005)
7. Murakami, H., Ura, Y., Kataoka, Y.: Assigning Library Classification Numbers to People on the Web. In: Banchs, R.E., Silvestri, F., Liu, T.-Y., Zhang, M., Gao, S., Lang, J. (eds.) AIRS 2013. LNCS, vol. 8281, pp. 464–475. Springer, Heidelberg (2013)
8. Murakami, H., Ueda, H., Kataoka, S., Takamori, Y., Tatsumi, S.: Summarizing and Visualizing Web People Search Results. In: Proceedings of the Second International Conference on Agents and Artificial Intelligence (ICAART 2010), vol. 1, pp. 640–643. INSTICC Press (2010)

9. Kimura, R., Oyama, S., Toda, H., Tanaka, K.: Creating Persnal Histories from the Web using Namesake Disambiguation and Event Extraction. In: Baresi, L., Fraternali, P., Houben, G.-J. (eds.) ICWE 2007. LNCS, vol. 4607, pp. 400–414. Springer, Heidelberg (2007)

10. Geocoding Tools & Utilities, `http://newspat.csis.u-tokyo.ac.jp/geocode/`

Gestures as Interface for a Home TV Digital Divide Solutions through Inertial Sensors

Stefano Pinardi and Matteo Dominoni

D.I.S.Co. University of Milano-Bicocca, Italy
{pinardi,dominoni}@disco.unimib.it

Abstract. Seniors are the fastest growing segment of populations not only in many parts of Europe, but also in Japan and the United States. ICT technologies are not very popular among many elderly and also are not designed around their cultural necessities and ergonomic needs. The risk is that in the very near future this growing segment will be digitally isolated, in a society that is more and more based on ICT as infrastructure for service, and communications.

Easy Reach Project proposes an ergonomic application to break social isolation through social interaction to help the elderly to overcome barrier of the digital divide. This paper focuses its attention on the development of the technology and algorithms used as Human Computer Interface of the Easy Reach Project, that exploits inertial sensors to detect gestures.

Many experimental algorithms for gesture recognition have been developed using inertial sensors in conjunction with other sensors or devices, or by themselves, but they have not been thoroughly tested in real situations, they are not devoted to adapt to the elderly and their way of executing gestures. The elderly are not used to modern interfaces and devices, and – due to aging – they can face problems in executing even very simple gestures.

Our algorithm based on Pearson index and Hamming distance for gestures recognition has been tested both with young and elderly, and was shown to be resilient to changes in velocity and individual differences, still maintaining great accuracy of recognition (97.4% in user independent mode; 98.79% in user dependent mode). The algorithm has been adopted by the Easy Reach consortium (2009-2013) to pilot the human machine gesture-based interface.

Keywords: Sensors, Inertial Sensors, Gestures, Human Machine Interfaces, Ambient Intelligence, Assisted Living, Elderly, Social, Digital Divide, Home, TV.

1 Introduction

The goal of this work is to develop an innovative and low cost HMI (Human Machine Interface) to exploit gestures to control a TV Set for smart applications. The HMI interface is part of the "Easy Reach" project, partially supported by AAL JP (Ambient Assisted Living Joint Program). The Easy Reach Project has the aim to help the elderly to interact with each other using ICT technology to break social isolation while at the same time coping with the ergonomic and cultural difficulties related to aging and

A. Moonis et al. (Eds.): IEA/AIE 2014, Part II, LNAI 8482, pp. 360–368, 2014.

the digital divide [1]. Seniors are the fastest growing part of the population in Europe [2], in Japan and in the United States: on one side ICT technologies are not popular among many elderly and on the other ICT applications are not designed around their mentality in order to solve their needs [1]. The risk is that in the very near future the elderly - who are becoming the most numerous segment of the population - will be digitally isolated in a society that is more and more based on ICT as infrastructure both for service, and communications. Easy Reach proposes a solution for social interaction, which is easy from the ergonomic point of view, that help the elderly break the barrier of the digital divide barrier. We use a device with inertial sensors that is handled by the user, to interpret the gestures that pilot the interface. This paper focus his attention about these algorithms and their accuracy.

Young   Elderly

Fig. 1. Gestures: young vs elderly. In black the ideal gesture in red the actual execution. More variations, noise and errors are present in elderly gestures.

The general purpose is to identify gestures considering the time feature, normalizing acquisitions, and making the system robust to variations in gesture execution, introduced both by individual variance and senile deterioration, while maintaining a high accuracy. As we shall see, the proposed method achieves an accuracy of 97.39% on the dataset 1 (8 gestures) in user *independent* mode and an accuracy of 98.79% in user *dependent* mode on the same dataset (both with a threshold of 58%), exploiting only inertial sensors information. The algorithm has been adopted by the Easy Reach consortium to pilot the "Easy Reach" Human Computer Interface.

2   Former Works

One of the first dynamics gesture recognition systems based on HMM and inertial sensors was created by Hofmann et al. in the mid-90s [3]. The application included the use of discrete Hidden Markov Models to reduce complexity, but the recognition algorithm took hours to arrive to its end. The work of Mäntylä et al. [4] is one of the first that uses only accelerometers to recognize both static and dynamic gestures using a sensor box installed on a portable device. The algorithms used for the recognition exploit the HMM and SOM (Self Organizing Map) of Kohonen, the first for dynamics gesture, the second for static ones. The recognition accuracy of the dynamic gestures is quite high, around 97% on average; it is to emphasize that this system use a dataset based on only two people[1].

---

[1] It must be highlighted that results presented must be weighed according to dimension, complexity and cardinality of the dictionaries.

In more recent works Schlömer et al. [5] created a classifier for four distinct gestures using a Nintendo Wiimote exploting a K-means, HMM, and Bayes classifier pipeline, reaching an accuracy of 89.72% on average (84.0%-93.4%). Prekopcsák [6] has an accuracy of 97.4% using an HMM, and 96.0 % using SVM (Support Vector Machines) interpreting the accelerometers data of a Sony Ericcson W910i. These accuracies are high, unfortunately very few is explained about the gestures used, a part from the fact that they involve the wrist and the arm, that is not very informative. Also, these results are obtained on datasets of only four different users. One interesting result is obtained by [7] which has an accuracy of 99.2% on average, using a Bayesian network model on a dataset of 13 different classes of gestures created by 15 people. These results are influenced by an appropriate choice of form of the gestures, adaptably modified to make them easily discernible from each other. For example, the number 7 and 1 are designed to be easily separable, and the number 4 is written not in the usual natural way. Another notable work is [8] that identifies the feature in the frequency domain. The authors reach an accuracy of 98.93 % in a group of 4 gestures, and an accuracy of the 89.29% in a set of 12 gestures using a method called FDSVM (Frame-based Descriptor and multi-class SVM) in user independent mode. In the document it is also shown that DTW (Dynamic Time Warping) and Naive Bayes have lower accuracy than FDSVM in groups of gestures of larger cardinality. Regarding recent papers that describe recognition methods based only on inertial data, we can report the work of Kratz and Rohs [9] that have the accuracy of 80% using 10 gestures created by 12 users, and Chen et al. [10] with an accuracy of 98.8% in user dependent mode using only inertial sensors (implicit) and 85.24 % in user independent mode using only the inertial sensors.

3   Euler Angles

Inertial sensors usually provide two different types of data i) acceleration, angular velocity, magnetic north, or ii) Euler angles. The first type of information is useful to determine the strength and quality of a gesture, for example, to verify the presence of tremors, apparent forces, peaks (e.g. for the identification of falls). This type of data are particularly suited to recognize the dynamic aspects of an action and its quality (cfr. [4] [5] [11]). In this work we use only the Euler angles to detect and classify the executed gestures.

3.1   Individual Variance and Noise

The execution of a gesture is conditioned by three factors:

• Thermal noise: each sensor produces a Gaussian noise due to temperature and electromagnetic fluctuations.
• Position of the sensor: slightly differently modifications of device handling or sensor position sensitively affect data of tilting and positioning of the sensor.
• Model noise (articulated body complexity): the body varies and changes from day to day and react differently all the time, this introduces a significant change in the execution of any gesture. The same gesture repeated in consecutive times, even by the same person, with the same device positioned in the same way, always looks different (see Fig. 2). It is the same effect we have when one person signs his name on paper.

Fig. 2. On the left: same gesture executed by different people. Right: the same gesture executed by the same person. Every gesture is different even if repeatedly executed by the same person.

4   Correlation Methodologies

To correlate gestures we test three different algorithms. The first exploits a Pearson correlation on Euler Angles' Yaw and Pitch. The other two are Hamming and Levenshtein distance. To increase the accuracy we have combined Pearson alternatively with one of the other two algorithms. The correlation algorithms are used both during the extraction phase of the centroids[2], and the recognition phase. Any new gesture is compared with the previously extracted centroids, and the higher scored centroid is referred to as "the recognized gesture". If we do not pass a certain score of confidence (see 5.1, Rejection Threshold) the gesture is consider invalid, this is interpreted as a poorly performed gesture or as a non-intentional gesture.

4.1   Pearson Correlation

This measure expresses a linearity correlation between the covariance of two random variables and the product of their standard deviations. The coefficient range in the interval [-1, 1], where 1 indicates a complete correlation between the two variables, and -1 indicates the random variables are inversely related. The higher the correlation, the more probable it is that two gestures are reciprocally similar.

4.2   Hamming and Levenshtein Distance

The Hamming and Levenshtein distance are the well know algorithms for string comparison, used to calculate the grade of difference of two gesture. The algorithm divides the signals in sub-segments, the segmentation algorithm find the local maxima and minima to determine beginning and end of every segment, then it builds up a string in which the four combinations of the Yaw and Pitch directions (up-up; up-down; down-up; down-down) are associated to an equivalent symbol (A,B,C,D) . Then, we use Hamming or Levenshtein to calculate the minimum number of changes required to transform one signal into the other: the resulting value reflects the edit distance between two gestures, i.e. to their "geometrical similarity".

4.3   Pearson-Hamming and Pearson-Levenshtein

To increase the accuracy we combine Pearson with Hamming and Levenshtein. Given a gesture, at first we "eliminate" all the dictionary centroids that are under the

---

[2] We assumes that gestures of the same class follow a Gaussian distribution with similar variance: the centroid is the gesture that has the greater intra-class similarity.

rejection threshold (see section 5.1) in the Pearson correlation, then we extract the best gesture assuming the Levenshtein or Hamming distance. In case of even results, we consider valid the gesture-centroid having the higher Pearson correlation.

5   Datasets

We carried out our tests on two data sets. The first one formed by 8 gestures inspired by the commands of a video player. The second one that contains 14 gestures representing the digits "0 to 9 ", and the symbols "plus", "minus", "multiply", and "divide". These two datasets have been created by 8 people between the ages 22 and 75, that performed every gesture 7 times. It was decided to acquire data without allowing the user to familiarize with the device, to simulate the same condition in which the application will be used.

In section 5.1 we present the results of tests by varying the rejection threshold (both in user dependent and user independent mode) to measure the effect on the accuracy. In section 5.3, more detailed results are presented in the form of a Confusion Matrix using the rejection threshold of 58%. In the Conclusions (see paragraph 6) we discuss our results.

Dataset 1 – Multimedia Player. The first dataset is the smallest one and is formed by a set of 8 natural gestures to interact with a video player. Legends are: (-1)-Rejection class; 1-Play, 2-Stop, 3-Previous, 4-Next, 5-Volume Up, 6-Volume Down, 7-Rewind, 8-FastForward.

Fig. 3. The eight gestures of dataset-1 ordered 1 to 8 from left to right

Dataset 2 – Numbers and Operations. The second dataset contains 14 gestures. Legends are: (-1)-Rejection class; 1-One, 2-Two, 3-Three, 4-Four; 5-Five, 6-Six, 7-Seven, 8-Eight, 9-Nine, 10-Zero, 11-Plus, 12-Minus, 13-Multiply, 14-Divide.

Fig. 4. The 14 gestures ordered 1 to 14 left to right. Leftmost "1-one", rightmost "14-Divide".

5.1   Rejection Threshold Test

We state that a gesture is recognized only if it passes a given threshold T. The threshold value must be at least $T > 1 / \|Dictionary\|$ to perform better than a random algorithm; a too high value creates too many false negatives. A reasonable number of false negative is acceptable (e.g. less than 2%) if it contributes to diminish the numbers of false positives, but we do not want the user to repeatedly redo gestures, as we prefer the use of interface to be easy and "natural".

A first test have been done on the given two datasets, changing the threshold value in a range of 48% to 68%, adding 5% at each iteration. Below we can see the most significant results related to the maximum (68%), medium (58%), and minimum (48%) value of the threshold T.

5.2   Results

The results below denote (highlighted in green) that the best accuracies are reached using a combination of Pearson and Hamming using a threshold T = 68%, in user *dependent* mode.

Table 1. Accuracies varying the rejection threshold T. Used combinations: P: Pearson; P-L: Pearson-Levenshtein; P-H: Pearson-Hamming.

Threshold T	Algorithm	Set 1 Usr-Indep.	Set 1 Usr-Dep.	Set 2 Usr-Indep.	Set 2 Usr-Dep.
68%	*P*	0.9596	0.9677	0.8708	0.9262
	P-L	0.9516	0.9778	0.8642	0.9288
	P-H	0.9616	1.0	0.9064	0.9433
58%	*P*	0.9596	0.9677	0.8708	0.9262
	P-L	0.9717	0.9778	0.8554	0.9380
	P-H	0.9738	0.9879	0.9288	0.9578
48%	*P*	0.9596	0.9677	0.8708	0.9262
	P-L	0.9838	0.9778	0.8906	0.9407
	P-H	0.9516	0.9798	0.9301	0.9450

The values in accuracy, in user *independent* mode, show a trend towards higher values as the threshold is decreased. The user *dependent* tests, instead show an opposite behavior as regards the two datasets: in fact, in the dataset 1, the recognition accuracy tends to decrease, while in the dataset 2 these values tend to increase.

On the basis of these results, we chose the Pearson-Hamming combination using a rejection threshold of 58% to promote a better balance in the dataset1 and dataset2, and in both user modes (*dependent* and *independent*), as long as we prefer maintain a certain flexibility in the dictionary choice.

5.3   Tests on Dataset1 and Dataset2 with Rejection Threshold of 58%

In this section we present more in detail the test results using dataset1 and dataset2 once fixed the threshold T = 58%. Results are presented in the form of a Confusion Matrix. Ground truth (GT) are in columns, while rows represent what has been actually recognized (R). In diagonal the correct results (true positive, in green), outside the diagonal the invalid recognitions: in the leftmost column the false negatives (in blue), out of this column and not in the diagonal the false positives (red).

Table 2. Dataset 1: Pearson-Hamming. Left user-independent; right user-dependent mode.

GT\R	-1	1	2	3	4	5	6	7	8
1	0	70	0	0	0	0	0	0	0
2	0	1	54	0	0	1	2	0	0
3	0	0	0	52	0	1	2	0	0
4	0	0	0	0	58	1	0	0	0
5	0	0	0	0	0	58	0	0	0
6	3	0	0	0	1	0	55	0	0
7	0	0	0	0	0	0	0	69	0
8	1	0	0	0	0	0	0	0	67

GT\R	-1	1	2	3	4	5	6	7	8
1	0	70	0	0	0	0	0	0	0
2	0	0	58	0	0	0	0	0	0
3	0	0	0	53	0	0	0	2	0
4	0	0	0	0	58	0	0	0	1
5	0	0	0	0	0	55	0	1	2
6	0	0	0	0	0	0	59	0	0
7	0	0	0	0	0	0	0	69	0
8	0	0	0	0	0	0	0	0	68

In **dataset 1** (8 gestures) the best combination using threshold T=58% in user *independent* mode is "Pearson and Hamming". *Accuracy* is 98.79%, precision is 98.17%, recall is 99.178%. We have only 4 false negatives out of 496 instances (cfr. Table 2, left). In user *dependent* mode the *accuracy* is 98.79%, precision is 98.79%, recall is 100%. We do not have any false negative (cfr. Table 2, right).

Table 3. Dataset 2: Pearson – Hamming. User-dependent mode.

GT\R	-1	1	2	3	4	5	6	7	8	9	10	11	12	13	14
1	0	75	0	1	0	0	0	0	0	0	0	0	0	0	0
2	0	2	45	1	0	0	0	0	0	0	0	1	0	0	0
3	0	0	1	47	0	0	0	0	0	0	2	0	0	1	0
4	0	1	0	0	49	0	0	0	0	0	0	0	0	0	0
5	0	0	0	0	0	50	2	0	0	0	0	0	0	0	0
6	0	0	0	0	1	0	49	0	0	0	0	0	0	0	0
7	0	1	1	2	0	0	0	49	0	0	0	0	0	1	0
8	2	0	0	0	0	0	0	0	47	0	0	0	0	1	0
9	0	0	0	0	0	1	0	0	0	42	0	3	0	1	0
10	0	0	0	0	0	0	0	1	1	0	59	0	0	0	0
11	0	0	0	0	0	0	0	0	0	0	0	58	0	0	0
12	0	0	0	0	0	0	0	1	0	0	0	0	50	0	1
13	0	0	0	0	0	0	0	0	0	0	0	0	0	52	2
14	0	0	0	0	0	0	0	0	0	0	0	0	0	0	55

In **dataset 2** (14 gestures) the best combination using a threshold T= 58% in user *independent mode* is the "Pearson and Hamming", again. The *accuracy* is 92.885%, precision is 94.631%, recall is 98.052%. In user *dependent* mode the *accuracy* is 95.784%, precision is 96.037%, recall is 99.725% (see Table 3 for user dep. mode)

6   Conclusions

Comparing our algorithm to others using only inertial systems in user *dependent* mode our recognitions algorithms obtain a better accuracy. The work of Kratz and Rohs 2010

[9] achieves an accuracy of 80% over a 10-gestures dictionary. Our algorithm appears also better in comparison with Chen et al. [10]. In the user *independent* mode their ranking algorithm obtains an accuracy of 85.24% without using optical sensors, against our accuracy of 97.4%. In user *dependent* mode their error rate is much lower, and their accuracy is 98.8% using only the inertial sensors, against our 98.79% (but using threshold 58%). Using a rejection threshold of 68% in user *dependent* mode we reached an accuracy of 100%, on the dataset1 of 8 gestures (see Table 1).

Our accuracy results can be also compared to systems that use more complex classification methods such as SVM , SOM and HMM that are in principle more effective, but do not allow a recalculation at "real time " of the class representative (centroid). Wu et al. [8] show that their classification algorithm has good results using a set of gestures that are very similar to our dataset. They have an accuracy of 99.38% in user *dependent* tests on a dataset of only 4 gestures, against our 100% reached using P-H with a threshold 68% on dataset1 (8 gestures) in user *dependent* mode (see Table 1). The classifier of Wu et al. has an accuracy of 95.21% on a dataset of 12 gestures, against our 95.78% of accuracy on the dataset2 (but with 14 gestures) in user *dependent* mode (see Table 3). Even in user *independent* tests our algorithm outperforms Wu et al. results: we reach an accuracy of 92.88 % on the dataset2 (14 gesture) against the 89.29% obtained by Wu et al. on their dataset of 12 gestures.

In conclusion, our classification algorithm obtains good results and performs better compared to analog recognition systems based only on inertial sensors and also gives better results than most of the multimodal works we found in the literature, reaching accuracies useful for a commercial device. The algorithm have quick time of reaction (less 0.1 sec), it is flexible, and resilient both to individual variations and to differences introduced by ageing. The algorithm has been adopted by the Easy Reach consortium to pilot the human machine gesture-based interface of the Easy Reach Project.

References

1. Bisiani, R., Merico, D., Pinardi, S., Dominoni, M., Cesta, A., Orlandini, A., Rasconi, R., Suriano, M., Umbrico, A., Sabuncu, O., Schaub, T., D'Aloisi, D., Nicolussi, R., Papa, F., Bouglas, V., Giakas, G., Kavatzikidis, T., Bonfiglio, S.: Fostering Social Interaction of Home-Bound Elderly People: The EasyReach System. In: IEA/AIE 2013, Amsterdam (2013)
2. Hans-Helmut, K., Dirk, H., Thomas, L., Steffi, G.R.-H., Matthias, C.A., Herbert, M., Vilagut, G., Ronny, B., Josep, M.H., Giovanni, D.G., Ron, D.G., Viviane, K., Jordi, A.: Health status of the advanced elderly in six european countries: Results from a representative survey using EQ-5D and SF-12. Health and Quality of Life Outcomes 2010 143(8), 143 (2010)
3. Hoffman, F.G., Heyer, P., Hommel, G.: Velocity Profile Based Recognition of Dynamic Gestures with Discrete Hidden Markov Models (1996)
4. Mäntylä, V.M., Mäntyjärvi, J., Seppänen, T., Tuulari, E.: Hand gesture recognition of a mobile device user. IEEE (2000)
5. Schlömer, T., Poppinga, B., Henze, N., Boll, S.: Gesture Recognition with a Wii Controller. In: Proceedings of the Second International Conference on Tangible and Embedded Interaction, Bonn (2008)
6. Prekopcsák, Z.: Accelerometer Based Real-Time Gesture Recognition. Poster (2008)

7. Cho, S.J., Oh, J.K., Bang, W.C., Chang, W., Choi, E., Jing, Y., Cho, J., Kim, D.Y.: Magic Wand: A Hand-Drawn Gesture Input Device in 3-D Space with Inertial Sensors. In: Proceedings of the 9th Int'l Workshop on Frontiers in Handwriting Recognition (2004)
8. Wu, J., Pan, G., Zhang, D., Qi, G., Li, S.: Gesture Recognition with a 3-D Accelerometer. In: Zhang, D., Portmann, M., Tan, A.-H., Indulska, J. (eds.) UIC 2009. LNCS, vol. 5585, pp. 25–38. Springer, Heidelberg (2009)
9. Kratz, S., Rohs, M.: A $3 Gesture Recognizer – Simple Gesture Recognition for Devices Equipped with 3D Acceleration Sensors. ACM (2010)
10. Chen, M., AlRegib, G., Juang, B.: A new 6D motion gesture database and the benchmark results of feature-based statistical recognition (2011)
11. Pinardi, S., Bisiani, R.: Movements Recognition with Intelligent Multisensor Analysis, A Lexical Approach. In: Proceedings of the 6th Int. Conf. on Intelligent Environments, Kuala Lumpur (2010)
12. Gupta, S., Morris, D., Patel, S.N., Desney, T.: SoundWave: Using the Doppler Effect to Sense Gestures. Redmond (2012)
13. Xu, R., Zhou, S., Li, W.J.: MEMS Accelerometer Based Nonspecific-User Hand Gesture Recognition. IEEE Sensors Journal (May 5, 2012)
14. Kratz, L., Saponas, T.S., Morris, D.: Making Gestural Input from Arm-Worn Inertial Sensors More Practical. ACM (2012)
15. XSens, XM-B Technical Documentation (2009)
16. STMicroelectronics, LIS331DLH - MEMS digital output motion sensor ultra low-power high performance 3-axes "nano" accelerometer (2009)
17. Zhou, S., Dong, Z., Li, W.J., Kwong, C.P.: Hand-Written Character Recognition Using MEMS Motion Sensing Technology. In: Proceedings of the 2008 IEEE/ASME International Conference on Advanced Intelligent Mechatronics (2008)
18. Keir, P., Elgoyhen, J., Naef, M., Payne, J., Horner, M., Anderson, P.: Gesture-recognition with Non-referenced Tracking. In: Proceedings of the 2006 IEEE Symposium on 3D User interfaces (2006)
19. Fihl, P., Holte, M., Moeslund, T., Reng, L.: Action Recognition using Motion Primitives and Probabilistic Edit Distance (2006)
20. Pylvänäinen, T.: Accelerometer Based Gesture Recognition Using Continuous HMMs. In: Marques, J.S., Pérez de la Blanca, N., Pina, P. (eds.) IbPRIA 2005. LNCS, vol. 3522, pp. 639–646. Springer, Heidelberg (2005)
21. Tuulari, E., Ylisaukko-oja, A.: SoapBox: A Platform for Ubiquitous Computing Research and Applications. In: Mattern, F., Naghshineh, M. (eds.) PERVASIVE 2002. LNCS, vol. 2414, pp. 125–138. Springer, Heidelberg (2002)
22. Vogler, C., Sun, H., Metaxas, D.: A Framework for Motion Recognition with Applications to American Sign Language and Gait Recognition. IEEE (2000)
23. Gavrila, D.M.: The Visual Analisys of Human Movement: A Survey. Academic Press (1998)
24. Wang, J.S., Chuang, F.C.: An Accelerometer-Based Digital Pen With a Trajectory Recognition Algorithm for Handwritten Digit and Gesture Recognition. IEEE (2011)
Choi, E., Bang, W., Cho, S., Yang, J., Kim, D., Kim, S.: Beatbox Music Phone: Gesture-based Interactive Mobile Phone using a Tri-axis Accelerometer. IEEE (2005)

Differential Diagnosis of Erythemato-Squamous Diseases Using Ensemble of Decision Trees

Mohamed El Bachir Menai and Nuha Altayash

Department of Computer Science,
College of Computer and Information Sciences,
King Saud University
P. O. Box 51178, Riyadh 11543, Saudi Arabia
{menai,nstayash}@ksu.edu.sa

Abstract. The differential diagnosis of erythemato-squamous diseases (ESD) in dermatology is a difficult task because of the overlapping of their signs and symptoms. Automatic detection of ESD can be useful to support physicians in making decisions if the model gives comprehensible explanations and conclusions. Several approaches have been proposed to automatically diagnosis ESD, including artificial neural networks (ANN) and support vector machines (SVM). Although, these methods achieve high performance accuracy, they are not attractive for dermatologists because their models are not directly usable. Decision trees can be converted into a set of if-then rules, which makes them particularly suitable for rule-based systems. They have been already used for the diagnosis of ESD. In this paper, we investigate the performance of boosting decision trees as an ensemble strategy for the diagnosis of ESD. We consider two decision tree models, namely unpruned decision tree and pruned decision tree. The experimental results obtained on UCI dermatology data set show that boosting decision trees leads to a relative increase in accuracy that attains 5.35%. Comparison results with other related methods demonstrate the competitiveness of the ensemble of unpruned decision trees. It performs 96.72% accuracy, which is better than those of some methods, such as genetic algorithms and K-means clustering.

1 Introduction

Psoriasis is a common dermatology disease that affects 2% to 3% people from the total population around the world [6]. Erythemato-squamous diseases (ESD) include six groups of skin diseases that share some psoriasis signs and symptoms with redness (erythema) causing loses in cells (squamous). ESD are psoriasis, seborrheic dermatitis, lichen planus, pityriasis rosea, chronic dermatitis and pityriasis rubra pilaris. They all share the clinical features of erythema and scaling, and many histopathological features as well. The exact causes of these diseases remain in general unknown but they could be a combination of hereditary and environmental factors that can affect people in any ages.

The differential diagnosis of ESD is a difficult task in dermatology, which depends on the analysis of features obtained from the evaluation of both clinical

A. Moonis et al. (Eds.): IEA/AIE 2014, Part II, LNAI 8482, pp. 369–377, 2014.

and histopathological features [7], and thus requires a detailed observation and experience to be correctly diagnosed. For example, pityriasis rubra pilaris is a rare disease that is usually mistakenly diagnosed in its first stages as psoriasis.

Medical decision support systems model the diagnosis task as a classification problem to assist physicians in making reliable and accurate decisions. As a result, several methods have been proposed for the classification of ESD, including genetic algorithms (GAs), support vector machines (SVM), artificial neural networks (ANN), K-means clustering, instance-based learning, decision tree learning, and other hybrid methods. Comprehensibility and interpretability are key factors of the usefulness of those methods. Indeed, methods that provide physicians with high diagnosis accuracy, but without explanations and conclusions, are not practically useful to physicians.

Decision tree learning is among the most practical methods for inductive inference. Moreover, decision trees can be converted to a set of if-then rules to improve their readability [11], which makes them suitable for rule-based systems, particularly in the medical domain. The purpose of this study is to investigate the performance of diagnosis models for ESD based on single decision tree and ensemble of decision trees (or forests) using boosting [5] as an ensemble strategy. The decision tree models considered are unpruned decision tree and pruned decision tree. The rest of this paper is organized as follows. The next Section describes the differential diagnosis of ESD. Section 3 outlines some related work. Section 4 describes the diagnosis models to be examined. Section 5 presents and discusses some experimental results. Section 6 concludes this work and outlines some future research directions.

2   Differential Diagnosis of ESD

A differential diagnosis of ESD is a systematic diagnostic procedure used by physicians to evaluate the differences between the diseases in order to identify the specific class of ESD in a patient. This procedure involves both a clinical and histopathological diagnostics for the evaluation of 12 clinical features and 22 histopathological features [6] described in Table 1.

The clinical diagnostic can distinguish between some ESD. For instance, the erythema and scaling are high in psoriasis and low in chronic dermatitis. Psoriasis, lichen planus and pityriasis rosea have the koebner phenomenon. Lichen planus has itching and polygonal papules where follicular papules are clinical feature for pityriasis rubra pilaris. Oral mucosa is usually seen in lichen planus, where the knee, elbow and scalp are usually preferred location of psoriasis. Family history is important in psoriasis. Pityriasis rubra pilaris is mostly seen during childhood [6]. The histopathological diagnostic consists in analyzing tissue samples under microscope to distinguish more accurately ESD. For instance, melanin incontinence is a diagnostic feature for lichen planus, fibrosis of the papillary dermis is for chronic dermatitis, exocytosis may be seen in lichen planus, pityriasis rosea and seborrheic dermatitis. Acanthosis and parakeratosis can be seen in all the diseases in different degrees. Follicular horn plug and perifollicular

Table 1. Features of erythemato-squamous diseases

Clinical Features	Histopathological Features
f_1: Erythema	f_{12}: Melanin incontinence
f_2: Scaling	f_{13}: Eosinophils in the infiltrate
f_3: Definite borders	f_{14}: PNL infiltrate
f_4: Itching	f_{15}: Fibrosis of the papillary dermis
f_5: Koebner phenomenon	f_{16}: Exocytosis
f_6: Polygonal papules	f_{17}: Acanthosis
f_7: Follicular papules	f_{18}: Hyperkeratosis
f_8: Oral mucosal involvement	f_{19}: Parakeratosis
f_9: Knee and elbow involvement	f_{20}: Clubbing of the rete ridges
f_{10}: Scalp involvement	f_{21}: Elongation of the rete ridges
f_{11}: Family history (0 or 1)	f_{22}: Thinning of the suprapapillary epidermis
f_{34}: Age (linear)	f_{23}: Spongiform pustule
	f_{24}: Munro microabcess
	f_{25}: Focal hypergranulosis
	f_{26}: Disappearance of the granular layer
	f_{27}: Vacuolisation and damage of basal layer
	f_{28}: Spongiosis
	f_{29}: Saw-tooth appearance of retes
	f_{30}: Follicular horn plug
	f_{31}: Perifollicular parakeratosis
	f_{32}: Inflammatory monoluclear infiltrate
	f_{33}: Band-like infiltrate

parakeratosis are strong features of pityriasis rubra pilaris. Clubbing of the rete ridges and thinning of the suprapapillary epidermis are strong features of psoriasis.

3   Related Work

Güvenir et al. [6] introduced a Voting Feature Interval classification algorithm (VFI5) for the differential diagnosis of ESD. The algorithm VFI5 classifies a new ESD instance based on real-valued voting, where the feature weights are learned using a GA. VFI5 achieved an accuracy of 96.2% on the UCI dermatology data set [2], which has been donated by the same authors. Güvenir and Emeksiz [7] proposed an expert system that makes a classification decision for the differential diagnosis of ESD based on decisions made by a nearest neighbor classifier, naïve Bayes classifier, and VFI5. Fidelis et al. [4] used a GA to discover comprehensible rules for the classification of ESD, that achieved an accuracy of 95%. Nanni [12] used a random subspace ensemble of SVMs and a feature selection method for the diagnosis of ESD, that achieved an accuracy of 97.22%. Übeyli [16] introduced a method based on the implementation of multiclass SVM with the error correcting output codes (ECOC) for the diagnosis of ESD and compared it with recurrent neural network (RNN) and a multi-layer perceptron neural network (MLPNN). SVM with ECOC was the outperforming method achieving an accuracy of 98.32%, while RNN and MLPNN reached an accuracy of 96.65% and 85.47%, respectively. A comparative study [10] of SVM with Gaussian kernel function, SVM with polynomial kernel function and RBN neural network for the diagnosis of ESD shows that SVM with polynomial kernel function learn about 10 times faster than SVM with Gaussian kernel function.

Xie and Wang [20] proposed a diagnosis model for ESD based on SVM and a new hybrid feature selection, called improved F-score and sequential forward search. This model achieved 98.61% classification accuracy with 21 features. Übeyli and Güler [19] proposed a new method for the diagnosis of ESD based on an adaptive neuro-fuzzy inference system that achieved an accuracy of 95.5%. Übeyli and Doğdu [18] applied K-means clustering to the diagnosis of ESD and reported an accuracy of 94.22%. In [17], Übeyli proposed to diagnose ESD using a combined neural network, that achieved an accuracy of 97.77%. Parthiban and Subramanian [13] proposed a co-active neuro-fuzzy inference system and GA (CANFIS) for the diagnosis of ESD, that achieved an accuracy of 96.65%. Karabatak and Ince [9] proposed a new feature selection method based on association rules (AR) and neural network for the diagnosis of ESD. This method achieved an accuracy of 98.61% and is considered among the best performing methods. Polat and Güneş [14] proposed a novel hybrid intelligent method based on C4.5 decision tree learning algorithm [15] and one-against-all approach for multi-class classification problems. Its application to the classification of ESD dermatology data gave an accuracy of 96.71%. To the best of our knowledge, the best model proposed for the diagnosis of ESD was proposed by Abdi and Giveki [1]. It is based on particle swarm optimization (PSO), SVM and AR. AR are used to select the optimal feature subset from the original feature set. Then a PSO based approach for parameter determination of SVM is developed to find the best parameters of kernel function. This model achieved 98.91% classification accuracy using 24 features of the ESD.

4   Ensemble of Decision Trees

The differential diagnosis of ESD has been addressed using various classification methods, as shown in the previous Section. The best performing methods are based on SVM, ANN, and their variants. Indeed, their overall results concerning the classification accuracy are very satisfactory, but the classification models generated by such methods are not easily comprehensible by physicians, since low-level information is involved in such models. Physicians need to understand the model's results and to put them together with their knowledge to validate the clinical hypothesis and to make a well-informed decision.

Which classification method to use to obtain a comprehensible model with a high predictive accuracy? Decision tree learning results in decision trees that can be translated into comprehensible if-then prediction rules (if <conditions on the predicting features> then <predicted pathology>). The C4.5 decision tree learning algorithm [15] has been already used as a component of a hybrid method for the diagnosis of ESD, achieving 96.71% accuracy [14]. The results that are generated by a committee (or ensemble) of decision trees in diagnosing ESD should be significantly better than anyone of them by its own.

In ensemble learning, several base classifiers are generated and combined to get a single classifier. Ensemble learning strategies include bagging and boosting. A bagging algorithm trains the base classifiers from bootstraps of the training data.

The combined classifier is obtained by majority voting of the ensemble of base classifiers. Boosting algorithms iteratively learn weak classifiers based on a distribution of training examples, weigh them according to their accuracy, and add them to a final strong classifier. The training examples are then reweighed, such as the weights of correctly classified examples are decreased and those of misclassified examples are increased to let the weak learners in the next iteration focus more on currently misclassified examples.

We propose to investigate the performance of ensembles of decision trees for the differential diagnosis of ESD using boosting as an ensemble strategy and C4.5 [15] as the base learner algorithm that generates decision trees. Two models are considered based on pruning or not pruning decision trees. In unpruned decision trees, there is no handling of overfitting problem. In pruned decision trees, each decision node is a candidate for pruning. The sub-trees or nodes under this decision node are removed if and only if the resulting pruned tree has the same or better performances compared to the previous one. When a pruning decision is made, then this decision node is set to a leaf node with the label of most common classification in the training set associated to it.

5   Experimental Evaluation

The programs are implemented using Java programming language and WEKA [8]. The UCI dermatology data set [2] used in this study is available online (http://archive.ics.uci.edu/ml/datasets/Dermatology). It consists of 366 records of clinical and histopathological patient cases distributed as shown in Table 2.

Table 2. Class distribution of erythemato-squamous diseases

Class code	Class name	number of instances
1	Psoriasis	112
2	Seboreic dermatitis	61
3	Lichen planus	72
4	Pityriasis rosea	49
5	Chronic dermatitis	52
6	Pityarisis rubra pilaris	20

A 10-fold cross validation method is used to estimate the average classification error (or the average accuracy). Sensitivity and specificity are also calculated for each class, but the results are averaged over the six classes.

$$errorRate = \frac{falsePositives + falseNegatives}{truePositives + falsePositives + trueNegatives + falseNegatives} \tag{1}$$

$$accuracy = \frac{truePositives + trueNegatives}{truePositives + falsePositives + trueNegatives + falseNegatives} \tag{2}$$

$$sensitivity = \frac{truePositives}{truePositives + falseNegatives} \tag{3}$$

$$specificity = \frac{trueNegatives}{trueNegatives + falsePositives} \tag{4}$$

Our first experiment involves detecting and preventing the overfitting problem by using a training and validation set approach (withhold one-third of the available data for the validation set, and use the other two-thirds for training).

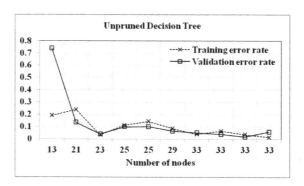

Fig. 1. Variation of the training and validation error rates with the size of the unpruned decision tree (number of nodes)

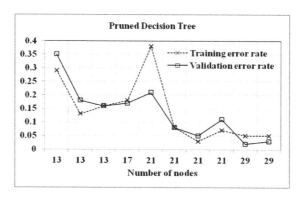

Fig. 2. Variation of the training and validation error rates with the size of the pruned decision tree (number of nodes)

Figures 1, 2 show the variation of the error rate measured over both training and validation data with the size of the decision tree (number of nodes). Different models are considered: the decision tree is allowed to overfit the data, which gives the unpruned decision tree (Figure 1); the reduced-error pruning method is used to prevent overfitting and results in a pruned decision tree (Figure 2). In the two cases, the results do not show an evidence of overfitting the data: a decrease of the training error rate and the validation error rate first decreases, then increases. This indicates that the data do not contain noise or a relatively low level of noise. The smallest validation error rates are achieved by an unpruned decision tree of 33 nodes (Figure 1, error rate of 1.8%, 25 rules) and a pruned decision tree of 29 nodes (Figure 2, error rate of 2%, 22 rules). In the next experiment, we test the performance of ensemble of decision trees over training and validation data. The unpruned decision tree is used as a base learner.

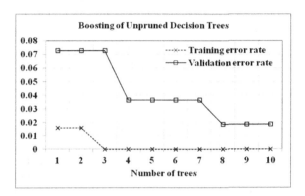

Fig. 3. Variation of the training and validation error rates with the size of the ensemble of unpruned decision trees

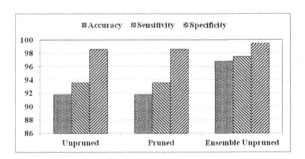

Fig. 4. Performance comparison of unpruned decision tree, pruned decision tree, and ensemble of unpruned decision trees

Figure 3 shows the variation of training and validation error rates of ensembles of unpruned decision trees with the number of trees. It indicates that the minimum validation error rate is obtained with an ensemble of 8 decision trees, while a zero training error is obtained with ensemble of 3 decision trees and more.

The bar chart in Figure 4 compares the performance of the diagnosis models considered in this study over test data: unpruned decision tree, pruned decision tree and ensemble of 8 unpruned decision trees. Accuracy, sensitivity and specificity results are shown. It is noticeable that the ensembles of decision trees outperform single decision trees. The best accuracy (96.72%), sensitivity (97.43%) and specificity (99.42%) are obtained with the ensemble of unpruned decision trees. In comparison to a single tree, the ensemble of unpruned decision trees achieves a relative increase of 5.35% in accuracy.

The comparison of the results with those of related work on ESD shows that ensemble of unpruned decision trees outperforms VFI5 [6], GA [4], MLPNN [16], an adaptive neuro-fuzzy inference system [19], and a K-means clustering method [18]. However, our proposed method is outperformed by SVM based methods [12,16,20] and a combined neural network [17].

6   Conclusion

In this paper, we have examined the performance of three models for the differential diagnosis of ESD based on single decision trees and ensemble of decision trees. The experimental results show that boosting decision trees can achieve a significant increase of the performance in comparison with single decision tree, especially when unpruned decision tree is used as the base learner. However, the number of rules tends to augment proportionally to the size of the ensemble. Ensembles of unpruned decision trees outperform the other models and some of the state-of-the-art models, achieving an average accuracy of 96.72%. Some models based on SVM and ANN remain the best models in terms of accuracy, but their lack of transparency makes them less useful models in practice.

Open research directions may include the investigation of other ensemble strategies, such as combination of boosting and bagging. Another important avenue of research is the reduction of the number of rules obtained by eliminating equivalent rules and replacing set of rules by more general rules.

References

1. Abdi, M.J., Giveki, D.: Automatic detection of erythemato-squamous diseases using PSO-SVM based on association rules. Engineering Applications of Artificial Intelligence 26(1), 603–608 (2013)
2. Bache, K., Lichman, M.: UCI machine learning repository (2013)
3. Breiman, L.: Bagging predictors. Machine Learning 24(2), 123–140 (1996)
4. Fidelis, M.V., Lopes, H., Freitas, A.: Discovering comprehensible classification rules with a genetic algorithm. In: Proceedings of the 2000 Congress on Evolutionary Computation, vol. 1, pp. 805–810 (2000)
5. Freund, Y., Schapire, R.E.: Experiments with a new boosting algorithm. In: International Conference on Machine Learning, pp. 148–156 (1996)
6. Güvenir, H.A., Demiröz, G., Ilter, N.: Learning differential diagnosis of erythematosquamous diseases using voting feature intervals. Artificial Intelligence in Medicine 13, 147–165 (1998)

7. Güvenir, H.A., Emeksiz, N.: An expert system for the differential diagnosis of erythemato-squamous diseases. Expert Systems with Applications 18(1), 43–49 (2000)
8. Hall, M., Frank, E., Holmes, G., Pfahringer, B., Reutemann, P., Witten, I.H.: The WEKA data mining software: An update. SIGKDD Explor. Newsl. 11(1), 10–18 (2009)
9. Karabatak, M., Ince, M.C.: A new feature selection method based on association rules for diagnosis of erythemato-squamous diseases. Expert Systems with Applications 36(10), 12500–12505 (2009)
10. Kecman, V., Kikec, M.: Erythemato-squamous diseases diagnosis by support vector machines and RBF NN. In: Rutkowski, L., Scherer, R., Tadeusiewicz, R., Zadeh, L.A., Zurada, J.M. (eds.) ICAISC 2010, Part I. LNCS, vol. 6113, pp. 613–620. Springer, Heidelberg (2010)
11. Mitchell, T.M.: Machine Learning, 1st edn. McGraw-Hill, Inc., New York (1997)
12. Nanni, L.: An ensemble of classifiers for the diagnosis of erythemato-squamous diseases. Neurocomputing 69(7-9), 842–845 (2006)
13. Parthiban, L., Subramanian, R.: An intelligent agent for detection of erythemato-squamous diseases using co-active neuro-fuzzy inference system and genetic algorithm. In: International Conference on Intelligent Agent Multi-Agent Systems, IAMA 2009, pp. 1–6 (2009)
14. Polat, K., Güneş, S.: A novel hybrid intelligent method based on C4.5 decision tree classifier and one-against-all approach for multi-class classification problems. Expert Systems with Applications 36(2, pt. 1), 1587–1592 (2009)
15. Quinlan, J.R.: C4.5: Programs for machine learning. Morgan Kaufmann Publishers Inc., San Francisco (1993)
16. Übeyli, E.D.: Multiclass support vector machines for diagnosis of erythemato-squamous diseases. Expert Systems with Applications 35(4), 1733–1740 (2008)
17. Übeyli, E.D.: Combined neural networks for diagnosis of erythemato-squamous diseases. Expert Systems with Applications 36(3, pt. 1), 5107–5112 (2009)
18. Übeyli, E.D., Doğdu, E.: Automatic detection of erythemato-squamous diseases using k-Means clustering. Journal of Medical Systems 34(2), 179–184 (2010)
19. Übeyli, E.D., Güler, I.: Automatic detection of erythemato-squamous diseases using adaptive neuro-fuzzy inference systems. Computers in Biology and Medicine 35(5), 421–433 (2005)
20. Xie, J., Wang, C.: Using support vector machines with a novel hybrid feature selection method for diagnosis of erythemato-squamous diseases. Expert Systems with Applications 38(5), 5809–5815 (2011)

Establishing the Relationship between Personality Traits and Stress in an Intelligent Environment

Marco Gomes, Tiago Oliveira, Fábio Silva, Davide Carneiro, and Paulo Novais

Department of Informatics, University of Minho
{marcogomes,toliveira,fsilva,dcarneiro,pjon}@di.uminho.pt

Abstract. Personality traits play a key role in the shaping of emotions, moods, cognitions, and behaviours of individuals interacting in a virtual environment. The personalities one exhibits reflect one's perception of the world and are demonstrated in the act of communication. Thus, the evaluation of a message can be changed due to stress and mood variations. Being able to identify the degree of relationship between one's personality characteristics and one's current stress state can thus facilitate the communication process. In particular, in this paper it is studied the correlation between some personality traits and the stress levels exhibited by users' interactions. To do so a novel approach was followed in which an intelligent environment is used to support the stress recognition process providing important personality related information. An experiment has been designed for the purpose of addressing the estimation of relevant aspects of interactions that occur in a rich sensory environment. Outputs from the experiment, such as the relation between personality characteristics and stress, can be used to maximize the benefits of virtual environments and its applications in fields such as learning, medicine or conflict resolution.

Keywords: Intelligent Environment, Stress, Emotions, Personality.

1 Introduction

As a society we nowadays experience a sense of proximity that stems from the unprecedented development in communication technologies over the last two decades. It is possible to communicate instantly with virtually anyone in the globe, leading to the feeling of a so-called "Global Village". However, this proximity is shallow and artificial as new communication mechanisms become increasingly poor, lacking many of the traditional Human aspects of communication. Addressing this subject, the lack of the non-verbal elements between interlocutors limits successful communication in Virtual Environments (VEs). In that sense, communicating in the absence of this information can be difficult [1]. To address it, preponderant factors in human communication must be covered, such as stress and personality traits. The use of a combination of multiple Artificial Intelligence techniques and, more particularly Ambient Intelligence techniques, can help to fill this gap.

Different personality types process and communicate information differently. Each personality type has a different impact in an individual's response to stress. Stress influences not only the efficiency of the communication but also the quality of its perception

A. Moonis et al. (Eds.): IEA/AIE 2014, Part II, LNAI 8482, pp. 378–387, 2014.

or even the relationship between the individuals communicating. Emotions significantly influence our social skills, our motivation and our actions in general. Likewise, the emotional state of one's interlocutor is very important to understand the true meaning of their words. Considering that people are individuals and behave individually, once you have identified the personalities and their communicative behaviours, an intelligent environment can potentiate more efficient communication mechanisms. The approach followed looks at how individual's behaviours change under stress and emotional stimuli. Specifically, we study the interaction of individuals with devices commonly used to communicate in virtual settings: keyboard and mouse. Preliminary work allowed to identify features extracted from interaction with these devices that are significantly affected by stress [2]. This allows us to identify and study behavioural changes associated to specific states of an individual such as stress, fatigue or emotional arousal. To accomplish this, one must understand how the patterns of interaction, the relationship between personality traits and stress can affect the human communication. Knowing how to deal with these dimensions will can improve our communication with others and our understanding of ourselves.

2   Multimodal Approach to Stress and Personality Dimensions

In this section, we look at the multi-dimensionality of stress and personality. Indeed, given the multiplicity of factors that influence human experience and the nature of this work, a single-modality approach to measure the effects of stress and personality would not be suited. Therefore, we propose the use of a multi-modal one. Indeed, it can be stated that human behaviour can be understood as an all-encompassing spectrum of what people do including thinking and feeling, which are influenced, for example, by culture, attitudes or emotions. Moreover, it is commonly accepted that people are all different from each other because they have different personalities, they are inside specific cultures, and they have distinct individual histories.

All behaviour can be considered as situational, i.e., when the situation or event changes, the behaviour changes too: the intention to carry out a certain action or the fashion in which we d it can be changed by the present circumstances. In the context of an intelligent environment, environmental factors can have a significant influence on shaping individual factors (e.g., personality types, cognitive maturity, emotional status and social experiences) and play a significant role in one's life. Human behaviour exhibits how a human being reacts intrinsically (determined by personality) and externally (stress responses) facing the stimuli that occurs in their environment.

Concerning *stress*, a consensual definition is still an open discussion in the scientific community. The main reasons for this lie in its multi-dimensionality and subjective nature, which led to multiple interpretations. And, with so many factors that can contribute to stress, it can be rather difficult to define it. Scientists have circumvented this problem by defining it empirically. To this end, researchers start to focus upon cognitive and behavioural consequences of stress, and stress became viewed as a mind-body, psychosomatic, or psycho-physiologic phenomenon. An interpretation of this phenomenon could refer stress as a psycho-physiologic arousal response occurring in the body as result of stimuli, and these stimuli becomes a stressor by virtue of the cognitive interpretation of the individual. This interpretation of stress is used in this paper to analyse stress in virtual environment scenarios.

In what concerns *personality*, it determines how a person reacts intrinsically to certain events. In a certain way, it can be seen like a long-term emotion, because it is formulated in the early years of life and remains almost unchanged during lifetime. The Big-Five Personality (BFP) theory also called Five Factor Model (FFM)[3] is a common definition of personality traits that indicate that is possible to separate them into five properties. Although psychological and physiological adverse ambient conditions can produce significantly changes in a person, the authors steer this topic to a very specific set of variables: sound, temperature and luminosity are studied as external factors that affect well-being and mood states.

Other researchers have also researched and debated the influence of such variables in the impact of mood change in people [10]. In a noisy environment people have diverging sensitivity to the volume of the sound. For instance, people that are sensitive to sound have a predisposition to become more anxious when the volume rises. Others, if provoked, can even become aggressive or hostile. In hot places psychological changes occur if the temperature increases, such as irritability, fatigue and discomfort. Very high values can also provoke hostility or violent attitudes. Assessing this type of information from the environment allows to estimate and even foresee the kind of emotional reactions that may arise, through a classification of mood states.

Different emotional reactions can be deduced in part by the personality structure of each person, represented in this study by the BFP theory and a mood simulator reacting to environment configurations. Taking into consideration the aforementioned stress and personality dimensions, we follow a multi-modal approach to analyse how they are related with each other.

3   An Intelligent Environment to Enhance Emotional and Stress Recognition

One of the main characteristics of Intelligent Environments is the key role that the user plays as the focus of the entire process. In other words, the process starts by collecting data about the user and the environment in which the user is situated, and it finishes by acting intelligently for the benefit of the user. Therefore, an intelligent environment with diverse devices and functionalities was built, aiming to compile and provide information about the user's context and state to the applications being used. This environment is composed of hard sensors (i.e., hardware sensors, in the traditional sense) and soft sensors (i.e., software sensors such as sensors for the interaction with the mouse and the keyboard), independent systems (with logic and intelligent services) and an integration framework.

The two most interesting components, from the point of view of this paper and described ahead in more detail, are the Stress Recognition and the Mood Recognition modules. The first classifies, in real-time, the level of stress of a user; the second provides an outlook of the user's mood using information about the personality traits previously provided. Based on this, a behavioural and contextual analysis is performed, applying Artificial Intelligence techniques in order to detect patterns and relationships among the selected features (see Sec. 4). A more detailed description of each of these modules is provided in the two following sub-sections.

3.1 Stress Recognition Module

The quantification of human stress can improve the way people behave and communicate in a virtual environment. Having this information, a system may correctly monitor how each issue or event is affecting each particular individual. The developed Stress Recognition Module is based in previous work [2] and relies on an approach of Behavioural Biometrics: a field of science that seeks to study the particularities of the behaviour of each individual in a given situation [13]. It can be used, for example, to accurately determine the identity of people for access control, through their behaviour (e.g. how they talk, how they walk, how they use the computer).

This field is very closed to the more traditional field of Biometrics, that relies on unique physical characteristics (e.g. hand size, fingerprint), independent of human behaviour but unique for each individual. However, Behavioural Biometrics looks at behavioural characteristics (e.g. the typing rhythm on a keyboard, the way the user looks at the screen of the computer). These characteristics change with factors such as mood, fatigue, stress or others [6]. When one knows how each individual behaves in each scenario or condition, one can then identify the individual from their behaviours or from changes in these behaviours. This is the key idea behind Behavioural Biometrics.

For this study Keystroke Dynamics and Mouse Dynamics were selected to this end. From between more than two dozen features that can be extracted in a non-invasive and transparent way, the following were selected from these two types of behavioural biometrics:

- *Time between Keys (TBK):* time spent between the use of two keys, that is, the time between events KEY_UP and KEY_DOWN row. Unit: millisecond
- *Key Down Time (KDT)* - time spent since the key is pressed down and is released later, in other words, time since the event KEY_DOWN and KEY_UP consecutively. Unit: milliseconds
- *Absolute Sum of Angles (ASA)* - absolute sum of the angles when turning left or right pointer during its travel. This measure tries to find just how much the mouse "turned", regardless of the direction to which it turned. Unit: degree
- *Mouse Velocity (MV)* - Mouse Velocity - velocity at which the cursor travels. The distance travelled by the mouse (in pixels) between a $C1$ coordinate $(x1, y1)$ and $C2(x2, y2)$ corresponding to $time1$ and $time2$ over travel time (in milliseconds). Unit: Pixel/Milliseconds
- *Mouse Acceleration (MA):* acceleration of the mouse at a given time. The acceleration value is calculated using the mouse velocity on movement time. Unit: Pixel/Milliseconds
- *Distance Pointer To Line Between Clicks (DPTLBC)* - between each two consecutive clicks, measures the distance between all the points of the path travelled by the mouse and the closest point in a straight line (that represents the shortest path) between the coordinates of the two clicks.

3.2 Mood Recognition Module

Personality traits, beliefs, mood or the state of the physical environment have influence on the emotional arousal of people. In a computational system it is still a challenging process to acquire accurate and rich information in this domain but there are some approaches proposed in the research community that have had broad acceptance. In regard to the representation of a personality this work uses an OCEAN structure similar to the approach adopted by the ALMA framework to represent the personality of people and initiate mood states [5]. In this theory personality is defined by a set of variables:

$$personality(o,c,e,a,n)$$
$$let\ o,c,e,a,n \in [-1,1]$$

These o,c,e,a,n variables represent five personality traits which mean Openness, Conscientiousness, Extraversion, Agreeableness and Neuroticism, respectively. The process of acquiring personality traits from real people is carried out through their answers to questionnaires used by psychologists in psychological studies. In this particular case it was used the summarized version of the Newcastle Personality Assessor (NPA) as a reference questionnaire [8].

Contrary to the personality, which is regarded as almost static during people's lifetimes, mood is a temporary state of the human mind, can last for minutes, hours or even days. It is seen as medium-term emotion. Given that personality also influences mood, the preliminary definition of personality traits, initial mood assessments can be implemented considering the predominant mood state for each personality. Individual personality parameters can also be taken into consideration for mood updates over time for each person.

The mood recognition module here described is responsible for assessing the current mood of each person using their personality structure and the perception of environmental variables such as sound, temperature and luminosity. Mood representation is based on the PAD space computationally designed by Gehbard in his work ALMA [5]. A person's mood is represented using the three variables that define PAD space, P, A and D for actual pleasure, arousal and dominance respectively. These variables are in the domain $[-1,1]$ to create the representation of the eight moods that are possible: hostile, exuberant, disdainful, docile, bored, anxious, dependent and relaxed. Other variables included on Mood are the *emotionalWeight* to represent the weight of the emotions related with each personality and *const* which represents the velocity of the Mood update.

Everyone is different and, just as literature suggests, people with high trend of neuroticism (n) have more probability to be emotionally negative i.e., they have predisposition to feel emotions like hate, anger or distress more often than people with low score of this trend. In other hand extraversion (e) awakens positive feelings more often, such as joy, love or gratitude.

As a consequence, the calculation of the weight is carried out through the signal of emotion: if it is higher than *emotionalWeight* $= (e+1.0)*const$ else *emotionalWeight* $= (n+1.0)*const$. This variable is in the range $[0,2]$ turning *emotionalWeight* positive to facilitate future calculations. The constant *const* is used to reduce or increase the velocity when mood is crossing from state to state. In this work the value used was 0.1.

One of Mood's parametrized constructors receives as parameters personalities' values o, c, e, a, n. To initiate mood in relation to personality in the *PAD* space an weighted reference was used as in:

$$P = 0.59a + 0.19n + 0,21e;$$
$$A = -0.59n + 0.30a + 0,15o;$$
$$D = 0.60e - 0.32a + 0.25o + 0.17c;$$
$$le\, to, c, e, a, \in [-1, 1]$$

This creates an initial mood state different from user to user, giving them a mood foundation that allows software agents to deal with different emotion types during the life time. These variables represent the mood state relating it with emotions triggered by events. The events are triggered by the environment's sound, luminosity and temperature conditions according to which the final mood is calculated. This final mood state is the predicted value for the exposure of people to those environment conditions after an infinite amount of time. The transition between mood states is assessed by applying an increment in the mood variables in the order and direction of the predicted final mood state after an event is triggered. An expression that makes the update of the current value with the displacement times the weight of emotion is used to calculate the displacement vector for each portion of time, as in :

$$\overrightarrow{M_{actual}M_{final}} = M_{final} - M_{actual} \qquad (1)$$
$$M_{actual} = M_{actual} + \overrightarrow{M_{actual}M_{final}} * emotionalWeight \qquad (2)$$

The vector $\overrightarrow{M_{actual}M_{final}}$ is divided in small pieces that are relative to *emotionalWeight* and this causes mood state to move more or less slowly, depending on each person's personality, over the time, thus preventing the instant passage from one state to another which does not happen in real life frequently.

Emotions are mapped in the PAD space using the OCC model [9] for emotion representation and ergonomic studies on the effect of the state temperature, luminosity and sound in emotion states [10]. From these studies and the current assessment of environment state, a final mood state can be calculated mapping the emotion derived and the current mood state using the PAD space and equation 2. The equation assures that the mood state does not react instantly to new emotions but it is gradually affected by them.

4   Practical Experiment

As stated before, the main objective of this research work was to identify which and how human psychological traits are correlated to stress and how they can be pointed out. To demonstrate the relationship discussed in the previous sections, an experiment was set up in which we tried to estimate all the relevant aspects of the interaction between the individual and their environment that occur in a rich sensory environment (where the contextual modalities were monitored). This environment was equipped with sensors and devices that acquire different kinds of information from the user in a non-intrusive way (using the multimodal approach described in Sec. 2)[1] . To have a psychological profile of the individual under study we have chosen a small questionnaire, the Newcastle Personality Assessor (NPA), referenced in scientific literature.

The participants of the proposed experiment were volunteers from our institution. Eight individuals participated, both female and male, aged between 20 and 32. All these individuals were familiar with the devices used. Therefore, the interaction with them was not an obstacle. The first step of the experiment was to ask the volunteers to fill in the NPA individual questionnaire. The following step was the monitoring of the individuals' interaction with their devices (where the stress recognition module was installed), within the set environment. While the user conscientiously interacts with the system, a parallel and transparent process takes place in which the contextual and behavioural data is sent in a synchronized way to the platform (a framework deliberately designed to handle with stress and mood recognition). The platform, upon converting the sensory data into useful information, allows for a contextualized analysis of the users' data.

In previous studies, the features mentioned in Sec. 3.1 were used in successful attempts to detect the stress of an individual or, to be more precise, to identify the transition of an individual to a higher state of arousal [11]. Considering the hypothesis that the occurrence of stressed states may be related with some personality traits, it could be possible to establish an association between the features tested in the experiment and the answers collected from the NPA questionnaire. Since the objective is to classify the personality traits of an individual according to his interaction patterns with a machine, a special type of Bayesian Networks was used, the Naïve Bayes [7]. It is a simpler probabilistic model that assumes strong independence between the evidence variables. In the following section a more detailed description will be provided on the construction of said classifiers.

4.1 Dataset Construction, Pre-processing and Attribute Selection

In order to construct the Naïve Bayes classifiers for the personality traits, each instance of the datasets used to train them would have to possess data from all of the six interaction features, meaning that there would have to be a correspondence between all the records. However, keyboard and mouse activities may not be simultaneous or happen with the same frequency, resulting in log datasets with different sizes. The solution for this problem was to calculate the mean values of the features for each individual every half hour, during the length of the experiment. Using this method, and after a selection of the valid records, it was possible to obtain a dataset with 138 instances with values of the individuals' key down time, time between keys, mouse velocity, mouse acceleration, the distance pointer to line between clicks and the absolute sum of degrees between clicks. Then, results of each individual's NPA questionnaire were added to this dataset. These results consist of the categories the subjects got for Openness, Conscientiousness, Extraversion, Agreeableness, Neuroticism and the final emotion in their personalities (as determined as described in Sec. 3.2).

---

[1] The following sensors have been used in the experiment:computer keyboard, mouse, temperature, luminosity, and sound sensors.

The full set variables consisted of the following: *TBK, KDT, MV, MA, ASA, DPTLBC, Openness, Conscientiousness, Extraversion, Agreableness, Neuroticism* and *Emotion*.

Bayesian models do not perform well in continuous data. Therefore, it was necessary to *discretize* the values of the interaction features. Using *Weka*, and more specifically the *PreProcess* tab of the Explorer interface, a supervised filter called *Discretize* was applied do the dataset in order to convert these values into intervals [12]. In order to select the variables with the strongest dependence relationship, the dataset underwent an attribute selection phase, supported by the resources in the *Select Attributes* tab of the *Explorer* interface. The *ClassifierAttributeEval* was selected as the attribute evaluator. Accordingly, the Naïve Bayes was set as the target classifier and the parameter folds was set to 12 for 12-fold cross validation. Only the three best features were selected for each personality trait. The results showed that the *DPTLBC* is always selected as one of the top three features, which reveals that the behaviour of this variable is more permeable to the characteristics of an individual's personality. On the other hand, the *KDT* is never selected, which eventually means that it is the feature that reflects less the personality traits.

4.2 Model Construction and Assessment

The Naïve Bayes classifiers were learned using subsets of the dataset containing all the attributes. The learning phase was executed using *R* software, more specifically the functionalities provided by the *bnlearn* library. 12-fold cross validation was performed in order to assess the model with the classification error [12] as the loss function. In the attribute selection phase, the top three features for Neurotiscism and Emotion were *DPTLBC, MV, ASA* and *DPTLBC, MV, MA* , respectively. The mean classification errors for the classifiers using these features were 0.08018 (for Neuroticism) and 0.09596 (for Emotion). The resulting model is depicted in the following diagram 1.

The Neuroticism and Emotion classifiers are, eventually, the most important. Neuroticism is a personality trait characterized by anxiety, moodiness and worry. The individuals that score high on Neuroticism usually respond poorly to stressors and manifest a change in behaviour when submitted to them. The Neuroticism categories present in the model has *high, low* and *medium low*. By conditioning the features in the Neuroticism classifier, it is possible to detect significant changes in its behaviour. In the Emotion classifier, the analyzed categories were *anxious, bored* and *relaxed*. When looking at it and performing the same type of conditioning for the features mouse velocity, mouse acceleration and the distance of the pointer to the line between each two consecutive clicks, it is revealed that the anxious class has higher probabilities for the central intervals of these features. For low intervals of *DPTLBC, MV* and *ASA* it is possible to see that the classifier assigns higher probabilities to the *high* class of Neuroticism. This indicates that more precision may be associated with high score in Neuroticism. This is also in line to what is expect as a tendency for the evolution of an individual in a stressed state [11]. The emotions expressed in the upper and lower boundaries of the features are usually *bored* and *relaxed*.

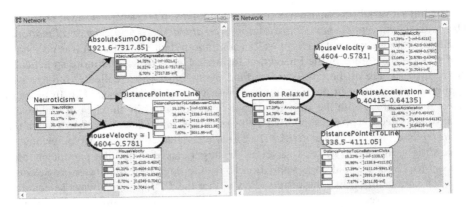

Fig. 1. The resulting Bayes model for the Neuroticism and Emotion networks

5 Conclusions

In order to develop an intelligent system that supports human communication in regular or demanding circumstances, a computational support is required that describes the richness of a human's emotional state in relation to communication and the environment. To this end, this paper presented significant advantages for understating how psychological and stress dimensions can be related in certain circumstances and how they can be identified using everyday devices.

Using a non-intrusive approach an experiment under different parameter settings have been performed. This experiment pointed dependence relationships between the features being monitored and the main personality traits. These relationships, especially those regarding *Neuroticism*, have the potential to enable the characterization of individuals through their device interaction patterns. The identification of a neurotic personality is an indicator that a person may not deal well with potentially stressful situations, which may impair their communication capabilities.

The findings presented herein may help to identify the degree of relationship between one's personality characteristics and his current stress state, maximizing the benefits of the communication within virtual environment applications in fields such as learning, medicine or conflict resolution. The main contribution of this work is in the identification of situations in which actions are required in order to smooth processes and not let people loose efficiency when they get ensnared by their own circumstances.

Acknowledgements. This work was developed in the context of the project CAM-CoF - Context-aware Multimodal Communication Framework funded by ERDF - European Regional Development Fund through the COMPETE Programme (operational programme for competitiveness) and by National Funds through the FCT - Fundação para a Ciência e a Tecnologia (Portuguese Foundation for Science and Technology) within project FCOMP-01-0124-FEDER-028980.

References

1. Bell, M.: Toward a definition of virtual worlds. Journal for Virtual Worlds Research 1(1) (2008),
   `http://journals.tdl.org/jvwr/index.php/jvwr/article/view/283`
2. Carneiro, D., Castillo, J.C., Novais, P., Fernández-Caballero, A., Neves, J.: Multimodal behavioral analysis for non-invasive stress detection. Expert Syst. Appl. 39(18), 13376–13389 (2012), `http://dx.doi.org/10.1016/j.eswa.2012.05.065`
3. Costa, P., MacCrae, R., Psychological Assessment Resources, Inc.: Revised NEO Personality Inventory (NEO PI-R) and NEO Five-Factor Inventory (NEO FFI): Professional Manual. Psychological Assessment Resources (1992),
   `http://books.google.pt/books?id=mp3zNwAACAAJ`
4. Friedman, N., Geiger, D., Goldszmidt, M.: Bayesian network classifiers. Machine Learning 29(2-3), 131–163 (1997)
5. Gebhard, P.: ALMA: A layered model of affect. In: Proceedings of the Fourth International Joint Conference on Autonomous Agents and Multiagent Systems, pp. 29–36 (2005)
6. Jain, A., Bolle, R., Pankanti, S.: Introduction to biometrics. In: Jain, A., Bolle, R., Pankanti, S. (eds.) Biometrics, pp. 1–41. Springer, US (1996),
   `http://dx.doi.org/10.1007/0-306-47044-6_1`
7. Jensen, F.V.: Bayesian networks. Wiley Interdisciplinary Reviews: Computational Statistics 1(3), 307–315 (2009)
8. Nettle, D.: Personality: What Makes You the Way You are. OUP, Oxford (2007),
   `http://books.google.pt/books?id=vk1kXARv1-IC`
9. Ortony, A., Clore, G., Collins, A.: Cognitive Structure of Emotions. Cambridge University Press (1988)
10. Parsons, K.: Environmental ergonomics: A review of principles, methods and models. Applied Ergonomics 31(6), 581–594 (2000),
    `http://dx.doi.org/10.1016/s0003-68700000044-2`
11. Rodrigues, M., Gonçalves, S., Carneiro, D., Novais, P., Fdez-Riverola, F.: Keystrokes and clicks: Measuring stress on e-learning students. In: Casillas, J., Martínez-López, F.J., Vicari, R., De la Prieta, F. (eds.) Management Intelligent Systems. AISC, vol. 220, pp. 119–126. Springer, Heidelberg (2013),
    `http://dx.doi.org/10.1007/978-3-319-00569-0_15`
12. Witten, I., Frank, E., Hall, M.: Data Mining: Practical machine learning tools and techniques, 3rd edn. Morgan Kaufmann (2011)
13. Yampolskiy, R.V., Govindaraju, V.: Behavioural biometrics: A survey and classification. International Journal of Biometrics 1(1), 81 (2008),
    `http://www.inderscience.com/link.php?id=18665`

Behavior Modeling and Reasoning
for Ambient Support: HCM-L Modeler

Fadi Al Machot, Heinrich C. Mayr, and Judith Michael

Application Engineering Research Group, Alpen-Adria-Universität Klagenfurt, Austria
{fadi.almachot,heinrich.mayr,judith.michael}@aau.at

Abstract. This paper introduces the architecture and the features of the HCM-L Modeler, a modeling tool supporting the Human Cognitive Modeling Language HCM-L and a comprehensive reasoning approach for Human Cognitive Models based on Answer Set Programming. The HCM-L tool has been developed using the ADOxx® meta modeling platform and following the principles of the Open Modeling Initiative: to provide open models that are formulated in an arbitrary, domain specific modeling language, which however is grounded in a common ontological framework, and therefore easily to translate in another language depending of the given purpose.

Keywords: Modeling Tool, Modeling Language, Behavior Modeling, Ontology, Model Mapping, Knowledge Base, Ambient Assistance, Reasoning, Answer Set Programming.

1   Introduction

Health is essential for individuals as well as for the society. Many everyday activities help to keep healthy: Washing hands after using the bathroom, body care, preparing healthy food and doing sports. However, when getting older, such activities in general become harder.

Ambient Assisted Living (AAL, [1]) focusses on softening or even compensating the effects of ageing. AAL research was as a key topic within the 7th Framework Programme of the European Union; and again, health, demographic change and well-being are important topics in the new funding program Horizon 2020. The range of related research projects is wide and covers mobility support, smart homes, fall protection, cognitive games, healthcare, and much more.

To assess the ability of a person doing her or his daily activities, scales like the Physical Self-Maintenance Scale or the scale of Instrumental Activities of Daily Living (IADL) [2] may be used. Using the former, e.g., the abilities in toileting, grooming, or bathing may be assessed; the IADL rates, among others, the competences in food preparation or the responsibility for the personal medication.

To support such activities is the main objective of the Human Behavior Monitoring and Support (HBMS[1] [3]) project. The key idea here is to use models of a person's

---

[1] Funded by Klaus Tschira Stiftung gGmbH, Heidelberg.

A. Moonis et al. (Eds.): IEA/AIE 2014, Part II, LNAI 8482, pp. 388–397, 2014.
© Springer International Publishing Switzerland 2014

former individual target-oriented behavior including its context as a knowledge basis for deriving support services when the person's abilities are reduced or (temporarily) forgotten.

Consequently, behavior modeling as well as reasoning for deriving support information from models are key topics of HBMS. As standard modeling languages like UML or BPMN proved to be suboptimal for the given purpose [4] the Human Cognitive Modeling Language (HCM-L) was designed as a lean set of concepts and constructs for integrated human behavior and context modeling. According to the OMI (Open Modeling Initiative [5]) HCM-L is a domain specific language based on common fundamentals, and thus allows for using models as knowledge carriers. It should be intuitively to understand by users of the AAL domain, in particular e.g. by psychologists or care givers.

HCM-L Modeler is a comprehensive tool for creating, managing and transforming models based on HCM-L. It was developed using the ADOxx® meta modeling platform [6]. Given the comprehensiveness of HCM-L, the Modeler might be useful for many application purposes beyond the borders of HBMS.

Modeling and reasoning are closely related and widely used in expert systems. In particular, intensive research is undertaken in the context of Semantic Web[2] [7,8,9]. Ontologies based on Semantic Web provide concise high-level semantic representations. However, parsing larger ontologies for recognition and heuristic problem tasks does not perform well with limited hardware resources [10,11,12].

Other supervised learning methods like Support Vector Machine, Artificial Neural Networks, Decision Trees, Bayesian Networks and Hidden Markov Models need training data that are quite difficult to extract from human cognitive models.

Within this paper we are targeting a reasoning approach for real time intelligent user support based on Answer Set Programming.

The paper is organized as follows: Section 2 introduces exemplarily the main HCM-L concepts and their representation using the HCM-L modeler. Section 3 discusses the architecture of the modeler and outlines its main features and functionality. Section 4 presents the overall architecture and the components of the proposed reasoning approach and a list of the obtained results. The paper closes with a conclusion and an outlook on future work (section 5).

2   HCM-L and Modeler

As Ambient Assistance in healthcare and other AAL domains mainly deals with supporting activities these are in focus when building ontologies (conceptualization) and modeling. Consequently, using HCM-L means modeling (the sequence of) activities together with their pre- and post-conditions first, and then adding models of the activities' personal, environmental, social, and spatio-temporal context [13].

As this paper addresses the HCM-L Modeler and reasoning, we introduce the underlying modeling language by means of an example instead of a systematic language definition as given in [4]. The graphical elements were chosen following the principles for designing effective visual notations [14].

---

[2] http://www.w3.org

Figure 1 shows a simplified example of the HCM-L key concept *behavioral unit* describing how a person, let's call her Paula, may achieve a given *goal*, here: "get ready for leaving the house" in the morning. As the reader might easily deduce from the figure, a behavioral unit represents an aggregate of *operations* which together lead to reaching a goal in daily life. As such, behavioral units correspond to use cases in business process modeling and can be broken down into different steps: the operations and the flows between these.

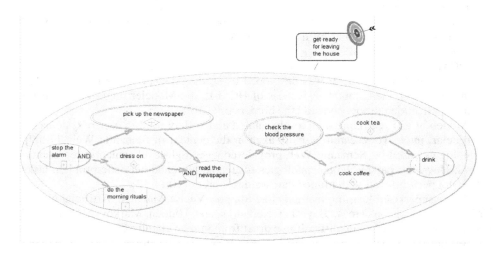

Fig. 1. Behavioral unit ‚morning activity'

Paula starts with stopping the alarm of the alarm-clock. Every morning, in an arbitrary order, she picks up the newspaper from the doormat in front of the entrance door, dresses on and does her morning rituals in the bathroom. After having done all that, she reads the newspaper in the kitchen. After 20 Minutes of reading, she checks her blood pressure. Depending on the result, she cooks and drinks tea (high blood pressure) or coffee (lower pressure).

Possible beginning and *successful ending* operations are just marked implicitly. The former have no *incoming flow*, the latter no *outgoing flow*. If one of the successful ending operations (there might be more than one) is executed, the goal is fulfilled.

Operations can have simple or complex pre - and post-conditions; in this case they are grey-shaded and have condition compartments. Precondition (AND) of operation 'read the newspaper' defines, that this operation can only be executed, if all three predecessor operations are executed completely. How these operations are performed will be defined in the *instruction* attribute of each operation. That Paula has to check the blood pressure at least 20 minutes after she has started to read the newspaper will be defined in the *time space* attribute.

The diamond at the bottom of an operation symbol like in 'check blood pressure' or 'dress on' indicates that the resp. operation is seen and modeled as a behavioral unit again. Clicking on the diamond thus opens the related model (see figure 2): To check the values, Paula takes the blood pressure monitor out of the top draw of her drawer cabinet in the kitchen, secures the cuff, presses the start button and after some seconds hears a beep signal and can record the results on her tablet PC.

The modeling granularity depends on the resp. objectives. In the context of HBMS, the behavioral unit models will be established by transforming and integrating results from sensor/video observations of concrete sequences of actions.

HCM-L modeler also offers a possibility to condense/expand a sequence of operations (square in the bottom of an operation symbol) for enabling clarity of larger graphs. This, however, is a pure syntactical aid and does not define a hierarchy of behavioral units.

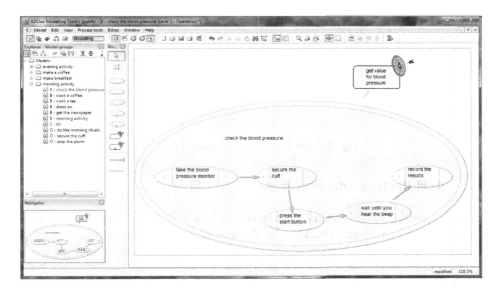

Fig. 2. Behavioral Unit ‚check the blood pressure'

Operations are performed within a personal, environmental, social, and spatio-temporal context [13]. The main HCM-L concepts for context modeling are *thing* and *connection* to describe arbitrary concrete or abstract objects, also persons, or relationships between things, respectively. Figure 3 shows a selection of the context of the "check the blood pressure" operations.

In this example, Paula is modeled as a person (thing box with smiley) with her left and right hand (aggregation). The blood pressure monitor is modeled as a thing, which again is made up of things, e.g. the start/stop button. The blood pressure monitor lays in the drawer cabinet which is located inside the kitchen (a *location thing*, flagged by the map symbol).

There are several relationships between operations and their context: *calling* (a thing initiates an operation), *participating* (things contribute to or are manipulated by an operation), and *executing* (a thing performs an operation).

The main HCM-L concepts are summarized in the meta model as discussed in [4].

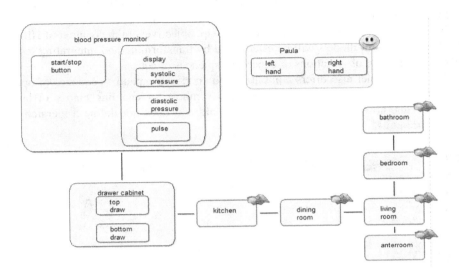

Fig. 3. Some elements of the context model

3   Architecture and Features

HCM-L Modeler is a comprehensive modeling tool for HCM-L including syntax, semantics and consistency checking. In the next development stage it will support complex scenarios, model optimization and advanced reasoning techniques.

HCM-L Modeler was developed using the meta modeling platform ADOxx® (www.ADOxx.org) which implements the upper three layers of the OMG Meta Object Facility (MOF) [15,16,17]. Figure 4 illustrates the overall architecture. On the third MOF layer the ADOxx® Meta^2 Model defines constructs such as Class, Relation Class, Endpoint, or Attribute.

The Meta^{1}Model (on the MOF M2 layer) corresponds to the meta model of the language, a modeling tool is to be developed for, i.e. in our case the HCM-L meta model. For that purpose, the corresponding ADOxx® tier provides the concept of library, representing a collection of meta models conforming to the Meta^{2}Model and formulated in the ADOxx® Library Language (ALL).

Models on MOF M1 layer (HCM-L models) are stored in a model repository, i.e. a generic model storage configured by a meta model library [16]. They can be exported in two formats, the ADOxx® Description Language (ADL) or in XML [17].

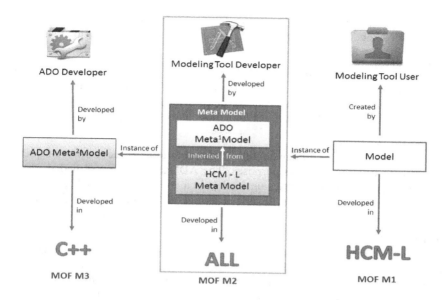

Fig. 4. HCM-L in the ADOxx® meta model hierarchy

We now proceed to the description of some key features and their implementation.

Objects and Relations

ADOxx® offers different types of meta constructs [18], out of which we selected three for the HCM-L modeler:

- *D_Construct*: Super class for a graph-base pre-defined meta model.
- *D_Container*: Container class providing the relation *is_inside*, i.e. O *is_inside* C means that the x/y coordinates of object O lie within the drawing area container C.
- *D_Aggregation*: Inherits from D_Container, hence also provides the *is_inside* relation; in addition, it enables a self-defined drawing area, e.g. a resizable rectangle.

Relation class is a construct that is used as a template for creating directed relations between objects; consequently, a relation class is defined between classes. As the relations are a directed, they have a from-side and a to-side.

Visualization

For the definition of the graphical elements ADOxx® provides the GRAPHREP grammar language which can be used to specify all graphical properties (e.g., shape and color) of a modeling language element.

Comprehensive mathematical support for drawing curves and polygons is provided. Furthermore, the developer is supported in controlling the coordinates of element's names, break-line properties and font types.

Consequently, using the HCM-L modeler, all objects and relations can be connected in a flexible way. The breaking lines of object's names give the possibility to resize the elements for an optimal visual form. This flexibility is a very important advantage of HCM-L where different models and sub-models are required for complex scenarios, e.g., the visualization of different levels of an aggregation hierarchy.

Traceability
Traceability relationship types are modeled as instances of the Meta^{2}Model construct "Relation Class". HCM-L modeler uses *Interref Relation Classes* that can be used to connect elements of different models. The definition of such traceability relationships is supported by the so-called *notebook*, and thus can be done by the user.

Querying
An important function of human cognitive modeling tools is the support of querying the model for different analysis purposes. The ADOxx® platform provides the AQL query language that allows queries on models in a style similar to SQL12 [17]. AQL queries can be pre-defined by the developer or may be formulated manually by a user; for that purpose, HCM-L modeler provides an interactive assistant.

Consistency Check
A major issue in modeling processes is the fact that comprehensive consistency checks are difficult, in particular for inexperienced users. However, inconspicuous mistakes in the logic may affect the whole model: contradictory semantics reduce the performance of reasoning processes and yield invalid results. For the HCM-L modeler we considered two main consistency issues: (1) using the right syntax of logical operators and (2) consistent naming of model elements throughout the whole model.

As an example, after clicking on the button "pre-defined queries", HCM Modeler yields a menu of different consistency checks for every model and sub model.

Transformation
ADOxx® offers the possibility to import and export models in a generic XML format. This feature is adopted by HCM-L modeler in order to allow transforming models to other formats, as used e.g. by inference or reasoning tools.

Reasoning Support
Both, model and rule based reasoning approaches for ambient support require the extraction of different features out of the given overall model. HCM-L modeler, among others, offers the possibility to calculate the frequency of specific activities based on the user history: every operation is supported by a percentage value which will be used for reasoning purposes (see next section).

In addition to that, HCM-L modeler calculates for each operation the "importance value" based on the user history, and "cost value" based on the similarity between the current user profile and other users. In the next section these values will be discussed in detail.

4   Ambient Support by Reasoning Based on HCM-L Models

For ambient support using conceptual cognitive models different requirements have to be covered, e.g., reasoning over time, constraints, and optimization.

Answer set programming (ASP) [19,20] is a form of declarative programming oriented towards difficult (primarily NP-hard) search problems. It is based on the stable model (answer set) semantics of logic programming. In ASP, time is usually represented as a variable the values of which are defined by an extensional predicate with a finite domain. Finite temporal intervals can be used to reason in ambient support. Optimization is indicated via maximization and minimization. Adding a constraint to a logic program P affects the collection of stable models of P in a very simple way: it eliminates the stable models that violate the constraint.

Figure 5 shows the components of the proposed ambient support reasoning approach. First we use the HCM-L modeler to model and design the conceptual cognitive model, and then we generate the required reasoning parameters that will be added to the model in our HCM-L tool automatically. Finally, we export the model in XML format as input for the answer set programming solver (i.e., its knowledge base).

Fig. 5. Components of the ambient support reasoning approach

The purpose of our reasoning module is to support users (patients or old people) in choosing the next operation (activity) when desired. In the case depicted in Fig. 1, the reasoning system would have to propose Paula on of the operations (1) picking up a newspaper, (2) doing morning rituals or (2) dressing on.

For that purpose, an optimization problem is to be solved based on three priority measures: (1) the importance of performing an operation according to the user history; (2) the cost value of choosing an operation based on the similarity between the current user profile and other users; (3) the time when the operation should be performed. Consequently, an operation is represented in our knowledge base as follows:

```
operation(Id).
operation_time(OperationId,Time).
user_hist_importance(OperationId,ImportanceValue).
cost(OperationId,CostValue).
bad_timing(OperationId).
user_current_time(Time).
```

The cost value is calculated based on the determined set of similar users. A common measure for such similarity is Pearson's correlation coefficient [16]. Based on the user history, a matrix R is established consisting of the scores of operations the users perform. The score is incremented by 1 each time the user chooses the particular operation. Dividing this score by the total number of operations a particular user is doing per day gives the probability of choosing that specific operation. Consequently, the cost value is the complementary probability of the resp. score probability.

The user_hist_importance is chosen to be the average value of performing the given operation over the last 20 days; it also could be fed into the model by the modeler.

To decide about timing we define two cases for the given example: if the operation is done in the morning then it is considered as good timing otherwise as bad.

The Optimization Process Based on ASP

To find the optimal solution for our optimization problem, we consider the reasoning values of the operations discussed previously:

```
1. #maximize[operation(X):user_hist_importance(X,Y)=Y @3].
2. #minmize[operation(X):cost(X,Y):user_hist_importance(X,Z)=Y/Z@2].
3. #minimize {bad_timing @1}.
```

Line 1–3 contribute optimization statements in inverse order of significance, according to which we want to choose the best operation. The most significant optimization statement (line 1) gives the main priority to the habits' history of the user. Line 2 is to minimize the cost per operation with respect to the importance of user's history. Line 3 serves to minimize the number of operations with bad timing.

Obtained Results

To check our ASP-based reasoning approach, we performed several tests on an embedded platform. In particular, we used pITX-SP 1.6 plus board manufactured by Kontron[3]. It is equipped with a 1.6 GHz Atom Z530 and 2GB RAM. We use Clingo[4] as ASP solver which is an incremental ASP system implemented on top of clasp[3] and Gringo[3] solvers [20]. Clingo is written in C and runs under Windows and Linux. We measured the execution time of the ASP solver on our embedded platform. The knowledge base consists of 10, 30 and 40 facts and supported by the previous optimization rules as discussed above. The overall execution time was between 0.4 and 0.6 seconds. I.e., that the reasoning system can run on smart phones and support the user in real time. 40 as a maximum number of facts mean that the user could choose between up to 8 operations which are much more than usual in everyday situations.

5   Conclusion and Future Work

As has been shown, HCM-Modeler is a powerful and comprehensive tool for developing, managing and exchanging models written in HCM-L. As HCM-L focuses on behavior and its context, HCM-Modeler might be used for a wide variety of applications in Ambient Assistance, healthcare and other process-oriented domains.

The Answer Set Programing paradigm proved to be an appropriate solution for solving heuristic problems. Furthermore, we showed that ASP allows solving such problems in real time which is important for the given application domain.

Currently we are designing usability experiments with end-users in order to reveal improvement potential for the modeler's interface.

References

1. Steg, H., et al.: Europe Is Facing a Demographic Challenge. In: Ambient Assisted Living Offers Solutions. VDI/VDE/IT, Berlin (2006)
2. Lawton, M.P., Brody, E.M.: Assessment of older people: Self-maintaining and instrumental activities of daily living. Gerontologist 9, 179–186 (1969)

---

[3] http://www.kontron.com
[4] http://potassco.sourceforge.net

3. Michael, J., Grießer, A., Strobl, T., Mayr, H.C.: Cognitive Modeling and Support for Ambient Assistance. In: Mayr, H.C., Kop, C., Liddle, S., Ginige, A. (eds.) UNISON 2012. LNBIP, vol. 137, pp. 96–107. Springer, Heidelberg (2013)

4. Michael, J., Mayr, H.C.: Conceptual Modeling for Ambient Assistance. In: Ng, W., Storey, V.C., Trujillo, J.C. (eds.) ER 2013. LNCS, vol. 8217, pp. 403–413. Springer, Heidelberg (2013)

5. Karagiannis, D., Grossmann, W., Höfferer, P.: Open Model Initiative: A Feasibility Study. Dpmt. of Knowledge Engineering, University of Vienna (2002), http://www.openmodels.at

6. Karagiannis, D.: Business Process Management: A Holistic Management Approach. In: Mayr, H.C., Kop, C., Liddle, S., Ginige, A. (eds.) UNISON 2012. LNBIP, vol. 137, pp. 1–12. Springer, Heidelberg (2013)

7. Kofod-Petersen, A., Mikalsen, M.: Context: Representation and Reasoning. Special Issue of the Revue d'Intelligence Artificielle on "Applying Context-Management" (2005)

8. Moody, D.: The "Physics" of Notations: Toward a Scientific Basis for Constructing Visual Notations in Software Engineering. IEEE Trans. Software Eng. 35, 756–779 (2009)

9. OMG Meta Object Facility (MOF) Core Specification Version 2.4.1: http://www.omg.org/spec/MOF/2.4.1/PDF/ (last visited on September 29, 2013)

10. Schwarz, H., Ebert, J., Lemcke, J., Rahmani, T., Zivkovic, S.: Using expressive traceability relationships for ensuring consistent process model refinement. In: Proc. Int. Conf. on Engineering of Complex Computer Systems (ICECCS), pp. 183–192 (2010)

11. Fill, H.G., Karagiannis, D.: On the conceptualisation of modelling methods using the ADOxx meta modelling platform. Enterprise Modelling and Information Systems Architectures 8(1) (2013)

12. ADOxx® Software Engineer Tutorial Package: http://www.adoxx.org/live/tutorial (last visited on September 29, 2013)

13. Baral, C.: Knowledge Representation, Reasoning and Declarative Problem Solving. Cambridge University Press (2003) ISBN 978-0-521-81802-5

14. Turaga, R., et al.: Machine recognition of human activities: A survey. IEEE Transactions on Circuits and Systems for Video Technology 18(11), 1473–1488 (2003)

15. Batsakis, S., Petrakis, E.G.M.: Sowl: Spatio-temporal representation, reasoning and querying over the semantic web. In: Proc. 6th Int. Conf. on Semantic Systems. ACM (2010)

16. King, B., et al.: Statistical reasoning in the behavioral sciences. Wiley.com (2011)

17. Batsakis, S., Petrakis, E.: SOWL: Spatio-temporal representation, reasoning and querying over the semantic web. In: Proc. 6th Int. Conf. on Semantic Systems. ACM (2010)

18. Sensoy, M., et al.: OWL-polar: A framework for semantic policy representation and reasoning. In: Web Semantics: Science, Services and Agents on the WWW 2012, pp. 148–160 (2012)

19. Gentner, D., et al.: Cognitive Approaches for the Semantic Web. Dagstuhl Reports 2.5, 93–116 (2012)

20. Al Machot, F., et al.: Real time complex event detection for resource-limited multimedia sensor networks. In: Proc. 8th IEEE Int. Conf. on Advanced Video and Signal-Based Surveillance (2011)

21. Gebser, M., et al.: A user's guide to gringo, clasp, clingo, and iclingo (2008)

Apply the Dynamic N-gram
to Extract the Keywords of Chinese News

Ren-Xiang Lin[*] and Heng-Li Yang

Dept. of MIS, National Chengchi University, Taiwan
{102356507,yanh}@nccu.edu.tw

Abstract. The explosive growth of information on the Internet has created a great demand for new and powerful tools to acquire useful information. The first step to retrieve information form Chinese article is word segmentation. But there are two major segmentation problems that might affect the accuracy of word segmentation performance, ambiguity and long words. In this paper, we propose a novel character-based approach, namely, dynamic N-gram (DNG) to deal with the two above problems of word segmentation and apply it to Chinese news articles to evaluate the accuracy of N-gram. The evaluation result indicated most of the readers agreed that dynamic N-gram approach could extract meaningful keywords. Even in different news categories, the keywords extraction results still have no significant difference. The primary contribution of this approach is that dynamic N-gram helps us to extract the most meaningful keywords in different types of Chinese articles without considering the number of grams.

Keywords: Chinese Word Segmentation, Dynamic N-gram, Information Retrieval.

1 Introduction

The explosive growth of information on the Internet has created a great demand for new and powerful tools to acquire useful information [1]. Research interest in Chinese information retrieval (CIR) has increased as a result of the large growth rate of online Chinese literature. The first step to retrieve information form Chinese article is word segmentation which also plays an important role in many Chinese language processing tasks such as text mining, question answering, and machine translating [2]. Chinese text consists of strings of ideographic Chinese characters and punctuations. Unlike English uses space as word delimiter, Chinese text has no delimiters to mark word boundaries other than the punctuations. This makes it difficult for computers to process Chinese linguistic data unless it was previously separated into individual words [3]. Therefore, the goal of Chinese word segmentation (CWS) is to segment a plain Chinese text to a sequence of meaningful words before linguistic analysis.

There are two major segmentation problems that might affect the accuracy of word segmentation performance, ambiguity and long words [4, 5]. The primary source of

---

[*] Corresponding Author.

A. Moonis et al. (Eds.): IEA/AIE 2014, Part II, LNAI 8482, pp. 398–406, 2014.

the ambiguity is out-of-vocabulary (OOV) in a Chinese article. Typical examples of OOV words include named entities such as organization names, person names, and location names. Those named entities may be very long and a difficult case when a long term consists of some meaningful words.

In this paper we propose a novel character-based approach, namely, dynamic N-gram (DNG) to deal with the two above problems of word segmentation. In order to evaluate the accuracy of dynamic N-gram, we apply the dynamic N-gram in five Chinese news articles to extract the keywords which are meaningful enough to represent the articles. Then, we invite volunteers to read the five articles and evaluate the extent of those keywords representing the articles.

2   Literature Review

According to the categories of Chinese segmentation which Foo and Li [6] elaborated, the basic approaches of Chinese segmentation can be roughly divided into two groups, namely, character-based approaches and word-based approaches. The character-based approaches include single character-based approach and multi character-based approaches; the word-based approaches include statistic-based approaches, dictionary-based approaches, and hybrid approaches as shown in Fig. 1.

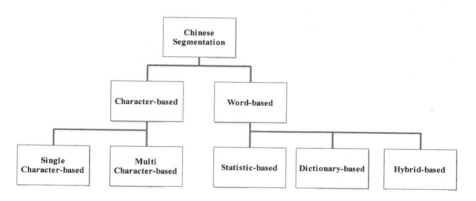

Fig. 1. Basic approaches of Chinese segmentation

2.1   Character-Based Approaches

The single character-based approach is a simple method to divide the context of Chinese into single characters. For example, the sentence of Chinese which is "國立政治大學" (National Chengchi University) could be divided into "國", "立", "政", "治", "大", and "學". Some studies have consistently shown that the information retrieval result using single-character indexing is significantly worse than using other segmentation approaches [7].

Multi character-based (or call N-gram) approaches split Chinese context into strings containing two, three or more characters. The approach which splits Chinese context into strings containing two characters which is called 2-gram (bigram). For instance, the bigram segments a sentence "國立政治大學" (National Chengchi University) into "國立", "立政", "政治", "治大", and "大學". Compared with the single character-based approach, the multi character-based approaches consistently yield superior CIR results [8].

The most outstanding advantage of these character-based approaches is the simple concept and ease of application that leads to other advantages such as reduced search costs, increased time benefits in the indexing and querying process.

2.2   Word-Based Approaches

Statistic-based approaches collect the statistic information such as occurrence frequency of characters or words in training documents to create a statistic model. By using the model to predict the probability of character combination, for example, Sproat and Shih [9] create a simple statistic model by calculating the occurrence frequency of words in a document. If the word frequency is over the criteria pre-defined by the model, the word will be segmented.

The dictionary-based approach creates a word dictionary that prior to being indexed and segments the texts which are matched in the dictionary [10]. For example, there are indexed words in a dictionary such as "國立" and "政治大學". By matching a sentence "國立政治大學" (National Chengchi University) with the dictionary, the sentence will be segmented as "國立" and "政治大學".

Hybrid-based approaches combine the statistic-based approaches and dictionary-based approaches. By using the created dictionary and statistic information such as word frequency to determine the text segmentation [11], hybrid-based approaches attempt to get a better precision of text segmentation.

3   Dynamic N-gram

Although the character-based approaches own advantages such as simple concept, ease of application, reduced search cost, and increased time benefits, first we have to decide the number of grams before applying these approaches and different number of grams will lead to different results of texts segmentation. For instance, the sentence "國立政治大學" (National Chengchi University) will be segmented into "國立", "立政", "政治", "治大", and "大學" by using 2-gram approach. However, through 3-gram approach, the segmentation results are "國立政", "立政治", "政治大", and "治大學". When using the 2-Gram approach, the segmentation results which have meaning in Chinese are "國立" (national), "政治" (chengchi), and "大學" (university), but there is no meaningful segmentation results through the 3-Gram approach. However, the correctly and most meaningful result is "國立政治大學", but it only could be segmented by 6-Gram approach. In order to overcome the determination of N, this study proposes an improved method of character-based approach, namely, dynamic N-gram (DNG). The dynamic N-gram approach does not need to set the N value, it will be determined by the algorithm. Table 1 is an example of an article in Chinese. The idea of dynamic N-gram as follows:

Table 1. Example of Chinese article

Title	第一屆台灣資訊管理夏季研討會
	(Meaning: The First Taiwan Summer Workshop on Information Management)
Context	由台灣國立政治大學商學院服務創新研究中心與資管系主辦的「第一屆台灣資訊管理夏季研討會」(TSWIM2013)，將於 2013 年 7 月 15 日至 17 日，在台灣台北舉行。TSWIM 致力於推動台灣與其他國家之間的資訊系統(IS)學者及博士研究生之研究合作。此 次 TSWIM 研討會將邀請台灣與國外對當前尖端的資訊系統與管理議題有研究興趣的學者，進行相關範疇(含括個人、企業、組 織與社會)的討論與互動交流。
	(Meaning: The First Taiwan Summer Workshop on Information Management (TSWIM 2013) will be held in Taipei, Taiwan on July 15-17, 2013. It will be organized and hosted by Center For Service Innovation and MIS Department, College of Commerce, National Chengchi University. TSWIM promotes research collaboration between Information Systems (IS) scholars and PhD students in Taiwan and other countries. The main purpose of TSWIM is to provide a forum for discussion and interaction among Taiwanese and International scholars with research interests in the cutting edge issues of information systems and management on individuals, businesses, organizations, and societies.)

3.1   Preprocessing

In the preprocessing step, we replace all punctuations such as parentheses and comma with hash sign (#) in the article. The article after preprocessed is shown in Table 2.

Table 2. Chinese article after preprocessing

Title	第一屆台灣資訊管理夏季研討會
Context	由台灣國立政治大學商學院服務創新研究中心與資管系主辦的#第一屆台灣資訊管理夏季研討會##TSWIM2013## ·········

3.2   Create the Candidate Characters Pool

After preprocessing, we count the occurrence frequency of each character in the article and add those characters whose frequency are over than one to candidate characters pool. Because the characters whose frequency are less than or equal to one could not be considered a high frequency word, we consider the characters whose frequency are less than or equal to one are unnecessary. The created candidate characters pool is shown in Table 3.

Table 3. Candidate characters pool

Character	Frequency	Character	Frequency	Character	Frequency
第	2	管	4	T	3
一	2	理	3	S	4
屆	2	夏	2	W	3
台	7	季	2	I	4
灣	6	研	7	M	3
資	5	討	4		
訊	4	會	4	....	

3.3   Create the Words Pool through Candidate Characters Pool

Then, by retrieving the article, if a character (A) in the article existed in the candidate characters pool, we continue to exam the second character (B) immediately appeared after first character (A). In case the second character (B) also exists in the candidate characters pool, we combine the character A and B into a word (AB).   We continue in the same way until the next character (C) does not exist in the candidate characters pool. The combined words will be added to the words pool according to the words position and frequency in the article to calculate their word weight (as formula 1). If a word appears in the article title, its position would be 3, and its title-frequency would be the times it appears in the title; otherwise, both position and title-frequency would set to 1. After calculating the words weight, we remove the words whose weight are less than or equal to one from words pool and keep the top 5 weight words. By doing this, the kept words might be the most meaningful words in the article. The created words pool are as shown in Table 4.

$$\text{Weight}_{word(i)} = \text{ContextFrequency}_{word(i)} \times \text{Position}_{word(i)} \times \text{TitleFrequency}_{word(i)}$$

$$\text{Position}_{word(i)} = \begin{cases} 3, word(i) \in title \\ 1, else \end{cases} \quad \text{TitleFrequency}_{word(i)} = \begin{cases} frequency, word(i) \in title \\ 1, else \end{cases} \tag{1}$$

Table 4. Words pool

Word	Frequency	Weight	Word	Frequency	Weight
第一屆台灣資訊管理夏季研討會	2	6	研究	4	4
一屆台灣資訊管理夏季研討會	2	6	TSWIM	3	3
屆台灣資訊管理夏季研討會	2	6	SWIM	3	3
台灣資訊管理夏季研討會	2	6	WIM	3	3
灣資訊管理夏季研討會	2	6	IM	3	3
資訊管理夏季研討會	2	6	....		

3.4   Merge Words by Longest Match

Finally, we use the longest match rule to merge the similar words in words pool. When merging, the longest word will be kept and the weight of longest word will plus the weight of the similar words. The merged words which also are the keywords of the article are as shown in Table 5.

Table 5. Merge words

Word	Frequency	Weight
第一屆台灣資訊管理夏季研討會(Meaning: The First Taiwan Summer Workshop on Information Management)	2	93
研究 (Meaning: Research)	4	4
TSWIM	3	12

3.5   Program Code

```
'Preprocessing
Title.replace(punctuations, "#")
Context.replace(punctuations, "#")
 'Create the Candidate Characters Pool
For Each s As Char In Context.ToCharArray
        If TempCharactersPool.Contains(s) Then
            TempCharactersFrequency.Item(TempCharactersPool.IndexOf(s)) += 1
        Else
          If s <> '#' Then
            TempCharactersPool.Add(s)
            TempCharactersFrequency.Add(1)
          End If
        End If
Next 'Count the each characters occurrence frequency
For i As Integer = 0 to TempCharactersFrequency.Count - 1
        If TempCharactersFrequency.Item(i) > 1 Then
            CharactersFrequency.Add(TempCharactersFrequency.Item(i))
                CharactersPool.Add(TempCharactersPoo.Item(i))
        End If
Next 'Remove the character which frequency is less than 1
'Create the Words Pool
For i As Integer = 0 To Context.ToList.Count - 1
        If CharactersPool.Contains(Context.ToList.Item(i)) Then
          Dim Word As String = Context.ToList.Item(i)
          For j As Integer = i + 1 To Context.ToList.Count - 1
                If CharactersPool.Contains(Context.ToList.Item(j)) Then
            Word += Context.ToList.Item(j)
                Else
```

```
                If Not TempWordsPool.Contains(Word) And Word.Length > 1 Then
                    TempWordsPool.Add(Word)
                    TempWordsFrequency.Add(Regex.Matches(Context, Word).Count)
                End If
                Exit For
                End If
            Next
        End If
    Next 'Dynamic N-gram
    For i As Integer = 0 To TempWordsPool.Count – 1
        Dim W = TempWordsFrequency.Item(i) *
            Math.Max(1, 3 * Regex.Matches(Title, TempWordsPool.Item(i)).Count)
        TempWordsWeight.Add(W)
    Next 'Calculate the weight of each word
    For i As Integer = 0 to TempWordsWeight.Count - 1
        If TempWordsWeight.Item(i) >= Top5Weight Then
            WordsWeight.Add(TempWordsWeight.Item(i))
            WordsFrequency.Add(TempWordsFrequency.Item(i))
                WordsPool.Add(TempWordsPool.Item(i))
        End If
    Next 'Remove the word which weight is less than Top 5
    'Merge Words by Longest Match
    For i As Integer = 0 To WordsPool.Count - 1
        For j As Integer = 0 To WordsPool.Count - 1
            If i <> j And
                WordsPool.Item(i).Contains(WordsPool.Item(j)) And
                WordsPool.Item(i).Length > WordsPool.Item(j).Length Then
                WordsPool.RemoveAt(j)
                WordsWeight.Item(i) += WordsWeight.Item(j)
            End If
        Next
    Next 'Merge Words; The merged words in WordsPool are the most meaningful
keywords in the article
```

4   Evaluation

We applied the dynamic N-gram approach in five different Chinese article categories (art, technology, health, politics, and entertainment) to extract the keywords from the articles. In order to evaluate the dynamic N-gram approach, we invited 31 volunteer readers to read the five articles and evaluate the extent of those keywords meaningfully represented the articles. The result of reader evaluation is as shown in Table 6. The result indicates the rating mean is 3.826 (scale is 1 to 5) in five different Chinese article categories, 4.065 in art category, 3.903 in technology category, 3.516 in health category, 3.871 in politics category, and 3.774 in entertainment category. All are significant greater than the middle 3. Furthermore, the ANOVA test provides preliminary evidence, p value is .495 which means there is no significant difference in five categories when applying the dynamic N-gram approach to extract article keywords.

Table 6. Reader evaluation

Category	All	Art	Technology	Health	Politics	Entertainment
Rating Mean (n = 31)	3.826	4.065	3.903	3.516	3.871	3.774

ANOVA	Sum of Squares	Mean Square	df	F	p value
	5.071	1.268	4	.852	.495

5 Conclusion

Today, the rapid growth information on the Internet has created an urgent demand for quick and precise way to acquire meaningful information. There are many approaches and researches trying to achieve the goal. One of these approaches, N-gram approach owns advantages such as simple concept, ease of application, reduced search cost, and increased time benefits, but first we have to decide the number of grams before applying these approaches and different number of grams will lead to different results of texts segmentation.

In this paper, we proposed an improved character-based approach, dynamic N-gram and evaluated the accuracy when applied in Chinese news articles. The evaluation result indicated most of the readers agreed the keywords that extracted by dynamic N-gram approach. Even in different news categories, the keywords extraction results still have no significant difference. The first contribution of this approach is the dynamic N-gram could help us to extract the most meaningful keywords in different type Chinese articles without considering the number of grams.

The second contribution is the major segmentation problems, ambiguity and long words have been overcome by using dynamic N-gram. Because the dynamic N-gram could automatically decide the number of grams, the long words still could be segmented and avoids ambiguity by longest match that makes sure the reserved words have clearly meaning for readers.

Although this study proposed a useful approach to extract keywords from Chinese article, there are still some inherent limitations. First, the dynamic N-gram approach did not solve the abbreviation problems. Second, if the context existed a lot of hybrid English and Chinese characters, the English keywords extraction accuracy would be poor. It is suggested that the subsequent studies could be conducted to solve these two primary limitations of this approach.

References

1. Zhang, M., Lu, Z., Zou, C.: A Chinese Word Segmentation Based on Language Situation in Processing Ambiguous Words. Information Sciences 162, 275–285 (2004)
2. Fu, G., Kit, C., Webster, J.J.: Chinese Word Segmentation as Morpheme-based Lexical Chunking. Information Sciences 178, 2282–2296 (2008)

3. Tsai, R.T.: Chinese Text Segmentation: A Hybrid Approach Using Transductive Learning and Statistical Association Measures. Expert Systems with Applications 37(5), 3553–3560 (2010)
4. Haizhou, L., Baosheng, Y.: Chinese Word Segmentation. Language, Information and Computation, 212–217 (1998)
5. Sun, X., Zhang, Y., Matsuzaki, T., Tsuruoka, Y., Tsujii, J.: Probabilistic Chinese Word Segmentation with Non-local Information and Stochastic Training. Information Processing and Management 49(3), 626–636 (2013)
6. Foo, S., Li, H.: Chinese Word Segmentation and Its Effect on Information Retrieval. Information Processing & Management 40, 161–190 (2004)
7. Tong, X., Zhai, C., Milic-Frayling, N., Evans, D.A.: Experiments on Chinese Text Indexing—CLARIT TREC-5 Chinese Track Report. In: TREC (1996)
8. Kwok, K.L.: Lexicon Effects on Chinese Information Retrieval. In: Proc. of 2nd Conf. on Empirical Methods in NLP, pp. 141–148. ACL (1997)
9. Sproat, R., Shih, C.: A Statistical Method for Finding Word Boundaries in Chinese Text. Computer Processing of Chinese and Oriental Languages 4(4), 336–351 (1990)
10. Wu, Z., Tseng, G.: Chinese Text Segmentation for Text Retrieval: Achievements and Problems. Journal of the American Society for Information Science 44(9), 532–542 (1993)
11. Sproat, R., Shih, C., Gale, W., Chang, N.: A Stochastic Finite-state Word-segmentation Algorithm for Chinese. Computational Linguistics 22(3), 377–404 (1996)

Burst Analysis of Text Document for Automatic Concept Map Creation

Wan C. Yoon, Sunhee Lee, and Seulki Lee[*]

Department of Knowledge Service Engineering, Korea Advanced Institute of Science
and Technology, Daejeon, Republic of Korea
{wcyoon,sunheelee,seulki.lee}@kaist.ac.kr

Abstract. In this paper, we propose a new method to extract relationships between words based on the burst analysis for creating a concept map from a document. Concept maps are graphical representation showing the relationships among concepts. An automatically generated concept map shows the whole picture of a certain domain or a document and helps people to understand it. A traditional approach to capture the association relationship between concepts uses co-occurrence of words. In this approach, the highly frequent words usually have strong relation with other words. However, these relations do not necessarily describe the content precisely. Instead of counting co-occurrence of words, the proposed method analyses burst interval of a word for detecting a topic word in a particular period and captures the relation between burst intervals. The case study shows that the proposed method outperforms the co-occurrence method in ranking meaningful relation-ships highly.

Keywords: Concept map, Burst analysis, Word relation.

1   Introduction

Concept maps are graphical representation of knowledge showing the relationships among concepts. A concept map consists of nodes and links. The nodes represent concepts and the links describe the relations between concepts. Since Novak developed a concept map grounded on Ausubel's assimilation theory [2] at the beginning of the 1970s, there have been many researches proving its usefulness in various fields. Concept maps are used as knowledge assessment tools [10], cooperative learning tools [13], and tools that visualize information and knowledge [3]. Usually, a concept map is constructed manually. However, it is a time-consuming process and constructing a concept map describing a whole picture of domain knowledge is hard to do by some experts. Thus, the problem of the automatic extraction of concept maps from documents is raised.

Creating a concept map from text data generally has two steps: keyword extraction and relation extraction. The automatically generated concept maps usually do not have labels on links but have the association relations between concepts. A traditional

---

[*] Corresponding Author.

A. Moonis et al. (Eds.): IEA/AIE 2014, Part II, LNAI 8482, pp. 407–416, 2014.

approach to capture the association relation between two words is the analysis of co-occurrence of words in a specified window such as a document, sentence, and paragraph. In this approach, the highly frequent words usually have strong relation with other words, and these relations do not necessarily describe the content precisely. In this paper, we propose the method which infers the association relation from the burst intervals of words.

A burst interval is a period when occurrence of a certain word is denser than in other periods. If a word has a burst interval, it is likely that the word is a topic word in a particular period. Burst analysis of a text stream is usually used in the research of temporal text summarization [4, 14]. In this study, we observed burst intervals of words in a document and their temporal relationships. Then, the strength of the relationship between words is calculated by a weighted sum of the relation patterns of burst intervals.

We conducted a case study for comparing the concept maps constructed from the co-occurrence based method and burst analysis based method. It shows that meaningful relationships are ranked highly in the proposed method, whereas the relation with high frequency is ranked highly in a co-occurrence method.

This paper is organized as follows. Section 2 discusses the related work. Section 3 describes the proposed method. Section 4 explains about a case study and its result. Section 5 concludes this study.

2   Related Work

Relation extraction between two concepts for creating a concept map is generally performed for two alternative types of relations; semantic relation and association relation. A semantic relationship between two concepts is represented as a verb phrase between two noun phrases. Oliveira et al. [11] tagged a text file using the external lexicon (WordNet base) and built predicates that map relations between two concepts from parsing sentences in a document. Two other studies, by Valerio et al. [15] and Zouaq et al. [16], also found verb phrases in the sentences for extracting labeled relationships of a concept map. These studies need natural language processing techniques to understand a text, and the type of relationship is limited (e.g. 'is-a' and 'have'). In this study, we extract the association relations between concepts.

An association relation represents psychological relationship between two words that are conceptually related but not equivalent such as a synonym. One of traditional approach of association relation extraction for creating concept maps is co-occurrence analysis. Co-occurrence represents simultaneous occurrence of two words in a specified window such as the same document, sentence, phrase, etc. It is based on a theory that a higher frequency of co-occurrence implies closer relationship between two words [12]. Clariana et al. [6] used the co-occurrences of terms in each sentence for calculating the weight of relationship for extracting concept maps. Chen et al. [5] constructed concept maps from academic articles in eLearning domain. They proposed the method of extracting relationship between the keywords which is directly proportional to the number of times two keywords appeared in the same sentence and is inversely proportional to the average distance between words when co-occurring. These studies use co-occurrence analysis of words, but the proposed method analyzes burst interval of a word for detecting a topic word in a period and their relationships. By using the proposed method, we can avoid the strong relationship caused by high word frequency.

3   Proposed Method

In this section, we describe the burst-based relation extraction method for creating a concept map from a document. It is assumed that representative keywords of a document are given. At first, the method detects bursts of each word through a document. Then, for inferring the relationship between words, the temporal relations of bursts are compared. The weight of relationship is a weighted sum of the relation patterns of burst intervals.

3.1   Burst Detection Algorithm

To find a burst interval of each word in a document, we adopt Kleinberg's [9] definition of burst and infinite-state automaton model. Kleinberg's algorithm identifies bursts in document streams. It assumes that the gap x_t in time between document i_t and i_{t+1} follows an exponential distribution $f_{i_t}(x_t) = \alpha_{i_t} e^{-\alpha_{i_t} x_t}$, and a stream of related documents is generated from a multiple-state automaton. Each state has its own parameter α_{i_t} in probability density function, and there is a probability for changing state.

Finding an optimal sequence of states is equivalent to finding a state sequence $Q = (q_{i_1}, \dots, q_{i_n})$ that minimizes the following cost function:

$$c(Q|X) = \left(\sum_{t=0}^{n-1} \tau(i_t, i_{t+1}) \right) + \left(\sum_{t=1}^{n} -ln f_{i_t}(x_t) \right) \tag{1}$$

where q_{i_1} is a state of a document i_t, and $X = (x_1, \dots, x_n)$ is the gap sequence between documents. $\tau(i_t, i_{t+1})$ denotes the cost of a state transition from state q_t of i_t to state q_{t+1} of i_{t+1}.

We apply Kleinberg's burst detection algorithm to a single document. A sequence of sentences is used as input for Kleinberg's algorithm instead of a sequence of documents. Therefore, a burst interval of a word is an intensive period with its high frequency in a sequence of sentences. All the other parts remain the same as in [9]. After performing burst detection, we get the sequence of bursts of each word. Let $B_w = (B_{w,1}, \dots, B_{w,k})$ denote the bursts of a word w, and let $B_{w,i}$ denote the i-th burst of the word w.

3.2   Temporal Relation Patterns between Bursts

After detecting the burst intervals of each word, the next step is to compare the burst intervals between words and to extract the relation from them. Burst interval continues from starting sentence index until ending sentence index, so bursts from two words can have several temporal relation patterns. We assume that not only the bursts which occur together but also the bursts which occur successively have some information about the relation of the words. We adopt Allen's [1] interval algebra for expressing temporal relation patterns between burst intervals. Allen's interval algebra is widely used in temporal reasoning and planning [7].

The key concepts of Allen's approach, interval algebra, are as follows. Temporal intervals are taken as primitives. Temporal relations, which may hold between intervals, are specified: before, equal, meets, overlaps, during, starts, finishes and their inverses. Table 1 shows temporal relation patterns of burst intervals. In this study, to measure the relatedness between two words, we assign a weight to each temporal relation. We induced these weights from comparing the result from the proposed method and domain expert's pairwise assessments about the relatedness between words. Then weights are selected from minimizing the sums of squared error (SSE) value between the expert's assessments and the result from the proposed method with given weights.

Table 1. Temporal relation patterns of burst intervals and weight adapted from Allen's interval algebra

Relation	Illustration	Weight
$B_{X,i}$ before $B_{Y,j}$	$---B_{X,i}---$ $\quad\quad\quad ---B_{Y,j}---$	0
$B_{X,i}$ equals $B_{Y,j}$	$------B_{X,i}------$ $------B_{Y,j}------$	10
$B_{X,i}$ meets $B_{Y,j}$	$---B_{X,i}---$ $\quad\quad ---B_{Y,j}---$	4
$B_{X,i}$ overlaps $B_{Y,j}$	$------B_{X,i}------$ $\quad\quad ------B_{Y,j}------$	2
$B_{X,i}$ during $B_{Y,j}$	$---B_{X,i}---$ $---------B_{Y,j}---------$	4
$B_{X,i}$ starts $B_{Y,j}$	$---B_{X,i}---$ $---------B_{Y,j}---------$	9
$B_{X,i}$ finishes $B_{Y,j}$	$\quad\quad ---B_{X,i}---$ $---------B_{Y,j}---------$	9

3.3   Measuring Relation Weight between Two Concepts

Synchronizing the Position of Burst Patterns. When we consider the start or the end of a burst, if we strictly count patterns based on a sentence index, the patterns of 'equals', 'meets', 'starts', and 'finishes' will be very rare. When we see the bursts from the real data, it is natural to say that two bursts start together even though their starting points are slightly different. For example, in Fig. 1, the second burst of the word 'X', $B_{X,2}$, starts at sentence index 150 and the first burst of the word 'Y', $B_{Y,1}$, starts at sentence number 146. Even though there is a gap of 4 sentences between the starting positions of $B_{X,2}$ and $B_{Y,1}$, two bursts seem to have 'starts' relation.

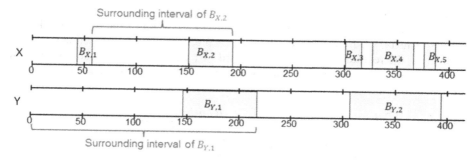

Fig. 1. Bursts of word 'X' and word 'Y'

To match the patterns of 'equals', 'meets', 'starts', and 'finishes' less strictly, we suggest the concept of the tolerance of a gap (TG). If a gap between two burst's starting or ending sentence index we want to compare is within the tolerance of a gap, then the gap is tolerable, and the two indexes are considered to be in the same position.

For calculating the tolerance of a gap, we consider two features: a length of a non-burst interval before a burst and a length of a current burst. The length of non-burst interval is important because if a burst occurs after relatively long non-burst interval, the burst seems to be more meaningful in a text. The length of a current burst is also important because a tolerance of a gap should be proportional to the length of a current burst. Surrounding interval of $B_{X,i}$, $SI(B_{X,i})$, is defined as the sum of previous non-burst interval and current burst interval (see Fig. 1). Therefore, we define a tolerance of a gap between $B_{X,i}$ and $B_{Y,j}$, $TG(B_{X,i}, B_{Y,j})$, and its rule as follows:

$$TG(B_{X,i}, B_{Y,j}) = \bigl(SI(B_{X,i}) + SI(B_{Y,j})\bigr) * \alpha \tag{2}$$

If $\Delta t\bigl(index\ of\ B_{X,i}, index\ of\ B_{Y,j}\bigr) < TG\bigl(B_{X,i}, B_{Y,j}\bigr),$
$$then\ synchronize\ the\ indexes. \tag{3}$$

where index of $B_{X,i}$ is starting or ending sentence index of $B_{X,i}$, Δt is the gap between two sentence indexes, and α is the parameter (e.g. 0.05).

In the case of $B_{X,2}$ and $B_{Y,1}$ in Fig. 1, the gap between starting indexes of two bursts is 4, since the burst of $B_{X,2}$ starts at 150 and the burst of $B_{Y,1}$ start at 146. If α is 5%, $TG(B_{X,2}, B_{Y,1})$ is $((91+38)+(145+65))*0.05=16.95$. Thus, the starting positions of two bursts $B_{X,2}$ and $B_{Y,1}$ are synchronized and have a temporal relation '$B_{X,2}$ starts $B_{Y,1}$'.

Assigning Weight to Burst Patterns. After synchronizing the starting and ending sentence indexes of bursts, proper weights are assigned to each temporal relation pattern of burst intervals. During assigning weights, some bursts can have more than one weight. Then the maximum weight is selected for each burst. For example, in Fig. 1, $\mathbf{B_{Y,2}}$ has three temporal relations with $\mathbf{B_{X,3}}$, $\mathbf{B_{X,4}}$, and $\mathbf{B_{X,5}}$. In this case, the maximum weight 9 ('starts' or 'finishes' relation) is selected as the relation weight of $\mathbf{B_{Y,2}}$. The relation weights of $\mathbf{B_{X,3}}$, $\mathbf{B_{X,4}}$, and $\mathbf{B_{X,5}}$ are respectively 9, 4, and 9.

Normalizing the Weights. After the weights assigned for bursts from the temporal relation patterns, they are normalized according to a portion of the total burst length and its density of frequency in the burst. Word relation is directional. When considering the relation of word 'X' and word 'Y', we can define a relationship with both directions. The normalized relation weight (NRW) from word X to word Y is defined as follows:

$$
\begin{aligned}
NRW_{X,Y} = \sum_i \bigg(& relation\ weight\ of\ B_{X,i} \times \frac{length\ of\ B_{X,i}}{total\ length\ of\ B_X} \\
& \times \frac{Word\ frequency\ in\ B_{X,i}}{length\ of\ B_{X,i}} \bigg) \\
= \sum_i \bigg(& relation\ weight\ of\ B_{X,i} \times \frac{Word\ frequency\ in\ B_{X,i}}{total\ length\ of\ B_X} \bigg)
\end{aligned} \tag{4}
$$

The first term in the equation represents a portion of the total burst length, and the second term is its density of frequency in the burst.

For example, if X= 'ontology' and Y= 'information' (Fig. 2), then two normalized relation weights are:

$$
NRW_{X,Y} = \left(9 \times \frac{28}{126}\right) + \left(4 \times \frac{20}{126}\right) + 0 + 0 + \left(4 \times \frac{21}{126}\right) + \left(9 \times \frac{12}{126}\right) = 4.16
$$

$$
NRW_{Y,X} = \left(9 \times \frac{7}{148}\right) + \left(4 \times \frac{6}{148}\right) + \left(4 \times \frac{13}{148}\right) + 0 + 0 + \left(9 \times \frac{7}{148}\right) = 1.36
$$

In this example,' ontology' is more related to 'information'.

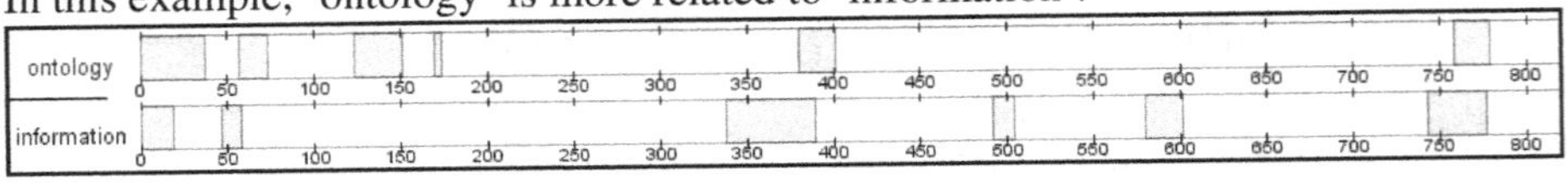

Fig. 2. Burst patterns of 'ontology' and 'information'

4   Case Study

We conducted a case study for comparing the concept maps constructed from the co-occurrence based relation extraction method and burst analysis based method. We used the document 'Ontology Visualization Method-A Survey [8]' which has 43 pages. When text content is extracted excluding title, figures, tables, references, etc. it has 783 sentences and 727 different words.

4.1   Keyword Extraction

For extracting representative keywords from a document, some preprocessing procedures were performed such as part-of-speech tagging, converting to lowercase, lemmatization, and stop words removal. NLTK (Natural Language Toolkit) [17] for Python was used for text processing. We treated only nouns as candidates for representative keywords of a document. Then, the word frequency was calculated.

We removed words that have a lower frequency rate than the frequency rate from general English. We assume that words that have less frequency rate in a document than its usual frequency rate in English language cannot represent a document precisely even though its frequency ranking in a document is relatively high. We compared a frequency rate to English word frequency data about the top 5,000 frequent words from Corpus of Contemporary American English (COCA) [18]. Finally, we selected top 20 frequent words as keywords of a document for this case study. The list of words is the following:

Table 2. List of extracted key words

Word (Frequency)
visualization (184), node (179), ontology (159), user (124), method (69), tree (61), technique (56), level (55), class (53), task (49), information (45), graph (40), space (39), tool (39), order (35), relation (34), instance (34), case (34), hierarchy (33), evaluation (32)

4.2   Method Implementation and Visualization

We selected top twenty frequent words from keyword extraction step. Then, we detected bursts of each word in the document using burst package in R [19] with arguments (s=2, gamma=0.25). Level 2 bursts are selected. For comparing the result with traditional co-occurrence approach, we obtained the co-occurrence counts using a window of ±3 sentences.

In the visualization of the concept map, the thickness of the edge represents the degree of relatedness. From the proposed method, we can extract two directional relation pairs from two words (i.e., $NRW_{X,Y}$ and $NRW_{Y,X}$ from words X and Y). For reducing the complexity of network visualization, the concept map shows only a greater relationship between two words if its relations are in a similar range. If there is a lot of difference between two relations, we defined it as an unbalanced relationship and represented it as a directed edge. In this case study, if ($NRW_{X,Y}$/ $NRW_{Y,X}$> 2, then it is unbalanced relationship, and represented as X → Y. This means word X is more related to word Y.

4.3   Results

From co-occurrence analysis and burst interval analysis of top twenty keywords, we extracted 380 relationships and 308 relationships respectively. Top 30 relationships from each analysis are shown in Table 3 and 4. From the proposed method, we can extract $NRW_{X,Y}$ and $NRW_{Y,X}$ for words X and Y. In co-occurrence based method, we can extract one relation between two words. For easily comparing the results of both analyses, we considered that there are two directional relation pairs between two words when we select relationships from co-occurrence analysis. We constructed concept maps with 20 representative words and their relationships. Fig. 3 (a) and (b) shows the concept map when the number of top edges in the concept map to be 21. If there are many relations in the map, the concept map is too complex to interpret it. Therefore, we choose most important relationships for the human analysis.

In the concept map from co-occurrence (Fig. 3 (a)), most links are related to 'visualization', 'ontology', and 'node' words. That is because three words are most frequent. When frequency of a word is high, it increases a chance to co-occur with other words. In the concept map from burst analysis (Fig. 3 (b)), highly connected words are 'ontology' and 'technique'. Common relationships (see Table 3 and 4) from both analyses represent well about the purpose of this article which is "to introduce several ontology visualization methods and techniques". High ranked relations among distinctive relationships from co-occurrence analysis (see Table 3) are related to the word 'user'. The word 'user' is distributed evenly over the text because there is no part explaining about the user. Therefore, its burst interval is short even though the frequency of word 'user' is high. High ranked relations among distinctive relationships from burst analysis (see Table 4) are (class, instance, weight: 4.065, unbalanced: →), (node, technique, weight: 3.560, unbalanced: →), and (ontology, instance, weight: 3.263, unbalanced: →). These are more meaningful than the relationships from co-occurrence analysis including the word 'user'. Their semantics highly accord with the content "an ontology is a formal explicit description of concepts, or classes in a do-main of discourse. Each class most probably has instances which are the actual data. In the visualization method, instances are represented as nodes." We can conclude that proposed method using burst analysis assigns more weight to a topic word in a particular period of time in a document and detects their temporal relationships, so meaningful relationships are ranked highly.

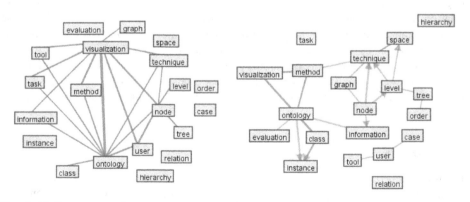

Fig. 3. (a) Concept map from co-occurrence analysis (relation threshold = 67, # of edges = 21) (b) Con-cept map from burst relation analysis (relation threshold =1.7, # of edges = 21)

Table 3. Top 30 relationships from co-occurrence analysis

Relation type	Relationship (word X, word Y, $co-occurrence_{X,Y}$)
Common relationships	(ontology, visualization, 347), (visualization, ontology, 347), (method, visualization, 159), (visualization, method, 159), (visualization, node, 142), (method, ontology, 106), (ontology, method, 106), (class, ontology, 91), (ontology, class, 91), (node, level, 89)
Distinctive relationships	(node, user, 160), (user, node, 160), (user, visualization, 153), (visualization, user, 153), (node, visualization, 142), (ontology, user, 137), (user, ontology, 137), (task, visualization, 124), (visualization, task, 124), (tool, visualization, 110), (visualization, tool, 110), (node, tree, 105), (tree, node, 105), (technique, visualization, 104), (visualization, technique, 104), (node, ontology, 96), (ontology, node, 96), (ontology, tool, 91), (tool, ontology, 91), (level, node, 89)

Table 4. Top 30 relationships from relation analysis of bursts

Relation type	Relationship (word X, word Y, $NRW_{X,Y}$, (unbalanced directionality))
Common relationships	(ontology, visualization, 4.197), (ontology, class, 4.139), (class, ontology, 4.065), (visualization, method, 3.173), (visualization, ontology, 3.121), (node, level, 2.967, →), (method, visualization, 2.811), (ontology, method, 2.482), (method, ontology, 1.945), (visualization, node, 1.659)
Distinctive relationships	(class, instance, 4.065, →), (node, technique, 3.560, →), (ontology, instance, 3.263, →), (ontology, information, 2.920), (node, information, 2.154, →), (level, technique, 2.092, →), (instance, class, 2.017, ←), (graph, node, 2.015), (method, technique, 1.906), (technique, space, 1.903), (user, case, 1.868), (user, tool, 1.868), (space, technique, 1.837), (graph, technique, 1.831), (ontology, evaluation, 1.803), (level, space, 1.782, →), (level, tree, 1.782), (tree, order, 1.742), (visualization, space, 1.665), (method, information, 1.661)

5   Conclusion

This study focus on creating a concept map of a document using burst intervals of words. When creating a concept map automatically, the traditional method measures the relation of words using their co-occurrence information. In that method, the highly frequent words usually have strong co-occurrence relation with other words. However, these relations do not necessarily describe the content precisely. The proposed method solves this problem by using burst intervals of words in a document and their temporal relationships. In this way, the burst of a word in which the word is a topic word in a specific section is highlighted even though it has relatively low word frequency. Moreover, the proposed method extract the relationships from not only the bursts which occur together but also the bursts which occur successively. The case study shows that the proposed method outperforms the co-occurrence method in ranking meaningful relationships highly. Therefore, a concept map constructed by proposed method implies more information about the document.

Acknowledgments. This work was supported by the Small and Medium Business Administration (SMBA). [Project No. S2096670].

References

1. Allen, J.F.: Maintaining knowledge about temporal intervals. Communications of the ACM 26, 832–843 (1983)
2. Ausubel, D.P., Novak, J.D., Hanesian, H.: Educational psychology: A cognitive view (1968)
3. Cañas, A.J., Carff, R., Hill, G., Carvalho, M., Arguedas, M., Eskridge, T.C., Lott, J., Carvajal, R.: Concept Maps: Integrating Knowledge and Information Visualization. Knowledge and Information Visualization, 205–219 (2005)
4. Chakrabarti, D., Punera, K.: Event summarization using tweets. In: Proceedings of the Fifth International AAAI Conference on Weblogs and Social Media, pp. 66–73 (2011)
5. Chen, N.-S., Kinshuk, W.C.-W., Chen, H.-J.: Mining e-Learning domain concept map from academic articles. Computers & Education 50, 1009–1021 (2008)
6. Clariana, R.B., Novak, R.: A computer-based approach for translating text into concept map-like representations. In: Proceedings of the First International Conference on Concept Mapping (2004)
7. Hirsch, R., Hodkinson, I.: Relation Algebras by Games. Gulf Professional Publishing (2002)
8. Katifori, A., Halatsis, C., Lepouras, G., Vassilakis, C., Giannopoulou, E.: Ontology visualization methods—a survey. ACM Comput. Surv. 39 (2007)
9. Kleinberg, J.: Bursty and Hierarchical Structure in Streams. Data Mining and Knowledge Discovery 7, 373–397 (2003)
10. McClure, J.R., Sonak, B., Suen, H.K.: Concept map assessment of classroom learning: Reliability, validity, and logistical practicality. Journal of Research in Science Teaching 36, 475–492 (1999)
11. Oliveira, A., Pereira, F.C., Cardoso, A.: Automatic reading and learning from text. In: Proceedings of the International Symposium on Artificial Intelligence, ISAI (2001)
12. Rijsbergen, C.J.V.: A theoretical basis for the use of co-occurrence data in information retrieval. Journal of Documentation 33, 106–119 (1977)
13. Stoyanova, N., Kommers, P.: Concept mapping as a medium of shared cognition in computer-supported collaborative problem solving. Journal of Interactive Learning Research 13, 111–133 (2002)
14. Subašic, I., Berendt, B.: From bursty patterns to bursty facts: The effectiveness of temporal text mining for news. In: Proc. ECAI (2010)
15. Valerio, A., Leake, D.: Jump-Starting Concept Map Construction with Knowledge Extracted From Documents. In: Proceedings of the Second International Conference on Concept Mapping, pp. 296–303 (2006)
16. Zouaq, A., Nkambou, R.: Building Domain Ontologies from Text for Educational Purposes. IEEE Transactions on Learning Technologies 1, 49–62 (2008)
17. NLTK (Natural Language Toolkit), http://nltk.org
18. The Corpus of Contemporary American English (COCA), http://corpus.byu.edu/coca/
19. Bursts: Markov model for bursty behavior in streams, http://cran.r-project.org/web/packages/bursts/index.html

Scale Invariant Multi-length Motif Discovery

Yasser Mohammad[1,2] and Toyoaki Nishida[2]

[1] Assiut University, Egypt
[2] Kyoto University, Japan
yasserm@aun.edu.eg, nishida@i.kyoto-u.ac.jp

Abstract. Discovering approximately recurrent motifs (ARMs) in time-series is an active area of research in data mining. Exact motif discovery was later defined as the problem of efficiently finding the most similar pairs of timeseries subsequences and can be used as a basis for discovering ARMs. The most efficient algorithm for solving this problem is the MK algorithm which was designed to find a single pair of timeseries subsequences with maximum similarity at a known length. Available exact solutions to the problem of finding top K similar subsequence pairs at multiple lengths (which can be the basis of ARM discovery) are not scale invariant. This paper proposes a new algorithm for solving this problem efficiently using scale invariant distance functions and applies it to both real and synthetic dataset.

1 Introduction

Discovering approximately recurrent patterns in timeseries is a basic problem in data mining and provides the basis for solving many real world problems (e.g. gesture discovery [10], any-time nearest neighbor algorithms [12], fluid imitation [8], etc). Consider a robot watching free hand motion of a human operator while operating another actor using gestures [10]. The ability to automatically discover recurring motion patterns allows the robot to learn important gestures related to this domain. Consider an infant listening to the speech around it. The ability to discover recurring speech patterns (words) can be of great value in learning the vocabulary of language. In both in these cases, and in uncountable others, the patterns do not recur exactly in the perceptual space of the learner. These cases motivate our search for an unsupervised algorithm that can discover these kinds of approximately recurring motifs (ARMs) in general time-series. Several algorithms have been proposed for solving this problem [7] [5] [1] [9].

A promising approach to solve the ARM problem is to use an algorithms that finds *exactly* the K timeseries subsequence pairs (called 2-motifs) of maximal similarity then use them as the basis for discovering recurrent patterns which by definition must have maximal similarity between its pairs. The naive algorithm for solving this problem exactly for a time series of length n and motifs of lengths between l_1 and $l_2 = l_1 + l$ has a time complexity of $O\left(n^2 K l\right)$. This quadruple complexity makes it impractical to apply this algorithm except for short timeseries, short motifs, and short motif length ranges.

A. Moonis et al. (Eds.): IEA/AIE 2014, Part II, LNAI 8482, pp. 417–426, 2014.

The simpler problem of exact discovery of the top 2-motif of a given length in a timeseries was defined by Mueen and Keogh [12] in 2009 and an efficient exact solution with amortized linear complexity was proposed (called the MK algorithm). This algorithm reduced the amortized time complexity from quadratic to linear which makes it practical to apply it to moderately long timeseries. The MK algorithm uses the Euclidean distance between zscore normalized subsequences as a dissimilarity measure. The main advantage of this distance function is that it is offset and scale invariant. It was also shown that it can provide a comparable performance to Dynamic Time Wrapping [4].

Mohammand and Nishida [9] proposed MK+ which is an efficient extension of MK to discover top K 2-motifs of a given length using the same distance function and showed that it outperforms iterative application of the MK algorithm. MK+ was further extended in [6] (MK++) to discover top K 2-motifs of a range of lengths but assuming that the distance between two subsequences of the timeseries cannot decreased with increased length. This assumption is true of the Euclidean distance and Euclidean distance between mean-shifted subsequences but is not true for zscore normalized subsequences. This means that MK++ cannot be used to discover scale-invariant 2-motifs which means that it cannot be a basis for scale invariant ARM discovery.

Recently, Mueen proposed MOEN [11] for solving the scale invariant version of the problem tackled in this paper. The main idea of MOEN is to calculate a lower bound on the distance between any two subsequences at length l given this distance at length $l-1$. Using this lower bound, it is possible to efficiently discover 2-motifs at different lengths. MOEN works with zscore normalized subsequences but the proposed algorithm can be applied to a more general class of distance functions.

This paper proposes two solutions to this problem: The first approach is to use a different normalization technique by dividing the mean shifted subsequence by its range (difference between maximum and minimum values) rather than standard deviation and using MK++ with minimal modification. We show that this renders the algorithm approximate but in most cases leads to exactly the same results as the exact algorithm. The second approach called sMD (for scale-invariant Motif Discovery) is to drive an incremental method to calculate any normalized distance function and then to use it to find motifs at all lengths in parallel leading to an exact 2-motif discovery algorithm.

The rest of the paper is organized as follows: Section 2 gives the problem statement. Section 3 describes MK and MK++ that form the basis of the proposed algorithm. Section 4 details the proposed incremental distance calculation method and section 5 gives the details of the proposed algorithm which is evaluated in section 6. The paper is then concluded.

2   Problem Statement

A time series $x(t)$ is an ordered set of T real values. A subsequence $x_{i,j} = [x(i) : x(j)]$ is a contiguous part of a time series x. In most cases, the distance between overlapping subsequences is considered to be infinitely high to avoid assuming that two sequences are matching just because they are shifted versions of each other (these are called trivial motifs [3]). In this paper we utilize the following definitions:

Definition 1. *2-Motif:* Given a timeseries x of length T, a motif length L, a maximum internal overlap $0 \geq wMO \geq 1$, maximum between-motifs overlap $0 \geq bWO \geq 1$, and a distance function $D(.,.)$; the top 2-motif is a pair of subsequences s_1, s_2 of length L with minimum distance compared with any other pair of subsequences in the time-series that have an overlap less or equal to wMO, the 2nd 2-motif is the pair of subsequences overlapping the top 2-motif no more than bMO that have the minimum distance compared with any other pair satisfying this overlapping condition. The Kth 2-motif is the pair of subsequences overlapping none of the top to the $K-1^{th}$ 2-motif more than bMO that have the minimum distance compared with any other pair satisfying this overlapping condition.

Using this definition, the problem statement of this paper can be stated as: *Given a time series x, minimum and maximum motif lengths (L_{min} and L_{max}), a maximum allowed within-motif overlap (wMO), and a maximum allowed between-motifs overlap (bMO), find the top K 2-motifs with smallest motif distance among all possible pairs of subsequences.*

3   MK and MK++ Algorithms

The MK algorithm finds the top 2-occurrences motif in a time series. The main idea behind MK algorithm [12] is to use the triangular inequality to prune large distances without the need for calculating them. For metrics $D(.,.)$ (including the Euclidean distance), the triangular inequality can be stated as:

$$D(A, B) - D(C, B) \leq D(A, C) \qquad (1)$$

Assume that we have an upper limit on the distance between the two occurrences of the motif we are after (th) and we have the distance between two subsequences A and C and some reference point B. If subtracting the two distances leads a value greater than th, we know that A and C cannot be the motif we are after without ever calculating their distance. By careful selection of the order of distance calculations, MK algorithm can prune away most of the distance calculations required by a brute-force quadratic motif discovery algorithm. The availability of the upper limit on motif distance (th), is also used to stop the calculation of any Euclidean distance once it exceeds this limit. Combining these two factors, 60 folds speedup was reported in [12] compared with the brute-force approach.

The inputs to the algorithm are the time series x, its total length T, motif length L, and the number of reference points N_r. The algorithm starts by selecting a random set of N_r reference points. The algorithm works in two phases: The first phase (called hereafter referencing phase) is used to calculate both the upper limit on best motif distance and a lower limit on distances of all possible pairs. During this phase, distances between the subsequences of length L starting at the N_r reference points and all other $T - L + 1$ points in the time series are calculated resulting in a distance matrix of dimensions $N_r \times (T - L + 1)$. The smallest distance encountered (D_{best}) and the corresponding subsequence locations are updated at every distance calculation.

The final phase of the algorithm (called scanning step hereafter) scans all pairs of subsequences in the order calculated in the referencing phase to ensure pruning most of the calculations. The scan progressed by comparing sequences that are k steps from each other in this ordered list and use the triangular inequality to calculate distances only if needed updating D_{best}. The value of k is increased from 1 to $T - L + 1$. Once a complete pass over the list is done with no update to d_{best}, it is safe to ignore all remaining pairs of subsequences and announce the pair corresponding to D_{best} to be the *exact* motif.

A better approach to discover the top K 2-motifs of a given length was suggested in [9] called MK+ that uses a single scanning rather than K-scanning runs. This approach can also be applied for every length to solve our problem.

Mohammad and Nishida [6] recently proposed an algorithm for solving the multi-length motif discovery problem (by iteratively running a modified version of MK) called MK++. The MK++ algorithm starts by detecting 2-motifs at the shortest length (L_{min}) and progressively finds 2-motifs at higher lengths. The algorithm keeps three lists: D_{bests} representing a sorted list of K best distances encountered so far and L_{bests} representing the 2-occurrence motif corresponding to each member of D_{bests}, and μ_{bests} keeping track of the means of the subsequences in L_{bests}. The *best-so-far* variable of MK is always assigned to the maximum value in D_{bests}. During the referencing phase, the distance between the current reference subsequences and all other subsequences of length L_{min} that do not overlap it with more than $wMO \times L_{min}$ points are calculated. For each of these distances (d) we apply the following rules in order:

Rule 1. If the new pair is overlapping the corresponding $L_{bests}(i)$ pair with more than $wMO \times L$ points, then this i is the index in D_{bests} to be considered

Rule 2. If *Rule1* applies and $D < D_{bests}(i)$, then replace $L_{bests}(i)$ with P.

Rule 3. If *Rule1* does not apply but $D < D_{bests}(i)$, then we search L_{bests} for all pairs $L_{bests}(i)$ for which *Rule1* applies and remove them from the list. After that the new pair P is inserted in the current location of L_{bests} and D in the corresponding location of D_{bests}

The main problem with MK++ is that it assumes that the distance function is nondecreasing which makes it inappropriate for scale-invariant distance functions.

4   Incremental Scale-Invariant Distance Calculation

We utilize the following notation: x_k is the k's element of the timeseries x where x is an ordered list of real numbers of length $L \geq l$. The symbols μ_x^l, σ_x^l,

mx_x^l, mn_x^l stand for the mean, standard deviation, maximum and minimum of x^l. The normalization constant r_x^l is assumed to be a real number calculated from x^l and is used in this paper to achieve scale-invariance by either letting $r_x^l = \sigma_x^l$ (zscore normalization), or $r_x^l = mx_x^l - mn_x^l$ (range normalization). The distance function (between any two timeseries x and y) used in this paper has the general form:

$$D_{xy}^l = \sum_{k=0}^{l-1} \left(\frac{x_k - \mu_x^l}{r_x^l} - \frac{y_k - \mu_y^l}{r_y^l} \right)^2 \tag{2}$$

This is an Euclidean distance between two subsequences $\bar{x}$ and $\bar{y}$, where $\bar{z}_k = (z_k^l - \mu_z^l)/r_z^l$. This means that it satisfies the triangular inequality which allows us to use the speedup strategy described in section 3. Nevertheless, because of the dependence of r_x^l and r_y^l on data and length, it is no longer true that $D_{xy}^{l+1} \geq D_{xy}^l$ and we cannot directly use MK++ [6]. Moreover, once any of these two values change, we can no longer use any catched values of $\bar{x}$ and $\bar{y}$.

We need few more definitions: $\alpha_x^l = r_x^{l-1}/r_x^l$, $\theta_{xy}^l = r_x^l/r_y^l$, $\Delta_k^l = x_k - \theta_{xy}^l y_k$, $^n\mu_{xy}^m = \mu_x^m - \theta_{xy}^n \mu_y^m$ and $\mu_{xy}^l = {}^l\mu_{xy}^l$. Notice that it is trivial to prove that the mean of the sequence $\left\langle \Delta_{xy}{}^l \right\rangle$ is equal to μ_{xy}^l.

The first contribution of this paper is a novel incremental formula for calculating scale invariant distances between time-series subsequences which is stated in the following theorem:

Theorem 1. For any two timeseries x and y of lengths $L_x > l$ and $L_y > l$, and using a normalized distance function D_{xy}^l of the form shown in Equation 2, we have:

$$D_{xy}^{l+1} = D_{xy}^l + \frac{1}{(r_x^l)^2} \begin{pmatrix} \left(\left(\alpha_x^l\right)^2 - 1 \right) \sum_{k=0}^{l-1} \left(x_k^2\right) + 2 \left(\theta_{xy}^l - \left(\alpha_x^l\right)^2 \theta_{xy}^{l+1} \right) \sum_{k=0}^{l-1} x_k y_k \\ + \left(\left(\alpha_x^{l+1}\theta_{xy}^{l+1}\right)^2 - \left(\theta_{xy}^l\right)^2 \right) \sum_{k=0}^{l-1} \left(y_k^2\right) + l\left(\mu_{xy}^l\right)^2 + \left(\alpha_x^{l+1}\Delta_l^{l+1}\right)^2 \\ - (l+1)\left(\alpha_x^{l+1}\mu_{xy}^{l+1}\right)^2 \end{pmatrix}$$

A sketch of the proof for Theorem 1 is:

$$D_{xy}^l = \sum_{k=0}^{l-1} \left(\frac{x_k - \mu_x^l}{r_x^l} - \frac{y_k - \mu_y^l}{r_y^l} \right)^2 = \left(\frac{1}{r_x^l}\right)^2 \sum_{k=0}^{l-1} \left(x_k - \mu_x^l - \theta_{xy}^l \left(y_k - \mu_y^l \right) \right)^2$$

$$\therefore D_{xy}^l (x,y) = \left(\frac{1}{r_x^l}\right)^2 \sum_{k=0}^{l-1} \left(\Delta_k^l - \mu_{xy}^l \right)^2$$

Notice that this is the form of a variance equation (since the mean of the sequence $\left\langle \Delta_{xy}{}^l \right\rangle$ is equal to μ_{xy}^l) and by simple manipulations we can arrive at the following equation:

$$D_{xy}^l = \left(r_x^l\right)^{-2} \left(-l\left(\mu_{xy}^l\right)^2 + \sum_{k=0}^{l-1} \left(\Delta_k^l\right)^2 \right) \tag{3}$$

From Equation 2, it follows that:

$$D_{xy}^{l+1} = \left(r_x^{l+1}\right)^{-2} \left(-(l+1)\left(\mu_{xy}^{l+1}\right)^2 + \sum_{k=0}^{l}\left(\Delta_k^{l+1}\right)^2\right) \tag{4}$$

Subtracting Equations 3 from Equation 4, using the definitions of α and θ given in this section and after some manipulations we get the equation in Theorem 1.

The important point about Theorem 1, is that it shows that by having a running sum of x_k, y_k, $(x_k)^2$, $(y_k)^2$, and $x_k y_k$, we can incrementally calculate the scale invariant distance function for any length l given its value for the previous length $l-1$. This allows us to extend the MK+ algorithm directly to handle all motif lengths required in parallel rather than solving the problem for each length serially as was done in MK++.

The form of D_{xy}^{l+1} as a function of D_{xy}^l is quite complicated but it can be simplified tremendously if we have another assumption:

Lemma 1. For any two timeseries x and y of lengths $L_x > l$ and $L_y > l$, and using a normalized distance function D_{xy}^l of the form shown in Equation 2, and assuming that $r_x^{l+1} = r_x^l$ and $r_y^{l+1} = r_y^l$, we have:

$$D_{xy}^{l+1} = D_{xy}^l + \frac{1}{(r_x^l)^2}\left(\frac{l}{l+1}\right)\left(\mu_{xy}^l - \Delta_l^l\right)^2$$

Lemma 1 can be proved by substituting in Theorem 1 noticing that given the assumptions about r_x^l and r_y^l, we have $\Delta_k^{l+1} = \Delta_k^l$ and $^{l+1}\mu_{xy}^{l+1} = {}^l\mu_{xy}^{l+1}$.

What Lemma 1 shows is that if the normalization constant did not change with increased length, we need only to use the running sum of x_k and y_k for calculating the distance function incrementally and using a much simpler formula. This suggests that the normalization constant should be selected to change as infrequently as possible while keeping the scale invariance nature of the distance function. The most used normalization method to achieve scale invariance is zscore normalization in which $r_x^l = \sigma_x^l$. In this paper we propose using the – less frequently used – range normalization ($r_x^l = mx_x^l - mn_x^l$) because the normalization constants change much less frequently. To support this claim we conducted two experiments. In the first experiment, we generated 100 timeseries pairs of length 1000 each using random walks and calculated the fraction of time in which either r_x^l or r_y^l changed using both zscore and range normalization. The zscore normalization constant changed 15.01% of the time while the range normalization constant changed only 0.092% of the time. In the second experiment, we used 50 timeseries representing the angles of wrist and elbow joints of an actor while generating free gestures as a real world dataset. The zscore normalization constant changed 34.2% of the time while the range normalization constant changed only 0.11% of the time. This suggests that just ignoring the change in the normalization constant would not affect the quality of returned 2-motifs even though it will render the algorithm approximate.

The formulas for incremental evaluation of the distance function given in this section assume that the change in length is a single point. Both formulas can be extended to the case of any difference in the length but proofs are much more involved and due to lack of space will not be presented.

5   Proposed Algorithm

The second contribution of this paper is to use the incremental normalized distance calculation formulas of Theorem 1 and Lemma 1 to extend the MK algorithm to handle the scale-invariant multi-length 2-motif discovery problem stated in section 2.

The first approach – as suggested by Lemma 1– is to use the range normalization and modify the calculation of distances in the D_{bests} list using the formula proposed in Lemma 1. Notice that during the scanning phase, the algorithm will decide to ignore pairs of subsequences based on the distances between them and reference points. We can either keep the exactness of the algorithm by recalculating the distance between the pair and all reference points at every length using the formula given in Theorem 1 or accept an approximate solution (that should not be much worse than the exact one by Lemma 1) and use the distances to reference points from previous lengths as lower bounds. In this paper we choose the second (approximate) method to maximize the speed of the algorithm. This approach is called MK++ for the rest of this paper. We will show that the proposed algorithm is faster than this *approximate* solution while being an exact algorithm.

The second approach is to use the formula in Theorem 1 and run the two phases of the MK algorithm in parallel for all lengths. The algorithm starts similarly to MK+ by calculating the distance between all subsequences of the minimum length and a randomly selected set of reference points. These distances will be used later to find lower bounds during the scanning phase. Based on the variance of the distances associated with reference points, these points are ordered. The subsequences of the timeseries are then ordered according to their distances to the reference point with maximum variance. These steps can be achieved in $O\,(nlogn)$ operations. The distance function used in these steps (D_{full}) uses Equation 2 for distance calculation but in the same time keeps the five running summation (x_k, y_k, $(x_k)^2$, $(y_k)^2$, and $x_k y_k$) needed for future incremental distance calculations as well as the maximum and minimum of each subsequence. After each distance calculation the structure S^l_{bests} is updated to keep the top K motifs at this length with associated running summations using the same three rules of MK++ (see section 3).

The next step is to calculate the S^l_{bests} list storing distances and running summations for all lengths above the minimum length using the function D_{inc} which utilized Theorem 1 to find the distances at longer lengths. The list is then sorted at every length. Both D_{inc} and D_{full} update the bsf variable which contains the best-so-far distance at all lengths and is used if the run is approximate to further prune out distance calculations during the scanning phase.

The scanning phase is then started in which the subsequences as ordered in the previous phase are taken in order and compared with increasing offset between them. If a complete run at a specific length did not pass the lower-bound test , we can safely ignore all future distance calculations at that length because by the triangular inequality we know that these distances can never be lower than the ones we have in S^l_{bests}. Scanning stops when all lengths are fully scanned.

During scanning we make use of Theorem 1 once more by using an incremental distance calculation to find the distances to reference points and between currently tested subsequences. If we accept approximate results based on Lemma 1, we can speed things up even more by not calculating the distances to reference points during the evaluation of the lower bound and by avoiding this step all-together if the distance at lower length was more than the current maximum distance in S^l_{bests}.

6   Evaluation

We conducted a series of experiments to evaluate the proposed approach to existing state of the art exact motif discovery algorithms. We evaluated MK++ (with the modifications discussed in section 5) and sMD proposed in this paper to the following algorithms: iterative application of the MK algorithm (iMK), the brute force approach of just comparing all possible pairs (using only bsf to prune calculations) at all lengths (called BF from now on), the brute-force algorithm but utilizing the incremental distance calculation proposed in section 4.

In the first experiment, we evaluated the five algorithms for scalability relative to the timeseries length. We used the EEG trace dataset from [12] and applied the algorithm to the first subsequence of length 1000, 5000, 10000, 15000, and 20000points. The motif range was 64 to 256 points and K was 15. Because it is always the case for all of these algorithms that execution time will increase with increased length, we did not evaluate any algorithm for lengths larger than the one at which its execution time exceeded one hour. Fig. 1-a shows the execution time in seconds of the six algorithms as a function of timeseries length. Notice that sMD, and BF++ outperform all other algorithms for even moderate lengths. The fact that the brute-force algorithm is better than iterative MK and even MK++ supports the effectiveness of the proposed incremental distance calculations because both iterative MK and MK++ cannot effectively utilize it.

The second experiment explored the effect of motif length range. We used the same dataset used in the first experiment and a fixed timeseries length of 1000 points. We used a minimum motif length of 50 and varied the maximum motif length from 54 to 249. K was 15 again and we stopped the execution in the same fashion as in the first experiment. Fig. 1-b shows the execution time in seconds of the six algorithms as a function of the motif length range. Again sMD, and BF++ outperform the other three algorithms.

In the final experiment, we tested the application of the proposed algorithm as the basis for ARM discovery by first deleting all 2-motifs of length l that are covered by 2-motifs of a higher length then combine 2-motifs at each length if either of their subsequences are overlapping more than a predefined threshold. We used the CMU Motion Capture Dataset available online from [2]. All the timeseries corresponding to basketball and general exercise and stretching categories were used. The occurrences of each recurring motion pattern in the time

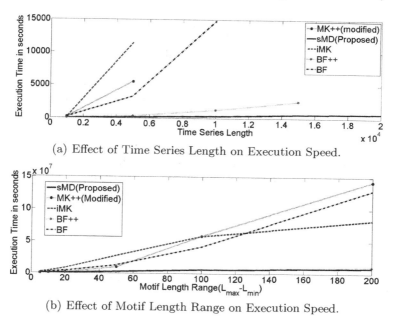

(a) Effect of Time Series Length on Execution Speed.

(b) Effect of Motif Length Range on Execution Speed.

Fig. 1. Comparing scalability of the proposed algorithm (sMD) with other exact motif discovery algorithms. See text for details.

series of the 20 available in this collection were marked by hand to get ground truth information about the locations of different motifs. The total number of frames in the 20 time series was 37788. Timeseries length ranged from 301 to 5357 points each. Before applying motif discovery algorithms, we reduced the dimensionality of each time series using Principal Component Analysis (PCA). To speedup PCA calculations, we used a random set of 500 frames and applied SVD to it then projected the whole time series on the direction of the first Eigen vector.

We applied sMD, MK++, PROJECTIONS [13], and MCFull [7] with a motif length between 100 and 300 to the 20 time series and calculated the accuracy and execution time for each of these five algorithms. The proposed algorithm achieved the highest accuracy (87% compared with 83% for MK++, 74% for MCFull, and 64% for the PROJECTIONS algorithm) and shortest execution time (0.0312 seconds per frame compared with 0.63 seconds for MK++, 3.2 for MCFull, and 10.3 seconds per frame for PROJECTIONS). These results show that the proposed algorithm is applicable to real-world motif discovery.

7   Conclusions

This paper presented an incremental formula for calculating scale invariant distances between timeseries. This formula was then used to design an algorithm for scale invariant multi-length exact motif discovery (called sMD). The proposed algorithm was evaluated against brute-force solution of the problem and two other motif discovery algorithms (MK++ and iterative application of the MK algorithm). The proposed algorithm is an order of magnitude faster than both of them for timeseries of moderate size (10000points).

The work reported in this paper opens several directions of future research. The most obvious direction is parallelizing the scanning phase of the algorithm. Another direction is integrating the proposed incremental distance calculation method, the lower bound used in MOEN and the pruning technique of the MK algorithm to provide even faster exact motif discovery. A third direction of future research is to utilize top-down processing (i.e. from higher to lower motif lengths) in conjunction with the bottom-up processing of the proposed algorithm to guarantee that 2-motifs found at every length are not overlapping those at higher lengths.

References

1. Chiu, B., Keogh, E., Lonardi, S.: Probabilistic discovery of time series motifs. In: ACM KDD 2003, pp. 493–498 (2003)
2. CMU: Cmu motion capture dataset, http://mocap.cs.cmu.edu
3. Keogh, E., Lin, J., Fu, A.: Hot sax: Efficiently finding the most unusual time series subsequence. In: IEEE ICDM, pp. 8–16 (November 2005)
4. Keogh, E., Kasetty, S.: On the need for time series data mining benchmarks: A survey and empirical demonstration. Data Mining and Knowledge Discovery 7(4), 349–371 (2003)
5. Minnen, D., Starner, T., Essa, I., Isbell, C.: Improving activity discovery with automatic neighborhood estimation. In: IJCAI (2007)
6. Mohammad, Y., Nishida, T.: Exact discovery of length-range motifs. In: Nguyen, N.T., Attachoo, B., Trawiński, B., Somboonviwat, K. (eds.) ACIIDS 2014, Part II. LNCS, vol. 8398, pp. 23–32. Springer, Heidelberg (2014)
7. Mohammad, Y., Nishida, T.: Constrained motif discovery in time series. New Generation Computing 27(4), 319–346 (2009)
8. Mohammad, Y., Nishida, T.: Fluid imitation: Discovering what to imitate. International Journal of Social Robotics 4(4), 369–382 (2012)
9. Mohammad, Y., Nishida, T.: Unsupervised discovery of basic human actions from activity recording datasets. In: Proceedings of the IEEE/SICE SII (2012)
10. Mohammad, Y., Nishida, T., Okada, S.: Unsupervised simultaneous learning of gestures, actions and their associations for human-robot interaction. In: Proceedings of IROS 2009, pp. 2537–2544 (2009)
11. Mueen, A.: Enumeration of time series motifs of all lengths. In: 2013 IEEE 13th International Conference on Data Mining (ICDM). IEEE (2013)
12. Mueen, A., Keogh, E.J., Zhu, Q., Cash, S., Westover, M.B.: Exact discovery of time series motifs. In: SDM, pp. 473–484 (2009)
13. Patel, P., Keogh, E., Lin, J., Lonardi, S.: Mining motifs in massive time series databases. In: IEEE International Conference on Data Mining, pp. 370–377 (2002)

Estimation of Missing Values in SNP Array

Przemyslaw Podsiadly

Institute of Computer Science, Warsaw University of Technology
Nowowiejska 15/19, 00-665 Warsaw, Poland
P.Podsiadly@ii.pw.edu.pl

Abstract. DNA microarray usage in genetics is rapidly proliferating, generating huge amount of data. It is estimated that around 5-20% of measurements do not succeed, leading to missing values in the data destined for further analysis. Missing values in further microarray analysis lead to low reliability, therefore there is a need for effective and efficient methods of missing values estimation.

This report presents a method for estimating missing values in SNP Microarrays using k-Nearest Neighbors among similar individuals. Usage of preliminary imputation is proposed and discussed. It is shown that introduction of multiple passes of kNN improves quality of missing value estimation.

Keywords: microarray, bioinformatics, SNP array, missing values.

1   Introduction

Single Nucleotide Polymorphism (SNP) is the simplest type of genetic polymorphism. As such SNP variants are not just mutations. They involve changes in only one nucleotide that occurred generations ago within the DNA sequence of a given species and are considered typical variations – they must be present in more than 1% of the population. In the case of a human genome, 50 million SNPs were identified, while unrelated individuals differ in about 0.1% of their 3.2 billion nucleotides. Measurements of SNP variations is a very good tool in biomedical research as SNPs are genetically stable throughout evolution and within population and high-throughput genotyping methods are available.

Microarray data sets often contain missing values for various reasons, such as similar intensity of the signal and the background, dust on the slides, not properly fixed probe on a chip, or the hybridization step not working properly. Unfortunately, missing data in genotyping SNP arrays is a common problem. For example, the published human chromosome 21 data in [5] contains about 20% undetermined markers. It is estimated that around 5-20% of measurements in high-throughput genotyping methods fail, and even a low percentage of missing SNPs leads to significantly lower reliability of down-stream analyses such as SNP-disease association tests [6]. Currently regenotyping the missing calls is not a practical alternative to dismissing individuals with missing values or disregarding failed markers. Missing value inference is therefore an important part of SNP array analysis.

A. Moonis et al. (Eds.): IEA/AIE 2014, Part II, LNAI 8482, pp. 427–435, 2014.

There are two types of missing SNP imputation: off-line and on-line. While off-line methods are designed to infer the missing SNP values before any further analysis is performed, on-line methods consider imputation of missing values simultaneously with the task of interest such as disease association study.

The aim of this report is to improve SNP inference accuracy and present a new method of iterative SNP imputation based on k Nearest Neighbors as an off-line method.

2   Basic Notions and Properties

A single microarray is responsible for measuring gene (or SNP) genotypes for a given individual. A single genotype (or variant) of a given SNP is represented as a single item from a set T of all possible nominal genotypes. An example set T presented below was taken from the data used in the experimental part of this paper:

$$T=\{AA, CC, GG, TT, AC, AG, AT, CG, CT, GT\}.$$

A single SNP array measures variants for a single individual i and is represented as a vector $g_i \in T^m$, where m is a number of investigated SNPs. In a given experiment with n individuals, a whole SNP dataset is represented as $G \in T^{n \times m}$:

$$G = \begin{pmatrix} g_1^T \\ \cdots \\ g_n^T \end{pmatrix} \in T^{n \times m}.$$

We name g_i a *gene profile* of an individual i.
A missing value of gene j for an individual i is denoted as follows:

$$g_i(j) = \emptyset, \text{where } 1 \leq i \leq n, 1 \leq j \leq m.$$

A sample matrix G is presented below.

	Gene1	Gene2	Gene3	Gene4	Gene5	Gene6
Individual 1	AA	CG	AT	CC	AC	Ø
Individual 2	AT	CC	AT	Ø	AA	CG
Individual 3	Ø	CC	TT	CG	Ø	CC
Individual 4	AT	GG	Ø	Ø	AA	CG
Individual 5	AT	CG	AA	CC	CC	GG

k Nearest Neighbors methods require a choice of similarity measure to be calculated between individuals. Similarity or distance measures need to take into consideration the fact that SNP data has nominal genotype values.

In this report, we use a simple match measure defined as the number of the genes on which individuals agree divided by the number of all genes:

$$simple\text{-}match(g_i, g_j) = \frac{the\ number\ of\ the\ genes\ on\ which\ g_i\ and\ g_j\ agree}{the\ number\ of\ all\ genes}$$

The comparison of vectors with missing values is a special case. In most papers it is assumed that if an element is missing in either or both vectors, they are considered not equal. In this report we consider two options:

1. All missing values are considered not matching (i.e. when either one or both are missing)
2. Missing values are counted as a ½ in the numerator, as there is a chance that the values match

For example, for individuals 2 and 4:

Individual 2	AT	CC	AT	Ø	AA	CG
Individual 4	AT	GG	Ø	Ø	AA	CG
	match	mismatch	missing	Missing	match	match

In variant 1, the similarity is as follows:

$$simple\text{-}match(g_2, g_4) = \frac{3}{6}$$

Meanwhile, variant 2 considers missing values differently:

$$simple\text{-}match(g_2, g_4) = \frac{3 + 2 * {}^{1}/_{2}}{6} = \frac{4}{6}$$

In variant 2, the similarity measure is always higher if missing values occur in the compared individuals.

3   Related Work

The problem of estimating missing values in DNA microarrays was explored mainly in the context of expression data. In this case, the genotype set T is substituted by a real set R, i.e. for each gene in place of a single genotype, a gene expression is measured. Gene expression roughly presents the activity level of a gene measured by an amount of produced RNA matching the DNA pattern.

The simplest approach to estimation of missing values was based on row average – a missing value was estimated using an arithmetic average of all the values for a given gene [1]. This approach provides a very simplistic method not requiring disregarding any values, but it does not take into account the structural similarities present in the data.

KNNimputed method, introduced in [3], tries to build upon discovery of similarities between genes. First $k<<m$ nearest neighbors are found for a given gene. A missing value is calculated using a weighted average where weights are based on similarity measures. In this method, the choice of similarity measure and how it influences weights is critical to effectiveness of this method. This aspect was developed further in [4], using different ways of weighting the neighbors, such as using least squares error minimization.

In the methods above, all the missing values are ignored during calculation of similarity and weighting.

4   New Method of Missing Values Estimation

In this report, a new method of SNP missing value estimation is presented. First, during preliminary imputation all missing values are substituted with most common variants of a given gene. Second, for each value originally missing, k nearest neighbors of the individual are found with their respective similarity. The missing value is substituted with a genotype voted by k nearest neighbors weighted with calculated similarity measures. When all missing values are inferred, the process is repeated, taking into account newly filled-in missing values.

The process iterates a number of times determined by a user or until the imputed values converge.

The algorithm is presented below:

1.   Store in *missingValuesList* pairs *(i,j)* identifying all missing values $g_i(j) = \emptyset$ in G.
2.   Preliminary imputation: for each missing value in *missingValuesList* calculate a most common value in a column:

$$g_i(j) = most\text{-}common(g_l(j) for\ all\ 1 \le l \le n)$$

3.   For each missing value identified by *(i,j)* in *missingValuesList*
   3.1.   Find k nearest neighbors of an individual i with indexes n_l where $l=1..k$ and calculated similarities $sim(g_i, g_{n_l})$
   3.2.   Calculate ranking of all gene variants $v \in T$ weighted by similarity:

$$p(v \in T) = \sum_{l=1}^{k} \begin{cases} sim(g_i, g_{n_l}), & if\ g_{n_l}(j) = v \\ 0, & if\ g_{n_l}(j) \ne v \end{cases}$$

   3.3.   The missing value is substituted by the most highly ranked value:

$$g_i(j) = argmax_v\ p(v)$$

4.   Repeat step 3 until no changes occur in the G matrix or the number of passes exceeds a threshold.

The preliminary imputation step of the algorithm is optional. The experimental part examines the influence of this step on the performance of the algorithm.

5   Example

To demonstrate the execution of the algorithm let us consider the sample matrix G presented above. The first step of this algorithm stores coordinates of all the missing values in a list:

$$missingValuesList = \{\ (1,6),(2,4),\ (3,1),(3,5),\ (4,3),\ (4,4)\ \}.$$

In the second step the most typical value for each gene (in columns) is calculated. The resultant matrix is presented below (previously missing values are presented in bold).

	Gene1	Gene2	Gene3	Gene4	Gene5	Gene6
Individual 1	AA	CG	AT	CC	AC	**CG**
Individual 2	AT	CC	AT	**CC**	AA	CG
Individual 3	**AT**	CC	TT	CG	**AA**	CC
Individual 4	AT	GG	**AT**	CC	AA	CG
Individual 5	AT	CG	AA	CC	CC	GG

In the third step of this algorithm, for each missing value in *missingValuesList*, k nearest neighbors are found. For example, for $(i,j) = (2,4)$, in order to find k nearest neighbors of vector Individual2, first similarities are calculated as follows:

$$simple\text{-}match(\text{Individual2, Individual1}) = 3\,/\,6$$
$$simple\text{-}match(\text{Individual2, Individual3}) = 3\,/\,6$$
$$simple\text{-}match(\text{Individual2, Individual4}) = 4\,/\,6$$
$$simple\text{-}match(\text{Individual2, Individual5}) = 2\,/\,6$$

For $k=3$, $kNN(\text{Individual2}) = \{\text{Individual4, Individual1, Individual3}\}$.
Next, the value distribution for gene 4 is calculated using the above weights:

$$p(CC) = 4/6 + 3/6 = 7/6$$
$$p(CG) = 3/6$$

The highest ranking is the value CC. As it is the same as already inferred, the number of changes does not increase. If a new value was different than previously imputed, it would replace the old value and the number of changes would be incremented.

When all the missing values are calculated in this way, the process is repeated if there were more than zero changes and maximal number of passes is not reached (if specified). If no changes were made, algorithm stops.

6   Experimental Results

The experiments were conducted on a dataset of 51 SNPs measured in 311 bovine samples. The data included 228 missing values. In order to measure the effectiveness of the algorithm additional artificial missing values were generated. For each $g_i(j)$ with a present value there was a 10% chance for it being marked missing for algorithm verification. For evaluation of the effectivess of the algorithm the number of correctly imputed values divided by all artificial missing values was used.

The experiments were conducted for k varying from 1 to 99. The influence of variant of *simple-match* was examined. Two variants of algorithm were used – with and without step 2 (preliminary imputation). Calculations were repeated 8 times for differently selected missing values.

The influence of preliminary imputation (true or false) is presented in Figure 1. Quality represents effectiveness of imputation, i.e. 65.02 means that 65.02% of all imputed values were estimated correctly.

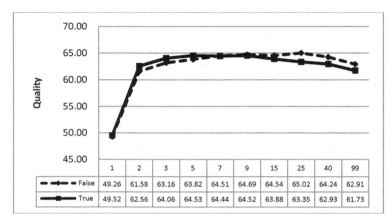

Fig. 1. Imputation quality in relation to $k=1..99$ and usage of preliminary imputation (True or False)

For $k<9$ the variant with preliminary imputation proves to be more effective. However with higher k the preliminary imputation causes the algorithm to settle in a local minimum with inferior results. This finding is confirmed by the number of passes presented in Figure 2.

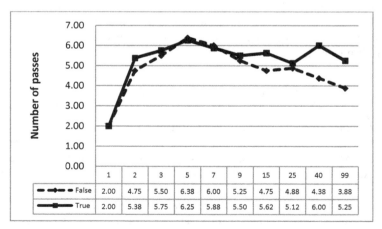

Fig. 2. Number of passes in relation to $k=1..99$ and usage of preliminary imputation (True or False)

Preliminary imputation causes the algorithm to look for a stable solution in more passes, as shown in Figure 2. In general, the number of passes decreases with growing

k with exception to $k<5$ due to finding a local minimum. Increasing k above 25 with preliminary imputation seems to generate more passes as a single missing value is influenced by bigger number of other missing values.

In order to investigate the influence of preliminary imputation on how fast a reasonable solution is found, the achieved quality is presented after each pass for two values of k (2 and 9, Figures 3 and 4 respectively).

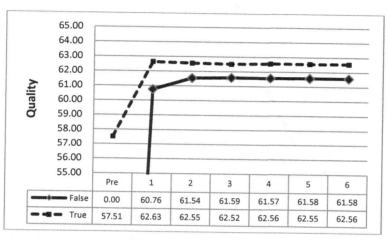

	Pre	1	2	3	4	5	6
False	0.00	60.76	61.54	61.59	61.57	61.58	61.58
True	57.51	62.63	62.55	62.52	62.56	62.55	62.56

Fig. 3. Imputation quality for $k=2$ after each pass in relation to the usage of preliminary imputation (True or False)

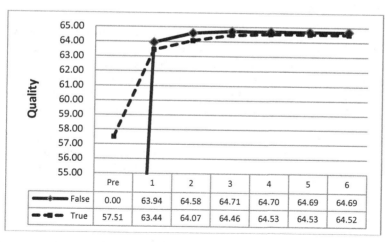

	Pre	1	2	3	4	5	6
False	0.00	63.94	64.58	64.71	64.70	64.69	64.69
True	57.51	63.44	64.07	64.46	64.53	64.53	64.52

Fig. 4. Imputation quality for $k=9$ after each pass in relation to the usage of preliminary imputation (True or False)

For lower values of k the algorithm using preliminary imputation performs better than straightforward kNN-based imputation. With higher values of k an inverse relation is observed. The figures show that additional passes improve imputation quality, especially the second pass.

Figure 5 presents the influence of the variant of similarity measure in case of algorithm with no preliminary imputation. The variant 2 of the measure proves to be a better suiting measure for data with missing values for $k>1$ and no preliminary imputation. When preliminary imputation is used, the effect is not observable.

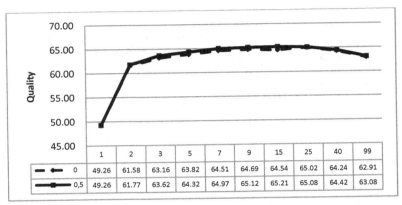

Fig. 5. Imputation quality in relation to $k=1..99$ and two variants of *simple-match* measure (0 – variant 1, 0.5 – variant 2) in case of no preliminary imputation

7   Conclusions

In this work, we presented a new method of SNP missing values estimation. We introduced preliminary imputation as a mean to roughly estimate values before kNN imputation starts, thus avoiding calculation of similarity between incomplete vectors. We proposed multiple passes of kNN imputation to improve effectiveness of the algorithm.

The experimental results show that additional passes of kNN imputation improve quality of estimation. Using just two passes in place of one improves the imputation significantly.

Preliminary imputation increases imputation quality for lower values of k ($k<9$), however, with higher values of k it introduces a bias that leads to suboptimal solutions.

When preliminary imputation is not used, it seems beneficial to use similarity measures that are aware of missing values.

References

1. Alizadeh, A.A., Eisen, M.B., Davis, R.E., Ma, C., Lossos, I.S., Rosenwald, A., Boldrick, J.C., Sabet, H., Tran, T., Yu, X., Powell, J.I., Yang, L., Marti, G.E., Moore, T., Hudson Jr., J., Lu, L., Lewis, D.B., Tibshirani, R., Sherlock, G., Chan, W.C., Greiner, T.C., Weisenburger, D.D., Armitage, J.O., Warnke, R., Staudt, L.M., et al.: Distinct types of diffuse large B-cell lymphoma identified by gene expression profiling. Nature 403, 503–511 (2000)

2. Han, J., Kamber, M., Pei, J.: Data Mining: Concepts and Techniques, 3rd edn. Morgan Kaufmann Publishers (2011)
3. Troyanskaya, O., Cantor, M., Sherlock, G., Brown, P., Hastie, T., Tibshirani, R., Botstein, D., Altman, R.B.: Missing value estimation methods for DNA microarrays. Bioinformatics 17(6), 520–525 (2001)
4. Kang, H., Qin, Z.S., Niu, T., Liu, J.S.: Incorporating Genotyping Uncer-tainty in Haplotype Inference for Single-Nucleotide Polymorphisms. Am. J. Hum. Genet. 74, 495–510 (2004)
5. Patil, N., et al.: Blocks of Limited Haplotype Diversity Revealed by High-Resolution Scanning of Human Chromosome 21. Science 294, 1719–1723 (2001)
6. Sinoquet, C.: Iterative two-pass algorithm for missing data imputation in SNP arrays. Journal of Bioinformatics and Computational Biology 7(5), 833–852 (2009)
7. Zezula, P., Amato, G., Dohnal, V., Batko, M.: Similarity Search: The Metric Space Approach. Springer (2006)

Configuring the Webpage Content through Conditional Constraints and Preferences

Eisa Alanazi and Malek Mouhoub

Department of Computer Science
University of Regina
Regina, Canada
{alanazie,mouhoubm}@cs.uregina.ca

Abstract. Configuring the webpage content to reflect the user desires is highly demanded in the era of personalization. The problem can be viewed as a preference-based constraint problem including a set of components forming the webpage along with the preferences. Our goal is then to locate each of these components such that the user preferences are maximized. Additionally, constraints might exist between different components of the given page. We investigate the problem of handling the web page content based on user preferences and constraints. Unlike previous attempts, we model the constraint part as an instance of the conditional CSP. This gives further expressive power to handle different relations among components. The preferences are expressed through the well-known CP-Nets graphical model.

1 Introduction

The large amount of information available today on the web brings many challenges to the Information Retrieval (IR) and Artificial Intelligence (AI) areas. Moreover, personalization is a key component in today's successful websites. In order to be effective, these websites are required to provide the visitors with the information they need without the complexity of finding useful information. Our goal is to provide the websites owners with a configuration tool with which they can easily manage and personalize their website contents according to their requirements and satisfaction. Providing such a tool is of great interest to different websites such as online stores, news portals and blogs. To effectively handle the problem, the provided tool must be dynamic (i.e. different users will have different configurations) and expressive enough (allows the owner of the webpage to express different types of tasks and relations). The problem is an example of interactive configuration problems where preferences coexist with the constraints over different components. Therefore, although the domain of webpage content is addressed here, we believe our approach is applicable to other interactive configuration domains such as product configuration and scheduling. A fundamental task in managing webpage content is specifying conditional dependencies among components. In other words, the existence of some components in the final configuration depends upon other components values. For instance,

A. Moonis et al. (Eds.): IEA/AIE 2014, Part II, LNAI 8482, pp. 436–445, 2014.

advertisement X is conditionally constrained by the appearance of article Y in the webpage. We view the Conditional Constraint Satisfaction Problem (CCSP) [1] as a convenient tool to manage this type of relations. The problem of determining the best configuration for a webpage content will then be formalized through CCSPs and Conditional Preference networks (CP-nets) [2]. CCSPs and CP-nets are used to manage constraints and preferences respectively. To the best of our knowledge, this work represents the first application of CCSPs and CP-Nets to the webpage content problem. Modeling the problem using these two models brings intuitive and clear usage of users requirements and preferences in the web context. The CCSP is dynamic in the sense that depending on the set of initial assignments, two users might have different configurations. In addition, the CCSP provides an intuitive method to handle conditional constraints that are naturally present within the webpage content problem. The CCSP helps us to answer the question of *what* to represent when given a set of conditional dependencies and constraints. The paper is structured as follows. Background and related work are provided in the next two sections. The problem is discussed along with different examples in Section 4. Section 5 addresses the problem of finding optimal configurations and updating the webpage. Experiments on randomly generated preferences and constraints instances are reported in Section 6. Section 7 lists concluding remarks and some future research directions.

2   Background

2.1   Preference Representations

Preferences are important when representing and reasoning about a set of outcomes in a decision making context. It gives us a way to distinguish between good and bad choices [2]. Preferences can be quantitative or qualitative. Qualitative preferences have no numeric values associated with them. Conditional Preferences networks (CP-nets) [2] is a well-known graphical model for reasoning and representing conditional qualitative preferences based on the ceteris paribus interpretation. More precisely, a CP-net represents qualitative preferences statements including conditional preferences such as: *"I prefer A to B when X holds"*. A CP-net works by exploiting the notion of preferential independency based on the *ceteris paribus* (with all other things being without change) assumption. A CP-net can be represented by a directed graph where nodes represent features (or variables) along with their possible values (variables domains) and arcs represent preference independencies among features. Each variable X is associated with a ceteris paribus table (denoted as $CPT(X)$) expressing the order ranking over different values of X given the set of parents $Pa(X)$. An outcome for a CP-net is an assignment for each variable from its domain. Given a CP-net, the users usually have some queries about the set of preferences represented. One of the main queries is the best outcome given the set of preferences. We say outcome o_i is better than outcome o_j if there is a sequence of worsening flips going from o_i to o_j [2]. A Worsening flip is a change in the variable value to a less preferred value according to the variable's CPT.

2.2   Constraint Representations

The Constraint Satisfaction Problem (CSP) [3] is a well-known framework for managing and solving constraint problems. More formally, a CSP consists of a set of variables each defined on a set of possible values (variable domain) and a set of relations restricting the values that each variable can take. A solution to a CSP is a complete assignment of values to variables such that all the constraints are satisfied. A Conditional Constraint Satisfaction Problem (CCSP) [1] is a formalism to handle constraint problems where a set of variables can be added or removed during the search space. Unlike typical CSPs, CCSPs have a set of optional variables that can participate in the final solution if they satisfy a particular type of constraints called activity constraints [4]. Therefore, each variable in the CCSP is associated with an activation status: active or not active (present or not present in the problem to solve). The set of activity constraints governs the participation of optional variables in the solutions. More formally, a CCSP is represented as tuple $< V, V_I, D, C_c, C_a >$ where V is the set of all variables, V_I is the set of initial active variables (i.e. set of variables that are initially active) and D is the set domains of the possible values for each variable. C_c and C_a represent the set of compatibility and activity constraints respectively. A compatibility constraint is similar to a typical hard constraint in CSPs [5]. Activity constraints have an inclusion or exclusion conditions. An inclusion constraint $A = a \xrightarrow{activate} B$ means that B is activated if a is assigned to the active variable A. Similarly, the exclusion constraint takes the form $A = a \xrightarrow{deactivate} B$ and will deactivate the active variable B if a is assigned to the active variable A.

3   Related Work

Effectively handling website content can be viewed as a decision problem where the system needs to choose the best viewable content for users. Only a few of the existing approaches adopt the qualitative decision theory as a tool to handle the personalization process [6,7,8,9]. In addition, the problem of intuitively representing content preferences and constraints in one framework has not been addressed in the literature. We believe that the CCSP is suitable for website content as it has been used as an intuitive tool to solve different problems in product configuration and synthetic tasks [10,1]. The closest work to ours is the CPML system proposed in [9]. This system handles webpage content through preferences represented by the webpage author (i.e. webmaster). However, the work does not address how to handle constraints in addition to preferences. In [6] an approach to personalize multimedia messages has been proposed based on TCP-nets. The authors consider a case where constraints assert some feasible presentations. Another close work is [8] where the authors described a framework to handle presentation components based on CP-nets and geometric constraints.

4   Managing Webpage Content

We view the webpage content problem as an optimization problem. That is the webmaster/website owner indicates his requirements and preferences over different set of components forming the page. Then, the system automatically generates and allocates each component such that the content is optimized for each user. The optimization part of the problem comes from the fact that most webpages content co-exists with constraints corresponding to user and webmaster requirements. In addition, the webmaster can specify different dependencies among components. In this work, we focus on constraints and preferences which will result in dynamic generations of webpages based on different viewers behaviours.

4.1   Problem Formulation

More formally, the webpage content problem is represented as follows given a CCSP $\langle V, V_I, D, C_c, C_a \rangle$ and a CP-Net N. The set of variables V corresponds here to the components forming a given webpage P. Each variable can take one value from the boolean domain (1: show, 0: hide) stating if the corresponding component is displayed or not. We assume here that $(N \subseteq V_I)$. That is the set of variables in N is a subset of the initial variables. This guarantees that for any two possible configurations, N can be consulted to determine which one is better. Compatibility and activity constraints (C_c and C_a respectively) are defined to properly configure the different components. For instance, the appearance of an article A in a news portal might be conditionally constrained by the value (appearance) of another component B.

Example 1. Let us consider a news portal page containing a set V of eight components including five articles $\{A_1, A_2, A_3, A_4, A_5\}$ and three advertisements $\{R_1, R_2, R_3\}$. The initial variables for the problem are $V_I = \{A_1, A_3, A_4, R_3\}$. We assume that the webmaster has specified the following requirements (constraints).

1. A_1 and A_2 cannot be displayed at the same time.
2. R_2 must be displayed when A_2 is part of the page.
3. R_1 and R_2 cannot be displayed at the same time.

In addition, the set of activity constraints are as follows.

- a_1: $A_1 = 1 \xrightarrow{activate} R_1,$
- a_2: $A_3 = 1 \xrightarrow{activate} R_2$
- a_3: $A_3 \xrightarrow{activate} A_2.$

Finally, the following webmaster preference relations are provided.

1. A_1 is always preferred to be shown.
2. When A_1 is not part of the webpage, the webmaster prefers A_4 to be shown.
3. If A_4 is part of the page, the webmaster prefers to display A_3.
4. if A_3 is displayed, the webmaster prefers R_3 to be displayed.

According to the above information, the constraints part is represented by the constraint network shown in Figure 1a. $N = \{A_1, A_3, A_4, R_3\}$ and the CPTs for these preference variables are shown in Figure 1b.

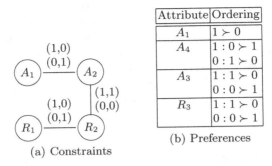

Attribute	Ordering
A_1	$1 \succ 0$
A_4	$1 : 0 \succ 1$
	$0 : 1 \succ 0$
A_3	$1 : 1 \succ 0$
	$0 : 0 \succ 1$
R_3	$1 : 1 \succ 0$
	$0 : 0 \succ 1$

(a) Constraints

(b) Preferences

Fig. 1. Constraint and Preference Attributes

4.2 Solving Method

Once a given webpage content is represented with a CCSP and a CP-Net as shown in example 1, the next task is to find a complete assignment of values to all the active components $c_i \in V$ satisfying all the constraints and maximizing all the preferences. In other words, the goal here is to find an optimal configuration (outcome) satisfying all the constraints and not dominated by other feasible configuration. Solving such a problem can be done using a backtrack search algorithm. In order to improve the efficiency of this search method, arc consistency can be used before and during search in order to remove earlier some possible conflicting values [11,12]. If the underlying CP-net structure forms an acyclic graph, then we can simply use the sweep-forwarding procedure [2]. The procedure in Figure 2 illustrates our solving approach for finding feasible configurations. In the next section, we study the problem of finding optimal configuration.

5   Finding Optimal Configurations

Initially, all users will have the same configuration (i.e content). Then the webpage is adapted according to the viewer activities. Adapting webpages to the viewer behaviour results in different changes to component values. For instance, if the user shows an interest in displaying an article X, the tool generates the optimal configuration given the event is true (i.e. $X = 1$). Variables are ordered according to the CP-net structure from top to bottom. Each time an event occurs, the tool generates optimal configurations consistent with the given information.

1. Sort the domains of all preference variables according to the CPTs.
2. Run the AC algorithm on the CCSP to reduce the search space.
3. Extend the partial assignment by assigning a new value to a selected active variable v.
4. Activate any variable due to this new assignment.
5. Run AC between v and the set of non assigned active variables.
6. If at least one domain of these non assigned variables becomes empty then assign another value to v, deactivate the variables according to this change, and go to 4. If there is no value to assign to v then backtrack to the previously assigned value.
7. Go to 3 until a complete assignment is found or the search space is exhausted without success.

Fig. 2. Procedure to find a consistent assignment

5.1   Updating Page Content

Example 2. Consider Example 1 again. According to the webmaster expertise, the initial portal page will show either A_1 or A_1 and R_1. Now, assume the user wants to display A_3. This corresponds to the partial assignment $A_3 = 1$. Trying to extend this assignment will result in a webpage with the following information $\{A_1 = 1, A_4 = 0, A_3 = 1, R_3 = 1, R_2 = 0, A_2 = 0, R_1 = 1\}$ which changes the webpage significantly.

The problem here is what to update for webpage P when an event e occurs. Events can be any triggers to change the webpage content. Usually, they form a series of clicks in different webpage components. An event is simply a pair (A, a) where A is variable and a is the value to be assigned to this variable. For instance, $e = (X, 1)$ means the event of the viewer is interested in showing article X. The challenge is: given an event e, how to transform the current webpage P to another webpage $\mathbf{P}$ reflecting e. The new updated webpage must be consistent with the constraints and should meet the preference statements represented in the CP-net N. First, we propagate the event e to the constraint network of P. If there is no solution for this propagation process, then, we quit and return to the initial page. Otherwise, the result of the propagation is a partial assignment $A = a$. After propagating the event value, we start traversing N in a top to bottom fashion trying to extend a to form a complete solution. Algorithm 1 shows the necessary steps to update webpage content. Note that in some cases, we are required to remove some variables from $\mathbf{P}$. In particular, when event e makes variable X inactive as in the following example.

Algorithm 1. *update(variables* v, *event* e, *constraints* con)

$C \leftarrow con$
let d_e be the value of variable e
$sol \leftarrow (e, d_e)$
for each $v_i \in v - \{e\}$ with value d_{v_i}
  if v_i becomes irrelevant
    remove v_i
  if any $c \in C_c$ becomes relevant
    add c to C
  apply activity constraints
  if (C, v_i, d_{v_i}) is consistent
    add (v_i, d_{v_i}) to *sol*
end
if *sol* is complete
  return *sol*

5.2 Example Analysis with Different Events

Example 3. Consider the initial webpage to be the vector $(1,0,0,0,1,0)$, where the values 1 and 0 correspond to show and hide respectively, and the event $A_2 = 1$ (that is the user wants to view article A_2). Propagating this event to the network in Example 1 will result in the partial assignment $(A_1 = 0, A_2 = 1)$ where R_1, as part of P, should be removed now since it is no longer an active variable according to a_1.

We consider Example 1 with some user events. First, we need to determine the initial configuration for the new users. Let $V = N = \{A_1, A_4, A_3, R_3\}$; we have $A_1 = 1$ and thus we execute a_1 and add R_1 to V when the process reaches $A_3 = 0$, a_3 is executed and A_2 is added to V. Given the parents values for R_3 we conclude with $R_3 = 0$. Thus, the initial webpage will contain $s_1 = (1,0,0,0,0,0)$ and when $R_1 = 1$ the solution is $s_2 = (1,0,0,0,1,0)$. Initial pages will display either s_1 or s_2. The algorithm response for different events is shown in Figure 3 where a component v is not displayed in the updated page if its status is hide (i.e. $v = 0$).

6   Experimentation

The goal of the conducted experiments is to analyze the required time to find the optimal configuration and to update the webpage. A special solver is implemented to randomly generate different CCSP and CP-net instances. The solver was implemented in Java under the NetBeans 7.0.1 environment. The experiments are conducted on a Mac OS X operating system version 10.7.5 with 2.4 GHz Intel Core i5 processor and 4 GB of DDR3 memory. We run the solver 100 times and fix the constraint tightness (percentage of allowed tuples for each constraint) and density (percentage of possible pairs between variables) to 50%.

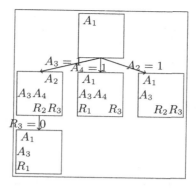

Fig. 3. Different events propagation for example 1

In each iteration we get n variables from the constraint network and randomly generate binary CP-nets where $0 < n \leq 100$. For each value of n, we randomly generate 5 different preference networks and take the average time needed to find the optimal configuration. Figure 4 shows the average time (in milliseconds) required to find an optimal configuration for the underlying CP-net and CCSP. The other part we experimented is the time needed to update the webpage. This problem is stated as follows. Given a partial assignments a we are interested in complete configurations $\mathbf{x} = xa$ that are consistent with the constraints and with minimum worsening flips. Figure 5 shows the different events along with the average time needed to update the webpage. We fix the number of events to be 50% of the original components in the webpage. For instance, if a given webpage has 10 components, we perform 5 different events (clicks) on that page.

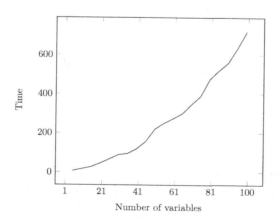

Fig. 4. The average time to find an optimal configuration

Number of Events	5	10	20	30	40	50	100
Time Needed	1.5	5.0	16.0	26.0	46.5	62.0	220.5

Fig. 5. Time required to update the webpage

7   Conclusion and Future Work

This work is part of an ongoing research project investigating the applicability of constraint problems under preferences to online and web applications [13,14]. The webpage content is treated as an interactive configuration problem where users have preferences over different components forming the page. To our knowledge, this work represents the first attempt to apply conditional constraint satisfaction together with CP-Nets to the web page content problem. Unlike previous attempts, we model the set of constraints as an instance of the conditional CSP. Thanks to our proposed method, the webmaster can now configure web content pages using compatibility and conditional constraints in addition to the set of qualitative preferences. In order to improve the time efficiency of our constraint solving approach, we are planning to integrate variable ordering heuristics based on learning and metaheuristics [15,16]. One possible extension of this work is to represent the set of qualitative preferences as a trade-off CP-net (TCP-net) [17]. The TCP-net is a generalization of the CP-Net by handling importance-tradeoff information between network attributes. Another possible extension is to handle the composite variables in addition to conditional ones [18,19,20]. A variable is composite if its domain contains other variables. Having composite and conditional variables is common in configuration problems and gives an intuitive meaning to different configuration tasks. Finally we intend to generalize our work by considering spatial constraints when configuring the web content. In this regard, we will rely on the dynamic spatio-temporal techniques we have proposed in the past [21,22].

Acknowledgements. Eisa Alanazi's research is supported by the Ministry of Higher Education, Saudi Arabia.

References

1. Mittal, S., Falkenhainer, B.: Dynamic constraint satisfaction problems. In: AAAI, pp. 25–32 (1990)
2. Boutilier, C., Brafman, R.I., Domshlak, C., Hoos, H.H., Poole, D.: Cp-nets: A tool for representing and reasoning with conditional ceteris paribus preference statements. J. Artif. Intell. Res (JAIR) 21, 135–191 (2004)
3. Dechter, R.: Constraint processing. Elsevier Morgan Kaufmann (2003)
4. Mouhoub, M., Sukpan, A.: Managing conditional and composite csps. In: Canadian Conference on AI, pp. 216–227 (2007)
5. Gelle, E., Sabin, M.: Solving methods for conditional constraint satisfaction. In: IJCAI 2003, pp. 7–12 (2003)

6. Brafman, R.I., Friedman, D.A.: Adaptive rich media presentations via preference-based constrained optimization. In: Preferences (2004)
7. Kuppusamy, K.S., Aghila, G.: Live-marker: A personalized web page content marking tool. CoRR abs/1202.2615 (2012)
8. Brafman, R.I., Domshlak, C., Shimony, S.E.: Qualitative decision making in adaptive presentation of structured information. ACM Trans. Inf. Syst. 22, 503–539 (2004)
9. Domshlak, C., Brafman, R.I., Shimony, S.E.: Preference-based configuration of web page content. In: IJCAI, pp. 1451–1456 (2001)
10. Sabin, M., Freuder, E.C., Wallace, R.J.: Greater efficiency for conditional constraint satisfaction. In: Rossi, F. (ed.) CP 2003. LNCS, vol. 2833, pp. 649–663. Springer, Heidelberg (2003)
11. Mackworth, A.K.: Consistency in networks of relations. Artificial Intelligence 8, 99–118 (1977)
12. Alanazi, E., Mouhoub, M.: Arc consistency for CP-nets under constraints. In: FLAIRS (2012)
13. Mohammed, B., Mouhoub, M., Alanazi, E., Sadaoui, S.: Data mining techniques and preference learning in recommender systems. Computer & Information Science 6 (2013)
14. Jiang, W., Sadaoui, S.: Evaluating and ranking semantic offers according to users'interests. Journal of Electronic Commerce Research 13 (2012)
15. Mouhoub, M., Jashmi, B.J.: Heuristic techniques for variable and value ordering in CSPs. In: [23], pp. 457–464
16. Abbasian, R., Mouhoub, M.: An efficient hierarchical parallel genetic algorithm for graph coloring problem. In: [23], pp. 521–528
17. Brafman, R.I., Domshlak, C.: Introducing variable importance tradeoffs into CP-nets. In: UAI, pp. 69–76 (2002)
18. Mouhoub, M., Sukpan, A.: Conditional and composite temporal CSPs. Applied Intelligence 36, 90–107 (2012)
19. Mouhoub, M., Sukpan, A.: Managing dynamic CSPs with preferences. Applied Intelligence 37, 446–462 (2012)
20. Mouhoub, M., Sukpan, A.: Managing temporal constraints with preferences. Spatial Cognition & Computation 8, 131–149 (2008)
21. Mouhoub, M.: Reasoning about Numeric and Symbolic Time Information. In: The Twelfth IEEE International Conference on Tools with Artificial Intelligence (ICTAI 2000), pp. 164–172. IEEE Computer Society, Vancouver (2000)
22. Mouhoub, M.: Dynamic path consistency for interval-based temporal reasoning. In: Applied Informatics, pp. 393–398 (2003)
23. Krasnogor, N., Lanzi, P.L. (eds.): Proceedings of the 13th Annual Genetic and Evolutionary Computation Conference, GECCO 2011, Dublin, Ireland, July 12-16. ACM (2011)

Layered Context Inconsistency Resolution for Context-Aware Systems

Been-Chian Chien and Yuen-Kuei Hsueh

Department of Computer Science and Information Engineering
National University of Tainan
Tainan, Taiwan, R.O.C.
bcchien@mail.nutn.edu.tw

Abstract. Ubiquitous computing or ambient intelligence initiates the era of integrating information techniques to build computing environments for serving users anytime and anywhere. For a context-aware system with large number of users, incorrect contexts are possibly caused by either imprecise noisy signals or the contradiction among context definitions. The incorrect context may cause context inconsistency and lead a context-aware system to bad performance. In this paper, the layered context inconsistency resolution is proposed. The layered scheme combines the prevention strategy and the detect-resolve strategy to accomplish an efficient and effective inconsistency context resolution. The proposed context model includes three layers: sensor layer, event layer, and service layer. All contexts defined on different layers apply specific strategies to resolve the problem of inconsistent contexts. The experimental results show that the proposed scheme provides an effective and efficient paradigm to improve the quality of context-aware application for smart living space.

Keywords: Ambient intelligence, context-aware, context inconsistency, inconsistency prevention, inconsistency detection and resolution.

1 Introduction

Ambient intelligence reveals a vision of future living environment, where people are surrounded by different sensing devices which are sensitive to their needs, anticipatory of their behavior and personalized to their requirements [1]. In decades, many researches on pervasive computing have drawn the technical possibility in developing ubiquitous computing environments. Context-aware computing is a pervasive computing paradigm in which application services are derived and provided by *context* information that characterizes the environmental conditions. The gathered contexts generally include user-centric context like habit and location of users and environmental context such as temperature, light, and surrounding devices. The techniques of context-awareness have been widely applied to the construction of a smart space, tour guiding and health care systems.

The research issues of context-aware systems include system frameworks, context representations, context reasoning and management, and security of services. Since

A. Moonis et al. (Eds.): IEA/AIE 2014, Part II, LNAI 8482, pp. 446–455, 2014.

contexts are the main operating elements of a context-aware system, it is an important task to capture contexts correctly. However, contexts are intrinsically fuzzy, easily inaccurate and changeful rapidly in practical applications. The incorrect contexts may lead to the contradiction among contexts of application service in the system, which is called context inconsistency. The consequence of occurring inconsistent context is that the effectiveness of a system and quality of application services are degraded. Most of the previous researches used the strategy of applying consistency constraints to detect and resolve inconsistency contexts. However, it is difficult to find all problematic contexts and identify their consistency constraints completely in real applications. Furthermore, detecting inconsistency by a large set of consistency constraints is also an overhead for the system.

In this paper, we proposed the layered context inconsistency resolution that combines the prevention strategy and the detect-resolve strategy to accomplish an efficient and effective inconsistency context resolution. First, we propose a context model which is divided into three layers: the sensor layer, the event layer, and the service layer. All contexts thus are defined as sensor context, event context, and service context according to their generating occasion and purpose. Then, the developed prevention strategy on the sensor layer involves specific constraints and rules for each sensor context to prevent the possible context error and inconsistency caused by imprecise sensing devices and environmental noises. For the contexts in the event layer, the freshness-persistence resolution are proposed and used to detect and resolve the inconsistency event contexts. The experimental results show that the prevention methods can effectively correct and resolve most of the inconsistency contexts caused by inaccurate contexts in the sensor layer. The inconsistency constraints in the event layer thus can detect and resolve the inconsistency contexts effectively and improve the quality of services in a context-aware system.

2   The Previous Work

Context inconsistency was firstly introduced by Henricksen et al. in 2004 [2] by analyzed and summarized several reasons of mismatching between sensed contexts and real-world activities. However, the resolution for imprecise context was not discussed in this article.

Generally, the approaches of resolving inconsistency context can be divided into two categories: the heuristic approach and the database approach. The difference between the heuristic and the database approaches is the information used for resolution. Before determining which context should be deleted while inconsistency occurs, the heuristic resolution considers more data, e.g. sampling time, frequency of context, and context data types. However, the heuristic approach cannot guarantee to delete the inconsistent context all the time.

The heuristic approaches are mainly proposed and studied by Bu et al. [3] and Xu [4]. Bu et al. used ontology to describe the context inconsistency detection rules for representing and detecting context inconsistency. This method applies the relationships between two classes including transitivity, inclusion, and exclusion to construct detection rules. The detection rules are then used to detect if the instances of context in ontology violate the rules of relationships. The merit of this approach is that the rules are able to be inferred easily. However, it is hard to represent complicated consistency constraints by relationships of ontology.

For the database approach of context inconsistency resolution, Xu et al. proposed a solution based on database systems in 2005 [5]. The traditional resolution of inconsistency in a database system considers two operations to perform: *Accept* or *Reject*. The operation *Accept* is to accept the last context and delete the previous old context causing inconsistency. The operation *Reject* is to reject the last context causing inconsistency and preserve the previous context for maintaining the consistency. In 2008, Xu [6] proposed a declarative constraint language to represent consistency constraints and detect context inconsistency. The proposed constraint language is based on first-order logic. Further, it supports user-defined functions. This approach can represent complicated rules easily, but it lacks of unified description and representation for user-defined functions. Mayrhofer [7] proposed the division of reactions for a context-aware system: *Reactive* and *Proactive*. The *Reactive* reaction will be triggered according to the sensing context and the situation of environments. The *Proactive* reaction will respond proper context actively to avoid inconsistency by predicting the attention and changing of users.

3   The Three-Layer Context Model

The context model is essential for a context-aware system to progress the context processing. For managing contexts in a structural paradigm, the produced contexts follow the design of the three-layer context model in CADBA [8]. The three-layer context model classifies all the contexts in the system into three classes of context: sensor context, event context, and scenario context. With the layered concept, contexts generated by sensors and users in a context-aware system are stratified as three layers called the *Sensor layer*, the *Event layer*, and the *Service layer*, as Fig. 1. The definitions of context in different context layers are declared as follows.

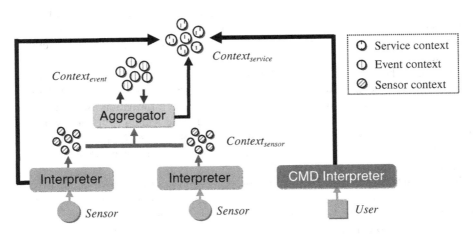

Fig. 1. The three-layer context model

1) *Sensor context (Context$_{sensor}$)*: The sensor context is the essential raw information triggered by sensors and interpreted by context interpreters. It is defined as

$$Context_{sensor} = Interpret\{ \ sensor_i \mid sensor_i \in D_j \ and \ D_j \subseteq \boldsymbol{D} \ \},$$

where *sensor$_i$* is a kind of sensors in the type of devices D_j, and $\boldsymbol{D}$ is the set of all types of devices in a context-aware system.

2) *Event context (Context$_{event}$)*: This event context is aggregated by different sensor contexts. The following definition defines the behavior of aggregating various actions of sensors into the event context.

$$Context_{event} = Aggregate\{ \bigcup_{j=1}^{m} Context_{sensor_j} \}.$$

3) *Service Context (Context$_{service}$)*: The service context is another high-level context containing not only the sensor context but also one event context at least. The service context is defined as

$$Context_{service} = Aggregate\{ \bigcup_{i=1}^{n} Context_{event_i} \} \cup \{ \bigcup_{j=0}^{m} Context_{sensor_j} \}.$$

4   Layered Context Inconsistency Resolution

As the context model presented in Section 3, the contexts in distinct layers are generated from various sources. The sensor context is formed by interpreting raw signals of sensors. The signals from sensor devices are usually full of noise. Hence, the sensor context may change and expire rapidly due to unstable signals. While the sensor context expire soon as time goes on, the incorrect event context may be easily derived since the event context is aggregated by the sensor context. For this reason, developing specific context inconsistency resolution for each context layer is natural and realistic. The layered framework of applying inconsistency resolution strategies is shown as Fig. 2. The low-level consistency module at *Sensor layer* uses prevention strategy to overcome the inaccurate problem in the sensor contexts. The high-level consistency module at *Event layer* then applies the strategy of inconsistency detection and resolution to resolve the wrong event contexts. Finally, the conflict-free module at *Service layer* considers the conflict issue that a device is activated by two services at the same time.

The Low-Level Inconsistency Prevention Strategy. The proposed prevention strategy [9] includes five context inconsistency constraints according to different kinds of sensor data types: binary data, numerical data and nominal data. The constraints are list by their data types in the following.

A. Binary data type. Three context inconsistency prevention constrains are supported: the range constraint ($\hat{C}_R$), the least persistence constraint ($\hat{C}_{LP}$), and the change-point constraint ($\hat{C}_{CP}$).

(1) The range constraint $\hat{C}_R(t_{min}, t_{max})$:

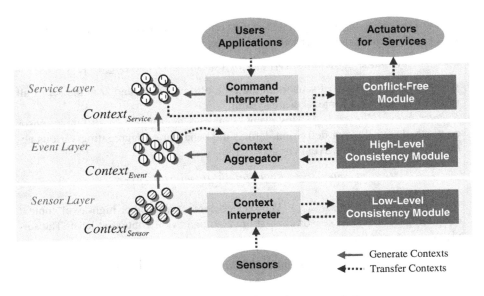

Fig. 2. The layered context inconsistency resolution

This constraint contains two parameters: t_{min} and t_{max}. t_{min} is the minimal sensing time and t_{max} is the maximal sensing time of the context. The consistency prevention rule for the constraint $\hat{C}_R(t_{min}, t_{max})$ is described as

> **if** $[\, t - start\_time(x_{signal} = 1)\,] \leq t_{min}$ **then** $x_{context}(t) = \text{ON}$;
> **if** $[\, t - start\_time(x_{signal} = 1)\,] > t_{max}$ **then** $x_{context}(t) = \text{OFF}$;

where x is a sensor, x_{signal} and $x_{context}(t)$ represent the raw signal's value and the context for the sensor x at current sampling time t, respectively. The function $start\_time()$ returns the start sampling time for signals or context of the sensor x. The prevention rule of the $\hat{C}_R$ constraint can be used in fixing the context of sensors with regular responding time. While such a sensor's signal is triggered, the context should be turned "ON" between the time t_{min} minimally and t_{max} maximally.

(2) The least persistence constraint $\hat{C}_{LP}(t_{least})$:

The least persistence constraint $\hat{C}_{LP}$ contains only one parameter t_{least} which is the least persistent time for the signals having to be sensed as "1" before the sensor context "ON" is triggered. The context is kept "OFF" if the raw signal was not sensed as "1" for t_{least} time length continuously. The consistency prevention rule for the least persistence constraint $\hat{C}_{LP}(t_{least})$ is described as

> **if** $[\, t - start\_time(x_{signal} = 1)\,] \geq t_{least}$ **then** $x_{context}(t) = \text{ON}$
>   **else** $x_{context}(t) = \text{OFF}$;

(3) The change-point constraint $\hat{C}_{CP}$:

The change-point constraint $\hat{C}_{CP}$ referred to the state correction method proposed by Kasteren et al. in 2008 [10]. Here, we modified the method to be the consistency prevention rule for the change-point constraint $\hat{C}_{CP}$. The rule is described as follows.

$$\textbf{if } [\, x_{signal}(t) \textbf{ xor } x_{signal}(t\text{-}1) \,] = 1 \quad \textbf{then} \quad x_{context}(t) = \text{ON}$$
$$\textbf{else} \quad x_{context}(t) = \text{OFF;.}$$

where $x_{signal}(t)$ is the raw signal's value at current sampling time t for the sensor x. The context of sensor x is set to "ON" while the signal is changing at the sampling point only.

B. Numerical data type. For numerical sensing data, we propose a context inconsistency prevention constraint called the bound constraint and denoted as $\hat{C}_B$. The bound constraint $\hat{C}_B(v_{max})$ contains one parameter v_{max} which is the maximal variation of signal values. The rule of consistency prevention for the constraint $\hat{C}_B(v_{max})$ is described as

$$\textbf{if } [\, x_{signal}(t) - x_{context}(t\text{-}1) \,] > v_{max} \quad \textbf{then} \quad x_{context}(t) = x_{signal}(t\text{-}1) + v_{max} \, ;$$
$$\textbf{if } [\, x_{context}(t\text{-}1) - x_{signal}(t) \,] > v_{max} \quad \textbf{then} \quad x_{context}(t) = x_{signal}(t\text{-}1) - v_{max}$$
$$\textbf{else} \quad x_{context}(t) = x_{signal}(t) \, ;$$

Since the context returns a value with numerical data type, the sensor's signal exceeding the maximal variation v_{max} is not allowed. The continuous sampling signals violates the bound constraint $\hat{C}_B(v_{max})$ will be restricted by the rule of consistency prevention.

C. Nominal data type. For nominal sensor data, we propose a context inconsistency prevention constraint named the Class Constraint, $\hat{C}_C$. Before applying the class constraint $\hat{C}_C$, a weighted transition graph G has to be built by training the historical signal sequence of an activity. The rule of consistency prevention for the class constraint $\hat{C}_C$ is described as

$$\textbf{if } [(x_{context}(t\text{-}1), x_{signal}(t)) \notin G] \quad \textbf{then} \quad x_{context}(t) = C \text{ with max}\{(x_{signal}(t\text{-}1), C) \in G\}$$
$$\textbf{else} \quad x_{context}(t) = x_{signal}(t);$$

where $\text{max}\{(x, y) \in G\}$ represents the edge (x, y) with the maximal weight in the weighted transition graph G, x and y are the context of signals.

The High-Level Inconsistency Detect-Resolve Strategy. This layer uses the traditional detection constraints [2] and resolution strategy to rectify inconsistent contexts. We give a freshness-persistence resolution(FP) approach based on context persistence and context freshness. Context persistence is defined as the existing duration of contexts in a system. Context freshness is the activation time since a context being started up. The main function of *FP* inconsistence resolution is to find the set of suspect contexts involved in the inconsistent context. The contexts with possible inconsistency then are called off from the system to avoid activating incorrect services. Of course, the cancelled contexts may affect the other correct services in the system. Hence, the design of algorithm should gain a good performance in such a manner. For describing the algorithm of FP resolution approach, we define the symbols first:

X : The set of all current contexts in the system.

X_C , X_{IC} : The set of consistent contexts and inconsistent contexts, respectively.

X^S_{IC} : The set of all sensor contexts contained in the inconsistent contexts.

x^S : The sensor context in X.

x^E : The event context which is composed of contexts x^S or other x^E.

x_{IC} : The detected inconsistent context in the current system.

$Cxt(x)$: The set of contexts of which the context x is composed.

$persist(x)$: The persistence of the context x, which being classified as *Sensor* or *Event*.

$fresh(x)$: The context freshness of the context x.

The *FP* inconsistence resolution algorithm is list as follows:

Algorithm $\hat{C}_{FP}$ (X, x_{IC})

1. $X_C = \varnothing$;
2. $X^S_{IC} = \{ x \mid x \in Ctx(x_{IC})$ and $persist(x) = Sensor$, for all $x \in X \}$;
3. **if** $X^S_{IC} = \varnothing$
4. $X^S_{IC} = \{ x \mid x \in Ctx(x_{IC})$ and $persist(x) = Event$, for all $x \in X \}$;
5. **if** $|X^S_{IC}| > 1$
6. $X^S_{IC} = \{ x \mid x \in Ctx(x_{IC})$ and $max_x\{fresh(Ctx(x_{IC}))$, for all $x \in X \}$;
7. $X_{IC} = \{ x \mid X^S_{IC} \cap Ctx(x) \neq \varnothing$, for all $x \in X \}$;
8. $X_C = X - X_{IC}$;
9. **return** X_C ;

5   Experiments and Evaluation Results

Two experiments were designed to evaluate the effectiveness of the proposed context inconsistency resolving approach for different layers. The first experiment uses two datasets, SD and KD, to evaluate the three prevention constraints for binary data type in sensor context layer. The second experiment evaluates and compares the performance of context inconsistence resolutions being applied in distinct context layers.

Experiment 1. The first test dataset uses the synthesized dataset (SD) generated by the Synthetic Data Generator in [11]. The generator simulates a living space equipped with 10 sensors. The data record the values of sensors and annotate the corresponding activities. Totally 7 activities are annotated and 28 days recorded data in the dataset. The second dataset, Kasteren dataset (KD) [10], contains the data streams that a man living alone in a three-room apartment where 14 state-change sensors were installed. The data are also collected for 28 days.

To evaluate the performance of the proposed prevention methods, each dataset was split into 27 days data for training and 1 day data for testing. The evaluation model uses Hidden Markov model (HMM) [12]. After the model was built by training data, the testing data were evaluated in cross-validation by two measures: the time slice accuracy (Acc_T) and the class accuracy (Acc_C). The two measures are defined as

$$Acc_T = \frac{\sum_{t=1}^{N} | HMM(t) = L(t) |}{N} , \qquad Acc_C = \frac{1}{C} \sum_{c=1}^{C} \left(\frac{\sum_{n=1}^{N_c} | HMM(n) = L(n) |}{N_c} \right) ,$$

where N is the total number of sampling time slices. N_c is the number of sampling time slices belonging to the labeled activity c and C is the number of possible activities. The $HMM(t)$ is the inferred activity and $L(t)$ is the labeled real activity in the dataset at sampling time t. The $|HMM(t)=L(t)|$ is the number of activities which $HMM(t)$ and $L(t)$ is identical at sampling time t.

The evaluation results for the three binary data constraints, $\hat{C}_R$, $\hat{C}_{LP}$, and $\hat{C}_{CP}$, are shown in Table 1. The time slice accuracy and the class accuracy of $\hat{C}_R$ are not good since the sampling rate in raw data is much larger than noises. The time slice accuracy and the class accuracy of $\hat{C}_{LP}$ get much gain and have the best results. Since the $\hat{C}_{LP}$ constraint accepts the context if signals can be kept for t_{least} time, $\hat{C}_{LP}$ can filter out the noise effectively when the noise rate are high and the t_{least} is large enough. The $\hat{C}_{CP}$ constraint also is not suitable for the case of large sampling time and unstable signals.

Table 1. The average accuracy of inconsistency prevention constraints

	Raw		$\hat{C}_R$		$\hat{C}_{LP}$		$\hat{C}_{CP}$	
	Acc_T	Acc_C	Acc_T	Acc_C	Acc_T	Acc_C	Acc_T	Acc_C
SD	37.09	19.95	33.02	15.94	88.67	80.82	33.37	17.97
KD	55.97	14.30	57.45	15.36	54.95	28.13	57.26	15.29

Experiment 2. The test dataset here uses Siafu simulator [13] written by M. Martin to generate various environmental contexts including 91 locations, 100 people, 8 activities and 3 surrounding contexts. The parameters of simulation are the same to the ones in [6]. The dataset contains 57,600 records in 24 hours with 150 seconds sampling rate. Each record consists of four sensor context information and one event context information. The sensor contexts include personal name, location, user's profile, and surrounding contexts. The event context is activity. The number of total contexts is 288,000 (57,600×5). In addition to the above generated contexts, 8 services: S1, S2, S4, S5, S6, S8, S14, S15, are defined by the system and they will be activated while the defined contexts occurring.

In such a simulation environment, 12 (C1 to C12) consistency constraints are used to keep the correctness of services. The constraints define the consistency rule of contexts. As Table 2, the constraints C1, C2, C3, and C4 belong to the event layer; the other constraints C5 to C12 can be applied in the sensor layer. Further, the C5 takes $\hat{C}_{S_B}$ as the resolution approach since it is a numerical type used to determine location. The C6 to C12 are nominal type, so they use the $\hat{C}_{S_C}$ resolution approach. The $\hat{C}_{FP}$ is applied by C1 to C4 to resolve the inconsistency in the event layer.

The three measures, the service activation rate (SAR), $precision$, and $recall$ are used to evaluate the performance of the context inconsistency resolution, which are defined as follows:

$$SAR = \frac{CS_{activated}}{S_{all}} \quad , \quad precision = \frac{CS_{activated}}{S_{activated}}, \quad recall = \frac{CS_{activated}}{CS_{all}} ;$$

where $CS_{activated}$ is the number of activated services after resolving correctly; CS_{all} is the number of correct services; $S_{activated}$ is the number of all activated services after resolving; S_{all} is the number of activated services without resolving inconsistency.

Table 2. The constraints for detecting and resolving inconsistent contexts

Constraints	Context Layer	Resolutions
C1, C2, C3, C4	Event layer	$\hat{C}_{FP}$
C5	Sensor layer	$\hat{C}_{S_B}$
C6, C7, C8, C9, C10, C11, C12	Sensor layer	$\hat{C}_{S_C}$

In this experiment, some noises are mixed in the low-level sensors. Then, four strategies are used to test the activation of services. The results of evaluation on service activation rates and precision-recall rates are shown in Table 3 and Table 4, respectively. The 'Noise' means that no inconsistency resolution is used for resolving noisy contexts. The 'Sensor' indicates that only the low-level module in the sensor layer is applied. The 'Event' stands for using the high-level module in the event layer only. The 'Sensor+Event' takes the resolutions in both of the two layers.

Table 3. The service activation rate (SAR) for each service in different resolution strategies

Services No.	Noise	Sensor	Event	Sensor + Event
S1	100.00	96.49	100.00	91.54
S2	69.54	100.00	70.81	99.96
S4	94.27	100.00	71.61	100.00
S5	54.36	95.47	42.20	95.07
S6	73.96	99.75	73.50	99.83
S8	73.40	100.00	74.59	100.00
S14	73.96	99.75	73.50	99.83
S15	47.97	97.18	36.52	97.23
Average	73.43	98.58	67.84	97.93

Table 4. Service activation *precision* and *recall* for different resolution strategies

Measures	Noise	Sensor	Event	Sensor + Event
$CS_{activated}$	3701	7415	3209	7386
$S_{activated} - CS_{activated}$	4510	795	5001	824
$S_{activated} - CS_{all}$	2122	1028	1563	1024
Recall	63.56%	87.84%	66.89%	87.84%
Precision	45.08%	90.32%	39.01%	89.96%

6   Conclusion

In this paper, we propose the layered context inconsistency resolution for context-aware systems. The context inconsistency constraints include five prevention rules in sensor layer and the inconsistency resolution in event layer. The experimental results show that the prevention constraints in sensor layer can improve service activation rate effectively. The multi-layer resolution can further decrease the number of wrong activation of services although the number of activated correct services is also reduced.

Obviously, the layered context inconsistency resolution is much better than the traditional inconsistency resolution in the event layer.

Acknowledgment. This research was supported in part by National Science Council of Taiwan, R. O. C. under contract NSC 99-2221-E-024-015.

References

1. Feng, L., Apers, P.M.G., Jonker, W.: Towards Context-Aware Data Management for Ambient Intelligence. In: Galindo, F., Takizawa, M., Traunmüller, R. (eds.) DEXA 2004. LNCS, vol. 3180, pp. 422–431. Springer, Heidelberg (2004)
2. Henricksen, K., Indulska, J.: Modeling and Using Imperfect Context Information. In: The 2nd IEEE International Conference on Pervasive Computing and Communications, pp. 33–37 (2004)
3. Bu, Y., Chen, S., Li, J., Tao, X., Lu, J.: Context Consistency Management Using Ontology Based Model. In: Grust, T., et al. (eds.) EDBT 2006. LNCS, vol. 4254, pp. 741–755. Springer, Heidelberg (2006)
4. Xu, C., Cheung, S.C., Chan, W.K., Ye, C.: On Impact-Oriented Automatic Resolution of Pervasive Context Inconsistency. In: The 6th Joint Meeting of the European Software Engineering Conference and the ACM SIGSOFT International Symposium on the Foundations of Software Engineering, pp. 569–572 (2007)
5. Xu, C., Cheung, S.C.: Inconsistency Detection and Resolution for Context-Aware Middleware Support. In: The 4th Joint Meeting of the European Software Engineering Conference and the ACM SIGSOFT International Symposium on the Foundations of Software Engineering, pp. 336–345 (2005)
6. Xu, C.: Inconsistency Detection and Resolution for Context-Aware Pervasive Computing. Ph.D dissertation, Department of Computer Science and Engineering, The Hong Kong University of Science and Technology, Hong Kong (2008)
7. Mayrhofer, R.: Context Prediction based on Context Histories: Expected Benefits, Issues and Current State-of-the-Art. In: The 1st International Workshop on Exploiting Context Histories in Smart Environments, pp. 31–36 (2005)
8. Chien, B.C., Tsai, H.C., Hsueh, Y.K.: CADBA: A Context-aware Architecture based on Context Database for Mobile Computing. In: The International Workshop on Pervasive Media, Joint with the Sixth International Conference on Ubiquitous Intelligence and Computing, pp. 367–372 (2009)
9. Chien, B.C., Hsueh, Y.K.: Initiative Prevention Strategy for Context Inconsistency in Smart Home. In: The 2011 IEEE International Conference on Granular Computing, pp. 138–143 (2011)
10. van Kasteren, T., Noulas, A., Englebienne, G., Kröse, B.: Accurate activity recognition in a home setting. In: The 10th International Conference on Ubiquitous Computing, pp. 1–9 (2008)
11. Szewcyzk, S., Dwan, K., Minor, B., Swedlove, B., Cook, D.: Annotating Smart Environment Sensor Data for Activity Learning. Methods of Information in Medicine 48(5), 480–485 (2009)
12. Rabiner, L.R.: A tutorial on hidden Markov models and selected applications in speech recognition. Proceedings of the IEEE 77(2), 257–286 (1989)
13. Siafu: An Open Source Context Simulator, http://saifusimulator.courceforge.net/

The Stowage Stack Minimization Problem with Zero Rehandle Constraint

Ning Wang[1], Zizhen Zhang[2,*], and Andrew Lim[3,**]

[1] Department of Management Sciences, City University of Hong Kong,
Tat Chee Ave, Kowloon Tong, Hong Kong
[2] School of Mobile Information Engineering, Sun Yat-sen University,
Tangjia Bay, Zhuhai, Guangdong, China
zhangzizhen@gmail.com
[3] School of Management and Engineering, Nanjing University,
Nanjing, Jiangsu, China

Abstract. The stowage stack minimization problem with zero rehandle constraint (SSMP-ZR) aims to find a minimum number of stacks to accommodate all the containers in a multi-port voyage without occurring container rehandles. In this paper, we first give the integer models of the SSMP-ZR (with uncapacitated and capacitated stack height). Next, heuristic algorithms are proposed to construct solutions to the SSMP-ZR. The theoretical performance guarantee of the algorithms is then discussed. To evaluate the actual performance of the algorithms, we conduct experiments on a set of instances with practical size. The results demonstrate that our heuristic approaches can generate very promising solutions compared with the random loading solutions and integer programming solutions by CPLEX.

Keywords: container ship stowage planning, stack minimization, zero rehandle, constructive heuristic.

1 Introduction and Literature Review

The stowage stack minimization problem (SSMP) investigates a stowage planning problem when carriers have the obligation to ship all the given containers in different ports, with the objective to utilize the fewest number of stacks on the ship. A ship calls a sequence of ports and containers are loaded and unloaded at each port by cranes in a last-in-first-out (LIFO) manner. If a container i is stacked above another container j and container i is to be discharged at a later port than container j (overstowage), then container i should be moved to another stack before container j is retrieved. This operation is known as *rehandling*. Generally speaking, if at most s rehandles are allowed throughout the journey, it is denoted as s-rehandle constraint. In this paper, we focus on the SSMP with zero rehandle constraint, abbreviated as SSMP-ZR. The SSMP-ZR is a good

---

* Corresponding Author.
** Andrew Lim is currently on no pay leave from City University of Hong Kong.

A. Moonis et al. (Eds.): IEA/AIE 2014, Part II, LNAI 8482, pp. 456–465, 2014.

starting point for general SSMP, as it exhibits fundamental properties that can be extended to the SSMP.

The SSMP-ZR is relative with container ship stowage planning problems (CSSP). Historically, the first study on the CSSP was done by Webster & Van Dyke [12]. The problem they addressed was quite simplistic and their methods were not extensively tested to indicate good and robust performance. Avriel et al. [3] developed a suspensory heuristic to deal with the CSSP with the objective to minimize overstowage. Similarly, Salido et al. [10] and Gümüş et al.[7] also solved the problem with the assistance of heuristic thought. Apart from heuristics, other techniques for combinatorial optimization, such as meta-heuristic, tree search, are widely used. Wilson & Roach [13] and Ambrosino et al. [1] applied tabu search for the optimization. Genetic algorithm was adopted by Dubrovsky et al. [4] and Imai et al. [8]. However, for the SSMP-ZR, it is rarely discussed in the literature. To the best of our knowledge, only Avriel et al. [2] and Jensen [9] discussed the SSMP-ZR and its connection with the chromatic number of circle graphs ([11]) when the stack height is unlimited.

In the rest of the paper, a formal description of the problem, together with some properties is addressed in Section 2. In Section 3, we propose constructive heuristic algorithms to find approximated solutions of the problem. The performance guarantees of the algorithms are also provided. The computational experiments in Section 4 illustrate the results generated by the heuristic algorithms and other methods. Section 5 gives some closing remarks.

2   Problem Description and Properties

A ship starts its journey at port 1 and sequentially visits port 2, 3, ..., P (port P could be the same as port 1 in a circular route). There are N standard containers to be shipped along the journey. At each port, the containers destined to the current port are discharged, and the containers stowed at the port are loaded to the ship. Each container i is characterized by its shipping leg $O(i) \rightarrow D(i)$, indicating that the container is loaded at port $O(i)$ and discharged at port $D(i)$. All the loading and discharging ports of N containers are known in advance.

We assume that the ship is empty initially and it has enough capacity to accommodate all the containers along the journey. The ship spaces are divided into several stacks. Each stack can hold at most H vertically piled containers, i.e., the height limit of every stack is H. When H is bounded, we call the SSMP-ZR as the capacitated SSMP-ZR (CSSMP-ZR); when H is sufficiently large (e.g., $H \geq N$) or unbounded, we refer to it as the uncapacitated SSMP-ZR (USSMP-ZR). For the CSSMP-ZR, if the number of containers in a non-empty stack is less than H, the stack is termed as a *partial* stack; otherwise it is a *full* stack. The objective of the SSMP-ZR is to transport all the containers using the fewest number of stacks without rehandles.

A feasible solution of the SSMP-ZR is represented by a loading plan $\langle (c_{p,1}, s_{p,1}),$ $\ldots, (c_{p,T_p}, s_{p,T_p}) \rangle$ at each port p, where $c_{p,i}$ is the index of the container, $s_{p,i}$ is the index of the stack, and T_p is the number of outbound containers at port p.

$(c_{p,i}, s_{p,i})$ denotes a loading assignment of container $c_{p,i}$ to stack $s_{p,i}$. We also denote $L_{p,i}$ (for $i = 1, \ldots, T_p$) as the layout of the ship after the i-th loading operation at port p. Specially, L_{p,T_p} is the layout when the ship departs port p. Because no container rehandle occurs, we do not need to explicitly determine a concrete container unloading plan. For notational convenience, we use $L_{p,0}$ to denote the ship layout after unloading all the containers destined to port p.

The CSSMP-ZR can be mathematically formulated by the following integer programming (IP) model. For the USSMP-ZR, the model is similarly formulated except that Constraints (4) are removed.

$$(IP) \quad \min \; \sum_{s=1}^{S} y_s \tag{1}$$

$$\text{s.t.} \quad \sum_{s=1}^{S} x_{is} = 1, \; \forall 1 \le i \le N \tag{2}$$

$$\sum_{i=1}^{N} x_{is} \le M y_s, \; \forall 1 \le s \le S \tag{3}$$

$$\sum_{i:O(i)\le p\le D(i)-1} x_{is} \le H, \; \forall 1 \le s \le S, 1 \le p \le P \tag{4}$$

$$x_{is} + x_{js} \le 1, \; \forall 1 \le s \le S, 1 \le i,j \le N, O(j) < O(i) < D(j) < D(i) \tag{5}$$

$$x_{is}, y_s \in \{0,1\}, \; \forall 1 \le s \le S, 1 \le i,j \le N \tag{6}$$

In the above model, M is a sufficiently large number and S is the number of stacks on ship. x_{is} is a binary decision variable which is equal to 1 if container i is loaded to stack s, and 0 otherwise. y_s is a binary decision variable that is equal to 1 if stack s is used for stowage, and 0 otherwise. $O(i)$ and $D(i)$ indicate the shipping leg of container i. The objective (1) minimizes the number of stacks used. Constraints (2) ensure that each container i is loaded to only one stack. Constraints (3) guarantee that containers can only be loaded to used stacks. Constraints (4) assure that at each port p, the number of containers in stack s does not exceed H. Constraints (5) avoid overstowage requiring that container i is prohibited to block container j in the same stack when container j is retrieved.

Note that the solution to the IP model only reflects the assignment of containers to stacks. It can be converted to a complete loading plan as follows. At each port p, load containers to their corresponding stacks (given by x_{is}) by the decreasing order of their destinations. It is straightforward that the resultant layout from such a loading order does not include overstowage.

As for the complexity of the SSMP-ZR, both CSSMP-ZR and USSMP-ZR belong to NP-hard according to Avriel et al. [2]. However, for $H = 1$, the problem is solvable in polynomial time, whereas the running time is still unknown for other fixed H.

In Avriel et al. [2], the authors showed that USSMP-ZR is equivalent to the coloring of the circle graph (a circle graph is the intersection graph of a set of chords of a circle), i.e., the chromatic number of the circle graph is the answer of

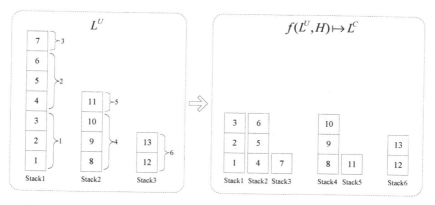

Fig. 1. An example of the transformation from L^U to L^C with $H = 3$

the corresponding USSMP-ZR instance. Therefore, conclusions on circle graphs coloring (see [5,6]) can be directly applied to the USSMP-ZR.

The USSMP-ZR is far from the real situation in the stowage planning, since stack heights of ships are restricted by operation cranes. However, given an instance, a feasible solution to the USSMP-ZR can be converted into a feasible solutions to the CSSMP-ZR. One method, amongst others, is to use the following transformation function.

Definition 1. *The transformation function $f(L^U, H) \mapsto L^C$ transforms an arbitrary layout L^U of the USSMP-ZR into a layout L^C of the CSSMP-ZR with height limit H: for each stack in the layout L^U, mark every H vertically adjacent containers bottom up with the same label. Containers marked with the same label compose a stack in L^C.*

Figure 1 illustrates an example of the layout transformation. Suppose that we have a feasible loading plan for the USSMP-ZR at port p, which results in a series of layouts $L^U_{p,0}, L^U_{p,1}, \ldots, L^U_{p,T_p}$. By applying the transformation function to $L^U_{p,i}$ ($\forall p = 1, \ldots, P, i = 0, \ldots, T_p$), we can generate a set of layouts $f(L^U_{p,i}, H)$. From this set of layouts, we can recover a feasible loading plan for the CSSMP-ZR.

3   Methodology

In this section, we talk about the heuristic algorithms to the USSMP-ZR and CSSMP-ZR, as well as the performance guarantee of the algorithms.

3.1   Heuristic Algorithms for the SSMP-ZR

We first provide an algorithm for the USSMP-ZR (see Algorithm 1). Given an instance I, the algorithm simply either loads a container to a stack or unloads a container from a stack until all the containers are processed.

Algorithm 1. A heuristic procedure for the USSMP-ZR

UHEURISTIC (I)

1 **for** each port $p = 1, 2, \ldots, P$
2   Unload containers destined at port p.
3   Sort containers at p by the decreasing order of their destinations.
4   **for** each container c at port p
5   S=currently non-empty stacks.
6            $c(s_i)$= the topmost container of stack $s_i \in S$.
7            $S' = \{s_i | s_i \in S \text{ and } D(c(s_i)) \geq D(c)\}$.
8   **if** $S' \neq \emptyset$
9                 $s_{min} = \arg\min_{s \in S'} D(c(s))$.
10   Load container c to stack s_{min}.
11   **else**
12   Load c to an empty stack.

Algorithm 1 generates a feasible loading plan $\langle (c_{p,1}, s_{p,1}), (c_{p,2}, s_{p,2}), \ldots, (c_{p,T_p}, s_{p,T_p}) \rangle$ for each port p, and a series of corresponding ship layouts $L_{p,0}^U, \ldots, L_{p,T_p}^U$ for the USSMP-ZR. As discussed in Section 2, we can apply the transformation function to obtain a feasible solution to the CSSMP-ZR. However, in this paper we devise another method (see Algorithm 2) to tackle the CSSMP-ZR.

Algorithm 2. A heuristic procedure for the CSSMP-ZR

CHEURISTIC (I)

1 **for** each port $p = 1, 2, \ldots, P$
2   Unload containers destined at port p.
3   Sort containers at p by the decreasing order of their destinations.
4   **for** each container c at port p
5   S=currently non-empty stacks with height smaller than H.
6            $c(s_i)$= the topmost container of stack $s_i \in S$.
7            $S' = \{s_i | s_i \in S \text{ and } D(c(s_i)) \geq D(c)\}$.
8   **if** $S' \neq \emptyset$
9                 $s_{min} = \arg\min_{s \in S'} D(c(s))$.
10   Load container c to stack s_{min}.
11   **else**
12   Load c to an empty stack.

3.2   Performance Guarantee of the Algorithms

We show that the heuristic algorithms have a performance guarantee if we regard the port number P as a fixed number. For ease of exposition, we first give some related notations.

- $\mathcal{U}^*$ & $\mathcal{C}^*$: the optimal solution to the USSMP-ZR and CSSMP-ZR instances, respectively.
- $\mathcal{U}$ & $\mathcal{C}$: the solution to the USSMP-ZR and CSSMP-ZR instances by the algorithm, respectively.
- $\mathcal{U}_p$ & $\mathcal{C}_p$: the number of stacks used at port p by the algorithm.
- N_p: the number of containers on the ship before its departure from port p.
- v_p: the number of loading ports that the ship has visited before it departs from port p. A port is called a loading port if there exists at least one container to be loaded. Clearly, $v_p \leq p, \forall p = 1, \ldots, P$.

Proposition 1. *For the USSMP-ZR instance, the heuristic approach generates a solution in which at most v_p stacks are used before the ship departs from port p, i.e. $\mathcal{U}_p \leq v_p, \forall p = 1, \ldots, P$.*

Proof. We use an inductive proof.

Basis: For $p = 1$, without loss of generality, we assume that port 1 is a loading port. The heuristic piles up containers in one stack in a way that the containers with later destinations are placed in lower tiers. We have $\mathcal{U}_1 = v_1 = 1$, and therefore $\mathcal{U}_1 \leq v_1$ holds.

Inductive step: Suppose that before the ship departs from port p, there are at most v_p stacks used, i.e., $\mathcal{U}_p \leq v_p$. Upon the ship arrives at port $p + 1$, it first unloads the containers from the ship. This process does not increase the number of stacks utilized. If there are containers to be loaded at port $p+1$, then $v_{p+1} = v_p + 1$, otherwise $v_{p+1} = v_p$.

(1) If there exist some containers to be loaded, according to our heuristic, they are placed on either the extant stacks or a new blank stack (with the containers of later destinations placed lower). Hence, $\mathcal{U}_{p+1} \leq \mathcal{U}_p + 1 \leq v_p + 1 = v_{p+1}$.

(2) If no container is loaded, $\mathcal{U}_{p+1} \leq \mathcal{U}_p \leq v_p = v_{p+1}$.

Based on the above two steps, we have $\mathcal{U}_p \leq v_p, \forall p = 1, \ldots, P$. Furthermore, we have

$$\mathcal{U} = \max_{p=1,\ldots,P} \mathcal{U}_p \leq \max_{p=1,\ldots,P} v_p = v_P \leq P$$

Proposition 2. *The CSSMP-ZR instance has a lower bound:*

$$\max_{p=1,\ldots,P} (\lceil \frac{N_p}{H} \rceil) \leq \mathcal{C}^*$$

Proof. $\lceil \frac{N_p}{H} \rceil$ is the least number of stacks to accommodate all the containers at port p, and thus, $\max_p(\lceil \frac{N_p}{H} \rceil)$ is the least number of stacks needed throughout the journey.

The lower bound can help to check the optimality of solutions obtained by our heuristic. If the solution is equal to the lower bound, we can conclude that the solution is optimal.

Proposition 3. *For the heuristic solution of the CSSMP-ZR instance, it holds*

$$\mathcal{C}^* \leq \mathcal{C} \leq \max_{p=1,\dots,P}(\lfloor \frac{N_p}{H} \rfloor + v_p) \leq \mathcal{C}^* + v_P$$

Proof. The first inequality obviously holds.

For the second inequality, $\mathcal{C} = \max_p \mathcal{C}_p$. Remind that $\mathcal{C}_p$ is the number of stacks in the layout L_{p,T_p}^C resulted from our algorithm. The number of partial stacks in the layout L_{p,T_p}^C is no greater than the number of stacks in the associated layout L_{p,T_p}^U (c.f., Definition 1). The number of stacks of L_{p,T_p}^U, from Proposition 1, is at most v_p, so the number of partial stacks in L_{p,T_p}^C is no greater than v_p. In addition, the number of full stacks in L_{p,T_p}^C is no greater than $\lfloor \frac{N_p}{H} \rfloor$. Therefore, $\mathcal{C}_p \leq \lfloor \frac{N_p}{H} \rfloor + v_p$ and $\mathcal{C} \leq \max_p(\lfloor \frac{N_p}{H} \rfloor + v_p)$ hold.

For the third inequality, $\lfloor \frac{N_p}{H} \rfloor \leq \lceil \frac{N_p}{H} \rceil$, and $\lceil \frac{N_p}{H} \rceil$ is the lower bound of the number of stacks used at port p by Proposition 2. Thus, $\lfloor \frac{N_p}{H} \rfloor \leq \mathcal{C}_p^*$. It then holds $\max_p(\lfloor \frac{N_p}{H} \rfloor + v_p) \leq \max_p(\mathcal{C}_p^* + v_p) \leq \max_p(\mathcal{C}_p^* + v_P) = \mathcal{C}^* + v_P$.

The above proposition indicates that our approximation algorithm has a constant performance guarantee if we regard the number of ports as fixed. In practice, the number of ports is generally much smaller than the number of stacks used along the journey. Hence, the gap between the optimal solution and our heuristic solution is relatively small, which demonstrates that our heuristic can generate promising solutions.

It is worth noting that the inequalities in Proposition 2 and Proposition 3 are also valid for the USSMP-ZR if we consider $H \to +\infty$. In addition, we can easily extend the bounds of the CSSMP-ZR to the capacitated stowage stack minimization problem allowing rehandles (CSSMP-R). Suppose that the optimal solution to the CSSMP-R is denoted as $\mathcal{R}^*$, then we have

$$\max_{p=1,\dots,P}(\lceil \frac{N_p}{H} \rceil) \leq \mathcal{R}^* \leq \mathcal{C}^* \leq \mathcal{C} \leq \max_{p=1,\dots,P}(\lfloor \frac{N_p}{H} \rfloor + v_p) \leq \mathcal{R}^* + v_P \leq \mathcal{C}^* + v_P$$

4   Experiments and Analysis

In this section, we will showcase the performance of our heuristic algorithms on a number of test instances. The heuristic was implemented in Java and the experiments were conducted on a computer with Intel Xeon processor clocked at 2.66 GHz and 4 GB RAM. The operating system of the computer is Linux. To provide a benchmark result, we also applied the ILOG CPLEX 12.4 with default setting to solve the IP model in Section 2.

As there is no test data available in the existing literature, we generated several sets of data by the following manner. There are 72 sets of instances in total, which are categorized by three parameters: the number of ports P, which is selected from $\{5, 10, 20\}$; the total number of containers to be shipped along the voyage N, which is selected from $\{50, 100, 200, 500, 1000, 5000\}$; the height limit H selected from $\{4, 8, 12, \infty\}$, where $H \to \infty$ indicates that the corresponding instance is the USSMP-ZR instance. Each set consists of 5 instances generated by different random seeds. In each instance, the origin $O(i)$ and the destination $D(i)$ of a container i are generated from a uniform distribution on the integers $1, 2, \ldots, P$ satisfying that $O(i) < D(i)$. All the data sets can be found at http://www.computational-logistics.org/orlib/topic/SSMP-ZR/index.html.

Table 1. Results of the instances with $N \in \{50, 100, 200\}$

N	P	H	LB	UB	CPLEX	Heuristic	Rand	N	P	H	LB	UB	CPLEX	Heuristic	Rand
50	5	4	8	9.6	8	8	9	100	10	12	5.2	10.4	5.8	7	8.6
50	5	8	4	6	4	4.6	5.2	100	10	∞	1	9	5	6.6	8.2
50	5	12	3	5	3	3.2	4	100	20	4	14.2	25	14.2	16	17.2
50	5	∞	1	4	3	3	4	100	20	8	7.4	19.6	8.2	10.8	13.2
50	10	4	7.8	12.6	8	9	9.8	100	20	12	5	19.2	7.8	10.2	12.2
50	10	8	4	10	5	6	7.2	100	20	∞	1	19	7.8	9.8	12.2
50	10	12	3	9.4	4.8	5.6	7.2	200	5	4	30.6	32.2	30.6	30.6	31.2
50	10	∞	1	9	4.8	5.6	7.2	200	5	8	15.6	17.4	15.6	15.6	17.4
50	20	4	7.2	19.6	7.6	9	10.2	200	5	12	10.4	12.4	10.4	10.4	12.2
50	20	8	3.8	19	5.8	7.8	8.6	200	5	∞	1	4	3	3	4
50	20	12	2.8	19	5.8	7.8	8.6	200	10	4	28.6	33.2	29.4	29.6	30.6
50	20	∞	1	19	5.8	7.8	8.6	200	10	8	14.4	19.2	15.6	16.4	18.6
100	5	4	15	16.8	15	15	16.2	200	10	12	9.8	15	11.4	12	14
100	5	8	7.6	9.6	7.6	7.6	9.2	200	10	∞	1	9	5	7.8	9
100	5	12	5.4	7.6	5.4	5.4	6.4	200	20	4	27.6	37.6	31	29.8	32
100	5	∞	1	4	3	3	4	200	20	8	14.2	25.2	19.4	17.6	20.4
100	10	4	14.8	19.6	14.8	15.8	16.6	200	20	12	9.4	21.6	15	13.8	17
100	10	8	7.6	12.6	7.6	9.2	11.2	200	20	∞	1	19	13.4	12.6	15.4

The experimental results are shown in Table 1 and Table 2. Table 1 summarizes the results of the instances with small number of containers, i.e. $N \in \{50, 100, 200\}$. There are 36 sets reported in Table 1. The columns with the heading LB, UB, CPLEX, Heuristic and Rand report the average lower bounds, upper bounds, CPLEX results, results by the heuristic algorithms, results by *random loading* on the five instances of each set, respectively. The lower bound of each instance is given by $\max_p \lceil \frac{N_p}{H} \rceil$, and the upper bound is given by $\max_p (\lfloor \frac{N_p}{H} \rfloor + p)$. The CPLEX results are obtained after 10 minutes of CPU time for each instance. We find that there are only 146 out of 180 instances are solved to optimality by CPLEX, while the other 34 instances have only feasible solutions. The heuristic approaches solve each instance in less than a second, to this end we do not report the running time. The Rand results are obtained by loading each container in a random order to a randomly selected available stack. Observe from Table 1 that the heuristic algorithms obviously outperform the random loading approach. Besides, they can generate very competitive solutions compared with CPLEX and LB.

Table 2. Results of the instances with $N \in \{500, 1000, 5000\}$

N	P	H	LB	UB	Heuristic	Rand	N	P	H	LB	UB	Heuristic	Rand
500	5	4	76.6	78	76.6	77.8	1000	10	12	46.2	50.4	46.6	55.2
500	5	8	38.6	40	38.6	40	1000	10	∞	1	9	8	9
500	5	12	26	27.4	26	28	1000	20	4	134.4	143.6	135.8	143.4
500	5	∞	1	4	3	4	1000	20	8	67.6	76.8	70.6	81.2
500	10	4	69.8	74.2	69.8	72	1000	20	12	45.2	54.4	48.8	59.6
500	10	8	35	39.4	35.6	40.6	1000	20	∞	1	19	17.6	19
500	10	12	23.6	28	25.2	29.4	5000	5	4	758.6	759.6	758.6	759
500	10	∞	1	9	8	9	5000	5	8	379.6	380.6	379.6	395.8
500	20	4	67.6	77.2	70.2	74.2	5000	5	12	253.2	254.2	253.2	270.4
500	20	8	34	43.6	38	44.2	5000	5	∞	1	4	3	4
500	20	12	22.8	33.2	27	33.4	5000	10	4	695.2	699.2	695.2	701.6
500	20	∞	1	19	15.8	18.6	5000	10	8	348	352	348	370.4
1000	5	4	152.2	153.8	152.2	152.8	5000	10	12	232.2	236.2	232.2	259.4
1000	5	8	76.6	78.2	76.6	79	5000	10	∞	1	9	8	9
1000	5	12	51	52.8	51	54.6	5000	20	4	664	672.8	664	686.8
1000	5	∞	1	4	3	4	5000	20	8	332.4	341.2	333.2	372.4
1000	10	4	137.6	141.8	137.6	141.6	5000	20	12	221.8	230.8	223.6	269.2
1000	10	8	69	73.2	69.2	76.6	5000	20	∞	1	19	18	19

Table 2 summarizes the results of the instances with large N, i.e. $N \in \{500, 1000, 5000\}$. The columns in this table are similarly defined as those of Table 1. In particular, the column with its heading CPLEX is removed, because we found that CPLEX ran out of memory for most instances of such large scales.

5   Conclusion

The stowage stack minimization problem with zero rehandle constraint problem (SSMP-ZR) is a good entry point for the investigation of minimum rehandle problem in container ship stowage planning. In this paper, we presented the integer programming model for the SSMP-ZR. Two simple heuristic approaches were then proposed to construct near-optimal solutions in a very short computional time. The experimental results show that our heuristic approaches generate very promising solutions on a variety of instances. In the near future, a more generalized version s-rehandle problem, or other extensions of the SSMP-ZR will be studied.

References

1. Ambrosino, D., Anghinolfi, D., Paolucci, M., Sciomachen, A.: A new three-step heuristic for the Master Bay Plan Problem. Maritime Economics & Logistics 11(1), 98–120 (2009)
2. Avriel, M., Penn, M., Shpirer, N.: Container ship stowage problem: Complexity and connection to the coloring of circle graphs. Discrete Applied Mathematics 103(1), 271–279 (2000)

3. Avriel, M., Penn, M., Shpirer, N., Witteboon, S.: Stowage planning for container ships to reduce the number of shifts. Annals of Operations Research 76, 55–71 (1998)
4. Dubrovsky, O., Levitin, G., Penn, M.: A genetic algorithm with a compact solution encoding for the container ship stowage problem. Journal of Heuristics 8(6), 585–599 (2002)
5. Garey, M.R., Johnson, D.S., Miller, G.L., Papadimitriou, C.H.: The Complexity of Coloring Circular Arcs and Chords. SIAM Journal Algebraic and Discrete Methods 1(2), 216–227 (1980)
6. Gavril, F.: Algorithms for a maximum clique and a maximum independent set of a circle graph. Networks 3(3), 261–273 (1973)
7. Gümüş, M., Kaminsky, P., Tiemroth, E., Ayik, M.: A multi-stage decomposition heuristic for the container stowage problem. In: Proceedings of the MSOM Conference (2008)
8. Imai, A., Sasaki, K., Nishimura, E., Papadimitriou, S.: Multi-objective simultaneous stowage and load planning for a container ship with container rehandle in yard stacks. European Journal of Operational Research 171(2), 373–389 (2006)
9. Jensen, R.M.: On the complexity of container stowage planning. Tech. rep. (2010)
10. Salido, M.A., Sapena, O., Rodriguez, M., Barber, F.: A planning tool for minimizing reshuffles in container terminals. In: IEEE International Conference on Tools with Artificial Intelligence (2009)
11. Unger, W.: On the k-colouring of circle-graphs. In: Cori, R., Wirsing, M. (eds.) STACS 1988. LNCS, vol. 294, pp. 61–72. Springer, Heidelberg (1988)
12. Webster, W.C., Van Dyke, P.: Container loading: a container allocation model I and II: introduction, background, stragegy, conclusion. In: Computer-Aided Ship Design Engineering Summer Conference (1970)
13. Wilson, I.D., Roach, P.A.: Principles of combinatorial optimization applied to container-ship stowage planning. Journal of Heuristics 5(4), 403–418 (1999)

The Multi-period Profit Collection Vehicle Routing Problem with Time Windows

Yubin Xie[1], Zizhen Zhang[1,*], Hu Qin[2], Songshan Guo[1], and Andrew Lim[3]

[1] Sun Yat-Sen University, Guangdong, China
zhangzizhen@gmail.com
[2] Huazhong University of Science and Technology, Wuhan, China
[3] Nanjing University, Nanjing, China

Abstract. This paper addresses a new variant of the vehicle routing problem with time windows that takes drivers' working periods into consideration. In this problem, each vehicle is dispatched to perform one route in a multi-period planning horizon. At the end of each period, each vehicle is not required to return to the depot but must stay at one of vertices for recuperation. We propose a tabu search algorithm to solve 48 test instances generated from Solomon's VRPTW instances. The computational results can serve as benchmarks for future researchers on the problem.

Keywords: vehicle routing problem; multiple working periods; tabu search.

1 Introduction

This paper examines a problem faced by a buying company that procures products from over one hundred suppliers across South China. The goods must be collected by the buying company upon the placement of product orders. Thus, each supplier has to make a collection request with the buyer when the ordered goods are ready. The buyer, in turn, schedules a fleet of vehicles to fulfill the collection requests. A weekly schedule is created to assign vehicles to collection requests for the upcoming week. Each vehicle only works during working hours, e.g., 9:30 am to 5:30 pm, and returns to the depot after completing all assigned collections, i.e., vehicles are not required to return to the depot every day. During the week, a vehicle generally travels to different locations and performs several collections per day, and finds overnight accommodation at its last/next collection site. The unfulfilled collection requests are outsourced to external agents.

We call this problem the *multi-period profit collection vehicle routing problem with time windows* (MP-VRPTW), which is formally described as follows. The MP-VRPTW is defined on a complete directed graph $G = (V, E)$, where $V = \{0, 1, \ldots, n\}$ is the vertex set and $E = \{(i, j) : i, j \in V, i \neq j\}$ is the edge set. Vertex 0 represents the depot location and $V_C = \{1, \ldots, n\}$ denotes the locations

---

* Corresponding Author.

A. Moonis et al. (Eds.): IEA/AIE 2014, Part II, LNAI 8482, pp. 466–475, 2014.

of n suppliers. Each supplier i is characterized by a location $i \in V_C$, a product quantity d_i, a required processing time s_i, and a time window $[e_i, l_i]$ within which the collection is allowed to be started. The profit gained from supplier i is directly proportional to d_i. Each edge $(i, j) \in E$ requires a non-negative travel time $t_{i,j}$, where the matrix $[t_{i,j}]$ satisfies the triangle inequality.

There are m vehicles available, each of which has a capacity Q and only works within a set $P = \{1, \ldots, w\}$ of w working periods. For any period $p \in P$, a_p and b_p ($a_p < b_p$) are its starting and closing times, and $b_p - a_p$ equals a constant T. A vehicle can arrive at vertex $i \in V_C$ prior to e_i and wait at no cost until the collection at supplier i becomes possible. All vehicles must leave the depot after e_0 ($e_0 = a_1 = 0$) and return to the depot before l_0 ($l_0 = b_w$). At the end of each working period, each vehicle is not required to return to the depot but has to stop traveling and stay at one of vertices. Moreover, all collections cannot be interrupted, i.e., if the collection at a supplier cannot be completed before the end of some period, it must be restarted in the later periods. Each vertex can be visited more than once while each collection can be performed by at most one vehicle, i.e., supplier locations can be used as *waypoints* (i.e., intermediate points on a route) while split collection is forbidden. The objective is to construct m vehicle routes, each starting from and ending at vertex 0, in the planning horizon consisting of w working periods such that the total collected profit is maximized. In reality, a_p should be greater than b_{p-1}, and the duration between a_p and b_{p-1} is the downtime for recuperation. Without loss of generality, we can assume the length of the downtime is extremely small by setting $b_{p-1} = a_p$ and imposing a break at time b_{p-1}.

The MP-VRPTW is NP-hard since it can be restricted to either the vehicle routing problem with time windows and a limited number of vehicles (m-VRPTW) [1] by setting $w = 1$, or the team orienteering problem with time windows (TOPTW) [2] by setting $Q = +\infty$, $w = 1$ and all $s_i = 0$. The TOPTW is extended from the team orienteering problem (TOP) [3] by imposing a time window on each vertex while the m-VRPTW is built on the TOPTW by involving vehicle capacity and customer service times.

The MP-VRPTW differs from the TOP or related vehicle routing problems in three main aspects. The first aspect is the existence of w disjoint working periods, within which vehicles are allowed to travel and/or perform collections. The majority of the existing vehicle routing models assume that vehicles are always in service within $[e_0, l_0]$ and do not need downtime for recuperation. However, this assumption is not valid for many practical applications that have to consider working hour regulations for drivers. The second one is the restriction that each vehicle must stay at one of vertices at the end of each working period. If the remaining time of the current period is insufficient for a vehicle to travel from its current vertex A to a certain vertex B, it can choose either waiting at vertex A until the start of the next period or traveling to a waypoint C that is closer to vertex B. The last one is to allow each vertex to act as a waypoint that can be visited more than once.

The most important characteristic of the MP-VRPTW that distinguishes it from other vehicle routing models is the consideration of multiple working periods. Recently, working hour regulations have received significant attentions from some researchers studying vehicle routing problems. Savelsbergh and Sol [4] studied a dynamic and general pickup and delivery problem in which lunch and night breaks must be taken into account. Xu et al. [5] applied column generation based solution approaches to solve a pickup and delivery vehicle routing problem that involves a set of practical complications, such as heterogeneous vehicles, last-in-first-out loading and unloading operations, pickup and delivery time windows, and working hour restrictions by the United States Department of Transportation. Similarly, Geol [6,7] and Kok et al. [8] investigated combined vehicle routing and driver scheduling problems under the European Union regulations for drivers. The multi-period planning horizon is related to the periodic vehicle routing problem (PVRP) [9]. In the PVRP, we need to make decisions on determining the visiting frequency for each customer and the route for each vehicle and each period. Unlike the MP-VRPTW, the PVRP requires that each vehicle must return to the depot at the end of each period.

2   Shortest Transit Time

In a complete graph that satisfies the triangle inequality, the shortest path from vertex i to vertex j must be edge (i, j). When working periods are imposed on the vehicles, edge (i, j) may be unusable in some situations and therefore the shortest transit time from vertex i to vertex j may be greater than $t_{i,j}$. The simplest such situation can be encountered when $t_{i,j} > T$. To move from vertex i to vertex j, a vehicle has to use some waypoints and the transit time may cross several periods. We illustrate this situation in Fig. 1, where a vehicle departs from vertex i at the beginning of a certain period.

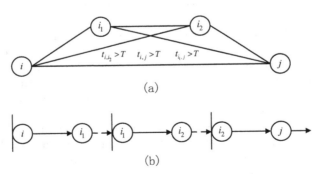

Fig. 1. (a) The vertex locations. (b) The route associated with the shortest transit time from vertex i to vertex j, where vertices i_1 and i_2 are waypoints.

Unlike the classical VRP models, in the MP-VRPTW the shortest transit times from vertex i to other vertices are affected by the departure time (denoted by dt_i)

of the vehicle. Therefore, we define $\hat{t}_{i,j}(dt_i)$ as the shortest transit time from vertex i to vertex j with departure time dt_i. If dt_i is the opening time of a certain period, e.g., $dt_i = a_p$, the notation for the shortest transit time is simplified to $\hat{t}_{i,j}$. Further, we define $ceil(dt_i)$ as the closing time of the period within which dt_i lies, i.e., if $a_p \leq dt_i \leq b_p$, then $ceil(dt_i) = b_p$. If $ceil(dt_i) - dt_i \geq t_{i,j}$, a vehicle can travel across edge (i,j) in the current period and thus $\hat{t}_{i,j}(dt_i) = t_{i,j}$. Otherwise, it has to either wait at vertex i until the start of the next period or travel to some waypoint. We illustrate these situations in Fig. 2.

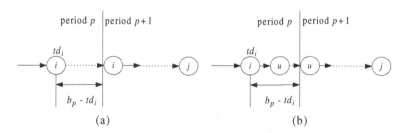

Fig. 2. (a) Wait at vertex i. (b) Travel to a waypoint u.

It is easy to observe that a waypoint can only be positioned as the last or the first vertex in the trip of some period. More precisely, if a vehicle travels to a waypoint u, it must stay at u during the downtime. Taking $N(dt_i) = \{u \in V | t_{i,u} \leq ceil(dt_i) - dt_i\}$ to be the set of all vertices that can be used as waypoints for vertex i, the value of $\hat{t}_{i,j}(dt_i)$ can be calculated by:

$$\hat{t}_{i,j}(dt_i) = \begin{cases} t_{i,j}, & \text{if } ceil(dt_i) - dt_i \geq t_{i,j}; \\ ceil(dt_i) - dt_i + \min_{u \in N(dt_i) \cup \{i\}}\{\hat{t}_{u,j}\}, & \text{otherwise.} \end{cases} \tag{1}$$

The above expression shows that computing any $\hat{t}_{i,j}(dt_i)$ requires $O(n)$ time given the values of all $\hat{t}_{i,j}$, which can be computed prior to applying any algorithm to solve the problem.

For each vertex u, we assume the values of $\hat{t}_{i,u}$ are known in advance. If the last waypoint between i and j is u, the corresponding shortest transit time, denoted by $\hat{t}_{i,j}^u$, can be obtained by:

$$\hat{t}_{i,j}^u = \begin{cases} \hat{t}_{i,u} + t_{u,j}, & \text{if } ceil(\hat{t}_{i,u}) - \hat{t}_{i,u} \geq t_{u,j}; \\ ceil(\hat{t}_{i,u}) + t_{u,j}, & \text{if } ceil(\hat{t}_{i,u}) - \hat{t}_{i,u} < t_{u,j} \leq T; \\ +\infty, & \text{otherwise.} \end{cases} \tag{2}$$

Obviously, the value of $\hat{t}_{i,j}$ can be achieved by $\hat{t}_{i,j} = \min_{u \in V}\{\hat{t}_{i,j}^u\}$. We can apply an algorithm modified from the Dijkstra's algorithm [10] to compute all $\hat{t}_{i,j}$. This modified Dijkstra's algorithm employs expression (2) as the extension function and has a time complexity of $O(n^2)$. Since we need to compute the shortest transit time between any pair of vertices, the total time complexity for all $\hat{t}_{i,j}$ is bounded by $O(n^3)$.

3   Tabu Search Algorithm

Tabu search algorithm [11] has been successfully applied to a wide variety of vehicle routing problems. In our proposed tabu search algorithm, we employ several local search operators adapted from classical edge-exchange operators, namely *2-opt, Or-opt, 2-opt*, Relocate* and *Exchange* [12]. The most noteworthy characteristic that distinguishes these adapted operators from their classical counterparts is the procedure of checking the feasibility of the modified solution. For example, after performing an edge-exchange operation on a VRPTW solution, we can check the feasibility of the resultant solution in $O(1)$ time (it is assumed that for each vertex the latest arrival time that does not lead to the violation of the time windows of all successive vertices has been calculated in a preprocessing step). However, for a modified MP-VRPTW solution, we may require up to $O(n^2)$ time to check its feasibility due to the re-computation of the shortest transit times associated with the affected vertices.

In our algorithm, we represent each vehicle route by a sequence of supplies whose goods have been collected. Fig. 3 shows an example solution, where routes r_1, r_2, r_3 and r_4 include the collected suppliers and U contains the leftover suppliers. Note that all waypoints are not displayed in this solution representation, and there may exist waypoints and/or breaks between two consecutive suppliers.

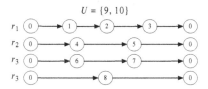

Fig. 3. An example solution used in the tabu search algorithm

The pseudocode of our tabu search algorithm is given in Algorithm 1. At the beginning of the algorithm, we initialize both the best solution S_{best} and the current solution S by an arbitrary initial solution S_0. In each iteration, we invoke the local search procedure with tabu moves (function *local_search*) and set S' to be the best solution found by this procedure. The best known solution S_{best} is updated by S' if possible. Subsequently, the search process is diversified by perturbing the best solution S' found in this iteration by function *perturb*. The above process is repeated until the perturbation procedure is consecutively performed *maxPerturbation* times without improving on S_{best}.

3.1   Local Search with Tabu Moves

The pseudocode of the local search procedure with tabu moves is provided in Algorithm 2. The neighborhood structure is one of the most important components that determine the size of the search space and the quality of the final

Algorithm 1. Framework of the tabu search algorithm

1. $S_0 \leftarrow$ an initial solution;
2. $S_{best} \leftarrow S_0$ and $S \leftarrow S_0$;
3. $i \leftarrow 0$;
4. **while** $i \leq maxPerturbation$ **do**
5. $S' \leftarrow$ the best solution found by $local\_search(S)$;
6.   **if** S' is better than S_{best} **then**
7. $S_{best} \leftarrow S'$ and $i \leftarrow 0$;
8.   **else**
9. $i \leftarrow i + 1$;
10.   **end if**
11. $S \leftarrow perturb\ (S')$;
12. **end while**
13. **return** S_{best}.

solution. This procedure employs five neighborhood operators, namely 2-*opt*, *Or*-*opt*, 2-*opt**, *Relocate* and *Exchange*. We regard the set U as a dummy route that includes all uncollected suppliers. The classical versions of these operators applied to the VRPTW were detailedly described in [12]. Compared to the operators for the VRPTW, the adapted operators require more computational efforts to check the feasibility of the resultant solution, and to update the earliest and latest arrival times at the affected vertices.

Algorithm 2. The local search procedure with tabu moves

1. INPUT: the initial solution S;
2. The current best solution $S' \leftarrow S$ and $Iter \leftarrow 0$;
3. **while** $Iter \leq maxLocalIter$ **do**
4.   Apply the 2-opt, Or-opt, 2-opt*, relocate and exchange operators on S ;
5. $S \leftarrow$ the best allowable solution found by the above operators;
6.   **if** S is better than S' **then**
7. $S' \leftarrow S$ and $Iter \leftarrow 0$;
8.   **else**
9. $Iter \leftarrow Iter + 1$;
10.   **end if**
11.   Update the tabu list;
12. **end while**
13. **return** S'.

Fig. 4 illustrates the 2-opt and Or-opt operations. We assume that the earliest and latest arrival times (denoted by ea_i and la_i, respectively) at each supplier i are already known. We can derive the earliest departure time (denoted by ed_i) of each supplier by:

$$ed_i = \begin{cases} ea_i + s_i, & \text{if } ceil(ea_i) - ea_i \geq s_i; \\ ceil(ed_i) + s_i, \text{otherwise.} \end{cases}$$

The 2-opt operation replaces edges $(i, i + 1)$ and $(j, j + 1)$ with edges (i, j) and $(i + 1, j + 1)$, and then reverses the directions of all edges between $i + 1$ and j. The resultant route r' shown in Fig. 4(b) must be feasible if its subroute $(j, j - 1, \ldots, i + 1, j + 1, \ldots, 0)$ is feasible. To check the feasibility of r', we need to re-calculate the earliest arrival time (ea'_k) at each supplier k in subroute $(j, j - 1, \ldots, i + 1, j + 1)$. If ea'_k is less than e_k or within $[e_k, l_k]$ for each supplier, this subroute must be feasible. If ea'_{j+1} in r' is less than or equal to la_{j+1} in r, subroute $(j+1,\ldots,0)$ must be feasible. We have $ea_{k+1} = ed_k + \hat{t}_{k,k+1}(ed_k)$, where $\hat{t}_{k,k+1}(ed_k)$ can be obtained in $O(n)$ time. Thus, it requires $O(n*n_s)$ time to check the feasibility of the route r' generated by a 2-opt operation, where n_s is the number of suppliers in subroute $(j, j-1,\ldots,i+1, j+1)$. Note that when dealing with the VRPTW, a 2-opt operation only requires $O(n_s)$ time to accomplish the feasibility checking. Moreover, after performing a 2-opt operation, the values of ea'_i and la'_i for all suppliers in r' can be updated in $O(n*|r|)$ time, where $|r|$ is the number of served suppliers in route r.

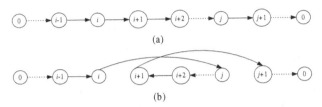

(a)

(b)

Fig. 4. (a) The original route r. (b) The resultant route r' after a 2-opt operation.

The exchange operation exchanges positions of two suppliers. Fig. 5(b) shows the resultant route after exchanging the positions of a supplier pair in the same route. The feasibility of this route can be checked by calculating ea'_k for all suppliers k in subroute $(j, i+1,\ldots,j-1, i, j+1)$. The resultant routes created by exchanging two suppliers from two different routes are shown in Fig. 5(c). In addition, this operation can also exchange a supplier in some route with a supplier in U.

Due to the space limitation, we cannot present the details of other modified operations in this paper. Tabu search algorithm employs a tabu list to prevent the search process from being trapped in local optima. In our algorithm, the tabu list stores edges that have been created within the previous ξ number of iterations. A move is forbidden if it attempts to remove the edges in the tabu list. The tabu restriction can be overridden if the aspiration criterion is fulfilled, i.e., we allow the tabu moves to be performed if the resulting solutions are better than the current best solution S'. The neighbors that are created by non-tabu moves or by overriding the tabu restriction are called *allowable*. All allowable moves are stored in a candidate list. The best candidate is performed to generate the next incumbent solution of the local search procedure. We terminate the procedure when *maxLocalIter* consecutive iterations are unable to improve on S'.

3.2   Perturbation

The perturbation procedure is a diversification scheme that helps the search process escape from local optima. Our perturbation procedure randomly removes some suppliers from the solution S' following the rule that the suppliers with

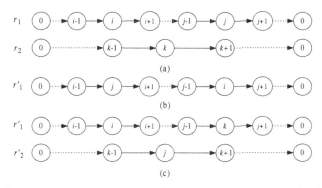

r_1 0 i-1 i i+1 j-1 j j+1   0

r_2 0 k-1 k k+1   0

(a)

r'_1 0 i-1 j i+1 j-1 i j+1   0

(b)

r'_1 0 i-1 i i+1 j-1 k j+1   0

r'_2 0 k-1 j k+1   0

(c)

Fig. 5. (a) The original routes r_1 and r_2. (b) The resultant routes r'_1 after exchanging suppliers i and j. (c) The resultants routes r'_1 and r'_2 after exchanging suppliers j and k.

smaller profits have more chance to be removed. Given a solution S', we sort the collected suppliers in a non-decreasing order of their profits, generating a supplier list $(v_1, v_2, \ldots, v_{n_s})$. The probability of removing the k-th supplier is determined by: $(p_{min} + p_\Delta \times \min\{N_{rep}, N_{max}\}) + (p_{max} - p_{min}) \times i/n_s$, where p_{min}, p_{max} and p_Δ are controlling parameters. The tabu search algorithm performs $maxPerturbation$ times of iterations, where the best solution found within each iteration is stored in a solution list. The number of times that the current solution S' appears in the solution list is counted by N_{rep}. The parameter N_{max} is used to set a upper bound for the probability. The introduction of N_{max} can help avoid the overly large probability, which would cause the process to degenerate into an undesirable multi-start method.

4   Computational Experiments

To evaluate our tabu search algorithm, we conducted experiments using the data set derived from Solomon's VRPTW instances [13]. The Solomon instances, each containing 100 customers, are divided into six groups, namely R1, C1, RC1, R2, C2, and RC2. We selected the first two instances in each Solomon instance group, for a total of 12 Solomon instances, and then generated four MP-VRPTW instances from each selected Solomon instance by taking the values of w and m from $\{3, 5\}$ and $\{7, 9\}$, respectively. The duration of each period is set to $T = (l_0 - e_0)/w$ and the vehicle capacity is set to 200. We executed the tabu search algorithm 10 times with different random seeds for each MP-VRPTW instance, for a total of 480 runs.

The tabu search algorithm was coded in C++ and compiled using the gcc 4.6.1 compiler. The algorithm was tested on a Dell server with an Intel Xeon E5430 2.66GHz CPU, 8 GB RAM and running Linux-CentOS-5.0 64-bit operating system. Computation times reported are in CPU seconds on this server. The parameters required by the algorithm are set as: $maxPerturbation = 5$, $maxLocalIter = 200$, $\xi = 100$, $p_{min} = 0.05$, $p_{max} = 0.30$, $p_\Delta = 0.1$ and $N_{max} = 5$.

Table 1. The computational results for the instances with $m = 7$

Instance	Total Profit	$w = 3$			$w = 5$		
		Max. Profit	Ave. Profit	Ave. Time (s)	Max. Profit	Ave. Profit	Ave. Time (s)
c101	1,810	1,400	1,399.0	25.4	1,240	1,239.0	26.0
c102	1,810	1,400	1,400.0	7.7	1,400	1,400.0	11.4
c201	1,810	1,400	1,400.0	7.4	1,400	1,400.0	7.8
c202	1,810	1,400	1,400.0	12.2	1,400	1,400.0	14.5
r101	1,458	891	889.1	36.0	885	879.1	51.7
r102	1,458	1,109	1,103.4	106.7	1,122	1,113.7	116.2
r201	1,458	1,400	1,400.0	6.1	1,400	1,400.0	7.4
r202	1,458	1,400	1,400.0	6.5	1,400	1,400.0	7.8
rc101	1,724	1,174	1,166.5	40.6	1,161	1,149.7	66.2
rc102	1,724	1,339	1,326.6	56.5	1,331	1,312.6	69.4
rc201	1,724	1,400	1,400.0	10.5	1,400	1,400.0	12.1
rc202	1,724	1,400	1,400.0	11.0	1,400	1,400.0	11.7

Table 2. The computational results for the instances with $m = 9$

Instance	Total Profit	$w = 3$			$w = 5$		
		Max. Profit	Ave. Profit	Ave. Time (s)	Max. Profit	Ave. Profit	Ave. Time (s)
c101	1,810	1,630	1,621.0	18.3	1,380	1,380.0	6.2
c102	1,810	1,750	1,748.0	30.3	1,620	1,620.0	9.8
c201	1,810	1,800	1,800.0	4.5	1,800	1,800.0	4.7
c202	1,810	1,800	1,800.0	5.1	1,800	1,800.0	4.8
r101	1,458	1,045	1,042.6	86.3	1,053	1,048.8	62.7
r102	1,458	1,247	1,239.3	111.9	1,264	1,259.9	92.4
r201	1,458	1,458	1,458.0	5.9	1,458	1,458.0	6.5
r202	1,458	1,458	1,458.0	6.6	1,458	1,458.0	7.0
rc101	1,724	1,391	1,380.2	78.3	1,371	1,362.4	91.6
rc102	1,724	1,567	1,556.4	75.9	1,535	1,523.1	86.6
rc201	1,724	1,724	1,724.0	5.1	1,724	1,724.0	5.5
rc202	1,724	1,724	1,724.0	5.7	1,724	1,724.0	4.5

The computational results are reported in Tables 1 and 2. The column "Total Profit" shows the total profit of all suppliers in each instance. Each block corresponds to a value of w and includes the maximum profit collected *Max. Profit*, the average profit *Ave. Profit* and the average computation time *Ave. Time* over ten executions. Since the MP-VRPTW is a new problem, there is no existing algorithm tailored for it. As a consequence, we cannot directly compare our algorithm with other approaches. These tables show that it is not very significant that the average computation times increase as the value of w.

5   Conclusions

In this paper, we studied a new vehicle routing problem in which each vehicle is dispatched to complete a route in a multi-period planning horizon. At the end of each period, each vehicle is not required to return to the depot but has to stay at one of the vertices for recuperation. We first studied the way of computing the shortest transit time between any pair of vertices with an arbitrary departure time. Next, we introduced several local search operators that are adapted from classical edge-exchange operators for the VRPTW. Finally, we integrated these new operators in a tabu search algorithm framework and conducted a series of computational experiments to evaluate the algorithm. The test instances and computational results reported in this article can serve as benchmarks for future researchers.

References

1. Lau, H.C., Sim, M., Teo, K.M.: Vehicle routing problem with time windows and a limited number of vehicles. European Journal of Operational Research 148(3), 559–569 (2003)
2. Vansteenwegen, P., Souffriau, W., Vanden Berghe, G., Van Oudheusden, D.: Iterated local search for the team orienteering problem with time windows. Computers & Operations Research 36(12), 3281–3290 (2009)
3. Vansteenwegen, P., Souffriau, W., Van Oudheusden, D.: The orienteering problem: A survey. European Journal of Operational Research 209(1), 1–10 (2011)
4. Savelsbergh, M., Sol, M.: Drive: Dynamic routing of independent vehicles. Operations Research 46(4), 474–490 (1998)
5. Xu, H., Chen, Z.L., Rajagopal, S., Arunapuram, S.: Solving a practical pickup and delivery problem. Transportation Science 37(3), 347–364 (2003)
6. Goel, A.: Vehicle scheduling and routing with drivers' working hours. Transportation Science 43(1), 17–26 (2009)
7. Goel, A.: Truck driver scheduling in the European union. Transportation Science 44(4), 429–441 (2010)
8. Kok, A.L., Meyer, C.M., Kopfer, H., Schutten, J.M.J.: A dynamic programming heuristic for the vehicle routing problem with time windows and European community social legislation. Transportation Science 44(4), 442–454 (2010)
9. Cordeau, J.F., Gendreau, M., Laporte, G.: A tabu search heuristic for periodic and multi-depot vehicle routing problems. Networks 30(2), 105–119 (1997)
10. Ahuja, R.K., Magnanti, T.L., Orlin, J.B.: Network flows: Theory, algorithms, and applications. Prentice Hall, Englewood Cliffs (1993)
11. Glover, F., Laguna, M.: Tabu Search. Kluwer Academic Publishers, Norwell (1997)
12. Bräysy, O., Gendreau, M.: Vehicle routing problem with time windows, part I: Route construction and local search algorithms. Transportation Science 39(1), 104–118 (2005)
13. Solomon, M.M.: Algorithms for the vehicle routing and scheduling problems with time window constraints. Operations Research 35(2), 254–265 (1987)

An Internet Core-Based Group-Trading Model

Pen-Choug Sun[*]

Department of Information Management, Aletheia University No. 32, Zhenli Street,
Tamsui Dist., New Taipei City, 25103, Taiwan, R.O.C.
au1159@au.edu.tw

Abstract. Coalitions, which allow traders to form teams in e-markets, can sometimes accomplish things more efficiently than individuals. Concepts and algorithms for coalition formation have drawn much attention from academics and
practitioners. However, most of coalitions are unstable and fall apart easily. The
core has become a popular solution, because it provides a way to find stable
sets. However its problems hinder researchers from applying it to a real world
market. This paper proposes a core-based group-trading model, which involves
bundle selling of goods, offering amount discount in group-buying in e-
markets. Its outcome is compared with a traditional market under the scenario
of a travel agent and is evaluated in terms of five criteria: the use of distributed
computing, the degree of computational complexity, incentive compatibility, efficiency and effectiveness.

Keywords: Broker, Core, Coalition, E-Market, Shapley Value.

1 Introduction

"Much of the retail sector's overall growth in both the US and the EU over the next five years
will come from the Internet," said the Forrester Research Vice President in March 2010 [1].
There are unceasing large potential profits for traders in Internet e-commerce. E-commerce
is "where business transactions take place via telecommunications networks." [2]. E-market,
also called e-marketplace, is a new trading model for e-commerce on the Internet in which
incentive compatibility, distributed computing, and less computational complexity, are all
highly relevant. Many traders gather together with different purposes and they can be
grouped in varied formations in an e-market. There are two parties, the sellers and the buyers, who generally are regarded as conflicting parties. When a seller gains more profit in a
transaction, this also means the buyer needs to pay more, and vice-versa. Is there any way to
benefit both parties at the same time and bring them to work together? There are many practical problems that need resolving, when a model is built for the real e-marketplace.

When customers and suppliers get together and work out deals, they seize every
opportunity to maximise their own profits. Forming coalitions on the Internet is an
effective way of striving to achieve their goals. Therefore, concepts and algorithms
for coalition formation have been investigated in both academics and practitioners [3].

---

[*] Corresponding Author.

A. Moonis et al. (Eds.): IEA/AIE 2014, Part II, LNAI 8482, pp. 476–486, 2014.

These research domains have approached the topic very differently. In Computer Science, the coalitions are formed in a precise structure to maximise the overall expected utility of participants. Since finding the coalition structure is NP-complete, researchers try to prescribe formation algorithms of less computational complexity [4]. On the other hand, the economics literature traditionally stresses the incentives of each selfish participant in a coalition [5]. The traders are self-interested to join a coalition only when it is to their own advantage [6]. A coalition with stability is in a condition when every member has the incentive to remain in the coalition.

Certain solution concepts for coalition problems related to stability notions have drawn much attention from researchers. The earliest proposed concept was called the stable set [7], which is a set with stabilities. Its definition is very general and the concept can be used in a wide variety of game formats. For years it was the standard solution for cooperative games [8]. However, subsequent works by others showed that a stable set may or may not exist [9], and is usually quite difficult to find [10]. The core has become a well-known solution concept because of its incentive compatibility and because the way it finds stable sets is more appropriate. It assigns to each cooperative game the set of profits that no coalition can improve upon [11]. The aim of this paper is to introduce a core-based model for e-markets. The concept of the core is described and its problems are discussed. Other solution concepts which are used to deal with these problems are introduced, and a new Core Broking Model (CBM) is proposed. The CBM aims to create a win-win situation for customers and providers in e-markets. The comparison between the results of the new model and a traditional market has been made.

2   The Core

The core plays an important role in the area of computer science and modern economics. The core was first proposed by Francis Y. Edgeworth in 1881, and it was later defined in game theory terms by Gillies in 1953. Individuals in a cooperative game are not only interested in maximising the group's joint efficiency, they are also concerned with their own profits. If they can gain better profit by working alone without involving others, they will not join a group. If a group can produce a better profit by excluding certain people, it will certainly form a new coalition without those individuals [12]. The core is a one-point solution concept for forming stable and efficient coalitions by calculating the profits of different possible coalitions. Consequently, it is useful to adopt it into the mechanism in an e-market.

In economics, the core indicates the set of imputations under which no coalition has a value greater than the sum of its members' profits [11]. Therefore, every member of the core stays and has no incentive to leave, because he/she receives a larger profit. A core in a cooperative game $<N, v>$ with n players, denoted as $C(v)$ of a characteristic function $v: 2^N \rightarrow \Re$, which describes the collective profit a set of players can gain by forming a coalition, is derived from the following function [13],

$$C(v) := \{ x \in \Re^n \mid x(N) = v(N), x(S) \geq v(S) \text{ for all } S \in 2^N \},$$

here N is the finite set of participants with individual gains $\{x_1, x_2, ..., x_n\}$. The $x(N) = \sum_{i \in N} x_i$ represents the total profit of each individual element in N by adding xi, which denotes the amount assigned to individual i, whereas the distribution of $v(N)$ can denote the joint profit of the grand coalition N. Suppose S, T are a set of pair wise disjoint nonempty subsets of coalition N. A cooperative game N is said to be convex, if

$$v(S)+v(T)\leq v(S\cup T)+v(S\cap T),$$

whenever S, T⊆N and for N∈2^N.

A convex core is always a stable set [14]. A game with x(N) = v(N) is regarded as efficient [13]. An allocation is inefficient if there is at least one person who can do better, though none is worse off. The definition can be summarised as "The core of a cooperative game consists of all un-dominated allocations in the game" [15].The profit of the allocations in a core should dominate other possible solutions, meaning that no subgroup or individual within the coalition can do better by deserting the coalition.

The core has been tested within some games in the previous work [16], with only one type of consumer and provider dealing with one kind of commodity. It simplifies the computational process of the core and helps us to understand the way the core works, although it seems a bit unrealistic. Undoubtedly, providing a way to find a stable set is one of the advantages of the core. It also gives that set incentive compatibility. In fact, the core is a Pareto efficiency [17], also called Pareto optimality. It is a central concept in Economics, proposed by an Italian economist, Vilfredo Pareto [18]. A Pareto improvement moves the current condition into a better situation that helps at least one person but harms no-one. And then, after some Pareto improvements, a situation will develop, called Pareto equilibrium, in which no Pareto improvement can be made at all. Pareto efficiency is important because it provides a weak but widely accepted standard for comparing economic outcomes. So its framework allows researchers to decide what is efficient.

Although it was proven that the core is always a stable set when the game is convex [14], a number of appealing sufficient conditions have to be identified to confirm whether it is stable or not [19]. The core may exist in different forms: unique-core game, multiple-core game or empty-core game [13]. Three problems of the core have to be solved in order to be applied in the real markets [20]. The first problem occurs when no stable set can be found in an empty-core game. It is a huge problem for everyone -- all the efforts will be in vain if no deal can be agreed with in a market. The second problem is high computational complexity [21]. The core tries to calculate the result of every combination of coalitions and finds the best outcome for a group and for the individuals. The process to find the core is NP-complete [22]. It is too complex to find a core in a large grand coalition. The third problem is that the core seems to deal with complete information only [16]. And these problems make it too difficult to apply in the marketplace.

3   Existing Solutions for the Core

Some solutions are introduced here for the problems of the core when it is applied in e-markets [23]. In cooperative games, the stability of cores is one of the most notorious open problems. It is important to have a stable coalition, for a game with an empty core is to be understood as a situation of strong instability, as any profits proposed to the coalition are vulnerable. Determining the stability of the core is NP-complete [24]. Attempts to solve the above problem seemed to have come to a dead-end, until a balanced game has been proven to have a stable core [25]. An important sufficient condition for a non-emptiness core, is that it is balanced. Let the game <N, v> be balanced. Then C(N, v)≠0 [26]. From the classical Bondareva-Shapley theorem, a cooperative game has a non-empty core if and only if it is balanced [27]. A game is balanced, if for every balanced collection B with weights $\{\lambda S\}_{S\in B}$ and the following holds,

$$v(N) \geq \Sigma_{S \in B} \lambda S \ v(S).$$

A collection B is called a balanced collection if

$\exists (\lambda S)_{S \in B} \ \forall i \in N \ \Sigma_{S \in B, \ S \ni i} \lambda S = 1$, where $(\lambda S)_{S \in B}$ is a vector of positive weights.

It is rather easy to check the stability of coalitions with simple calculations in the above formula.

To locate the core in a coalition is NP-complete [27]. It can be helpful to divide a grand coalition into several small sized sub-coalitions [28] but it is not practical. Some papers show the collaborations of players can effectively reduce the computational complexity [28]. A broker may persuade traders to cooperate together. Suppose n brokers gather m traders. Instead of m traders, the core now only needs to deal with n brokers. Since n is much smaller than m, the size of the coalition is therefore successfully reduced.

When providers collaborate together, it is very important to make sure all of them in the coalition can get what they deserve. Shapley value [14] presents a fair and unique way of solving the problem of distributing surplus among the players in a coalition, by taking into account the worth of each player in the coalition. Assume a coalition N and a worth function $v:N \rightarrow \mathfrak{R}$, with the following properties,

$$v(\emptyset)=0, \ v(S+T)=v(S)+v(T), \text{ whenever S and T are subsets of N.}$$

If the members of coalition S agree to cooperate, then v(S) describes the total expected gain from this cooperation. The amount that $player_i$ gains is,

$$\text{Shi}(N) = \Sigma S \subseteq (N \backslash \{i\})(|S|)!(|N|-|S|-1)!/|N|![v(S \cup \{i\})-v(S)].$$

It shows the fact that collaboration can only help and can never be harmful.

4   Core Broking Model (CBM)

This core-based model involves joint-selling of multiple goods, offering volume discount for group-buying coalitions in e-markets. Several providers conduct bundle selling together, while, on the other hand, many buyers form coalitions for the amount discount in the e-markets. The descriptions of its components, which are in the structure represented in Fig. 1, are as follows:

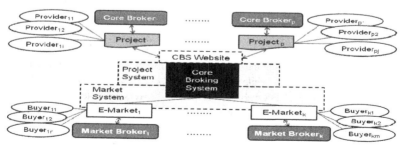

Fig. 1. The Structure of the CBM

- **Core-brokers:** the initiators of the group-trading projects.
- **Projects:** Each project has several sessions of group-trading in e-markets.
- **Providers:** The core-brokers invite them to provide products and services for customers.
- **Market-brokers:** who play the role of team members to help with the group-trading projects.
- **E-markets:** may be any existing online shopping avenues such as eBay or the market-brokers' own sites on which they can post projects and find customers.
- **Buyers:** the market-brokers' clients, who have been attracted to the projects.
- **The Core Broking System** (CBS) consists of three components as follows:
    - **CBS Website:** list of group-trading projects. It is a place where core-brokers and market-brokers meet together. Its resolution centre is designed to deal with any problem with a transaction. Members of the site can report and open a case for a problematic transaction.
    - **Project Subsystem:** a system specially designed to assist the core-broker in managing all the necessary tasks to assure quality outcomes.
    - **Market Subsystem:** by, the market-brokers can use it to perform transactions for a session on a project; purchase electronic coupons from the core brokers and sell them to their clients.

The system flow chart of the model is shown in Fig. 2 and was illustrated in one of my papers [29]. A brief description for the six stages in the process is as follows:

1) **Initiating** – In this stage, a core broker setups a proposal for a group-trading project, settle the project with some providers and lists the project on the CBS website.
2) **Commencing** – After recruiting several market brokers, the core broker officially begins sessions of group trading in the group-trading project.
3) **Gathering** – The market brokers attract buyers to their websites, combine the orders of buyers into market-orders and submit the market-orders to the core broker.
4) **Combining** – The core broker checks the stability of the coalitions, combines the coalitions together, decides the final prices for the items and sends acceptance notices to the market brokers.
5) **Closing** – When the buyers have paid for their purchases, they receive coupons. Finally, the brokers close the deal with their clients and the benefits of the participants are calculated.
6) **Terminating** – The core brokers backup the tables and analyse the transactions in the sessions of the group-trading project for the future use.

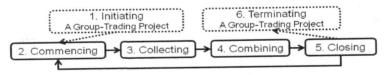

Fig. 2. System Flow Chart of the CBM

To ensure a healthy level of competition, the new model adopts brokers to prevent price makers. The brokers make possible the collaboration between the members of coalitions. Core-brokers are like project managers, while market-brokers act like salesmen in the CBM. The core-brokers initiate projects on the CBS website. The market-brokers promote products of the project on the appropriate shopping sites and form buyers' coalitions there. The core-brokers invite providers to perform joint-selling to increase the 'competitive advantage' [30]. They provide all the necessary information to the market-brokers for them to promote the product and market it. Each session has a starting and an ending date. The suggested duration for a session is usually one week. The iterative process is looping between stage 2 and stage 5. At the end of a session, the core brokers may choose to have a new session of trading or stop the project for good.

There are several assumptions used in the CBM [29]. The model manages the orders on a First Comes First Serves (FCFS) basis. There are many ways to decide how to distribute the items fairly. Other than FCFS, another option is to use the Shapley value, which is decided by the original amount ordered by each customer. An agreement may also be reached amongst the customers on the distribution over the conflicting issues through a multiple stage negotiation process [31].

After surveys to current markets, a fees system for the CBM including the commission for brokers and how the members of the CBS site pay their fees was constructed. Four kinds of fees: final value fee (FVF), handling fee, session fee and online store fee (OSF) are suggested. Session fee of £30.00 is the only fee the core-brokers need to pay for a project, when they list a session in a project on the site. An OSF of £24.50 is a suggested monthly fee for market-brokers, who wish to open an online store on the CBS site including purchase of domain name, server use, maintenance etc. A FVF is 5.5% of final selling price. When a FVF is paid by the providers, it is then divided into 3 portions. The core-broker takes 2%, the market-broker gains 3% and the CBS site keeps 0.5%. A handling fee from the buyers rewards the brokers, because they combine the orders and let the buyers get better discounts. It is 15% of the extra discount, after each of the brokers has processed the orders.

When buyers pay for their orders, they receive electronic coupons from the brokers. Each coupon has a unique ID to ensure that one coupon is redeemed only once. No extra fee for shipping is charged, when they claim the real products and services from the providers printed on the coupons. Alternative payment methods are used including bank transfers, PayPal and utility & debit cards. The results of surveys in Tulsa indicated that most sellers do not feel any risk associated with transactions with a reliable payment mechanism [32]. Bank transfers are regarded as secure and are a common and efficient way of making payments today. PayPal is an alternative safe way but it involves additional cost [33]. The providers and the core-brokers are assumed to receive money by bank transfer, while customers pay for their items via PayPal.

A prototype of the CBS was developed in C# under the development environment of Visual Studio 2008. It consists of one CBS website and two subsystems: the project subsystem and the market subsystem. A database was designed for the CBM so that the core-brokers and market-brokers can manage and store all the data they need to fulfil their tasks in the group-trading.

5   Evaluation

One possible way to evaluate the CBM is to use it in a real-world site and collect data from it. A CBS site has been developed, but the crux of this paper is to make a comparison between the CBM and a traditional market (will be referred to as TM). At this stage, a real-world test would be inappropriate; further it would be unlikely to produce the data set necessary for a rigorous testing, for instance, the costs of the products should be private data of the providers. In order to calculate the net profits of every provider, the costs of the products need to be given here. Therefore, a simulation system was developed to evaluate the CBM using the scenario of a travel agent and was used to produce outputs from the TM and the CBM.

Core-broker Ben created group-trading project S1: 'Summer Time around the Midlands' by integrating the products from the three providers offering inexpensive hotel rooms and low car rentals for economical travel in the Midlands. The purpose of the project is to enable sessions of bundle selling by integrating the resources of the providers. By offering volume wholesale discounts, customers may form groups in order to purchase items. Coupons can be chased and sent to the providers on them and exchanged into hotel rooms or car for the buyers to travel around the Midlands in the UK.

The simulation system was written in C# in the Visual Studio 2008 development environment. The results in this paper were produced in it on a common personal computer with Windows Vista. The system contains a Test Case Generator (TCG), a TM Simulator (TMS) and a CBM Simulator (CBMS). The TMS is based on the core concept and aims to find a core of the coalition in a TM. The CBMS is built to the pattern of the CBM and aims to find a bigger core of coalitions in multi-e-markets. The data generated by the TCG were put into the TMS and the CBMS at the same time. By examining the outputs of the TMS and CBMS, a comparison between the TM and the CBM using the following five criteria:

Table 1. Distributed Computing

Distributed Computing	CBM	TM
Multi Computers	Yes	No
Internet	Yes	Maybe
Multi e-Markets	Yes	No

1. **Distributed Computing:** To fit in a distributed computing environment, by nature, a distributed model requires involving multiple computers to be effective. Another two distributed contexts, namely Internet and multi-markets, are used here to assess the models too. The core is used to find a stable set of a coalition in a TM. This is normally done on one computer. It might cause extra complexity if this problem were to be solved by using multi computers. The TM might transfer into an e-market where it would allow customers to place orders via the Internet, but it is difficult to apply it to multiple e-markets in the Internet environment. Table 1 shows the results of the CBM and the TM in distributed computing.

2. **Computational Complexity:** can be expressed as big O notation, which is also a useful tool to analyse the efficiency of algorithms regarding the time it takes for them to complete their jobs. Assume there are p providers, b orders and g products in the market. There are at most 10 order lines in an order. The number of sub-coalitions is $(p+1)^b$. The fact that CBM has less computational complexity than the TM can easily tell by examining their big O notations. The TM and CBM are $O(n^n)$ and $O(n^2 \times 2^n)$ respectively.

A task, which uses Shapley value to decide the number of items for each provider, is proven to be a time-consuming job. It have nothing to do with the number of buyers, but mainly come from the number of providers, i.e. p. Assume p is limited within 20, its big O becomes $O(n^2)$ and can be classified as an algorithm with low computational complexity. The computational complexity can be dramatically lessened in the CBM by limiting the number of providers.

One interesting finding in this paper is that CBM's computational complexity will not become higher, when there are more orders in the e-markets, so market-brokers are free to take as many clients as they could in their e-markets. However, it may take time for them to negotiate with buyers if the size of a coalition becomes too big or the communications get too busy. In order to make sure brokers to collect the orders in time, one important principle is to limit the size of coalitions in e-markets within one group-trading session, and this needs to be applied in the model. It is usually the brokers who monitor the size of coalitions in their e-markets. All of the brokers have to follow this principle in order to prevent situations where a coalition becomes too large to handle.

The aim is to produce outcomes for a group-trading project on an average computer within a reasonable time. To achieve this, a core-broker should limit the number of suppliers to form a group-trading project. The suggested number of providers is 15.

3. **Incentive Compatibility:** It is crucial to give people incentives to participate in online trading. The effectiveness of both models is assessed by comparing the benefits for participants. In order to compare the incentive compatibilities, the assumption has been made that providers are willing to offer more volume discount to customers if a group of them purchases the same item from the same provider in the core, although it is rather unusual for them to give customers such discounts in the TM. If this is not done, there will be no means of comparing the two systems.

There are three incentives for traders: discounts, an equilibrium price and a fair distribution. The providers offer volume discounts to customers in the CBM. If the core may be empty and unstable, an equilibrium price will not be reached in a coalition. The CBM performs a stability check to make sure that there is a best price for traders. By using the Shapley value, the CBM can make a fair decision as to which items are allocated to which provider, even when customers' requirements are less. The CBM provides fair shares to customers and providers, but the TM does not. Fair distribution is crucial in teamwork. The providers might leave the team if the profits have been distributed unfairly, even though the profit they can get out of it is good. In the CBM, suppliers have a great chance to sell out their products. Even if the customers do not purchase all that is on offer, the suppliers still get their fair share. This will also give them satisfaction. Table 2 shows the CBM has higher incentive compatibility than the TM. The CBM created a win-win situation for all participants in a group-trading session.

Table 2. Incentive Compatibility

Incentive Compatibility	CBM	TM
Volume Discount	Yes	Yes
Equilibrium Price	Yes	Not Sure
Fair Share	Yes	No

4. **Efficiency:** It generally describes the intent to which some valuable resources are well used for an intended outcome. In this paper, the efficiency of the CBM and the TM are judged by the time and cost which they use to complete tasks. The CBM is more efficient than the TM in collecting the necessary data to find a stable set. It is a time-consuming job to collect the information of marginal utility functions, which describe the values that customers are willing to pay for the goods and the price the providers want to sell them for. Collecting such information can be a big problem in the TM. On the other hand, the inputs needed for the CBM, which are price lists and orders, are market information and can easily be collected by a core broker.

Table 3. Efficiency in the CBM and the Core

Efficiency \ Task	Less Time		Less Cost	
	CBM	TM	CBM	TM
Information Collecting	√	-	√	-
A Stable Set Finding	√	-		√

Table 3 shows that the CBM has better efficiency than the TM. By jointing providers together and combining orders from multiple e-markets, the CBM becomes more efficient than the TM in locating a stable set in a large coalition. Firstly, it hides the information of providers and customers to ensure that there are less information and quicker data transfer between e-markets. Secondly, the CBM successfully reduces the computational complexity and executing time in group-trading, but there are extra costs in doing so, including some expenses involving multiple e-markets and the commissions for the brokers.

Table 4. Effectiveness in the CBM and in the TM

Effectiveness	Normal Buyers		Demanding Buyers	
	CBM	TM	CBM	TM
More Net Discount	√	-	√	-
More Profit	√	-	√	-
More Net Profit	√	√	√	-
More Total Benefit	√	-	√	-

5. **Effectiveness:** It includes the discounts to customers, the profits for the providers and the total benefits. The net total benefits excluding the extra expenses to the brokers need to be checked to certain that it is worthy to spend the costs to perform group-trading in the CBM. There are effectiveness of data protection and fairness in the CBM. Market brokers can hide their customers' personal information before they submit orders to the core brokers. In a market, it is usual that if the customers get better discounts, then the providers receive less profit. To ensure the CBM has taken into account the interests of both customers and providers, the average discount that buyers can obtain and the average profit of suppliers are examined here. In this way, the CBM can be evaluated to see whether it gives more benefits than the TM to both customers and providers. Table 4 shows that the customers gain higher discounts in the CBM in both scenarios: the one with normal buyers, and the one with demanding buyers, who ask for extremely high discount for every item. Table 5 summarises the effectiveness in the CBM and the TM.

Table 5. Effectiveness

Effectiveness	Data Protection	Fairness	More Discount	More Profit	More Total Benefit	Commission
CBM	√	√	√	√	√	√
TM	-	-	-	-	-	-

The above comparison between the results of the CBMS and the TMS shows that the CBM is superior to the TM, in terms of five criteria: the use of distributed computing, the degree of computational complexity, incentive compatibility, efficiency and effectiveness. The results from the simulator also demonstrate that the CBM can bring more profits to providers and attract both demanding and normal customers to e-markets.

6   Conclusion

Because the customers can gain higher discounts, the CBM can really attract them to the e-markets and encourage them go through with their purchases, so that the providers can also earn more profits in it, even after part of the profits goes to the brokers as commission. A wise trader, no matter buyer or seller, will definitely choose to join the group-trading in the CBM rather than stay in the core of a TM. The above evaluation has proven that the CBM creates a win-win-win situation for buyers, sellers and brokers.

The new model has overcome a number of group-trading problems on the Internet. The main contribution of this research is the CBM, but during the process of creating this new model for group-trading in e-markets, three additional issues have emerged which also made a contribution to knowledge in this field: (a) the advantages and problems of the core (b) the check of stability for a coalition and (c) the use of brokers in group-trading.

All systems are capable of improvement and some issues with the CBM can be identified. There will be two main targets for future research. One main target will be to create more incentives for participants. Another target will be to expand the CBM by including particular e-markets and selling a great diversity of products and services on them.

Reference

1. Forrester Research, Inc. 'Forrester Forecast: Double-Digit Growth for Online Retai in the US and Western Europe'. CAMBRIDGE, Mass (2010) (February 3, 2011),
   http://www.forrester.com/ER/Press/Release/0,1769,1330,00.html
2. Lieberman, H.: Autonomous Interface Agents. ed. by Pemberton, S. Human Factors in Computing, 67–74 (1997)
3. Ferguson, D., Nikolaou, C., Yemini, Y.: An Economy for Flow Control in Computer Networks. In: Proceedings of the 8th Infocom, pp. 100–118. IEEE Computer Society Press, Los Alamitos (1989)
4. Moulin, H.: Cooperative Microeconomics: A game theoretic Introduction. Princeton university Press, Princeton (1995)
5. Shehory, O., Kraus, S.: Feasible Formation of coalitions among autonomous agents in non-super-additive environments. Computational Intelligence 15(3), 218–251 (1999)
6. Turban, E., et al.: Electronic Commerce: A Managerial Perspective. Prentice Hall, New Jersey (1999)
7. Neumann, J., Morgenstern, O.: Theory of Games and Economic Behaviour. Princeton University Press, Princeton (1944)
8. Owen, G.: Game theory. Academic Press, New York (1995)
9. Lucas, W.F.: A game with no solution. Bulletin of the American Mathematical Society 74, 237–239 (1968)
10. Lucas, W.F.: Von Neumann-Morgenstern stable sets. In: Aumann, R.J., Hart, S. (eds.) Handbook of Game Theory, pp. 543–590. Elsevier, Amsterdam (1992)
11. Gillies, D.: Some theorems on n-person games. Unpublished PhD thesis, Princeton University (1953)
12. Osborne, M., Rubenstein, A.: A Course in Game Theory. MIT Press, London (1994)
13. Curiel, I.: Cooperative Games and Solution Concepts. In: Cooperative Game Theory and Applications, pp. 1–17. Kluwer Academic, Netherlands (1997)

14. Shapley, L.: Cores of convex games. International Journal of Game Theory 1, 11–26 (1971)
15. McCain, R.A.: Cooperative Games and the Core, William King Server (October 11, 2005), http://william-king.www.drexel.edu/top/eco/game/game-toc.html
16. Sun, P., et al.: Core-based Agent for Service-Oriented Market. In: IEEE International Conference Proceedings of Systems, Man, and Cybernetics, SMC 2006, pp. 2970–2975. IEEE, Taipei (2006)
17. Sun, P., et al.: A Core Broking Model for E-Markets. In: Proceedings of The 9th IEEE International Conference on e-Business Engineering (ICEBE 2012) Held September 9-11 at Zhejiang University, pp. 78–85. IEEE Press, Hangzhou (2012)
18. Ng, Y.: Welfare Economics. Macmillan, London (1983)
19. Gellekom, J., Potters, J., Reijnierse, J.: Prosperity properties of TUgames. Internat. J. Game Theory 28, 211–227 (1999)
20. Sun, P., et al.: Extended Core for E-Markets. In: Isaias, P., White, B., Nunes, M.B. (eds.) Proceedings of IADIS International Conference WWW/Internet 2009 Held November 19-22 at Rome, Italy, pp. 437–444. IADIS Press (2009)
21. Sun, P.: A Core Broking Model for E-Markets. Unpublished PhD thesis, Coventry University (2011)
22. Faigle, U., et al.: On the complexity of testing membership in the core of min-cost spanning tree games. International Journal of Game Theory 26, 361–366 (1994)
23. Sun, P., et al.: Evaluations of A Core Broking Model from the Viewpoint of Online Group Trading. In: Proceedings of the IEEE International Conference on Industrial Engineering and Engineering Management (IEEM 2012) Held December 10-13 at Hong Kong Convention and Exhibition Centre, pp. 1964–1968. IEEE Press, Hong Kong (2012)
24. Conitzer, V., Sandholm, T.: Complexity of determining non-emptiness of the core. In: Proceedings of the 4th ACM Conference on Electronic Commerce, San Diego, CA, USA, pp. 230–231 (2003)
25. Jain, K., Vohra R.: On stability of the core (2006) (October 16, 2008), http://www.kellogg.northwestern.edu/%20faculty/vohra/ftp/newcore.pdf
26. Scarf, H.: The core of an N person game. Econometrica 38, 50–69 (1967)
27. Bondareva, O.: Some application of linear programming to cooperative game. Problemy Kibernetiki 10, 119–139 (1963)
28. Yamamoto, J., Sycara, K.: A Stable and Efficient Buyer Coalition Formation Scheme to E-Marketplaces. In: AGENT 2001, pp. 576–583 (2001)
29. Sun, P.-C., Yang, F.-S.: An Insight into an Innovative Group-Trading Model for E-Markets Using the Scenario of a Travel Agent. In: Bădică, C., Nguyen, N.T., Brezovan, M. (eds.) ICCCI 2013. LNCS, vol. 8083, pp. 582–592. Springer, Heidelberg (2013)
30. The New Oxford Dictionary of English, 3rd edn. Oxford University Press, Oxford (2010)
31. Sun, P.-C., Yang, F.-S.: An Insight into an Innovative Group-Trading Model for E-Markets Using the Scenario of a Travel Agent. In: Bădică, C., Nguyen, N.T., Brezovan, M. (eds.) ICCCI 2013. LNCS, vol. 8083, pp. 582–592. Springer, Heidelberg (2013)
32. Chao, K., et al.: Using Automated Negotiation for Grid Services. IJWIN, 141–150 (2006)
33. Leonard, L.: Attitude Influencers in C2C E-commerce: Buying and. Selling. Journal of Computer Information Systems 52(3), 11–17 (2012)
34. PayPal 'Transaction Fees for Domestic Payments.' (May 15, 2011), https://www.paypal.com/uk/cgi-bin/webscr?cmd=_display-receiving-fees-outside

Considering Multiple Instances of Items in Combinatorial Reverse Auctions

Shubhashis Kumar Shil and Malek Mouhoub

Department of Computer Science, University of Regina, Regina, SK, Canada
{shil200s,mouhoubm}@uregina.ca

Abstract. Winner determination in combinatorial reverse auctions is very important in e-commerce especially when multiple instances of items are considered. However, this is a very challenging problem in terms of processing time and quality of the solution returned. In this paper, we tackle this problem using genetic algorithms. Using a modification of the crossover operator as well as two routines for consistency checking, our proposed method is capable of finding the winner(s) with a minimum procurement cost and in an efficient processing time. In order to assess the performance of our GA-based method, we conducted several experiments on generated instances. The results clearly demonstrate the good time performance of our method as well as the quality of the solution returned.

Keywords: Winner Determination, Combinatorial Reverse Auctions, Genetic Algorithms.

1   Introduction

The main purpose in combinatorial auctions is to increase the efficiency of bid allocation especially when biding on multiple items [3], [13]. More precisely, the goal here is to minimize the procurement cost in an efficient running time [3], [13], [21] when determining the winner [7]. Winner determination is still one of the prime challenges in combinatorial auctions [3] and is identified as an NP-complete problem [13], [21]. Nevertheless, applying combinatorial auctions to a procurement scenario [11], [14], such as transportation services, travel packages and sales of airport time slots, is cost saving [13]. Many algorithms have been introduced to tackle combinatorial auctions, e.g. Sitarz defined Ant algorithms and Hsieh and Tsai developed a Langrangian heuristic method [13]. While Genetic Algorithms (GAs) have been applied successfully to solve many combinatorial optimization problems [13], such as quadratic assignment problem [18], job scheduling [15] and travelling salesman problem [20], not much work has been done to solve the problem of winner determination in the context of Combinatorial Reverse Auctions (CRAs) with GAs.

In [17], we proposed a GA based method, called GACRA, to solve winner determination specifically for CRAs by considering multiple items. In [17], we introduced two repairing methods, RemoveRedundancy and RemoveEmptiness, along with a

A. Moonis et al. (Eds.): IEA/AIE 2014, Part II, LNAI 8482, pp. 487–496, 2014.

modified two-point crossover operator. By using these two methods and an efficient crossover operator, we solved the problem of winner determination successfully and efficiently. Indeed, the goal of these two methods is to repair the infeasible chromosomes and the main task of the crossover operator is to ensure a diversity of solutions. Indeed, increasing diversity will allow bidders to get more chances to be selected. In [17], we have demonstrated that GACRA is better in terms of processing time and procurement cost when compared to another GA-based winner determination method [13]. Subsequently in [16], we conducted several statistical experiments and showed that GACRA is a consistent method since it was able to reduce the solution variations in different runs. Moreover, we described the any-time behaviour of GACRA in [16].

In this paper, we consider the procurement of a single unit of multiple instances of items and propose an improved GA-based method, that we call GAMICRA (Genetic Algorithms for Multiple Instances of Items in Combinatorial Reverse Auctions), by modifying the methods and operators of [16] and [17] as follows.

1. We change the configuration of chromosomes to represent the bids in the case of multiple instances of items. We define the configuration of the chromosomes as well as the length of the chromosomes based on the number of sellers, items and instances of items.
2. We modify the fitness function to consider multiple instances of items. The fitness function is kept simple enough in order not to perform a lot of calculation. This will help the processing time to remain reasonable.
3. We also improve RemoveRedundancy and RemoveEmptiness methods to solve the winner determination problem associated with multiple instances of items. RemoveRundancy guarantees that the selected number of instances for each item does not exceed the buyer's requirements. On the other hand, RemoveEmptiness ensures that the selected number of instances for each item is not less than that of the buyer's requirements.

With these modifications and using the modified two-point crossover defined in [17], we solve efficiently the problem of winner determination for multiple instances of items in CRAs. In order to evaluate the performance of our method in terms of procurement cost and processing time, we perform several experiments. The results we report in this paper clearly demonstrate that our method is prominent in terms of time and cost and also show that it does not suffer from the inconsistency issue.

In many e-commerce and resource allocation problems, real-time response to a very large problem is expected [8]. When solutions are needed quickly and problem instances are large, exact algorithms that produce solutions at the expense of consuming extended required time are not only inadequate but also become infeasible as instances become larger. In real life, certain domains may require high-quality and approximate solutions within a suitable running time [8]. Sometimes it is unnecessary to spend a lot of time to slightly improve the quality of the solution. For these reasons, in this research we consider GAs to solve the winner determination problem as they have the ability to produce good solutions from the very first generations.

The rest of this paper is organized as follows. In Section 2, our GAMICRA method is presented and then explained through an example. In Section 3, the experimental study we conducted to evaluate the processing time, procurement cost and the consistency of our method is reported. In Section 4, concluding remarks with some future research directions are given.

2   Proposed Method (GAMICRA)

In a CRA, there is one buyer and several sellers who compete according to the buyer's requirements. First the buyer announces his demand (multiple instances of items) in the auction system. Then, the interested sellers register for that auction and bid on a combination of items. In this paper, our research goal is twofold: (1) to solve the winner determination problem for multiple instances of items in the context of CRAs, and (2) to find the winners in a reasonable processing time and reduced procurement cost. For this purpose, we propose the method GAMICRA that utilizes two techniques, namely RemoveRedundancy and RemoveEmptiness, to repair infeasible chromosomes. In addition, we utilize the modified two-point crossover operator defined in [17] that provides a good distribution of the solutions and preventing a premature convergence. Figure 1 illustrates a possible scenario in CRAs. Here the buyer wants to purchase three instances of item A and five instances of item B. There are three sellers (S1, S2 and S3) who want to bid on items A and B. Therefore, the maximum instance between these two instances of items A (here 3) and B (here 5) is equal to 5. Hence, 3 bits are required to represent each item for each seller considering the following equation (1).

$$\max(insItemA, insItemB) = 2^{reqBit} \tag{1}$$

With 3 bits, we can generate $2^3 = 8$ combinations $\{000, \dots, 111\}$. We assume for this example that the number of chromosomes = 4, and the length of chromosomes = $2 \times 3 \times 3 = 18$ according to the following equation (2).

$$chromosomeLength = m \times n \times reqBit \tag{2}$$

Figure 2 presents the pseudo code of our GAMICRA method (Algorithm 1) while RemoveRedundancy and RemoveEmptiness procedures are reported respectively in Figures 3 and 4. In Algorithm 1, the bid price of each item for each seller is generated randomly from the interval [100, 1000] (see Table 1). Table 2 shows the initial chromosomes that are produced randomly.

The next steps of Algorithm 1 are to convert infeasible chromosomes into feasible ones with the help of RemoveRedundancy and RemoveEmptiness methods. For our example, the conversion of chromosome X1 from infeasible to feasible one is described below.

Table 1. Bid Price Generation

	S1	S2	S3
Bid Price of Item A	951	868	525
Bid Price of Item B	826	920	696

Table 2. Initial Chromosomes

Chromosome	S1		S2		S3	
	Item A	Item B	Item A	Item B	Item A	Item B
X1	010(2)	011(3)	110(6)	001(1)	001(1)	101(5)
X2	100(4)	100(4)	110(6)	011(3)	011(3)	101(5)
X3	010(2)	010(2)	010(2)	101(5)	000(0)	010(2)
X4	010(2)	101(5)	101(5)	010(2)	101(5)	110(6)

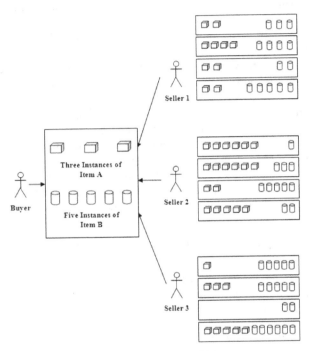

Fig. 1. An example in CRAs

Initially, chromosome *X1* is *010 011 110 001 001 101*. As shown above, the number of instances for Item A is 9 which is more than the buyer's requirement. The number of instances of Item B also exceeds the initial requirement. Hence, this chromosome is infeasible and needs to be repaired.

At first we will remove redundancy by using RemoveRedundancy method according to Algorithm 2 shown in Figure 3. After removing the redundancy, *X1* is now equal to *000 000 000 000 000 101*. Although the number of instances of Item B is the same as the buyer's requirement, chromosome *X1* is still infeasible as the number of instances of Item A is less than that of the buyer's requirement. In this case, we will apply the RemoveEmptiness method on this chromosome according to Algorithm 3 given in Figure 4. After removing the emptiness, *X1* becomes *000 000 000 000 011 101*. Now, chromosome X1 is a feasible one. By applying the same process to all the chromosomes, the initial chromosomes after remove redundancy and remove emptiness are as follows.

X1: 000 000 000 000 011 101
X2: 000 000 000 101 011 000
X3: 000 000 010 101 001 000
X4: 000 101 011 000 000 000

Algorithm 1 GAMICRA (m: number of bid items, n: number of sellers, δ: number of generations, α: crossover rate, β: mutation rate)

---

1. $t = 1$;
2. bidGenerator();
   //generates bid prices for each combination of bid items
   //for each seller
3. chromosomeGAMICRA();
   //generates initial chromosomes
4. RemoveRedundancy();
   //removes redundant instances of items
5. RemoveEmptiness();
   //adds instances of items as buyer's requirements
6. fitnessGAMICRA();
   //computes fitness values of chromosomes
do{
7. selectionGAMICRA();
   //selects chromosomes using gambling-wheel disk
   //method
8. crossoverGAMICRA();
   //generates child chromosomes from parent
   //chromosomes with modified two-point crossover
   //considering rate α
9. mutationGAMICRA();
   //mutates chromosomes considering mutation rate β
10. RemoveRedundancy();
11. RemoveEmptiness();
12. fitnessGAMICRA();
13. newChromosomeGAMICRA();
    //selects better chromosomes from both initial and
    //new chromosomes of each generation
}while $t \leq \delta$;
14. return winner(s);
    //returns winner(s) with minimum bid price in optimal
    //processing time

Fig. 2. Algorithm for GAMICRA

Algorithm 2 RemoveRedundancy (X: chromosome, m: number of items, n: number of sellers)

1. for each X do
2. { for each m do
3.   { if(redundantViolet())
      //item possesses redundant instances
4.   { generateRN();
         //generates random number between 1 and n
5.   convertZero();
         //converts bits for this seller to 0s
6.   }
7.   }
8. }

Fig. 3. RemoveRedundancy pseudo code

Algorithm 3 RemoveEmptiness (X: chromosome, m: number of items, n: number of sellers)

1. for each X do
2. { for each m do
3.   { if(emptinessViolet())
      //item possesses less instances
4.   { while number of instances are not exactly same
         as the buyer wants to buy do
5.   { generateRN();
6.   convertOne();
            //converts required bits for this seller to 1
7.   }
8.   }
9.   }
10. }

Fig. 4. RemoveEmptiness pseudo code

For evaluating the fitness value of each chromosome, the following fitness function is applied.

$$F(x_i) = 1 \Big/ {\sum_{N=1}^{n} \sum_{M=1}^{m} b_{NM} \times l_{NM} \times x_{NM}(C)} \tag{3}$$

$$/x_{NM}(C) \in \{0,1\}$$

Where $F(X_i)$ is the fitness value of chromosome X_i; b_{NM} is the bid price of item M submitted by seller N; l_{NM} is the number of instances of item M submitted by seller N; $x_{NM}(C)$ is 1 when the item combination C is selected for item M of seller N, and 0 otherwise.

Gambling Wheel Disk [7] is used in the selection operation. After selection, the following chromosomes are produced based on the fitness values calculated by using equation (3).

X1: 000 000 000 101 011 000 (Previous X2)
X2: 000 000 000 101 011 000 (Previous X2)
X3: 000 101 011 000 000 000 (Previous X4)
X4: 000 101 011 000 000 000 (Previous X4)

Then the modified two-point crossover is applied as shown in Figure 5(a). In our modified two-point crossover operation, the first child inherits the parent chromosomes between the crossover points in the forward direction as the basic two-point crossover, but the second child inherits it in reverse order to create a positive effect.

After crossover, the following chromosomes are generated.

X1: 000 000 000 101 011 000
X2: 011 000 000 101 000 000
X3: 000 101 011 000 000 000
X4: 000 101 011 000 000 000

Then the procedure will enter into the mutation operation as illustrated in Figure 5(b). For each chromosome, two unique random numbers are selected between 1 and n. Assume for this example, 1 and 3 are chosen, so the bit configurations of sellers S1 and S3 are interchanged. Algorithm 1 is repeated until the specified number of generations is completed. Thus, the winner(s) are determined based on the minimum bid price in an optimal processing time.

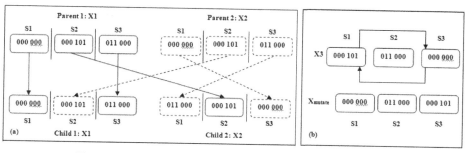

Fig. 5. (a) Modified two-point crossover (b) Mutation Operation

3   Experimentation

We have fully implemented GAMICRA in Java. This method is executed on an AMD Athlon (tm) 64 X2 Dual Core Processor 4400+ with 3.43 GB of RAM and 2.30 GHz of processor speed.

We use the following parameters and settings for all the experiments. The population size is 100 and we use the binary string for chromosomes encoding. The selection method is based on the Gambling Wheel Disk technique [7]. The crossover and mutation rates are 0.6 and 0.01 respectively. The running time in milliseconds and procurement cost (bid price) are averaged over 20 runs. The maximum number of generations is 50. The total number of sellers, items and instances are 500, 8 and 5 respectively unless indicated otherwise.

Figures 6(a), 6(b), 6(c), 6(d), 6(e) and 6(f) report the results of these experiments. Figure 6(a) shows how the running time varies with the increase of the number of generations. As we can notice our method can still run in a reasonable time (under the minute) even for a large number of iterations. Figure 6(b) shows how does the bid price improve with the increase of the number of generations, thanks to our modified two-point crossover initially defined in [14]. As we can easily see, this cost significantly improves in the first 30 iterations and stabilized afterwards. Figures 6(c) and 6(d) show how the running time increases when we augment the number of

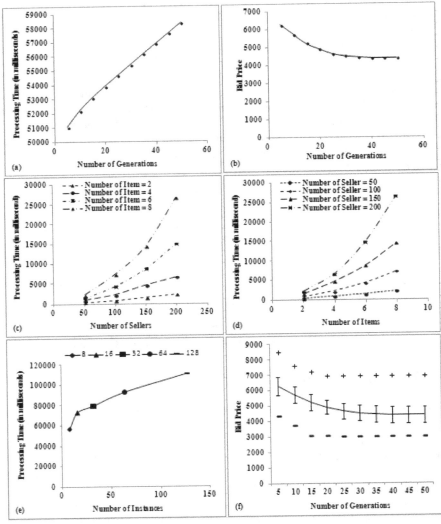

Fig. 6. Running time and procurement cost performances

sellers as well as the number of items. In this regard our method is capable of producing the solutions in less than 30 seconds even for large number of sellers and a good number of items. Figure 6(e) shows the time performance of our method when varying the number of instances from 1 to 128. Here too, GAMICRA successfully returns the solutions in a reasonable running time (under than 2 minutes in the case of 128 instances). Finally Figure 6(f) demonstrates the consistency of GAMICRA. The bid price is indicated by a solid line with the maximum and minimum values represented by + and - respectively. The error bars are depicted with a confidence level of 95%. From the figure, it is clearly notable that the variations of bid price in a particular generation over different runs are quite small. This means that GAMICRA is able to reduce the solution variations. It is also notable that the solution quality increases steadily over the generations. After 5 and 10 generations, our method produces an average bid price of 6267.55 and 5731.3 respectively. It then continues producing better solutions generation after generation. Furthermore, the variability of the solution quality over multiple runs improves over generations. The variability of the solutions fluctuates till the 25th generation and stabilizes afterwards. In addition, the minimum bid price remains constant after 25 generations. This suggests that the best solution found by GAMICRA might be the optimal solution.

4   Conclusion and Future Work

We propose a GA based method for winner determination with multiple instances of items in CRAs. Thanks to the RemoveRedundancy and RemoveEmptiness functions as well as the modified two-point crossover operator, GAMICRA can produce optimal solutions in a reasonable processing time and optimal procurement cost. Moreover, GAMICRA is a consistent method as it is able to reduce the solution variations over different runs.

As it is obvious that parallel GAs are capable of providing the solutions in a better running time [1], the future target of this research is to design our proposed method in parallel based on Parallel GAs (PGAs). Another future direction is to consider other bidding features besides 'Price', such as 'Seller Reputation', 'Delivery Time', 'Feedback from Other Buyers' and 'Warranty'. The winner will then be determined according to these criteria.

References

1. Abbasian, R., Mouhoub, M.: An Efficient Hierarchical Parallel Genetic Algorithm for Graph Coloring Problem. In: 13th Annual Genetic and Evolutionary Computation Conference (GECCO), pp. 521–528. ACM, Dublin (2011)
2. Avasarala, V., Mullen, T., Hall, D.L., Garga, A.: MASM: Market Architecture or Sensor Management in Distributed Sensor Networks. In: SPIE Defense and Security Symposium, Orlando FL, pp. 5813–5830 (2005)
3. Avasarala, V., Polavarapu, H., Mullen, T.: An Approximate Algorithm for Resource Allocation using Combinatorial Auctions. In: International Conference on Intelligent Agent Technology, pp. 571–578 (2006)

4. Das, A., Grosu, D.: A Combinatorial Auction-Based Protocols for Resource Allocation in Grids. In: 19th IEEE International Parallel and Distributed Processing Symposium (2005)
5. Easwaran, A.M., Pitt, J.: An Agent Service Brokering Algorithm for Winner Determination in Combinatorial Auctions. In: ECAI, pp. 286–290 (2000)
6. Goldberg, D.E., Deb, K.: A Comparative Analysis of Selection Schemes Used in Genetic Algorithms. Edited by G. J. E. Rawlins, 69–93 (1991)
7. Gong, J., Qi, J., Xiong, G., Chen, H., Huang, W.: A GA Based Combinatorial Auction Algorithm for Multi-robot Cooperative Hunting. In: International Conference on Computational Intelligence and Security, pp. 137–141 (2007)
8. Hoos, H.H., Boutilier, C.: Solving Combinatorial Auctions using Stochastic Local Search. In: Proceedings of the 17th National Conference on Artificial Intelligence, pp. 22–29 (2000)
9. Muhlenbein, H.: Evolution in Time and Space The Parallel Genetic Algorithm, pp. 316–337. Morgan Kaufmann (1991)
10. Mullen, T., Avasarala, V., Hall, D.L.: Customer-Driven Sensor Management. IEEE Intelligent Systems 21(2), 41–49 (2006)
11. Narahari, Y., Dayama, P.: Combinatorial Auctions for Electronic Business. Sadhana 30 (pt. 2 & 3), 179–211 (2005)
12. Nowostawski, M., Poli, R.: Parallel Genetic Algorithm Taxonomy. In: Third International Conference on Knowledge-Based Intelligent Information Engineering Systems, pp. 88–92 (1999)
13. Patodi, P., Ray, A.K., Jenamani, M.: GA Based Winner Determination in Combinatorial Reverse Auction. In: Second International Conference on Emerging Applications of Information Technology (EAIT), pp. 361–364 (2011)
14. Rassenti, S.J., Smith, V.L., Bulfin, R.L.: A Combinatorial Auction Mechanism for Airport Time Slot Allocation. The Bell Journal of Economics 13, 402–417 (1982)
15. Senthilkumar, P., Shahabudeen, P.: GA Based Heuristic for the Open Job Scheduling Problem. International Journal of Advanced Manufacturing Technology 30, 297–301 (2006)
16. Shil, S.K., Mouhoub, M., Sadaoui, S.: An Approach to Solve Winner Determination in Combinatorial Reverse Auctions Using Genetic Algorithms. In: The 15th Annual Genetic and Evolutionary Computation Conference (GECCO), pp. 75–76 (2013)
17. Shil, S.K., Mouhoub, M., Sadaoui, S.: Winner Determination in Combinatorial Reverse Auctions. In: Ali, M., Bosse, T., Hindriks, K.V., Hoogendoorn, M., Jonker, C.M., Treur, J. (eds.) Contemporary Challenges & Solutions in Applied AI. SCI, vol. 489, pp. 35–40. Springer, Heidelberg (2013)
18. Tate, D.M., Smith, A.E.: A Genetic Approach to the Quadratic Assignment Problem. Computers and Operations Research 22(1), 73–83 (1995)
19. Walsh, W.E., Wellman, M., Ygge, F.: Combinatorial Auctions for Supply Chain Formation. In: ACM Conference on Electronic Commerce, pp. 260–269 (2000)
20. Watabe, H., Kawaoka, T.: Application of Multi-Step GA to the Traveling Salesman Problem. In: Fourth International Conference on Knowledge-Based Intelligent Engineering Systems and Allied Technologies, Brighton, UK (2000)
21. Zhang, L.: The Winner Determination Approach of Combinatorial Auctions based on Double Layer Orthogonal Multi-Agent Genetic Algorithm. In: 2nd IEEE Conference on Industrial Electronics and Applications, pp. 2382–2386 (2007)

Constraint and Qualitative Preference Specification in Multi-Attribute Reverse Auctions

Samira Sadaoui and Shubhashis Kumar Shil[*]

Department of Computer Science, University of Regina, Regina, SK, Canada
{sadaouis,shil200s}@uregina.ca

Abstract. In the context of Multi-Attribute and Reverse Auctions (MARAs), two significant problems need to be addressed: 1) specifying precisely the buyer's requirements about the attributes of the auctioned product, and 2) determining the winner accordingly. Buyers are more comfortable in expressing their preferences qualitatively, and there should be an option to allow them describes their constraints. Both constraints and preferences may be non-conditional and conditional. However for the sake of efficiency, it is more suitable for MARAs to process quantitative requirements. Hence, there is a remaining challenge to provide the buyers with more facilities and comfort, and at the same time to keep the auctions efficient. To meet this challenge, we develop a MARA system based on MAUT. The proposed system takes advantage of the efficiency of MAUT by transforming the qualitative requirements into quantitative ones. Another benefit of our system is the complete automation of the bid evaluation since it is a really difficult task for buyers to determine quantitatively all the weights and utility functions of attributes, especially when there is a large number of attributes. The weights and utility functions are produced based on the qualitative preferences. Our MARA looks for the outcome that satisfies all the constraints and best satisfies the preferences. We demonstrate the feasibility of our system through a 10-attribute reverse auction involving many constraints and qualitative preferences.

Keywords: Constraint Specification, Qualitative Preference Specification, Winner Determination, Multi-Attribute Reverse Auctions, MAUT.

1 Introduction

Over the last decades, several companies have started using reverse auctions [21]. According to an annual benchmark e-sourcing survey of the Purchasing magazine, in 2006, 31% of participating companies employed reverse auctions [13]. Procurement software vendors, such as PurchasePro, IBM, DigitalUnion and Perfect, have also adopted reverse auctions [4]. In addition, there are numerous popular service providers of reverse auctions such as eBay.com, Priceline.com, Bidz.com, and AuctionAnything.com. A survey reported that buyers of reverse auctions achieved savings of

---

[*] Both authors contributed equally for the work presented in this paper.

A. Moonis et al. (Eds.): IEA/AIE 2014, Part II, LNAI 8482, pp. 497–506, 2014.
© Springer International Publishing Switzerland 2014

15% in cost and up to 90% in time [19]. On the other hand, multi-attribute auctions are considered one of the most valuable procurement systems in which bidders negotiate over the price along with other dimensions such as warranty, quality, sellers' reputation, delivery and payment terms, [5], [20]. This type of auctions guarantees higher efficiency than traditional auctions by exchanging information between the buyers' preferences and sellers' offers [4]. Multi-attribute auctions have been successfully utilized for the acquisition of a huge amount of specific products [4]. In this paper, we are interested in Multi-Attribute and Reverse Auctions (that we call MA-RAs). Several research works have been done in the context of MARAs, still two significant problems have to be solved: 1) eliciting precisely the buyers' requirements containing the constraints and preferences about the product attributes, and 2) determining the winner (the best bid) accordingly. Qualitative preferences and constraints are important features for any multi-attribute auction systems [8]. Indeed, the buyers would like to express their requirements qualitatively and in a friendly and interactive way, which is the most convenient. MARAs should provide these facilities to the buyers so that they can feel that their desires are satisfied at the maximum. Nevertheless, for the purpose of efficiency, it is more suitable for the auction systems to process quantitative requirements. Consequently, there is a remaining challenge to provide the buyers with more features and comfort, and to keep the auctions efficient at the same time.

Maximizing the satisfaction of the users may be achieved by considering precisely their preferences and interests [11]. [11] proposes a personalized matchmaking system that determines the best offer for a buyer by evaluating and sorting the sellers' offerings according to the buyer's precise interests. Preference elicitation is essential in interactive applications such as online shopping systems [2], [14]. [2] introduces an online shopping system that provides the users with the ability to specify their requirements (constraints and preferences) in an interactive way. [14] enhances the online shopping system developed in [2] with a learning component to learn the users' preferences, and therefore suggest products based on them. More precisely, the new component learns from other users' preferences and makes a set of recommendations using data mining techniques. Nowadays, there are several preference-aware commercial systems, such as Amazon, Netflix, SmartClient, Teaching Salesman and PExA [14]. Nevertheless, these systems do not allow the users to express qualitative preferences nor constraints. Formalizing users' preferences accurately also is very important in most decision support systems [6]. These systems rely on preferences to produce effective user models. For instance, [18] discusses and evaluates different methods to elicit preferences in negotiation support systems. [7] proposes a general interactive tool to obtain users' preferences about concrete outcomes and to learn utility functions automatically based on users' feedback. [12] develops a variant of the fully probabilistic design of decision making strategies. In [12], the elicitation of preferences is based on quantitative data. Preference elicitation plays an essential role in negotiations such as in automated market places and auctions [9]. Moreover, the acquisitions of high quality users' preferences are also significant for interactive Web services [10]. The quality of the returned results depends on the capability of the services to acquire the preferences. [10] examines the adaptation and personalization strategies during the elicitation process for Web services. Preferences and constraints can co-exist together in many domains [1], [20], and it is of great benefit to handle them together in many real-world applications. For instance, [8] employs MAUT to process quantitative preferences and Conditional Preference networks (CP-nets) to formalize qualitative preferences. [1] represents preferences as an instance of CP-nets and the constraints as a Constraint Satisfaction Problem (CSP). [1] also introduces a new algorithm to determine the best outcome based on the arc consistency propagation technique. It performs several experiments to show that the proposed approach is able to save substantial amount of time to generate the optimal solution. [16] introduces an algorithm that processes constraints and preferences and finds the optimal outcomes of a constrained CP-nets. [2] employs C-semiring to describe quantitative preferences, CP-nets for qualitative ones, CSP for constraints. This paper utilizes branch and bound method to provide the users with a list of outcomes.

This research proposes a MARA system where the buyer specifies his requirements containing constraints and qualitative preferences about the product attributes. Our system looks for the outcome that completely satisfies all the constraints and best satisfies the preferences. More precisely, we develop a MARA protocol to assist the buyer step by step in specifying conditional and non-conditional qualitative preferences as well as conditional and non-conditional constraints. The specification process is carried out through several friendly and interactive GUIs. The proposed auction system is based on MAUT [3], a widely used technique in multi-attribute decision-making. It takes advantage of the efficiency of MAUT by transforming the qualitative requirements into quantitative ones. Another major contribution of our auction system is the complete automation of the MAUT calculation. Indeed, it is a really challenging task for the buyers to determine quantitatively all the weights and all the utility functions of attributes, especially when there are lots of attributes for the auctioned product. The weights and utility functions are generated based on the qualitative preferences. We may also note that our system can handle multiple rounds to give the sellers a chance to improve their bids and compete better in the next rounds, allow any number of attributes to aid the buyer elicit his interests in a more precise way. Our protocol is semi-sealed because the bidders' offers are kept private during the auction, but at the end of each round, MARA reveals the overall MAUT utilities of the valid bids and statuses of all the submitted bids.

2   The Proposed MARA Protocol

When the buyer requests to purchase a product (e.g. a Television), our system first provides him with possible attributes for the product (generated from the product database). The buyer then selects the attributes he is interested in (e.g. Brand, Weight, Display Technology, Refresh Rate and Price). Subsequently, a reverse auction based on the buyer's attributes is launched. Next, the buyer will specify his requirements about the auctioned product.

A. Specifying Constraints on Attributes: first the buyer expresses hard constraints on the attributes of his choice. Constraints are described with the following syntax:

$$(condition_{a_1}) \, and/or, \ldots, and/or \, (condition_{a_i}) => constraint_{a_j} \qquad (1)$$

where *condition* and *constraint* both denote a relation between an attribute and its possible values: $rel(a_i, value(s) \, of \, a_i)$ such that $rel \in \{=, \neq, <, >, \leq, \geq\}$. When the condition clause in (1) is NULL, the constraint is non-conditional i.e. it has no dependencies on other attributes.

Assume the buyer submits the following two constraints regarding the TV auction:

(c1) NULL => Brand $\neq$ Panasonic
(c2) (Weight $\geq$ [5 - 5.9]) or (Display Technology = LED) and (Refresh Rate $\leq$ 120)
　　$\Rightarrow$　Price $\leq$ [1000 - 1499.99]

Here (c1) indicates that the TV brand must not be a Panasonic. (c2) means if the TV weight is greater than or equal to [5kg - 5.9kg] (i.e. not less than 5kg), or the display technology is LED, and the refresh rate is less than or equal to 120Hz, then the price must be less than or equal to [$1000 - $1499.99] (i.e. not greater than $1499.99). Since the precedence of 'and' is greater than of 'or', the condition clauses of (c2) is evaluated as: $condition_{Weight}$ or ($condition_{DisplayTechnology}$ and $condition_{RefreshRate}$).

B. Specifying Qualitative Preferences on Attributes: next the buyer should submit the qualitative importance levels for all the attributes. The system assigns equivalent quantitative values of these importance levels as shown in Table 1. Rank is a relative value of an attribute and M the number of attributes.

Table 1. Attribute Importance Levels in MARA

Qualitative Importance Level	Quantitative Importance Level	Rank
ExtremelyImportant (EI)	1	M
VeryImportant (VI)	0.75	M - 1
Important (I)	0.5	M - 2
Not VeryImportant (NVI)	0.25	M - 3

C. Specifying Qualitative Preferences on Attribute Values: the buyer can also specify qualitative preferences (liking) on the values of some attributes by using the following format:

$$(condition_{a_1}) \, and/or, \ldots, and/or \, (condition_{a_i}) => preference_{a_j} \qquad (2)$$

where $condition_{a_i} = rel(a_i, value(s) \, of \, a_i)$ such that $rel \in \{=, \neq, <, >, \leq, \geq\}$, and $preference_{a_j}$: $aj(value_{a_{j_1}}(liking), \ldots, value_{a_{j_n}}(liking))$

Values can be of two types: string and numeric. Table 2 gives the corresponding quantitative liking where rank is a relative value of an attribute value and N the number of values of an attribute. If the condition clause in (2) is NULL, then the preference is non-conditional.

Table 2. Attribute Value Liking in MARA

Attribute Type	Value	Qualitative Liking	Quantitative Liking	Rank
String		Highest (H)	1	N
		Above Average (AA)	0.8	N - 1
		Average (A)	0.6	N - 2
		Below Average (BA)	0.4	N - 3
		Lowest (L)	0.2	N - 4
Numeric		Highest (H)	1	
		Lowest (L)	0.2	

Suppose the buyer submits the following preferences for the TV auction:

(p1) NULL => Price ([800 - 899.99] (H), [1000 - 1499.99] (L))

(p2) (Refresh Rate > 120) => Brand (LG (BA), Panasonic (A), Sharp (A), Sony (AA), Toshiba (H))

(p1) expresses [800 - 899.99] as the highest preference and [1000 - 1499.99] as the lowest one for the attribute Price; (p2) means if the value of Refresh Rate is greater than 120, then the liking of each value of Brand is given. (p1) is an example of preference on numeric values whereas (p2) is on string values.

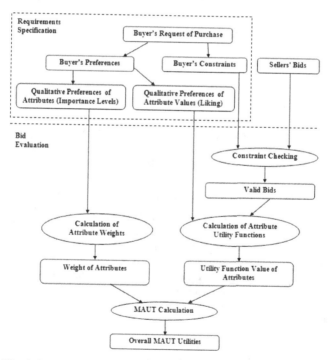

Fig. 1. Requirements Specification and Bid Evaluation in MARA

D. Evaluating Automatically the Bids: the buyer's requirements are displayed to the registered sellers who will compete by following the first sealed bid protocol. In

MAUT, each attribute a in a bid has a value denoted by v_a, and is evaluated through a utility function $U_a(v_a)$ [4]. An attribute has a weight which shows its importance to the buyer. The MAUT utility of a bid is calculated as the sum of all weighted utilities of the attributes [4]:

$$MAUT(bid) = \sum_{a=1}^{M} Weight_a \times U_a(v_a) \tag{3}$$

As illustrated in Fig. 1, after each round, the system deletes from the auction those bids that violate any constraint. For the qualified bids, MARA calculates automatically the weights and utility functions of all the attributes. As shown in Table 3, the weights are produced based on the buyer's importance levels. The utility function values are generated using the buyer's liking. If the buyer does not include a liking for an attribute value, then the system assigns 0 to the corresponding quantitative liking. In the case of conditional preferences, if a bid does not satisfy the condition clause, then the system assigns 0 to the quantitative liking of the attribute value for that bid. Finally, MARA informs the bidders about the overall utilities of the valid bids and statuses of all the submitted bids. In the next round, each seller is expected to place a bid with a utility higher than the one of the previous round. Bidding continues until the highest utility values of two consecutive rounds remain the same. Suppose now the following two bids have been submitted:

Bid_1: Brand = Panasonic; Display Technology = LCD; Price = 1000; Refresh Rate = 240; Weight = 4
Bid_2: Brand = Toshiba; Display Technology = Plasma; Price = 1200; Refresh Rate = 600; Weight = 4

The first bid does not satisfy the two constraints (c1) and (c3), however respects (c2) and (c4). In fact, this bid is deleted as soon as it violets (c1) and therefore there is no need to check the other constraints and calculate its MAUT utility value in order to save some processing time. On the other hand, the second bid satisfies all the constraints but does not guaranty to be the best bid.

Table 3. Attribute weight and utility function determination

Attribute Weight Calculation
$Weight_a = quanImpLevel_a \times rank_a \times weightRate$ $/\sum_{a=1}^{M} weight_a = 1$
Attribute Utility Function Calculation (String)
$U_a(v_a) = quanLikeliness_{v_a} \times rank_{v_a} \times utilityRate$
Attribute Utility Function Calculation (Numeric)
v_{aH} is the attribute value which has the highest preference v_{aL} is the attribute value which has the lowest preference $U_a(v_{aH}) = 1$ **if** $v_a \in]v_{aL}, v_{aH}[$ **then** $U_a(v_a) = (v_{aH} - v_a)/(v_{aH} - v_{aL})$ **else** $U_a(v_a) = 0$ such that $(v_a > v_{aH})$ or $(v_a < v_{aL})$ $U_a(v_{aL}) = $ lowest $U_a(v_a)/$ N $v_{aL}/v_{aH}/v_a$ is a value or an average value in case of range of values of a

Table 4. Weight and utility function value calculation for Brand

Weight calculation		Rank	Quantitative Importance Level	Weight
		4	0.75	0.27
Utility function value calculation	Attribute Value	Rank	Quantitative Liking	Utility Value
	LG	4	0.8	0.32
	Panasonic	3	0.6	0.36
	Sharp	2	0.4	0.24
	Sony	4	0.8	0.64
	Toshiba	5	1	1

Here, we illustrate the MAUT calculation for Bid_2. MAUT (Bid_2) = $Weight_{Brand} \cdot U_{Brand}$ (Toshiba) + $Weight_{DisplayTechnology} \cdot U_{DisplayTechnology}$ (Plasma) + $Weight_{Price} \cdot U_{Price}$ (1200) + $Weight_{RefreshRate} \cdot U_{RefreshRate}$ (600) + $Weight_{Weight} \cdot U_{Weight}$ (4). In Table 4, we show for example how the weight and utility function values are computed by MARA for the attribute Brand.

3   A 10-Attribute Reverse Auction

We fully developed our MARA system using the Belief-Desire-Intention model [17] and the agent-based simulation environment Jadex [15]. We designed MARA as 3-layer architecture. The presentation layer contains several GUIs to assist the buyer step by step in specifying the (non)conditional constraints as well as (non)conditional qualitative preferences. It also aids the sellers to submit their bids. The business layer contains the algorithms for constraint checking, weight, utility function value and MAUT calculation. The database layer stores all the information regarding the products and auctions. In this case study, the buyer wants to purchase a TV with ten attributes (given in Fig. 3).

The buyer can now submit the constraints for this auction:

(c1) NULL => Model Year $\neq$ 2011
(c2) NULL => Warranty $\geq$ 2
(c3) NULL => Refresh Rate $\geq$ 120
(c4) NULL => Screen Size $\geq$ [30 - 39]
(c5) (Refresh Rate $\leq$ 240) => Price $\leq$ [900 - 999.99]
(c6) (Brand = Panasonic) and (Resolution = 720p HD) => Weight $\leq$ [5 - 5.9]
(c7) (Brand = LG) or (Resolution = 1080p HD) => Screen Size $\leq$ [40 - 49]

Next, the buyer ranks the ten attributes according to their importance: Brand (VI), Customer Rating (NVI), Display Technology (I), Model Year (NVI), Price (EI), Refresh Rate (NVI), Resolution (I), Screen Size (VI), Warranty (NVI), and Weight (VI).

The following are examples of buyer's qualitative preferences on attribute values:

(p1) NULL => Price ([300 - 399.99] (H), [1000 - 1499.99] (L))
(p2) NULL => Refresh Rate (600(H), 120(L))

(p3) NULL => Brand (Bose (BA), Dynex (L), Insignia (BA), LG (AA), Panasonic (A), Philips (A), Samsung (A), Sharp (BA), Sony (AA), Toshiba (H))

(p4) NULL => Screen Size ([50 - 60] (H), [30 - 39] (L))

(p5) NULL => Model Year (2013(H), 2012(L))

(p6) NULL => Warranty (3(H), 2(L))

(p7) NULL => Customer Rating (5(H), 3(L))

(p8) (Price > [300 - 399.99]) and (Screen Size ≥ [40 - 49]) =>
     Display Technology (LCD (BA), LED (A), OLED (AA), Plasma (H))

(p9) (Refresh Rate ≥ 120) =>
     Resolution (1080p HD (H), 4K Ultra HD (AA), 720p HD (A))

(p10) (Screen Size ≥ [30 - 39]) => Weight ([4 - 4.9] (H), [6 - 7] (L))

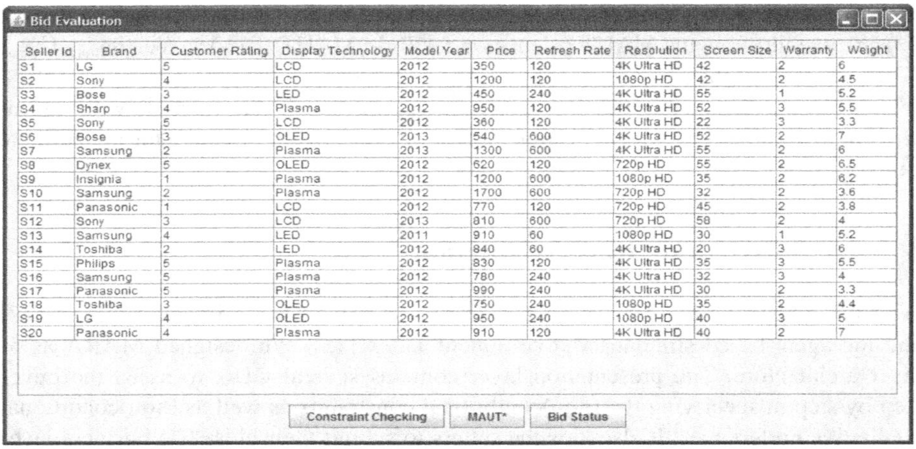

Seller Id	Brand	Customer Rating	Display Technology	Model Year	Price	Refresh Rate	Resolution	Screen Size	Warranty	Weight
S1	LG	5	LCD	2012	350	120	4K Ultra HD	42	2	6
S2	Sony	4	LCD	2012	1200	120	1080p HD	42	2	4.5
S3	Bose	3	LED	2012	450	240	4K Ultra HD	55	1	5.2
S4	Sharp	4	Plasma	2012	950	120	4K Ultra HD	52	3	5.5
S5	Sony	5	LCD	2012	360	120	4K Ultra HD	22	3	3.3
S6	Bose	3	OLED	2013	540	600	4K Ultra HD	52	2	7
S7	Samsung	2	Plasma	2013	1300	600	4K Ultra HD	55	2	6
S8	Dynex	5	OLED	2012	620	120	720p HD	55	2	6.5
S9	Insignia	1	Plasma	2012	1200	600	1080p HD	35	2	6.2
S10	Samsung	2	Plasma	2012	1700	600	720p HD	32	2	3.6
S11	Panasonic	1	LCD	2012	770	120	720p HD	45	2	3.8
S12	Sony	3	LCD	2013	810	600	720p HD	58	2	4
S13	Samsung	4	LED	2011	910	60	1080p HD	30	1	5.2
S14	Toshiba	2	LED	2012	840	60	4K Ultra HD	20	3	6
S15	Philips	5	Plasma	2012	830	120	4K Ultra HD	35	3	5.5
S16	Samsung	5	Plasma	2012	780	240	4K Ultra HD	32	3	4
S17	Panasonic	5	Plasma	2012	990	240	4K Ultra HD	30	2	3.3
S18	Toshiba	3	OLED	2012	750	240	1080p HD	35	2	4.4
S19	LG	4	OLED	2012	950	240	1080p HD	40	3	5
S20	Panasonic	4	Plasma	2012	910	120	4K Ultra HD	40	2	7

Constraint Checking   MAUT*   Bid Status

Fig. 2. Bid Submission in MARA

After the buyer's requirements have been submitted, the sellers can now start bidding. Assume 20 sellers participate and their first bids are depicted in Fig. 2. Five sellers, S2, S3, S5, S13 and S14, are disqualified as they do not satisfy some buyer's constraints. The remaining bids respect all the seven constraints. Fig. 3 shows the overall MAUT utilities of the valid bids and statuses of all the submitted bids for the first round. It also depicts S18 as the winner and the remaining 14 bids are challenged.

4   Conclusion and Future Work

The proposed auction system is able to assist the buyer in specifying non-conditional and conditional constraints as well as qualitative non-conditional and conditional preferences about the attributes of a product. To determine the winner efficiently, it converts qualitative requirements into quantitative ones. To generate the MAUT utility for each valid bid, MARA calculates automatically the attribute weights and utility functions. The latter are directly calculated from the buyer's preferences. In this paper, we explored experimentally the feasibility of MARA with a 10-attribute reverse auction involving 7 constraints, 10 preferences and 20 sellers.

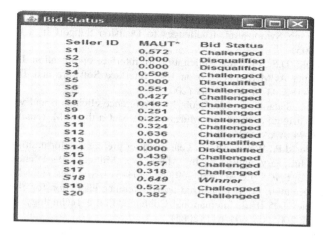

Fig. 3. Overall Utilities and Statuses of Bids

There are several promising future directions for this research work. In the literature, there are approaches to assign weights for the attributes, such as Simple Multi-Attribute Rating Technique (SMART) and Weight determination based on Ordinal Ranking of Alternatives (WORA) [5]. We can compare our weight calculation algorithm with these approaches in terms of processing time and best outcome. Another direction is to compare MAUT with other decision analysis techniques [3], such as Analytic Hierarchy Process and Conjoint Analysis, specifically in the domain of reverse auctions. Also, we would like to test the performance of our MARA system on a large dataset involving a large number of attributes, constraints, preferences and sellers. Our system allows the buyer to specify his preferences only qualitatively. To increase the acceptability of MARA, the buyer can be allowed to elicit qualitative preferences on some attributes as well as quantitative preferences on others.

References

1. Alanazi, E., Mouhoub, M.: Managing qualitative preferences with constraints. In: Huang, T., Zeng, Z., Li, C., Leung, C.S. (eds.) ICONIP 2012, Part III. LNCS, vol. 7665, pp. 653–662. Springer, Heidelberg (2012)
2. Alanazi, E., Mouhoub, M., Mohammed, B.: A preference-aware interactive system for online shopping. Computer and Information Science 5(6), 33–42 (2012)
3. Bichler, M.: An experimental analysis of multi-attribute auctions. Decision Support Systems 29(3), 249–268 (2000)
4. Bichler, M., Kalagnanam, J.: Configurable offers and winner determination in multi-attribute auctions. European Journal of Operational Research 160(2), 380–394 (2005)
5. Chandrashekar, T.S., Narahari, Y., Rosa, C.H., Kulkarni, D.M., Tew, J.D., Dayama, P.: Auction-based mechanisms for electronic procurement. IEEE Transactions on Automation Science and Engineering 4(3), 297–321 (2007)

6. Faltings, B., Pu, P., Zhang, J.: Agile Preference Models Based on Soft Constraints. In: AAAI Spring Symposium: Challenges to Decision Support in a Changing World, pp. 23–28 (2005)

7. Gajos, K., Weld, D.S.: Preference elicitation for interface optimization. In: Proceedings of the 18th Annual ACM Symposium on User Interface Software and Technology, UIST 2005, pp. 173–182. ACM, New York (2005)

8. Ghavamifar, F., Sadaoui, S., Mouhoub, M.: Preference elicitation and winner determination in multi-attribute auctions. In: Murray, R.C., McCarthy, P.M. (eds.) FLAIRS Conference, pp. 93–94. AAAI Press (2011)

9. Guo, Y., Müller, J.P., Weinhardt, C.: Learning user preferences for multi-attribute negotiation: An evolutionary approach. In: Mařík, V., Müller, J.P., Pěchouček, M. (eds.) CEEMAS 2003. LNCS (LNAI), vol. 2691, pp. 303–313. Springer, Heidelberg (2003)

10. Jannach, D., Kreutler, G.: Personalized user preference elicitation for e-services. In: Proceedings of the 2005 IEEE International Conference on e-Technology, e-Commerce and e-Service, EEE 2005, pp. 604–611 (2005)

11. Jiang, W., Sadaoui, S.: Evaluating and Ranking Semantic Offers According to Users' Interests. Journal of Electronic Commerce Research, JECR 13(1), 1–22 (2012)

12. Karny, M., Guy, T.V.: Preference elicitation in fully probabilistic design of decision strategies. In: 2010 49th IEEE Conference on Decision and Control (CDC), pp. 5327–5332 (2010)

13. Manoochehri, G., Lindsy, C.: Reverse auctions: Benefits, challenges, and best practices. California Journal of Operations Management 6(1), 123–130 (2008)

14. Mohammed, B., Mouhoub, M., Alanazi, E., Sadaoui, S.: Data Mining Techniques and Preference Learning in Recommender Systems. Computer and Information Science 6(4), 88-1-2 (2013)

15. Pokahr, A., Braubach, L., Lamersdorf, W.: Jadex: A BDI Reasoning Engine. In: Bordini, R.H., Dastani, M., Dix, J., El FallahSeghrouchni, A. (eds.) Multi-Agent Programming. Multiagent Systems, Artificial Societies, and Simulated Organizations, vol. 15, pp. 149–174. Springer, US (2005)

16. Prestwich, S., Rossi, F., Venable, K.B., Walsh, T.: Constraint-based preferential optimization. In: AAAI, vol. 5, pp. 461–466 (2005)

17. Seow, K.T., Sim, K.M.: Collaborative Assignment using Belief-Desire-Intention Agent Modeling and Negotiation with Speedup Strategies. Information Sciences 178(4), 1110–1132 (2008)

18. Shao-bin, D., Yu Qiang, F., Kexing, L., Xuncheng, S.: Preference elicitation of the NSS in the ecommerce. In: International Conference on Management Science and Engineering, ICMSE 2007, pp. 135–140 (2007)

19. Talluri, S., Narasimhan, R., Viswanathan, S.: Information technologies for procurement decisions: A decision support system for multi-attribute e-reverse auctions. International Journal of Production Research 45(11), 2615–2628 (2007)

20. Teich, J.E., Wallenius, H., Wallenius, J., Zaitsev, A.: A multi-attribute e-auction mechanism for procurement: Theoretical foundations. European Journal of Operational Research 175(1), 90–100 (2006)

21. Williams, C.L.: An overview of reverse auctions. In: Southwest Decision Sciences Institute Conference (2010)

Applications of Multivariate Time Series Analysis, Kalman Filter and Neural Networks in Estimating Capital Asset Pricing Model

An Zeng[1], Dan Pan[2,*], Yang Haidong[3], and Xie Guangqiang[1]

[1] Faculty of Computer, Guangdong University of Technology, China
[2] Batteries Plus LLC, Hartland, USA
pandan2006@gmail.com
[3] Pipe Material & Equipment Corp. of CNPC, China

Abstract. In modern finance theory, the Capital Asset Pricing Model (CAPM) is used to price an individual security or a portfolio. The model makes use of the relation between the systematic risk and the asset's expected rate of return to show how the market must price individual securities according to their security risk categories. In this study, traditional multivariate time series analysis, Kalman filter and neural networks are utilized to estimate the pricing model of a stock (YunNanBaiYao, YNBY) in Shenzhen Stock Exchange Market in China. From the case, we can see that the CAPM is valid in its theory, but there is still a room to improve the accuracy of pricing achieved with traditional regression and econometrics methods. Among those alternatives, Kalman filter and Neural networks seem to be promising and to deserve a try. Besides, it is indicated that how to combine various technical methods together to pricing a security or a portfolio could be worthwhile to research.

Keywords: Capital Asset Pricing Model; Kalman filter; Neural networks.

1   Introduction

In modern finance theory, the Capital Asset Pricing Model (CAPM), introduced by Jack Treynor, William Sharpe, John Lintner and Jan Mossin independently, is used for determining a theoretically appropriate required rate of return of an asset, if the asset is to be added to an already well-diversified portfolio, according to the asset's non-diversifiable risk. For individual securities, the model makes use of the relation between the systematic risk (beta or β) and the asset's expected rate of return to show how the market price individual securities in relation to their security risk category. Estimated Beta has been used in financial economics including testing of asset pricing theories, estimation of the cost of capital, evaluation of portfolio performance and calculation of hedge ratios for index derivatives, etc. Hence, improvements in the measurement of beta could be good for asset pricing, cash flow valuation, risk management, or making investment decisions, etc.

---

* Corresponding Author.

A. Moonis et al. (Eds.): IEA/AIE 2014, Part II, LNAI 8482, pp. 507–516, 2014.

It has been indicated that stock's beta coefficient may move randomly through time instead of remaining constant and beta is linked to leverage of the firm, which changes owing to changes in the stock price or decisions by the firm. There could be links between macro-economic / micro-economics factors and the firm beta, as illustrated in the work of paper [1,2]. When equity is interpreted as a call option on the assets of the firm, the beta of a stock is related to the beta of the firm's assets through a factor varying with the level of risk free interest rate. Hence, changes in interest rate should generate changes in beta. Moreover, beta is the covariance between the stock returns and index returns, scaled down by the variance of index returns. Bollerslev et al. [3] provided tests of the CAPM that imply time-varying betas. And there is considerable evidence that index volatility is time-varying according to paper [4]. This could lead to time-variation in β.

Unfortunately, the β estimation problem has still existed in the CAPM. For example, traditionally speaking, β is assumed to be constant. In this framework, β can be estimated as the slope parameter in a simple regression model fitted by ordinary least squares (OLS). Thus, problems arise when returns come from a non-normal distribution and/or are serially correlated. To violate the OLS assumptions results in inefficient estimates. Meanwhile, the regression specifications impose that the model parameters (e.g. β) are constant over the period. However, this is incompatible with a variety of economic arguments which predict time-varying beta. Empirical evidence in numerous studies indicates that betas are not constant over time [5~8]. Also, the length of a regression window is important and is likely to affect the results. But there is no theory which helps to determine the optimal length of the regression window.

Given that beta is time-varying, estimating β in the CAPM model has attracted enormous research interest owing to central roles of both the CAPM in modern finance theory and β in the CAPM model. Moreover, since the beta (systematic risk) is the only risk that investors should be concerned about, prediction of the beta value helps investors to make their investment decisions easier. The value of beta can also be used by market participants to measure the performance of fund managers through Treynor ratio. In addition, for corporate financial managers, forecasting the conditional beta accurately not merely benefit them in the capital structure decision but in investment appraisal.

So far, there have been a wide range of methods to estimate the time-varying beta, such as GARCH, Kalman Filter, and Artificial neural networks. This paper chooses YunNanBaiYao (sz000538), a stock in Shenzhen Stock Exchange in China and compares the applications of three methods (i.e. multivariate time series analysis, Kalman Filter and Artificial neural networks) in estimating the beta of the stock.

2   Related Works

Many techniques have been proposed to model and estimate time-varying beta. Three kinds of these approaches which have been widely utilized are as follows:

One strategy uses various GARCH models, such as standard bivariate GARCH, bivariate BEKK, and the bivariate GARCH-X according to paper[9,10]. These models use time varying second moments of the market and index returns to obtain time varying betas.

The second approach is based on the Kalman filter, which enables us to model beta as a time-series process, and hence allows us to test for beta constancy Bos & Newbold. The standard Kalman filter is applicable to the problem of the market model with a time-varying beta, under the assumptions of normality and homoscedasticity of the error. A series of papers has conducted these tests in various countries and reject beta constancy. The Kalman filter estimates a process by using a form of feedback control: the filter estimates the process state at some time and then obtains feedback in the form of (noisy) measurements. As such, the equations for the Kalman filter fall into two groups: time update equations and measurement update equations. The time update equations are responsible for projecting forward (in time) the current state and error covariance estimates to obtain the a priori estimates for the next time step. The measurement update equations are responsible for the feedback— i.e. for incorporating a new measurement into the a priori estimate to obtain an improved a posteriori estimate. The time update equations can also be thought of as predictor equations, while the measurement update equations can be thought of as corrector equations. Indeed the final estimation algorithm resembles that of a predictor-corrector algorithm for solving numerical problems.

The third strategy is a time-varying beta market model approach suggested by Schwert and Seguin [11]. This is a single factor model of stock return heteroscedasticity and it incorporates this into the market model equation which in its modified form provides estimates of time-varying market risk. The authors use the data over the sample period 1927-1986 to test monthly size-ranked US portfolios. It is indicated that the inability of previous studies to validate the CAPM model may be due to their failure to take into account the heteroscedasticity in stock returns.

3   Methodology

3.1   Source of Data

The data of stock prices was obtained from the Shenzhen Stock Exchange in China. For the validity of results, rates of return of YunNanBaiYao stock (YNBY, sz000538) for a five years' duration were used.

1   Time horizon: from January 1, 2004 to August 20, 2010. Here, the data from January 1, 2004 to December 31, 2008 (243 weeks in all) is used to acquire prediction model and the rest data (i.e. from January 1, 2009 to August 20, 2010, 83 weeks in all) is employed to evaluate the performance of prediction models;

2   Frequency: weekly. We use daily stock prices of YNBY to calculate its daily rates of return with the formula:

$$r_i = \frac{P_i - P_{i-1}}{P_{i-1}} \tag{1}$$

Here, P_i and P_{i-1} mean the closing price in day i and day (i-1), respectively. And then, weekly average rate of return r_j,

$$r_j = \frac{1}{n} \sum_{i \in week_j} r_i \tag{2}$$

Where n refers to the number of working days in week j. The method is used to calculate the weekly average rate of return of both YNBY and market portfolio.

3   Assume the proxy of required rate of return of the market portfolio is its actual rate of return.

3.2   Variables Used to Estimate Rate of Return of YNBY

$$R_t = \alpha + \beta R_{M,t} + \varepsilon_t \tag{3}$$

Here, α and β are the variables to be estimated with a variety of methods. Especially, β is used to measure the undiversified risk level of a stock in the stock market.

3.3   Three Methods for Rate of Return

3.3.1   Multivariate Time Series Model.

In this study, we make use of an Auto-Regressive Integrated Moving Average with eXtra / eXternal (ARIMAX) procedure in SAS to acquire the estimates of α and β. And then, the estimate of required rate of return of stock YNBY in week t can be calculated with the estimates of α and β.

$$\hat{R}_t = \hat{\alpha} + \hat{\beta} R_{M,t} \text{, } t = 1, 2, \dots, N; \tag{4}$$

The steps to obtain the above model are as follows.

1   Build the data set and plot graphs of Rt and RM,t against time; From the graphs, we can visually decide whether or not the time series are non-stationary. At the same time, if the two time series co-vary to some extent, we can consider to build the ARIMAX model.

2   To verify the time-series stationarity of variable Rt and RM,t, conduct ADF (Augmented Dickey-Fuller), PP (Phillips-Perron) or DW (Durbin-Watson) unit root test on them, respectively.

3   Build ARIMAX model. The spurious regression is prone to arise when a regression model on two non-stationary time-series is built unless there is not a cointegration relationship between the two series. ARIMA procedure can help intuitively judge the cointegration of the two variables, identify the order of the ARIMAX model and estimate the parameters of the model.

4   Forecast the out-sample values of the variable Rt when the time-series RM,t is given based on the acquired model and generate confidence intervals for these forecasts.

5   Evaluate the performance of the model with the comparison of the out-sample values $\hat{R}_t$ and the actual value R_t.

3.3.2   Kalmanfilter

Since the parameters, i.e. α and β, in the CAPM model are prone to be affected by certain stochastic factors, it might be better to fit in with the real life scenarios if we consider them to follow a certain stochastic model. In this paper, the RWM model (Random Walk Model) is employed to estimate the parameters. That is to say,

$$\alpha_t = \alpha_{t-1} + \mu_t; \tag{5}$$

$$\beta_t = \beta_{t-1} + \eta_t; \tag{6}$$

Here, $t = 1, 2, \ldots, N$.

The above formula can be merged as follows.

$$\begin{bmatrix} \alpha_t \\ \beta_t \end{bmatrix} = \begin{bmatrix} 1 & 0 \\ 0 & 1 \end{bmatrix} \begin{bmatrix} \alpha_{t-1} \\ \beta_{t-1} \end{bmatrix} + \begin{bmatrix} \mu_t \\ \eta_t \end{bmatrix} \tag{7}$$

Where μ_t and η_t are zero-mean white noise, and satisfy $\mu_t \sim N(0, Q_1)$ and $\eta_t \sim N(0, Q_2)$.

Thus, the formula $R_t = \alpha + \beta R_{M,t} + \varepsilon_t$ can be modified as follows.

$$R_t = \begin{bmatrix} 1 & R_{M,t} \end{bmatrix} \begin{bmatrix} \alpha_t \\ \beta_t \end{bmatrix} + \varepsilon_t \tag{8}$$

Here, ε_t is zero-mean white noise, and satisfies $\varepsilon_t \sim N(0, \Re)$.

If
$$x_t = \begin{bmatrix} \alpha_t \\ \beta_t \end{bmatrix}, \quad A = \begin{bmatrix} 1 & 0 \\ 0 & 1 \end{bmatrix}, \quad w_t = \begin{bmatrix} \mu_t \\ \eta_t \end{bmatrix}, \tag{9}$$

$$z_t = R_t, \quad H_t = \begin{bmatrix} 1 & R_{M,t} \end{bmatrix}, \text{ and } v_t = \varepsilon_t,$$

Then the state equation and the measurement equation can be expressed as follows.

$$x_t = Ax_{t-1} + w_t; \quad t = 1, 2, \ldots, N. \tag{10}$$

$$z_t = H_t x_t + v_t \tag{11}$$

$$E(w_t w_\tau') = \begin{cases} \begin{bmatrix} Q_1 & 0 \\ 0 & Q_2 \end{bmatrix} & t = \tau \\ 0 & t \neq \tau \end{cases}; \quad E(v_t v_\tau') = \begin{cases} \Re & t = \tau \\ 0 & t \neq \tau \end{cases}; \quad E(w_t v_\tau') = \begin{bmatrix} 0 \\ 0 \end{bmatrix} \tag{12}$$

With Kalman filter, we use the data from January 1, 2004 to December 31, 2008 (243 weeks in all) to obtain the optimal estimates $\hat{\alpha}_t$ and $\hat{\beta}_t$ of time-varied parameters αt and βt in the CAPM model. And then, we employ the data from January 1, 2009 to August 20, 2010 (83 weeks in all) to calculate the estimate $\hat{z}_t$ (i.e. $\hat{R}_t$) of required rate of return of YNBY based on the Kalman filter method and evaluate the performance of the model through comparing the estimate $\hat{R}_t$ and the actual value R_t.

3.3.3 Artificial Neural Networks

In this study, we use the three-layer neural network architecture with ten neurons in input layer, fifteen neurons in hidden layer and one neuron in output layer (Figure 1). Among the ten neurons in input layer, seven neurons represent the R_M from period t (current period) to period t-6 and three neurons the R from period t-1 to period t-3.

We use the MATLAB neural networks toolbox to build, train and test the neural networks. The neural network uses a hyperbolic tangent sigmoid transfer and activation function for the hidden layers, defined as '*tansig*' in MATLAB and for the output we use a linear layer which is defined as '*purelin*'. As the training function we used the Bayesian regularization, which is defined as '*trainbr*' in MATLAB.

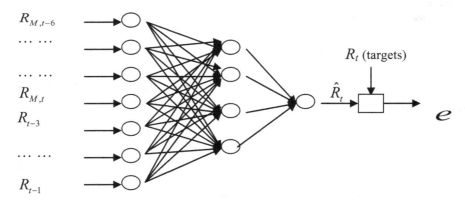

Fig. 1. The Architecture of Three Layer Neural Networks Used in the Study

One of the main advantages of the Bayesian regularization neural networks is that it provides estimation for the entire distribution of the model parameters instead of obtaining single "optimal" set of network weights, while over fitting problems can be avoided and the generalization performance can be improved. In the Bayesian regularization, we estimate the targets by minimizing the squared error of the output and simultaneously we include a regularized term, which is used to "penalize" the large values of weights in the network.

In this study, we firstly use the data from January 1, 2004 to December 31, 2008 (243 weeks) to train the neural networks. And then, we utilize the data from January 1, 2009 to August 20, 2010 (83 weeks) to calculate the estimate $\hat{R}_t$ of required rate of return of YNBY with the well-trained neural networks and to evaluate the performance of the model through comparing the estimates $\hat{R}_t$ and the target values R_t.

4   Results and Analysis

The three methods to forecast the rate of return are applied in stock YNBY.

4.1   Data Description

We have obtained the daily price of YNBY stock in Shenzhen Stock Exchange in the latest six and a half years from January 1, 2004 to August 20, 2010, as shown in Fig. 2. The daily rate of return and weekly average rate of return are shown in Fig. 3 and 4.

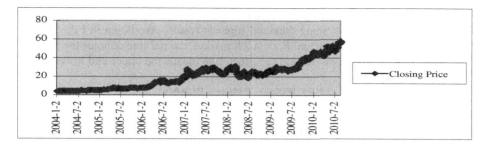

Fig. 2. Stock Price of YNBY

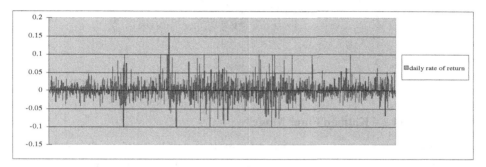

Fig. 3. Daily Rate of Return of YNBY

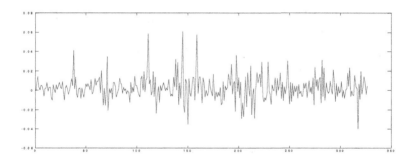

Fig. 4. Weekly Average Rate of Return of YNBY

4.2   Method 1 – Multivariate Time Series Model

As mentioned before, based on the first three steps, we can get the estimates of parameter α and β are 0.0019835 and 0.69806 (0.05 significance level) with the Conditional Least Squares Method, respectively. And the prediction model is obtained as follows.

$$R_t = 0.001983 + 0.698056 \times R_{M,t} + \varepsilon_t \quad_, \quad \varepsilon_t \sim N(0, 0.00011590) \tag{13}$$

Then we can forecast the values of variable R_t for the period from January 1, 2009 to August 20, 2010. The training data and forecast results are shown in Fig. 5. Here, the black line represents the actual R_t in stock market, the red line denotes the forecast R_t generated by the above model and the green line means the upper and lower bound with 95% confidence.

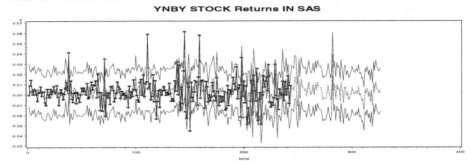

Fig. 5. Training Data and Forecast Results Achieved by Regression (RGR)

4.3   Method 2 – Kalman Filter Model

We use the rate of return in period from January 1, 2004 to December 31, 2008 to estimate the relevant parameters in Kalman Filter Model with the Least Squares Method.

Here,

$N=243$, $x_0 = \begin{bmatrix} 0.001619 \\ 0.640352 \end{bmatrix}$, $P_0 = \begin{bmatrix} 0.1 & 0 \\ 0 & 10 \end{bmatrix}$, $Q_1 = 0.2$, $Q_2 = 0.16$, $\Re = 800$. And

then, we make use of the Kalman Filter Model with the above parameters to forecast the weekly average rate of return of YNBY in the period from January 1, 2009 to August 20, 2010.

4.4   Method 3 – Artificial Neural Networks Model

We set the learning rate and the momentum rate as 0.5 and 0.2, respectively. In addition, the error target is set as 2.5e-2 and the maximum epochs are 200.

We employ the following formula to calculate the error between the forecasts $\hat{R}_t$ and the target values R_t.

$$e = \sqrt{\sum_{t=1}^{T} (\hat{R}_t - R_t)^2} \tag{14}$$

The forecasted weekly average rates of return are in Table I and shown in Fig. 6. Here, in order to compare with the forecasted results obtained by Neural Networks, we just keep the 77 values generated with the Kalman Filter Model.

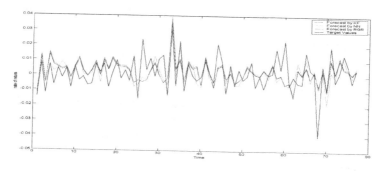

Fig. 6. Prediction Results from Three Methods: KF, NN and RGR Applied to the Out-sample Data

Table 1. Prediction results and errors from three methods: kalman filter (KF), neural networks (NN), and regression (RGR) applied to the out-sample data

Obs.	Forecast by KF	Forecast by NN	Forecast by RGR	Target Values
250	-0.011	-0.013	-0.012	-0.014
251	0.012	0.007	0.014	0.008
252	-0.004	-0.005	-0.004	-0.012
...	...	...	...	...
324	0.004	0.003	0.004	0.006
325	0.001	-0.001	0.001	-0.004
326	0.004	0.006	0.005	0.004
e	0.086	0.084	0.087	N/A

Table I gives the prediction results and errors from the three methods: Kalman filter (KF), Neural networks (NN), and Regression (RGR) when these methods are applied to the out-sample data (i.e., from January 1, 2009 to August 20, 2010). Owing to the NN requires the RM in previous 6 weeks, we focus on the period from Feb. 23, 2009 to Aug. 20, 2010 (77 weeks in all) to calculate the errors between the forecasts and the targets, shown as Table I. We can see that compared with the regression method, artificial neural networks has decreased the error by 4.15% and the Kalman filter has reduced the error by 1.73% when they are utilized to forecast the rate of return of stock YNBY in the period from February 23, 2009 to August 20, 2010.

5   Conclusion

We estimate the Capital Asset Pricing Model (CAPM) of a stock of the Shenzhen Exchange Stock Market, i.e. the firm "YunNanBaiYao" (YNBY). We examine if there are cointegration effects and conclude that the cointegration effects are definitely in the case of the stock YNBY. So we can build the regression model on R_t and $R_{M,t}$ without concerning the spurious regression problem.

Owing to various factors dynamically impacting the beta in the CAPM, we estimated the CAPM model for the stock with Kalman filter method and obtained the better forecasting results. In addition, we apply a multilayer neural network with Bayesian regularization and concluded neural networks could acquire the pattern hidden behind the seemingly random data and employ the acquired pattern to obtain much better forecast performance than the results obtained with multivariate time series analysis.

Also, we can say that the Capital Asset Pricing model is valid in its theory, but there is still a room to improve the prediction accuracy achieved with traditional regression and econometrics methods. Among those alternatives, Kalman filter and Artificial neural networks seem to be promising and to deserve a try.

Finally, we think if we could combine the advantages from a variety of time-series forecast methods, the prediction performance would be further enhanced. Thus, how to combine the methods together could be a next hot research topic. As a matter of fact, the predictor ensembles might be a feasible solution to the benefits maximization from the combination of various prediction methods.

Acknowledgments. This study was supported by the grants from the National Natural Science Foundation of China (61300107) and NSF of Guangdong (S2012010010212).

References

1. Rosenberg, B., Guy, J.: Prediction of beta from investment fundamentals. Financial Analysts Journal, 62–70 (1976)
2. Bos, T., Newbold, P.: An empirical investigation of the possibility of stochastic systematic risk in the market model. Journal of Business, 34–41 (1984)
3. Bollerslev, T., Engle, R.F., Wooldridge, J.M.: Capital asset pricing model with time-varying covariances. Journal of Political Economy, 116–131 (1988a)
4. Bollerslev, T., Chou, R.Y., Kroner, K.F.: ARCH modeling in finance: A review of the theory and empirical evidence. Journal of Econometrics 52(1-2), 5–60 (1992)
5. Blume, M.E.: Betas and their Regression Tendencies: Some Further Evidence. Journal of Finance 34, 265–267 (1979)
6. Abberger, K.: Conditionally parametric fits for CAPM betas, Technical report, Diskussionspapier des Center of Finance and Econometrics, Universität Konstanz (2004)
7. Ebner, M., Neumann, T.: Time-varying betas of German stock returns. Financial Markets and Portfolio Management, 29–46 (2005)
8. Eisenbeiß, M., Kauermann, G., Semmler, W.: Estimating beta-coefficients of German stock data: A nonparametric approach. The European Journal of Finance, 503–522 (2007)
9. Engle, C., Rodrigues, A.: Tests of International CAPM with Time-Varying Covariances. Journal of Applied Econometrics, 119–138 (1989)
10. Yun, J.: Forecasting Volatility in the New Zealand Stock Market. Applied Financial Economics, 193–202 (2002)
11. Schwert, G.W., Seguin, P.J.: Heteroscedasticity in stock returns. The Journal of Finance, 1129–1155 (1990)

Author Index